# SURVEY OF CLIMATOLOGY

17-95

# SURVEY OF CLIMATOLOGY

**JOHN F. GRIFFITHS**
**DENNIS M. DRISCOLL**
TEXAS A&M UNIVERSITY

CHARLES E. MERRILL PUBLISHING COMPANY
*A Bell & Howell Company*
Columbus   Toronto   London   Sydney

We acknowledge with gratitude the assistance of our colleagues of the Department of Meteorology, and Mrs. Dorothy Lorenz.

Cover photograph by Gill Kenny

Published by Charles E. Merrill Publishing Co.
A Bell & Howell Company
Columbus, Ohio 43216

This book was set in Optima and Benguiat
Cover Design Coordination: Will Chenoweth
Text Designer: Ann Mirels
Copy Editor: Jo Ellen Gohr
Production Coordination: Cynthia Brunk
Cartography: Cartographic Service Unit, College of
　　　　　Geosciences
　　　　　Texas A&M University

Copyright © 1982, by Bell & Howell Company. All rights reserved.
No part of this book may be reproduced in any form, electronic, or mechanical, including photocopy, recording, or any information storage and retrieval system, without permission in writing from the publisher.

Library of Congress Catalog Card Number: 81-81252
International Standard Book Number: 0-675-09994-3
Printed in the United States of America
1 2 3 4 5 6 7 8 9 10—86 85 84 83 82

*To our patient and understanding wives,
Joan and Anne*

# CONTENTS

## 1
## INTRODUCTION
1.1 What is climatology? *1*
1.2 What are the elements of climate? *1*
1.3 How have we learned about climatology? *2*
1.4 What is required of the reader of this text? *3*

## 2
## RADIATION AND THE ENERGY BALANCE
2.1 Introduction *5*
2.2 Solar radiation *5*
    2.2.1 General characteristics *5*
    2.2.2 Spatial and temporal variations of solar radiation *6*
2.3 The energy balance and net radiation *12*
    2.3.1 General considerations *12*
    2.3.2 Energy fluxes at the interface *15*
    2.3.3 Net radiation *17*
    2.3.4 Land versus water *17*
    2.3.5 Geographic and time variations of the energy balance components *19*

## 3
## TEMPERATURE AS A CLIMATIC ELEMENT
3.1 Introduction *29*
3.2 The physical basis of measurement and temperature *30*
3.3 Temperature in terms of the energy balance equations *31*
3.4 Time variations of temperature *32*
3.5 The spatial variation of temperature *39*
3.6 Continentality *42*
3.7 Other climatic aspects of temperature *44*

## 4
## CIRCULATION OF THE ATMOSPHERE AND OCEANS
4.1 Introduction *51*
4.2 Requirements for atmospheric and oceanic circulation *52*
    4.2.1 Conservation of mass, heat and moisture *52*
    4.2.2 Redistribution of heat and moisture *52*
    4.2.3 Conservation of angular momentum *52*
4.3 The basis of air movement *53*
    4.3.1 Forces affecting air movement *53*
    4.3.2 Rules for wind flow *55*
    4.3.3 The geostrophic approximation *55*
    4.3.4 The effects of friction *56*
4.4 Convergence and divergence *59*
4.5 Scales of atmospheric motion *60*
4.6 Surface circulation *62*
    4.6.1 Fundamental considerations *62*
    4.6.2 Centers of action *65*
    4.6.3 Transient disturbances on the synoptic scale *69*
    4.6.4 Climatological characteristics of transient disturbances *78*

- 4.7 Upper-air circulation  *84*
  - 4.7.1 The Hadley cells  *84*
  - 4.7.2 The zonal circulations of extratropical latitudes  *85*
- 4.8 Climatic influences of the oceans  *95*
  - 4.8.1 The transfer of heat by ocean currents  *96*
  - 4.8.2 Surface circulation and its climatic influences  *98*

# 5

# MOISTURE IN THE ATMOSPHERE, CLOUDS AND PRECIPITATION

- 5.1 Introduction  *103*
- 5.2 Phase changes  *103*
- 5.3 Measures of water vapor  *104*
- 5.4 The hydrologic cycle  *106*
- 5.5 Global variations  *108*
- 5.6 Evaporation  *111*
  - 5.6.1 Measurement  *112*
  - 5.6.2 Potential evapotranspiration  *113*
- 5.7 Water budgets  *114*
- 5.8 Clouds and precipitation  *116*
  - 5.8.1 Cloud formation and vertical motion  *116*
  - 5.8.2 Condensation  *125*
  - 5.8.3 The formation of precipitation  *126*
  - 5.8.4 Forms of precipitation  *127*
  - 5.8.5 Characteristics of precipitation  *127*

# 6

# ANALYTICAL METHODS IN CLIMATOLOGY

- 6.1 Introduction  *129*
- 6.2 Definitions  *130*
- 6.3 Descriptive statistics  *131*
  - 6.3.1 Numerical descriptors  *131*
  - 6.3.2 Graphical descriptors  *132*
- 6.4 Probability  *135*
  - 6.4.1 The binomial distribution  *135*
  - 6.4.2 The normal distribution  *137*
  - 6.4.3 The Poisson distribution  *138*
  - 6.4.4 Other theoretical distributions  *140*
- 6.5 Sampling and tests of statistical significance  *140*
  - 6.5.1 Nonparametric tests  *140*
  - 6.5.2 Parametric tests  *142*
- 6.6 Regression and correlation  *144*
  - 6.6.1 Bivariate linear regression and correlation  *145*
  - 6.6.2 Multivariate linear regression and correlation  *146*
- 6.7 Time series analysis  *147*

# 7

# PATTERNS OF THE CLIMATIC ELEMENTS

- 7.1 Introduction  *149*
- 7.2 Temperature  *149*
  - 7.2.1 Spatial patterns  *151*
  - 7.2.2 Temporal patterns  *157*
- 7.3 Precipitation  *157*
  - 7.3.1 Spatial patterns  *157*
  - 7.3.2 Temporal patterns  *162*
  - 7.3.3 Other aspects of hydrometeors  *166*
- 7.4 Air masses  *172*
- 7.5 Climate iterations  *174*
  - 7.5.1 Temperature  *176*
  - 7.5.2 Precipitation  *178*
  - 7.5.3 Climatic types  *180*
- 7.6 Climatic extremes  *180*
  - 7.6.1 Temperature  *181*
  - 7.6.2 Precipitation  *181*
  - 7.6.3 Other climatic extremes  *183*

# 8

# CLIMATES OF MACROSCALE AREAS AND THEIR CLASSIFICATION

- 8.1 Introduction  *187*
- 8.2 History of climate classification  *187*
- 8.3 Climate of the standard continent  *190*
  - 8.3.1 Temperature zones  *191*
  - 8.3.2 Precipitation zones  *191*
  - 8.3.3 Topographic and shape effects  *192*
- 8.4 Classification requirements  *192*
- 8.5 Marine and continental climates  *193*
- 8.6 Köppen's classification  *194*
- 8.7 Thornthwaite's classification  *197*
- 8.8 Other important classifications  *199*

# 9

# REGIONAL CLIMATES

- 9.1 Introduction  *205*
- 9.2 Hot, humid climates (A climates)  *205*
  - 9.2.1 Hot, wet climate (Af)  *206*
  - 9.2.2 Hot, short dry season (Am)  *207*
  - 9.2.3 Hot, wet and dry seasons (Aw)  *208*
- 9.3 Dry climates (B climates)  *209*
  - 9.3.1 Hot (low latitude) steppe (BSh)  *210*
  - 9.3.2 Cold (mid-latitude) steppe (BSk)  *210*
  - 9.3.3 Hot (low latitude) desert (BWh)  *211*
  - 9.3.4 Cold (mid-latitude) desert (BWk)  *212*
- 9.4 Mid-latitude climates (C climates)  *213*
  - 9.4.1 Mid-latitude, uniform precipitation, hot summer (Cfa)  *214*
  - 9.4.2 Mid-latitude, uniform precipitation, warm summer (Cfb)  *215*
  - 9.4.3 Mid-latitude, uniform precipitation, cool summer (Cfc)  *216*
  - 9.4.4 Mid-latitude, dry winter, hot summer (Cwa)  *217*
  - 9.4.5 Mid-latitude, dry winter, warm summer (Cwb)  *218*
  - 9.4.6 Mid-latitude, dry and hot summer (Csa)  *219*
  - 9.4.7 Mid-latitude, dry and warm summer (Csb)  *220*
- 9.5 The high latitude climates (D climates)  *221*
  - 9.5.1 High latitude, uniform precipitation, hot summer (Dfa)  *221*
  - 9.5.2 High latitude, uniform precipitation, warm summer (Dfb)  *222*
  - 9.5.3 High latitude, uniform precipitation, cool summer (Dfc)  *224*

# CONTENTS

- 9.5.4 High latitude, uniform precipitation, extremely cold winter (Dfd) *225*
- 9.5.5 High latitude, dry summer (Ds) *226*
- 9.5.6 High latitude, dry winter, hot summer (Dwa) *227*
- 9.5.7 High latitude, dry winter, warm summer (Dwb) *227*
- 9.5.8 High latitude, dry winter, cool summer (Dwc) *228*
- 9.5.9 High latitude, dry and extremely cold winter (Dwd) *229*
- 9.6 Cold climates (E climates) *229*
  - 9.6.1 Polar marine climate (EM) *230*
  - 9.6.2 Tundra climate (ET) *230*
  - 9.6.3 Frost climate (EF) *232*
- 9.7 Highland climates (H climates) *232*

# 10
## SMALL-SCALE CLIMATES

- 10.1 Introduction *235*
- 10.2 Mesoscale controls *235*
  - 10.2.1 Land-sea configuration *236*
  - 10.2.2 Topography *238*
  - 10.2.3 Lake effects *239*
- 10.3 Microscale controls *239*
  - 10.3.1 Bare surface characteristics *239*
  - 10.3.2 Topography *242*
  - 10.3.3 Vegetation *243*
  - 10.3.4 Constructions *245*

# 11
## CLIMATIC CHANGES

- 11.1 Introduction *247*
- 11.2 Assessing past climates *249*
- 11.3 The Precambrian Era *251*
- 11.4 The Paleozoic Era *252*
- 11.5 The Mesozoic Era *254*
- 11.6 The Tertiary Period *254*
- 11.7 1.8 million years before present to 100,000 years before present *255*
- 11.8 100,000 to 10,000 years before present *256*
- 11.9 10,000 to 1,000 years before present *258*
- 11.10 The last 1000 years *259*
- 11.11 The last 100 years *260*
- 11.12 The suspected causes of climatic change *260*
  - 11.12.1 Solar output *261*
  - 11.12.2 The sun-earth path *263*
  - 11.12.3 Atmospheric composition *263*
  - 11.12.4 Surface changes *264*
- 11.13 Modeling the climate *267*
- 11.14 Forecasting future climates *269*

# 12
## SYNOPTIC CLIMATOLOGY

- 12.1 Introduction *271*
- 12.2 Some aspects of synoptic meteorology *272*
- 12.3 The synthesis of synoptic weather maps *274*
  - 12.3.1 Extratropical latitudes *274*
  - 12.3.2 The tropics *281*
  - 12.3.3 Upper-air patterns *285*
- 12.4 Applications *293*

# 13
## CLIMATE, AGRICULTURE AND FORESTRY

- 13.1 Introduction *299*
- 13.2 Radiation and vegetation *299*
- 13.3 Temperature and vegetation *301*
- 13.4 Water and vegetation *302*
  - 13.4.1 Evaporation and evapotranspiration *303*
  - 13.4.2 Precipitation and drought *304*
- 13.5 Phenology *304*
- 13.6 Climate and crop yield *305*
- 13.7 Atmospheric modification *308*
  - 13.7.1 Irrigation *308*
  - 13.7.2 Mulches *308*
  - 13.7.3 Shelterbelts *309*
  - 13.7.4 Freeze protection *310*
  - 13.7.5 Reduction of evaporation *310*
  - 13.7.6 Cloud seeding *311*
- 13.8 Climate, plant diseases, and pests *311*
- 13.9 Climate and livestock *313*
- 13.10 Climate and forestry *314*

# 14
## CLIMATE, PEOPLE, AND ARCHITECTURE

- 14.1 Introduction *317*
- 14.2 Human biometeorology *317*
  - 14.2.1 Direct effects of the atmosphere on people *317*
  - 14.2.2 The body's heat balance *318*
  - 14.2.3 Indices of comfort *320*
  - 14.2.4 Clothing *322*
  - 14.2.5 Health *322*
  - 14.2.6 Food and diet *323*
- 14.3 Buildings *324*
  - 14.3.1 Climatic variables and building design *324*
  - 14.3.2 Urban climates *327*

# 15
## CLIMATE AND TRANSPORT, ENERGY, BUSINESS, AND LEISURE ACTIVITIES

- 15.1 Introduction *331*
- 15.2 Travel and transport *331*
  - 15.2.1 Water *331*
  - 15.2.2 Road *333*
  - 15.2.3 Rail *335*
  - 15.2.4 Air *336*
- 15.3 Power generation and transmission *337*
- 15.4 Utilities *337*
- 15.5 Industry *340*
- 15.6 Commerce *342*
- 15.7 Communications *346*
- 15.8 Tourism *346*
- 15.9 Recreation and entertainment *347*

# 1
# INTRODUCTION

**WHAT IS CLIMATOLOGY?**
**WHAT ARE THE ELEMENTS OF CLIMATE?**
**HOW HAVE WE LEARNED ABOUT CLIMATOLOGY?**
**WHAT IS REQUIRED OF THE READER OF THIS TEXT?**

## 1.1
### WHAT IS CLIMATOLOGY?

Climatology is the scientific study of climate. The word *climate* comes from the Greek word *klima*, which refers to the angle of incidence of the sun's rays. *Weather* is the instantaneous or short-term state of the atmosphere; when weather conditions are averaged or summed over periods of a year or longer, climate is the result. Climate is therefore the synthesis of weather. Climate consists of more, however, than just averages or representative values of the elements which comprise it; the variability of these elements is also important.

Climatology is a branch of meteorology, the science of the atmosphere, and is distinguished from its parent discipline by its emphasis on the behavior of atmospheric elements over relatively long time, but not necessarily large space, scales. Other disciplines have benefitted from, and contributed to, the advancement of climatology. Chief among these is geography, which considers that climate, as a factor of the physical environment, helps to explain the distribution of other natural phenomena and of people and their artifacts over the earth. Other fields concerned with phenomena directly influenced by climate are the agricultural sciences, especially those concerned with plants, animals, and insects; geology, which examines the way the atmosphere's erosive agents (wind, rain, and sun) have molded the earth's surface; and oceanography, which recognizes that the oceans and the atmosphere are a coupled system. Indeed, there is hardly a physical science that is not affected by the long-term state of the earth's atmosphere. The same is true, to a lesser extent, of the social sciences and humanities.

## 1.2
### WHAT ARE THE ELEMENTS OF CLIMATE?

The elements of climate are the characteristics of the atmosphere: temperature, precipitation, wind speed and direction, humidity, radiation, sunshine, cloudiness, and evaporation. How we determine these elements and how we apply our knowledge of them distinguish the five general subdivisions of climatology outlined in this book. In Chapters 2 through 5 the reasons for the spatial (place-to-place) and temporal (time-to-time) variations of the climatic elements, principally temperature and precipitation, are analyzed. This is *physical climatology*. Much of what we know in this area is based upon the principles of physics. Chapter 6 deals with analytical methods—chiefly statistical—in climatology. Chapter 7 links physical principles to patterns of the climatic elements, while Chapters 8 through 10 detail these patterns, which are based on instrumental observations of the climatic elements from records gathered over tens of years to as many as 200 years. This subdivision is called *climatography*, or *regional climatology*, and emphasizes the description and classification of climate rather than an analysis of its causes.

In Chapter 11 we examine the behavior of the climatic elements over even longer periods of time, back to millions of years ago. Of course, the conjectures about what the earth's climate was like are not based on instrumental observations, but rather on the consequences of climate—for example, the width of tree rings, the chemical composition of cores from deep-sea bottoms, and our observations of how glaciers changed the earth's surface. This is *historical climatology*, or *paleoclimatology*. In Chapter 12 we consider only certain characteristics, particularly those over the long term, of the free atmosphere (that part of it not directly influenced by the earth's surface properties) that are obtained from analyses of weather maps. This is called *synoptic climatology*, which deals with features such as pressure patterns (e.g., lows, highs), and weather fronts, and their consequences for temperature, precipitation, and wind speed and direction.

Chapters 13 through 15 describe the influence of the climatic elements on plants (especially crops and other agricultural commodities), animals, people, and buildings. This subdivision of climatology is generally called *applied climatology*, and its particular subdivisions include *agricultural climatology*, *biometeorology*, and *architectural climatology*.

# 1.3
## HOW HAVE WE LEARNED ABOUT CLIMATOLOGY?

The history of climatology cannot be separated arbitrarily from the history of meteorology, but there are certain landmarks in the development of climatology. During the eleventh century B.C., in China, under the Chou dynasty, details of precipitation, temperature, and storms were kept in official records. At about the same time the height of the Nile River flood was recorded each year in what is now Egypt.

Around 100 B.C. the Greek Kyrrhestes designed the Tower of the Winds in Athens, often mentioned as the earliest meteorological observatory. However, this tower was used mainly for astronomical purposes and for housing the city's official *clepsydra*, or water clock. During this time winds were thought to bring specific types of weather and thus were emphasized over other climatic elements. In Europe the conditions of the ice around Iceland before 1000 A.D., as reported by explorers, were recorded by monks. Weather records of a descriptive nature exist from the thirteenth century.

It was not until about the mid-seventeenth century that weather measurements were systematized. Apparently the first network of stations was created by the order of Ferdinand II of Tuscany (Italy). Seventy years later James Jurin proposed that the Royal Society of Great Britain should collect worldwide data. The first known climatic map appeared in 1683 when Edmund Halley published a chart of the winds, the basis of his study to explain trade winds and monsoon circulations.

Climate was defined in 1771, in the first edition of *Encyclopedia Britannica*, as "a space upon the surface of the terrestrial globe, contained between two parallels, and so far distant from each other that the longest day in one differs by half an hour from the longest day in the other parallel." There would thus be 48 climatic bands, 24 in each hemisphere. Interestingly, in 140 B.C. the Greek Hipparchus had used the concept of climatic zones that depend upon length of the daylight hours at the summer solstice, so even then sun control of climate was recognized. Thus, the *Encyclopedia Britannica* was using a definition about 1900 years old!

In 1790 a significant advance occurred when the Meteorological Society of Mannheim, Germany, established 39 weather-observing stations using standardized equipment in various countries, 4 of these in North America. From 1787 to 1844 John Dalton, an English chemist, operated a rain gage network in northwestern England. During this period the first rainfall map from this area, as well as some mean monthly temperature maps, was published. In 1825 New York established a state climatological network. However, these networks did not remain for long, and there are very few representative stations that have records spanning long periods.

By the early years of the nineteenth century, as more measurements became available, the first maps of mean annual temperature were developed, as were synoptic charts based on observations made in the last years of the eighteenth century. During the next decades mapping various elements, especially wind flow and pressure, was emphasized. Many European meteorological services were being formed and scientific analyses begun.

From these rather modest beginnings the present worldwide network of observing stations has grown. Of course, there are still gaps in this network, and especially over the oceans data are sparse. A particularly difficult problem to resolve is that most of the long-period stations have been affected by location changes or urbanization so that climatic changes may reflect these effects and not necessarily atmospheric variations. Remote sensing from satellites, a mid-twentieth century innovation, has assisted in filling some of these gaps. The process of obtaining readings has become more automated, so fewer observers are needed and measurements can be telemetered to central data processing facilities.

As more data became available and climatic classifications were developed in the late nineteenth to mid-twentieth century, the physical basis of climate became clearer and specialized books began to appear. The first general text was written in 1853 by Jean Charles Houzeau, but thirty-five years earlier Luke Howard had published his famous *Climate of London*. Lorin Blodget wrote the first *Climatology of the United States* in 1857, but it was Julius Hann's classic work *Handbuch der Klimatologie* (1883) that began a spate of books over the next century. Now, in the age of computers, weather scientists have begun to construct mathematical models of the processes that govern climate. Thus, from observation and description, through mapping and classification, through the beginnings of an understanding of the physical basis of climate, to mathematical models, the science of climatology has developed.

# 1.4
## WHAT IS REQUIRED OF THE READER OF THIS TEXT?

Curiosity and a desire to learn are necessary for students of any scientific discipline. This text is largely nontechnical and uses very few mathematical equations, except in Chapter 6, which deals with applications of statistics to climatology. Any student with a high school background in physics, mathematics, and the physical sciences in general should be able to master the material with little difficulty.

Attention to and conceptualization of illustrations is also important. Climatology makes extensive use of maps and graphs on which **isopleths**, lines of constant values of a given quantity, have been drawn. For example, **isotherms** are lines of constant temperature, and **isobars** are lines of constant pressure. On maps there are three variable quantities: latitude, longitude, and the climatological variable being shown, usually at a constant height. In graphical representations using isopleths, the quantities, or dimensions, are variable.

It is extremely important that these illustrations are visualized as three-dimensional. It may help to think of them in terms of their topographic counterpart, that is, as contour maps. In Figure 2.4, for example, there is a "ridge" centered in late June that decreases in elevation from 90° to about 20° N. Similarly, a flat "valley" from late March to late September at 90° S reaches its most northward extent in late June. A more direct analogy, and for this reason perhaps an easier one, would be the "contours" of a pressure map, where lows and highs become depressions and hills, respectively.

In this regard also it helps to visualize **gradients**, the rate of change of a variable across some dimension. Again in Figure 2.4, notice that the isopleths (in this case, lines of constant radiation received at the top of the atmosphere) are closest together from January to March and again from September to November at 90° S. This means that the rate of change of radiation with time is greatest during these months.

Each of the many isoplethed maps and graphs should be studied as it is encountered. Of course, nothing replaces experience, and one successfully visualizes and integrates all the information in such representations only after long experience and careful study.

# 2
# RADIATION AND THE ENERGY BALANCE

**INTRODUCTION**
**SOLAR RADIATION**
General Characteristics
Spatial and Temporal Variations of Solar Radiation
**THE ENERGY BALANCE AND NET RADIATION**
General Considerations
Energy Fluxes at the Interface
Net Radiation
Land versus Water
Geographic and Time Variations of the Energy Balance Components

## 2.1 INTRODUCTION

One of the controls of climate is solar radiation. The sun provides virtually all—some 99.97 percent—of the energy required for the physical processes taking place in the earth-atmosphere system. Spatial (geographic) and temporal (time) variations in shortwave (solar) radiation play a decisive role in determining the long-term means (arithmetic averages), as well as short-term variations, of the climatic elements. In this chapter we consider the physical basis for temperature and proceed from such nonterrestrial considerations as the geometry of earth-sun relationships, through atmospheric modifications of shortwave radiation, and then to the earth's surface, which, upon receipt of this radiation, acts as the "furnace" for heating the atmosphere. We use a one-dimensional perspective here, emphasizing the vertical fluxes of radiation and heat. In subsequent chapters we will consider how horizontal motions of the atmosphere influence the climatic elements.

## 2.2 SOLAR RADIATION

### 2.2.1 GENERAL CHARACTERISTICS

The spectrum of the sun's radiation is divided according to wavelength. Wavelength is usually measured in micrometers ($\mu$m), a unit of length equal to one ten-thousandth of a centimeter (0.0001 cm). Solar radiation that is meteorologically important has wavelengths ranging from about 0.15 $\mu$m to about 4.0 $\mu$m. Of this, 9 percent is in the **ultraviolet** range, with wavelengths ($\lambda$) equal to or less than 0.4 $\mu$m; 45 percent is in the **visible** range (0.4 $\mu$m $< \lambda <$ 0.74 $\mu$m); and 46 percent is in the **infrared** ($\lambda >$ 0.74 $\mu$m). One of the fundamental radiation laws states that the higher the temperature at which a substance radiates, the shorter are the wavelengths in which this radiation is emitted (Wien's law). Thus, because the sun is a relatively hot body, it radiates at relatively short wavelengths. The earth's surface and its atmosphere are cooler than the sun and thus radiate at longer wavelengths. Longwave radiation emitted by the earth's surface is called *terrestrial radiation*; that emitted downward by the atmosphere is atmospheric counterradiation.

The **flux density** of solar radiation, defined as the amount of radiant energy passing through a unit area in a unit of time, decreases with the square of distance from the sun. This is symbolized as

$$Q \propto \frac{1}{R^2}$$

where $Q$ is a general indicator for the flux density of solar radiation and $R$ is the earth-sun distance. For the earth at its mean distance from the sun this figure is 1.95 cal cm$^{-2}$ min$^{-1}$ (calories per square centimeter per minute). If we use the unit called the langley (ly), defined as 1 cal cm$^{-2}$, we get 1.95 ly min$^{-1}$. This is equivalent to 1360 watts per square meter (W m$^{-2}$). It should be understood that this figure refers to a quantity of radiant energy passing through an area of one square centimeter in one minute. Specifically, it is called the **solar parameter** ($S$) for the earth (also called the *solar constant*); that is, it is the flux density of solar radiation at the outer boundary of the earth's atmosphere that is received on a surface held perpendicular to the sun's direction at the mean distance between the earth and the sun. Notice that this definition implies that the value 1.95 would be less if the earth were at a distance greater than the mean earth-sun distance (and vice versa) and would be less if the receiving surface were other than perpendicular to the solar beam. This value also would be less if the measurement were made within the atmosphere or at the earth's surface, because some radiation is depleted by the atmosphere by absorption and backscattering to space.

Notice that the solar parameter has a "per area" quantity (per square centimeter) in its units. To find the total amount of solar radiation intercepted by the earth, we must multiply the solar parameter by the area over which this radiation is absorbed. The sun "sees" the earth as a disk, and the area of this disk is the intercept area. The total amount is thus the solar parameter times the intercept area, or

$$S\pi r^2$$

where $r$ is the mean earth radius, or $6.37 \times 10^8$ cm. This amount is spread unequally over the earth's surface, an area four times as great as the intercept area (since the surface area of a sphere is four times its cross-sectional area). Thus, the average amount available at the top of the atmosphere, per unit area and time, is one-fourth the solar parameter, or very nearly 0.50 ly min$^{-1}$.

Unless the earth-atmosphere system warms or cools over long periods of time, it must return to space as much energy as it receives from the sun. Because the latter amount is known, it is possible to calculate the mean temperature of this system, called the **planetary temperature.** This is done by means of an equation which relates temperature to radiation, as shown in the boxed explanation on this page. This temperature, just about 255K (K = °C + 273; see Figure 3.1 and accompanying explanation), is lower than the earth's mean surface temperature but is higher than the average temperature at which the atmosphere radiates.

### 2.2.2 SPATIAL AND TEMPORAL VARIATIONS OF SOLAR RADIATION

The fundamental causes of the spatial and temporal variation of climate are related to the altitude of the

---

**CALCULATION OF EARTH'S MEAN PLANETARY TEMPERATURE**

Earth's mean planetary temperature is calculated on the assumption that the earth-atmosphere system radiates as much energy to space as it receives from the sun, and thus, at least over long periods of time, does not warm or cool. The flux density of received radiation is given by the solar parameter, 1.95 cal cm$^{-2}$ min$^{-1}$, or 1360 W m$^{-2}$. This amount is spread out over an area four times the intercept area because the area of a sphere is four times that of a circle of the same diameter. So, the average amount reaching the outer atmosphere over a long period of time is 340 W m$^{-2}$. Of this amount 70 percent is absorbed and 30 percent reflected, so the amount actually available is reduced to

$$0.70(340 \text{ W m}^{-2}) = 238 \text{ W m}^{-2}$$

which is the flux density of radiation which must be emitted if the earth emits to space as much radiation as it receives from the sun.

The radiation emitted by a black body is related to its temperature by the Stefan-Boltzmann equation:

$$I = \epsilon \sigma T^4$$

where $\epsilon$ is the emissivity (assumed to be unity, although the earth-atmosphere system does not radiate exactly as a black body); $\sigma$ is the Stefan-Boltzmann constant, equal to $5.67 \times 10^{-8}$ W m$^{-2}$ K$^{-4}$; and $T$ is the temperature at which the earth-atmosphere system radiates. Given that we know $I$, $T$ can be solved as follows:

$$T = \left(\frac{I}{\epsilon\sigma}\right)^{1/4}$$

$$T = \left(\frac{238 \text{ W m}^{-2}}{5.67 \times 10^{-8} \text{ W m}^{-2} \text{ K}^{-4}}\right)^{1/4} = 254.5\text{K}$$

This is equivalent to $-18$°C (0°F), and is the overall temperature of the earth-atmosphere system. The mean earth surface temperature is higher, and the mean temperature of the atmosphere is lower, than 254K.

sun in the sky as seen by an observer at the earth's surface. Altitude here means the angular displacement between the sun and the horizon. In general, the warmest time of the day occurs when the sun is highest (altitude is greatest); the coolest time occurs just before sunrise. On an annual basis, for latitudes outside the tropics, the warmest months are those during which the average altitude of the sun is highest and vice versa. Actually, the coincidence is not exact and temperatures lag behind maximum and minimum sun altitudes; the reasons for this will be considered in Chapter 3. In our discussions we will not use the average altitude of the sun during a day, but the altitude of the sun at noon, i.e., the highest angle attained in a 24-hour period.

The relationships between sun altitude and solar radiation exist because the flux density—or more generally, the intensity—of solar radiation depends on the angle at which this radiation strikes the earth's surface. At low sun altitude a beam of radiation of a given cross-sectional area is spread out over the surface, while at high altitude that same radiation is relatively concentrated. This principle can be demonstrated by directing the beam of a flashlight directly downward to a tabletop. When the beam is perpendicular, the lighting is most intense because it illuminates a small, circular area. As the angle between beam and tabletop changes, the same amount of light is spread out over an increasingly larger area, and the intensity of illumination decreases. For the solar beam this principle is illustrated in Figure 2.1. Flux density of solar radiation also is reduced by the atmosphere itself as the solar beam penetrates it, which is considered later in this section.

This principle can be verified on the basis of experience. What is perhaps not as obvious is that the heating of the earth's surface by the sun depends on *duration* as well as intensity. The day-length period (sunrise to sunset) is important; obviously, the longer the sun is above the horizon at a given latitude, the more radiation is received and the more heat is produced.

### Orbital Characteristics

These two factors, intensity and duration of solar radiation, vary quite predictably at the outer limits of the earth's atmosphere and somewhat less so at the surface. To understand how these factors vary in time and with latitude, we must investigate the geometrical relationships between the sun and the earth. The manner in which the earth is oriented toward the sun—or, more particularly, its *orbital characteristics*—determines the large-scale, or global, features of climate and especially of temperature. These orbital characteristics are **revolution, rotation, inclination,** and

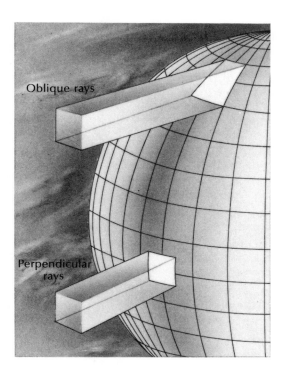

**FIGURE 2.1**
The angle at which the sun's rays strike the earth's surface determines the intensity of the solar radiation received there. This intensity is greatest when the solar beam is perpendicular (at right angles) to the surface. (From *An Introduction to Climate,* 4th ed., by G. T. Trewartha. Copyright © 1968 by McGraw-Hill Book Company. Used with the permission of McGraw-Hill Book Company.)

**parallelism.** The earth revolves around the sun in a nearly circular orbit, taking 365.25 days to complete one revolution. The earth-sun distance varies from 147 million km (at the *perihelion,* the first week in January) to 152 million km (at the *aphelion,* during the first week in July). Thus the changes from the mean distance of about 150 million km are quite small, about 1.7 percent of this mean distance. These differences have no significant implications for the present global climate.

The earth also rotates on its axis, counterclockwise when viewed from over the North Pole, with one rotation taking 24 hours. To an observer in either hemisphere, then, the sun rises at the eastern horizon and sets at the western horizon.

The earth's orbit around the sun describes a geometrical plane, called the **plane of the ecliptic.** The earth's axis is inclined 23.5° from a perpendicular to this plane. Furthermore, this inclination not only remains constant throughout a revolution, but occurs in such a way that the axis is parallel to itself at all times. This is the same as saying that the poles are always directed toward the same points in deep space, since the distance around the orbit is negligi-

**FIGURE 2.2**
Earth-sun relationships.
(From A. N. Strahler,
*Introduction to Physical
Geography.* New York: John
Wiley & Sons, 1965.)

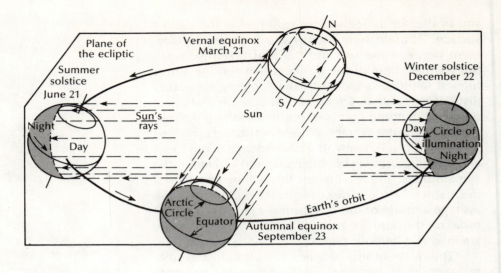

ble compared with most astronomical distances. Thus, the North Pole always points to within a few degrees of Polaris, the North Star.

The earth's orbital characteristics are illustrated in Figure 2.2. Try to imagine this diagram in motion, with the earth rotating and maintaining parallelism of its inclined axis as it revolves around the sun. It also will be helpful to remember the correct size and scale. If the earth-sun distance were scaled down to the length of a football field (100 yd, or about 90 m), the earth would be a sphere about one-third inch in diameter (0.8 cm); the sun, at the other goal line, would be 2 ft, 9 in. (85 cm) in diameter.

### Intensity and Duration

We can now associate these aspects of earth-sun geometry with the two characteristics of solar radiation stressed earlier, intensity and duration. To determine the variation of intensity, by latitude and time of the year, we must first follow the latitude at which the noon rays of the sun are perpendicular to the surface (actually, to a tangent at the surface). Notice in Figure 2.2 that there are two times of the year when the sun's rays are perpendicular at the equator. That is, an observer at the equator would see the sun at the zenith, or directly overhead, twice annually. These dates are March 21 and six months later, September 23. From March 21 to June 21 the latitude at which the noon rays of the sun are overhead moves into the Northern Hemisphere as far as 23.5° N. Then the sun "moves" back toward the equator, which it passes on September 23, and continues as far as 23.5° S, which it reaches on December 22. From there the progression is back toward the starting point on March 21. This latitudinal variation of the noon rays of the sun is the **solar declination.**

Figure 2.2 uses familiar names for these dates. Those in March and September are referred to as the **equinoxes;** in the Northern Hemisphere the vernal, or spring, equinox occurs in March and the autumnal, or fall, equinox in September. The solstices occur when the noon rays of the sun are farthest from the equator: at the summer solstice (June 21) and the winter solstice (December 22). It is important to note that the equinoxes and solstices occur as a consequence of earth-sun geometry and are unrelated to the dates when the earth is closest to and farthest from the sun.

At the equator, then, the earth's surface is heated most intensely at the time of the equinoxes. The farther one travels from the equator, the more removed the sun is from the zenith, so heating decreases with increasing latitude. At the poles the sun circles the horizon at the equinoxes. At the June solstice the Northern Hemisphere is inclined toward the sun and thus is heated more strongly than the Southern Hemisphere. At this time of year, and considering intensity only, radiation decreases with distance north and south from 23.5° N. North of 66.5° N, the sun never sets, while for the corresponding latitudes in the Southern Hemisphere the sun is not seen. The latitudinal reverse of these conditions occurs at the December solstice.

The latitudinal and temporal variation of the duration of solar radiation, or of day length, can be understood by referring to the **circle of illumination,** illustrated in Figure 2.2. This is a circle with a diameter equal to that of the earth, which separates the light and dark halves of the earth. At the equinoxes this circle passes through the poles, and thus every latitude experiences 12 hours of day and 12 hours of night (neglecting civil twilight and other astronomically defined phenomena that result in relatively minor changes in these lengths). This can be verified by counting the number of degrees of longitude, at any latitude, around the light half and dividing by 15. Since at all latitudes this longitudinal span is 180°, the day length is 12 hours.

A different latitudinal division of day and night applies at all other times of the year, the most extreme difference occurring at the solstices. On June 21 the circle of illumination divides the earth in such a way that all latitudes in the Northern Hemisphere have greater than 12 hours of day, and day length increases with increasing latitude in this hemisphere. This increase is from 12 hours at the equator to 24 hours at 66.5° N. From there to the pole the sun never sets.

These same conditions apply to the Southern Hemisphere at the December solstice, while the reverse conditions (i.e., a decrease in day length with increasing latitude and constant night poleward of 66.5° N) apply to the Northern Hemisphere at the December solstice and to the Southern Hemisphere at the June solstice (Figure 2.2). At the solstices and the equinoxes—and for that matter, at any time of the year—day length may be determined by counting the number of degrees of longitude around the lighted portion at any latitude. Figure 2.3 shows the relationship between the altitude of the noon sun and the radiation intensity and day length (duration) for both solstices in the Northern Hemisphere. A similar illustration for the two equinoxes would show the height of the noon sun decreasing uniformly from 90° at the equator to 0° at the North Pole. Day length would be 12 hours at all latitudes, although its definition is somewhat arbitrary at the North Pole, where the sun circles the horizon in 24 hours. At the equator day length is 12 hours throughout the year.

When both intensity and duration are combined, the following patterns of the spatial and temporal variation of solar radiation result. At the lowest latitudes the sun is never far from the zenith at noon, and day length is very nearly 12 hours throughout the year. This results in relatively stable receipts of solar radiation with no pronounced changes from month to month within the year. With an increase in latitude in the summer hemisphere (Northern Hemisphere in June, Southern Hemisphere in December), radiation intensity increases to a maximum at 23.5° at the solstices, then decreases to the poles, while day length increases to 66.5°. In the winter hemisphere both intensity and duration decrease over the entire 90° span (see Figure 2.3). At all other times of the year a situation intermediate between the equinoxes and solstices exists.

Solar radiation intensity and duration now can be portrayed graphically. First, we must look at the spatial and temporal variation in receipts of solar radiation outside the atmosphere, called **extraterrestrial solar radiation** (see Figure 2.4). Extraterrestrial radiation is equivalent to radiation that would be received at the surface in the absence of an atmosphere.

Several features of Figure 2.4 are important. Maximum receipts of solar radiation occur at the poles during their respective solstices. This may seem sur-

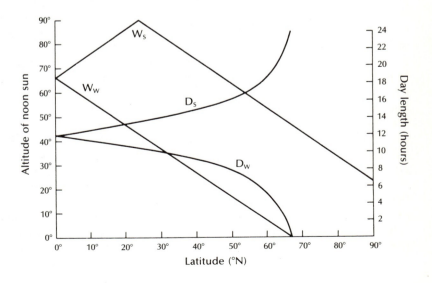

**FIGURE 2.3**
Variation of day length (*D*) and altitude of the sun at noon (*W*) with latitude in the Northern Hemisphere, for the summer (June) solstice (subscript *s*) and winter (December) solstice (subscript *w*).

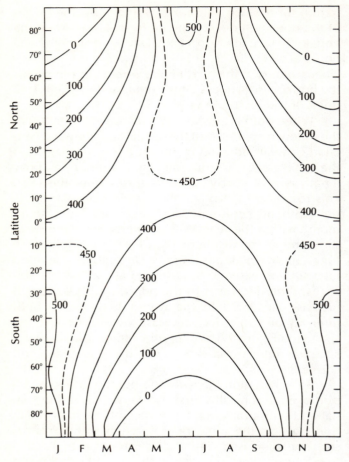

**FIGURE 2.4**
The daily variation of extraterrestrial solar radiation as a function of latitude and time of year. Units are watts per square meter averaged over a day. (From data prepared by Starley L. Thompson.)

prising, but it should be remembered that although the sun is fairly low in the sky, it never sets. Six months later, however, no radiation is received at the poles. At the equator values are uniformly high and there is only a small variation throughout the year. Maxima occur at the equinoxes, minima at the solstices. At all latitudes higher than 23.5° the range between minimum and maximum daily receipts during the year increases with increasing latitude. At 30°, for example, the range is about 250 W m$^{-2}$; at 60° it is about 450 W m$^{-2}$. Notice also that extraterrestrial radiation is "mirror-imaged" in the two hemispheres at the equinoxes, but that more radiation is received in the Northern Hemisphere at times between the March and September equinoxes, and more is received in the Southern Hemisphere between the September and March equinoxes. These statements ignore the relatively small differences that exist between hemispheres because of changing earth-sun distance.

In terms of its association with temperature, Figure 2.4 is misleading. Indeed, if we were to make a direct correspondence between extraterrestrial solar radiation and temperature, the warmest places on earth would be the North and South Poles at their respective solstices! The radiation that survives its passage through the earth's atmosphere and arrives at the surface is of more direct interest. This radiation, called **insolation** (a contraction of *in*coming *sol*ar radi*ation*), is quite variable. Unlike the values in Figure 2.4, which are virtually constant from year to year, radiation at the surface is affected differentially in time and space by the atmosphere. These factors will be considered quantitatively in Section 2.3.5. At this point it is sufficient to know that there are two factors chiefly responsible for this differential depletion of solar radiation (Figure 2.5). First, the depth of the atmosphere through which the solar beam must pass, the path length, varies with latitude. This path length is greater at high than at low latitudes, and thus the depletion due to air molecules generally increases with latitude.

Second, the various solid and liquid constituents of the atmosphere scatter, reflect, and absorb solar radiation. Clouds and nonwater suspended particulates such as ash, soot, and salt, all of which are highly variable in time and space, are responsible for marked variations in insolation. Water vapor, similarly highly variable, and other gases such as ozone and oxygen, also absorb some solar radiation.

Insolation can be represented by constructing a graph similar to Figure 2.4 if we simplify the effects of these factors. Figure 2.6 shows insolation for a clear atmosphere and assumes that 70 percent of extraterrestrial radiation is transmitted to the surface when the sun is at the zenith. For lower sun angles this percentage decreases, in keeping with the greater depth of the atmosphere the solar beam must penetrate. Thus, of the two factors we have just specified, the first is approximated and the second ignored.

### Insolation Patterns

We can use Figure 2.6 to study insolation patterns. Since on a global basis insolation is well correlated with mean air temperatures at the surface when the effects of elevation on temperature are ignored, this illustration reflects large-scale latitudinal variations of temperature. All of the specified features of Figure 2.4 also apply to Figure 2.6, with two important excep-

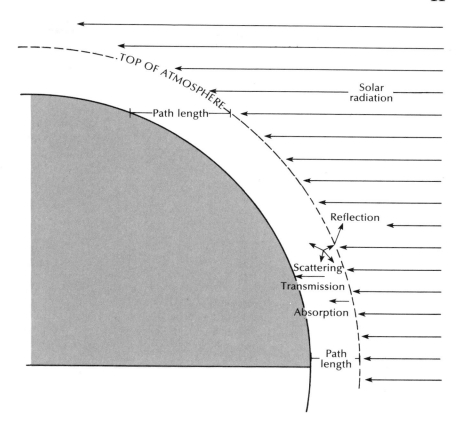

**FIGURE 2.5**
Differential atmospheric depletion of solar radiation with latitude, and absorption, reflection, scattering, and transmission of solar radiation within the atmosphere. The thickness of the atmosphere is greatly exaggerated.

tions. First, values are reduced overall, and the reduction of amounts increases with increasing latitude. Second, maximum receipts are no longer at the poles, but at about 30° in both hemispheres at their respective solstices. It is at these places and times, then, that maximum temperatures are reached, as will be noted again in Chapter 3.

Finally, a description of spatial and temporal variation of solar radiation must include an examination of the insolation that actually reaches the earth's surface and is therefore differentiated by longitude as well as by latitude. This aspect is represented on maps rather than graphs. Unlike Figures 2.4 and 2.6, which were constructed according to mathematical formulas that incorporate specified geometric principles, Figures 2.7, 2.8, and 2.9 represent averages of observed values. At about 800 weather stations around the globe measurements are made of *global radiation*, the shortwave radiation falling on a horizontal surface. When a record of sufficient length has been accumulated, averages (in the case of Figures 2.7, 2.8, and 2.9, monthly averages) are calculated. These values are then mapped and *isopleths* (lines connecting equal values) drawn. Because these are observed values, and because climate is at least somewhat variable from one averaging period to another, we should not interpret these values, or for that matter averages of other weather elements, too literally.

Figures 2.7, 2.8, and 2.9 show observed values of insolation for June and December, the months in which the solstices occur, and the total insolation received in an average year. These maps should be compared with Figure 2.6, which shows only variations in latitude and time and ignores longitudinal differences. The "real" world is represented by the maps. Here the variables are latitude and longitude; time does not vary.

There are other features of interest in these maps. First, all three show significant meridional (east-west) differences due chiefly to variations in cloudiness. These variations may be ascribed to the location of continents, since continents tend to be less cloudy than oceans; to semipermanent low- and high-pressure areas (see Chapter 4), since the former are generally associated with more widespread cloudiness than the latter; and to elevation. Second, zonal patterns (north-south differences, east-west similarities) prevail in middle and high latitudes but are not apparent in the tropics. Next, maximum values of insolation occur not at the very lowest latitudes but in the subtropics. On all three maps this can be ascribed to a lesser degree of cloudiness. In addition, in June

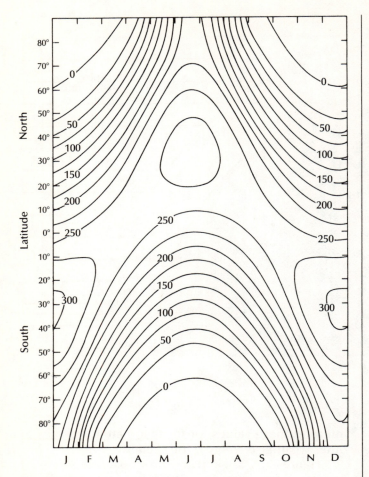

**FIGURE 2.6**
The daily variation of solar radiation at the earth's surface, as a function of latitude and time of year, assuming that 70 percent of extraterrestrial radiation is transmitted to the surface when the sun is at its zenith. (From data prepared by Starley L. Thompson.)

and December the maximum insolation values occur at about 30° (north and south, respectively); this can be seen clearly in Figure 2.6. The last features to examine are the north-south gradients of insolation, which have a large-scale similarity to gradients of temperature. They are most pronounced in the middle and high latitudes of the winter hemisphere.

Having explored the many aspects of the spatial and temporal variation of solar radiation, we should address the consequences of this variation for temperature at the earth's surface. Again, we can simplify our discussion by saying that large-scale (global) patterns of insolation are indicative of those of temperature. Figure 2.6 indicates that total annual insolation is greatest in the lowest latitudes and decreases poleward. The range of values is least in low latitudes and increases poleward. This is represented schematically in Figure 2.10 for an earth of uniform substance. Actual patterns of temperature will, of course, not be exactly zonal. The various earth surface substances (water, soil, snow, and ice) respond quite differently to insolation, and variations in cloudiness have already been noted. Still, Figure 2.10 shows the major features of interest—features which may be attributed to the spatial and temporal variation of solar radiation, which in turn are a consequence of earth-sun geometry.

# 2.3

## THE ENERGY BALANCE AND NET RADIATION

### 2.3.1 GENERAL CONSIDERATIONS

In the preceding section we examined the large-scale (in both space and time) variations of both extraterrestrial radiation and insolation. It should be clear now that climatic phenomena such as the seasons, low-latitude warmth, and high-latitude cold can be explained readily in terms of these variations. Now we will narrow our focus somewhat, to the physical processes taking place at the boundary between the atmosphere and the solid or liquid earth. Here, at the **interface,** most of the energy available to the earth-atmosphere system is produced. This production and transformation of energy to a great extent governs atmospheric motions and is responsible for variations in the elements that describe weather and climate. In this section, then, we will examine these energy transformations and exchanges at the interface.

The concept of the interface is critical to the following discussion and needs some elaboration. The interface is an infinitesimally thin boundary between the atmosphere and the earth's surface; that is, it is a discontinuity between two media differing greatly in density and other physical characteristics. This surface can be water, bare or vegetated soils, or even the roofs of buildings. Try to think of energy fluxes as occurring in such a way that there is no energy transfer across the interface. Therefore, fluxes either originate elsewhere and are directed toward the interface, or they originate at the interface and are directed away from it. Each of the separate energy transfer processes about to be detailed fits one or the other of these two categories.

Part of the solar radiation available at the outer limits of the atmosphere is transmitted to the interface

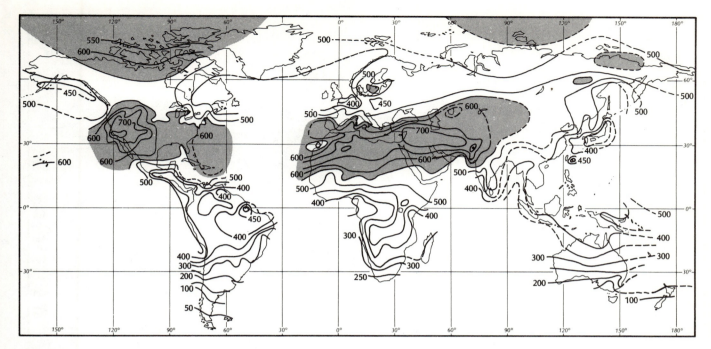

**FIGURE 2.7**
Average June insolation, in langleys per day. (After G. O. G. Löf, J. A. Duffie, and C. O. Smith, *World Distribution of Solar Radiation*. Report No. 21. University of Wisconsin Engineering Experiment Station, 1966.)

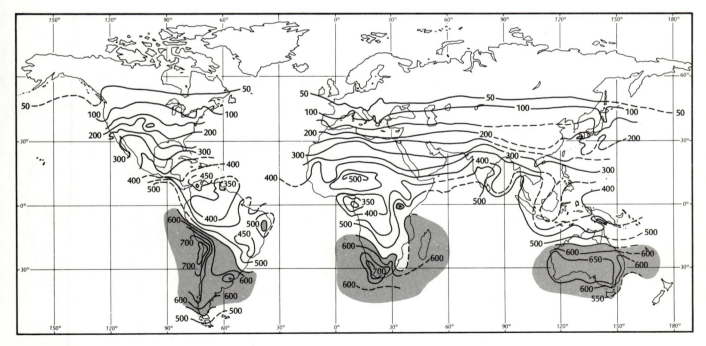

**FIGURE 2.8**
Average December insolation, in langleys per day. (After G. O. G. Löf, J. A. Duffie, and C. O. Smith, *World Distribution of Solar Radiation*. Report No. 21. University of Wisconsin Engineering Experiment Station, 1966.)

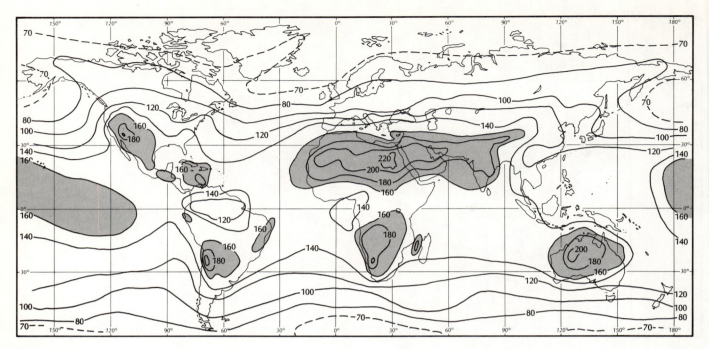

**FIGURE 2.9**
Average annual insolation, in kilolangleys. (After H. E. Landsberg. From E. Rodenwaldt and H. J. Jusatz, eds., *World Maps of Climatology*, New York: Springer-Verlag, 1965.)

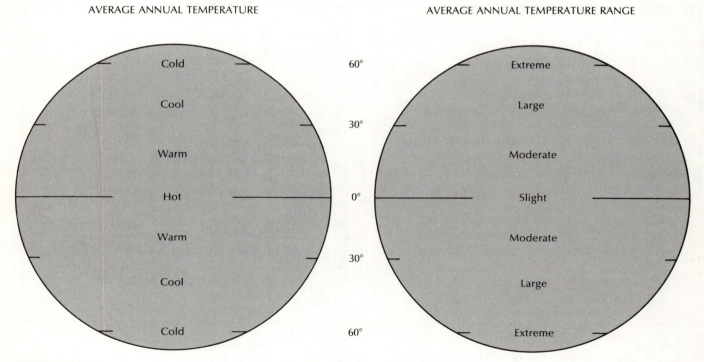

**FIGURE 2.10**
Average annual surface temperatures and mean annual range of temperature, as inferred from insolation.

# SEC. 2.3 | THE ENERGY BALANCE AND NET RADIATION

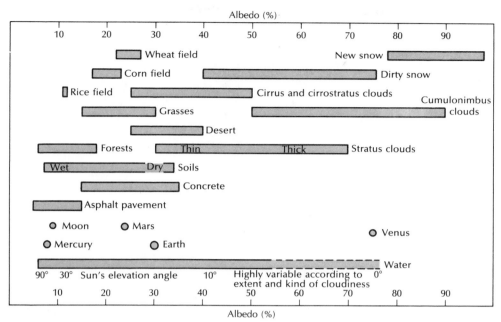

**FIGURE 2.11**
Albedos of common earth surface substances, clouds, and some planets.

(see Figure 2.5). For the earth as a whole, and for all times of the year, this proportion is about one-half; the remaining half is reflected from clouds or absorbed and scattered in the atmosphere. But not all of the radiation reaching the surface is available to take part in energy exchange. The various earth surface substances reflect a part of the insolation back to the atmosphere and space. The term **albedo** is used to indicate the reflected proportion. In general, lighter colored substances have higher albedos; that is, they reflect more radiation. Figure 2.11 indicates albedos for common earth surface substances, various types of clouds, and a few planets. Natural and artificial surfaces and crops generally have low albedos. New snow is highly reflective, but the albedo decreases as snow becomes dirty. Clouds have moderate to high albedos, depending largely on the type and thickness. Since water occupies almost three-quarters of the earth's surface, its albedo is very important. Although it varies depending on the sun's elevation angle, the albedo of water is quite low, around 6 to 10 percent for angles of 30° to 90°.

A distinction is made between **surface albedos**, which are highly variable from place to place, and **planetary albedos**. The latter can refer to the albedo of the entire earth-atmosphere system, which is about 0.30 (30 percent). This is an approximate measure of how bright our planet would appear to an observer seeing it from space. For comparison, the planetary albedos of some other planets as well are shown in Figure 2.11. Planetary albedo also can refer to the proportion of incident radiation reflected from a portion of the earth and atmosphere when viewed from space.

The unreflected portion of insolation is available for energy transformations at the interface. In general, increasing insolation raises the temperature of the absorbing substances at the air-land or air-water interface; decreasing insolation—or the absence of it, as at night—lowers the temperature. The schematic diagram in Figure 2.12 accounts for all the energy fluxes involved.

## 2.3.2 ENERGY FLUXES AT THE INTERFACE

Energy fluxes at the interface are accomplished by the processes of **radiation, convection,** and **conduction,** the three means of heat transfer. These processes are shown schematically in Figure 2.12. Insolation, composed of both direct beam ($Q$) and diffuse ($q$) radiation, is represented by the quantity ($Q + q$). As we have learned, not all of this insolation is available at the interface. The depletion due to surface albedo is given as ($1 - a$), where $a$ is the surface albedo. The proportion of insolation actually available at the interface is therefore ($Q + q$)($1 - a$). The earth's surface also radiates, but in longer wavelengths because earth surface substances are much cooler than the sun. The

**FIGURE 2.12**
Energy fluxes at the land-air and water-air interfaces.

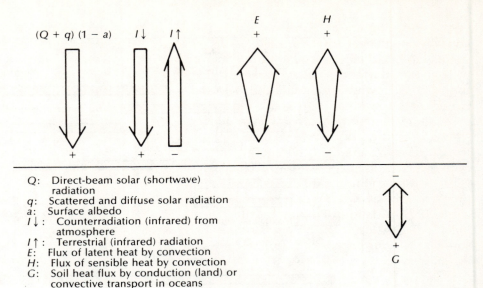

Q: Direct-beam solar (shortwave) radiation
q: Scattered and diffuse solar radiation
a: Surface albedo
$I\downarrow$: Counterradiation (infrared) from atmosphere
$I\uparrow$: Terrestrial (infrared) radiation
E: Flux of latent heat by convection
H: Flux of sensible heat by convection
G: Soil heat flux by conduction (land) or convective transport in oceans

symbol $I\uparrow$ (I for infrared) indicates terrestrial radiation directed upward. The atmosphere also radiates in the infrared, and the portion directed downward to the interface is important in energy fluxes. This quantity is shown as $I\downarrow$ in Figure 2.12.

The nonradiative processes are the convection of (1) sensible heat and (2) latent heat, and (3) the conduction and convection of heat into and out of the solid and liquid earth, respectively. Convection is very generally defined here as mass motions within a fluid (air or water) resulting in the transport and mixing of the properties of that fluid. This includes convection due to density differences within the atmosphere. Density differences result when the layer of air nearest the interface is heated, becomes less dense, rises, and is replaced by the layer of air above it. Convection also includes mixing induced by mechanical forces such as the frictional drag imposed when air moves over rough or irregular surfaces (in the atmosphere) and by wind-induced roughness, tides, and currents (in large water bodies). The convective transport of sensible heat into the atmosphere is represented by the symbol H in Figure 2.12.

The flux of latent heat proceeds in a similar fashion, except that the heat energy involved is that which is latent in both water vapor and liquid water, having been added to water by the process of melting, and to water vapor by the processes of melting and evaporation. Relatively large amounts of heat energy (about 600 cal g$^{-1}$ to effect evaporation and 80 cal g$^{-1}$ to effect melting) are required for these transformations, or *phase changes*. Thus, water in its vapor phase has an amount of heat which is proportional to its temperature, plus the latent heats required by both melting

and evaporation. The reverse phase changes—freezing and condensation—release heat. For the entire earth-atmosphere system melting and evaporation occur predominantly at the interface and thus remove heat from it. Freezing and condensation occur largely in the atmosphere, where they are a highly significant heat source. The contribution of latent heat is shown by E in Figure 2.12.

Unlike air and water, which are fluids and therefore capable of motions within themselves, heat is transferred in solid substances by the relatively slow process of conduction, which involves only molecular motion. At various times of the day and year, heat is conducted down into or up from the solid earth, that is, away from or toward the interface. Conduction in the solid, and convection in the liquid, earth are represented by G in Figure 2.12.

The plus and minus signs in Figure 2.12 are important, as is the ability to recognize which fluxes are radiative and which are nonradiative. To simplify the mathematics of Section 2.3.3, radiative fluxes directed toward the interface are designated as positive, and vice versa. The nonradiative fluxes that originate at the interface and are directed away from it are positive ($+H$, $+E$, $+G$); those directed toward the interface are negative ($-H$, $-E$, $-G$). Over periods of a year or longer both H and E are positive, but the reverse can occur over shorter time periods. During temperature inversions, for example, when temperature increases with height, H may be directed toward the interface, although such a stable atmosphere is usually not conducive to vertical circulations. Similarly, instead of evaporation or melting, condensation or freezing (of water vapor and liquid water, respectively) may occur here, producing energy gains instead of losses. The conduction of heat through solid earth substances such as soil, and its convection through water, varies with time of day and time of year. Heat tends to be propagated downward during the day and in the spring, and upward at night and during the fall.

There are minor heat exchange processes not shown in Figure 2.12. These include the flux of heat from the earth's interior; the energy used in photosynthesis; the dissipation of the mechanical energy of wind, waves, and tides and currents; heat transfer by precipitation (as when cold rain falls on a warm earth); the oxidation of biological substances (forest fires); and volcanic eruptions and artificial heat sources. For our purposes here, these processes are negligible.

### 2.3.3 NET RADIATION

Because for periods of approximately a day or multiples of a day energy gains must equal losses, the algebraic sum of all significant fluxes at the interface must be zero. Over shorter periods, and instantaneously, the fluxes are also zero if heat storage is accounted for. In other words, a decrease in one flux (or a net decrease in two or more fluxes) must be compensated by an increase in another flux (or a net increase in two or more other fluxes), and total positive and negative fluxes must balance. We may therefore write an equation for the summation of the fluxes:

$$(Q + q)(1 - a) + I\downarrow - I\uparrow + H + E + G = 0 \quad (2-1)$$

$H$, $E$, and $G$ can be either positive or negative quantities, depending on whether they are directed away from or toward the interface, respectively. As we noted at the beginning of this chapter, our perspective is only a vertical one. This formulation therefore excludes any horizontal fluxes and considers only fluxes in the vertical at the interface.

We now introduce a new concept, **net radiation** ($R$). This is defined as

$$R = (Q + q)(1 - a) + I\downarrow - I\uparrow \quad (2-2)$$

Notice that $R$ includes only radiative terms. Because energy gains must equal losses, we can rearrange Equation (2–1) to get

$$(Q + q)(1 - a) + I\downarrow - I\uparrow = H + E + G \quad (2-3)$$

By substituting Equation (2–2) into Equation (2–3), we get

$$R = H + E + G \quad (2-4)$$

Equation (2–4) indicates that the energy acquired by the radiation fluxes that comprise net radiation is utilized by the processes represented in the right-hand side of the equation. In general, net radiation is positive during the day, and thus the right-hand side also must be positive. This means that one or more of three processes will occur: (1) since heat is being generated at the interface, sensible heat warms the air ($+H$); and (2) evaporation or melting occurs if water (or solid $H_2O$) is present and if plants are transpiring moisture ($+E$); (3) a downward flux of heat warms the soil ($+G$). At night, and in some other restricted circumstances, $R$ is negative because the ($Q + q$) term vanishes and, in general, $I\uparrow$ exceeds $I\downarrow$. This means that the signs of the right-hand quantities are reversed. Thus sensible heat may be transferred toward the interface, condensation or freezing may occur, and there is an upward flux of heat in the surface substance to replace heat being lost from it at the interface. Notice that when $R$ is positive, the three right-hand terms are not necessarily all positive, but the entire quantity must be. The same reasoning applies when $R$ is negative.

These energy balance equations provide a valuable framework for expressing energy fluxes at the air-land and air-water interfaces. They can be applied to nearly all time-space scales, although we will see later in this chapter that the principal applications involve two extremes: annual averages of the components for continental and global areas, and hourly values for a day at particular sites. Furthermore, these equations can be applied in other than terrestrial circumstances. The energy balance at the interface of other astronomical bodies (where an interface exists) also can be calculated. On the moon, for example, the same energy balance applies, except that the $q$, $I\downarrow$, $H$, and $E$ terms vanish because the moon has no significant atmosphere.

### 2.3.4 LAND VERSUS WATER

The magnitude of the energy balance components and their variation with time and place are fundamental to the study of the physical basis of the earth's climate. Before we present examples for particular locations, however, we must examine the various substances comprising the earth's surface. If the earth were of a uniform substance then, just as Figure 2.10 suggests for temperature, the components of the energy balance equations would differ only latitudinally. But the earth's surface substances are not uniform. Nearly three-quarters of the surface is water, while land surfaces vary from exposed soil, to soil covered by vegetation ranging from sparse grasses to dense forests, to snow- and ice-covered areas of the high latitudes and high elevations. People sometimes impose their own surface materials; parking lots are an example.

Each of these substances has its own physical and thermal properties, and these influence the magnitude of the energy balance components. $I\uparrow$, $H$, $E$, and $G$ are especially dependent on these properties. We will not specify the influences of all these surface substances. We will distinguish between the properties of the most fundamental of these: water and land. Next to the latitudinal variation of insolation, the most significant control of climate is the way land and water differ in their response to heat energy produced at, or removed from, the interface.

Positive values of net radiation are accompanied by fluxes of sensible and latent heat by convection into the atmosphere, and by a flux of heat by conduction into the solid earth and by convection into the oceans. When water is present at the surface, either as open water or as soil or plant moisture, most of the net radiation is expended for evaporation and transpiration, although the rate at which these processes occur varies with the dryness of the overlying air and the extent of air movement. Over the oceans about 90 percent of the net radiation is used to evaporate water. With such large expenditures of radiation there is little left to promote fluxes into the atmosphere or the ocean, and the interface temperature is therefore less than it would be if no water were available to be evaporated. The same reduction of temperature applies also over moist land surfaces, but to a lesser degree because for all land areas $E$ is about one-half of $R$. This is discussed further in Section 2.3.5.

The relatively small proportion of $R$ not used in evapotranspiration—or all of $R$ when water is not present—is used to raise the temperature at the interface and promote heat fluxes away from it, if $R$ is positive. If $R$ is negative, the temperature falls and the fluxes are opposite in direction. This temperature rise or fall is strongly dependent on two thermal properties of the interface substances. First, the **heat capacity** (strictly, the volumetric heat capacity) of a substance indicates the ratio of heat absorbed (or released) to the corresponding temperature rise (or fall). Water has a high heat capacity because when large amounts of heat are added to it the observed temperature rise is comparatively small. Soils have varying heat capacities, but in general these range from only one-third to one-half that of water. Air has a very small heat capacity, about one three-thousandth that of water.

Second, how rapidly heat is transported away from or toward the interface depends on the ability of the substance to accomplish such fluxes. This varies considerably from air to soil to water, and the means of heat transport also varies. Both air and water are capable of motion within themselves, and thus heat transfer is accomplished by bulk motions of these fluids—that is, by convection. In air this transport is highly variable, but in general it increases with the rate at which temperature decreases with height. The more rapid this decrease, the faster heat is transferred away from the interface. On a diurnal basis heat ascends to heights of as much as 2000 m, although this will vary depending on the extent of surface heating. On an annual basis this flux extends to the tropopause (see Section 3.5), about 12 km in mid-latitudes, which is the top of the troposphere, or mixed layer of the atmosphere.

Heat produced at the interface over oceans is transported downward by convective overturnings and distributed through a considerable depth. This transport is highly variable, as with air, and depends on wind-induced roughness and tides and currents. On a diurnal basis this transport extends to about 40 m, and on an annual basis to about 150 m, in the mid-latitudes.

Heat transport by conduction through solid substances is described by the property **thermal conductivity.** This transport in soil occurs much less rapidly than by convection in air and water, and for this reason only a shallow layer of soil is heated. The depth varies with the water content of the soil, and heat penetration is deeper in wet soils. Representative diurnal and annual penetrations are 0.2 m and 4 m for dry sand, 0.5 m and 9 m for wet soil, and 0.5 m and 10 m for concrete.

The temperature rise or fall at the interface depends on these thermal properties: heat capacity and the substance's ability to transport heat. In air and water the specific mechanism is convection; in soil it is conduction. Using these terms, we can now explain the way in which these temperature changes occur. Consider first a given amount of net radiation available to the air-water interface. A substantial part of this will be used to evaporate water, but this amount will vary with the dryness of the overlying air and the extent of air movement. Most of the remaining net radiation will be used to warm the water, and this heat will be transported downward and diffused through a considerable depth. Because water has a high heat capacity, the temperature rise through this depth will be minimal. Thus, there will be little left to warm the air. In addition, solar radiation—the chief contributor to net radiation—can penetrate water and thus distribute its warmth through the topmost layer. When net radiation is negative over large water bodies, the temperature fall at the interface also is small. Representative diurnal ranges are only 0.2C° to 0.5C° (0.4F° to 0.9F°). Maximum annual ranges over deep ocean water are about 8C° to 10C° (14F° to 18F°).

Now consider the disposition of that same amount of net radiation at the air-soil interface. If water is present in the soil, part of the net radiation will be used to evaporate it, but this proportion is considerably less than that over an open water surface. The remaining net radiation, or all of it in the absence of soil water, will be used to warm the soil and air. Since soil has a relatively low heat capacity, and especially because heat is diffused through only a small depth in the soil, the temperature rise will be much more marked than at the air-water interface. This marked increase in temperature also means that relatively large amounts of heat will be transported upward by convection because of the great depth of the atmosphere through which heat can be distributed. On the other hand, the quantity of heat transported upward is limited by the very small heat capacity of air. However, the net result is still a relatively pronounced rise in the interface temperature. The maximum increase occurs with a completely dry, sandy soil, which conducts heat very slowly. The increase is less marked in other soils or when soil water is present.

When net radiation is negative, a situation that invariably results when $I\uparrow$ exceeds $I\downarrow$ in the absence of solar radiation, heat transfer away from the interface is through the $I\uparrow$ flux, and, when the surface has cooled to a temperature lower than that of the subsoil, the movement of soil heat is toward the interface. The loss of heat from the interface by terrestrial radiation is only slightly compensated by subsoil heat flux, however, and the interface cools quite rapidly compared with the similar situation over a large body of water. The resulting decrease in interface temperature, in general, is of the same magnitude as the increase in interface temperature when the net radiation is positive. Mean annual and diurnal temperature ranges tend to be at a maximum in the deep interior of continents, where oceanic influences are minimal.

The foregoing discussions can be summarized as follows: *Independent of the evaporation process, the observed rise or fall of temperature at the interface depends inversely on both the joint heat capacities of the interface substances and their ability to carry away heat from the point of generation, i.e., the interface.* Air and water have a relatively large combined heat capacity, and heat produced at their interface is rapidly carried away, especially into the water. Air and land have a small joint heat capacity by comparison, and although heat is transported readily into the atmosphere, its penetration into the soil is quite limited with respect to depth and to the quantity of heat. Thus, over both diurnal and annual heating cycles, the air-land interface temperatures show a greater range than those at the air-water interface. Even more tersely, *land heats and cools more rapidly than water.*

### 2.3.5 GEOGRAPHIC AND TIME VARIATIONS OF THE ENERGY BALANCE COMPONENTS

The geographic and time variations of the energy balance components can now be examined. We begin with a representation of typical diurnal variations of these components. Monthly averages are then used to portray the seasonal variations for different climates. Finally, annual averages for varying latitudinal belts are given.

Figure 2.13 shows the diurnal variation of $R$, $H$, $E$, and $G$ for two simulated land locations during a mid-latitude spring. Location A is in a humid climate, so it is assumed that surface water, either free standing or as plant or soil moisture, is readily available to be evaporated and transpired. Location B is a dry climate with negligible soil moisture.

The dependence of net radiation ($R$) on insolation is clearly shown at both locations. Typically, $R$ reaches a maximum at local noon and is zero shortly after sunrise and shortly before sunset, when insolation just balances the net longwave radiation ($I\downarrow - I\uparrow$). At other times, when insolation is zero, $I\uparrow$ exceeds $I\downarrow$ by a small amount and $R$ is negative. The energy used to evaporate and transpire water is relatively high where such water is available, and at A most of $R$ is used for this purpose. Notice that $E$ is about one-half of $R$ during daylight hours. $E$ is zero during most of the sunset-to-sunrise period, but becomes negative shortly before sunrise, when dew forms. At location B, $E$ is essentially zero throughout the 24-hour period.

The flux of sensible heat away from the interface ($H$) is of course much larger at B, and most of $R$ is used to accomplish this. Still, even at location A an appreciable part of $R$ is expended in $H$. This is a land surface, and although most of $R$ will be used to promote a flux of latent heat, the interface will warm during the daylight hours, and an upward flux of sensible heat also will occur.

The flux of $G$ into and out of the soil is the small-

**FIGURE 2.13**
Diurnal variations of the components of the energy balance for **(A)** humid, and **(B)** dry, mid-latitude areas.

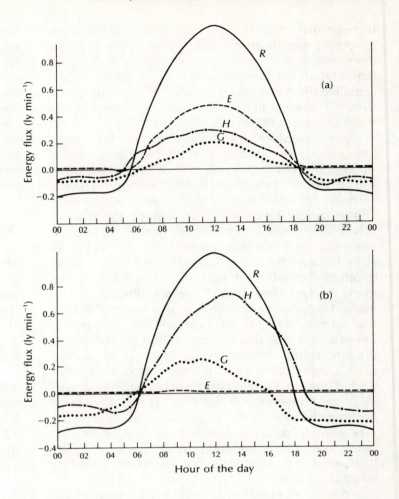

est of the three energy balance components. In the absence of expenditures for $E$, air gains, through $H$, some two to ten times as much energy as does the soil. We recall that this reflects the ability of the atmosphere to transport heat to great heights, compared with the depths to which it is conducted into the soil, even though the air's heat capacity is much less than that of the solid earth. It should also be noted that $G$ is highly variable, depending on the thermal conductivity of the soil. Furthermore, we can see in Figure 2.13 that the area between the zero line and positive values (heat flux into the soil) exceeds the area between the zero line and negative $G$ values. This means that during the spring a small net amount of heat is being accumulated in the soil each day in response to the annual cycle of insolation. In the fall the reverse occurs, and heat is taken from the soil. Over the course of a year $G$ must be zero for land surfaces. For the oceans, however, $G$ is both negative and positive depending on location. The reasons for this will be discussed in Section 4.8.

There are other factors that may affect the patterns seen in Figure 2.13. This figure shows the fluxes expected to occur under clear skies, with light to moderate winds. Cloud cover of any kind would produce reductions and fluctuations in net radiation, chiefly through changes in insolation. The $E$, $H$, and $G$ components would then vary in a more irregular fashion than is shown. As we have noted, we are emphasizing vertical exchanges. But in the presence of strong advection (i.e., high horizontal wind speeds), appreciable quantities of sensible heat may be carried into or out of the column of air above the interface. The same may be true for latent heat, as when "oasis" effects occur. In this situation very dry air is moved across an atypically humid area, and the expenditures of latent heat are greatly magnified, even to the point of exceeding the net radiation. A balance is then achieved, as of course it must be, by negative fluxes of sensible heat (toward the interface).

The scope of our inquiry is now enlarged, and we examine variations in the energy balance components

## SEC. 2.3 | THE ENERGY BALANCE AND NET RADIATION

over the year for particular areas around the globe. From Figures 2.6, 2.7, and 2.8 and Equation (2–2), we infer that net radiation will be at a maximum in cloud-free parts of the subtropics at the solstices in their respective hemispheres. This is primarily because the $(Q + q)$ term, which dominates Equation (2–2), is at a maximum under these conditions. Also, we expect $R$ to be higher over large water bodies than over land because water has a lower albedo and, at this time of year, lower radiating temperatures (and therefore smaller $I\uparrow$) than adjacent land areas. Surprisingly, the subtropical deserts have only moderately high values of net radiation. Here the $(Q + q)$ term is comparable to that over adjacent oceans, but the surface albedo is higher, as is the surface temperature. The lowest average monthly values of $R$ occur at high latitudes in winter, especially over snow- and ice-covered areas. Below-zero values at this time of year are generally found poleward of 40°.

We now examine the annual "march," or variation throughout the year, of the energy balance components $R$, $H$, $E$, and $G$ for locations typical of different climates. The smallest month-to-month variations in these components occur in tropical areas that are either very humid or very dry throughout the year. In very humid areas a far greater proportion of $R$ is used for evapotranspiration than for the transfer of sensible heat. The reverse is true for a very dry climate. In both $G$ is quite small. These conditions are shown in Figure 2.14(a) and (b).

A large part of the tropics experiences a wet and dry climate, with abundant rains during part of the year and drought during the remainder. The annual march of the energy balance components for such

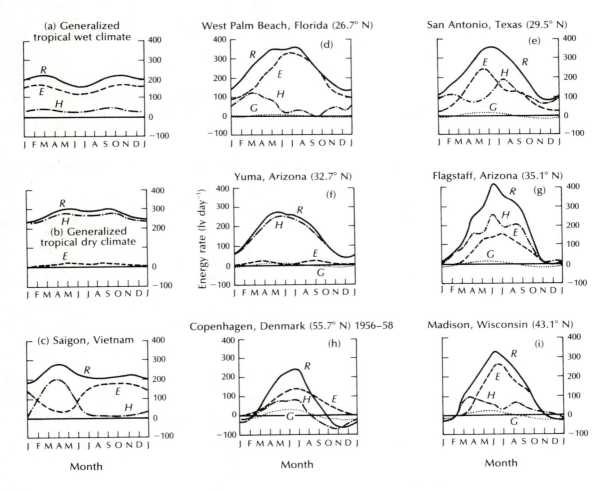

**FIGURE 2.14**
Average annual variation of the components of the energy balance at selected locations. (Reprinted from *Physical Climatology* by William D. Sellers by permission of The University of Chicago Press. Copyright © 1965.)

conditions is represented by Ho Chi Minh City (Saigon), Vietnam, in Figure 2.14(c). Its rainfall regime thus incorporates the characteristics of both tropical climates shown in Figure 2.14(a) and (b). We may infer from Figure 2.14(c) that a rather dry period exists from about January to May or June, when $E$ and $H$ are low and high, respectively. For the rest of the year the reverse occurs. Notice that $R$ increases during the dry period, when the sky is relatively free of clouds.

For latitudes other than the tropics, the annual variation of $R$ is more pronounced and the march of the components generally more irregular. Figure 2.14(d–i) shows the energy balance components for various climates in subtropical to moderately high latitudes. The variation of $R$ closely follows that of insolation but is not necessarily in the same proportion. A comparison of $R$ for Yuma and Flagstaff (representing hot-dry and cool-dry climates, respectively) shows that considerably higher values of $R$ are recorded at the latter. Part of this difference is due to elevation (Flagstaff is at 2100 meters, Yuma near sea level), because there is less of the atmosphere above Flagstaff to deplete solar radiation. This difference also is due to the much higher surface temperatures (and thus higher $I\uparrow$) at Yuma. This observation reinforces the point made earlier about moderately high values of $R$ in the subtropical deserts. Notice that midsummer values of $R$ are only slightly higher at Yuma than those at Copenhagen.

The amount and annual variation of expenditures for $E$ depend largely on the quantity of monthly precipitation and its distribution throughout the year. At West Palm Beach, representative of a humid subtropical climate, rainfall occurs year-round but is concentrated in the warm season; this is reflected in high $E$ values then. This is also the case in San Antonio, except that its monthly amounts and its annual total are only about one-half those of West Palm Beach, and the maximum occurs in late spring rather than in summer or early fall. Another reason why maximum $E$ tends to occur in summer is that a greater evaporating potential is associated with high temperatures (an explanation of this is given in Chapter 5). This is apparent at all stations except Yuma, where moisture for evapotranspiration is scarce throughout the year. There the slight maxima of $E$ in the spring and fall occur during and shortly after the light precipitation which falls then. At Copenhagen, which is representative of a cool, ocean-dominated climate and is at a fairly high latitude, $E$ is the only component to remain positive or near zero through the winter.

$G$ is clearly the smallest of the four terms. This

**TABLE 2.1**
Mean latitudinal values of the components of the energy balance equation for the earth's surface, in kilolangleys per year.

| Latitude Zone | Oceans | | | | Land | | | Earth | | | |
|---|---|---|---|---|---|---|---|---|---|---|---|
| | R | E | H | T | R | E | H | R | E | H | T |
| 80–90° N | — | — | — | — | — | — | — | −9 | 3 | −10 | −2 |
| 70–80 | — | — | — | — | — | — | — | 1 | 9 | −1 | −7 |
| 60–70 | 23 | 33 | 16 | −26 | 20 | 14 | 6 | 21 | 20 | 10 | −9 |
| 50–60 | 29 | 39 | 16 | −26 | 30 | 19 | 11 | 30 | 28 | 14 | −12 |
| 40–50 | 51 | 53 | 14 | −16 | 45 | 24 | 21 | 48 | 38 | 17 | −7 |
| 30–40 | 83 | 86 | 13 | −16 | 60 | 23 | 37 | 73 | 59 | 24 | −10 |
| 20–30 | 113 | 105 | 9 | −1 | 69 | 20 | 49 | 96 | 73 | 24 | −1 |
| 10–20 | 119 | 99 | 6 | 14 | 71 | 29 | 42 | 106 | 81 | 16 | 9 |
| 0–10 | 115 | 80 | 4 | 31 | 72 | 48 | 24 | 105 | 72 | 11 | 22 |
| 0–90° N | — | — | — | — | — | — | — | 72 | 55 | 16 | 1 |
| 0–10° S | 115 | 84 | 4 | 27 | 72 | 50 | 22 | 105 | 76 | 10 | 19 |
| 10–20 | 113 | 104 | 5 | 4 | 73 | 41 | 32 | 104 | 90 | 11 | 3 |
| 20–30 | 101 | 100 | 7 | −6 | 70 | 28 | 42 | 94 | 83 | 16 | −5 |
| 30–40 | 82 | 80 | 8 | −6 | 62 | 28 | 34 | 80 | 74 | 11 | −5 |
| 40–50 | 57 | 55 | 9 | −7 | 41 | 21 | 20 | 56 | 53 | 10 | −7 |
| 50–60 | 28 | 31 | 10 | −13 | 31 | 20 | 11 | 28 | 31 | 11 | −14 |
| 60–70 | — | — | — | — | — | — | — | 13 | 10 | 11 | −8 |
| 70–80 | — | — | — | — | — | — | — | −2 | 3 | −4 | −1 |
| 80–90 | — | — | — | — | — | — | — | −11 | 0 | −11 | 0 |
| 0–90° S | — | — | — | — | — | — | — | 72 | 62 | 11 | −1 |
| Earth as a whole | 82 | 74 | 8 | 0 | 49 | 25 | 24 | 72 | 59 | 13 | 0 |

SOURCE: Reprinted from *Physical Climatology* by William D. Sellers by permission of The University of Chicago Press. Copyright © 1965.

flux is downward in spring and summer and upward in fall and winter. Notice that the amplitude increases with latitude; this reflects the greater range of interface temperatures throughout the year. Other differences in $G$ among the stations in Figure 2.14 may be ascribed to the presence or absence of soil and the varying soil conductivities.

The flux of sensible heat into the atmosphere ($+H$) usually is greatest during the warm months (San Antonio), but may be at a maximum in late winter (West Palm Beach). Or there may be a double maximum as at Madison, which represents a continental climate with moderate, year-round precipitation. At higher latitude stations (Copenhagen, for example), $H$ is negative (directed toward the interface) during fall and winter. At Madison it is negative for a short time during the coldest part of the year.

The mean latitudinal values of the energy balance components at the interface are shown in Table 2.1. $G$ is not shown because, as noted, this flux is zero when summed over a year. For all but the highest latitudes $R$ is higher over water than over land. This is due to water's lower albedo and the somewhat higher mean annual temperature of land surfaces. Poleward of about 50° N and S the values are comparable. Over land and oceans, and for the earth as a whole, net radiation is rather uniform between 30° N and S; from there it decreases with distance poleward. This reflects the dominant influence of annual insolation totals. The relative magnitude of the $E$ and $H$ terms over land is due to the latitudinal distinction between deserts and more humid areas. For the 20°–30° spans in both hemispheres, the proportion of $R$ used for evapotranspiration is less than at other latitudes because these areas receive little precipitation. This proportion tends to increase both toward the equator and poleward of about 30° N and S as more humid areas are encountered. For all land areas $H$ and $E$ are about equal.

The quantity $T$ for the oceans and the earth is new and requires explanation. As we have learned, water, unlike land, is capable of motion within itself; thus quantities of heat can be transported laterally as well as vertically. The negative values of $T$ indicate that a latitudinal span receives a net import of heat from lateral transport, and positive values indicate a net loss of heat. The energy balance equation for any latitudinal span must still balance, of course. In middle and high latitudes the net radiation is augmented and the gains acquired by both net radiation and lateral transport are balanced by losses through $E$ and $H$. Low-latitude oceanic areas lose heat by lateral transport, and this is a drain on net radiation.

Evaporation over the oceans is, perhaps surprisingly, greater in the subtropics than in the tropics. This is because the subtropics experience higher net radiation and are drier than adjacent areas, and evaporation can therefore occur more rapidly than in the tropics, which are predominantly humid. In addition, fairly strong and persistent winds—the trade winds—dominate the surface air circulation in subtropical latitudes and thereby promote evaporation. Air movement in the tropics is less well developed. Poleward of about 30° $E$ values decrease rapidly in keeping with comparable reductions in net radiation. For the oceans about 90 percent of $R$ is used to evaporate water, as noted earlier.

The latitudinal variation of $H$ over the oceans is noteworthy because, unlike the other components, the highest values occur at high latitudes and vice versa. Sensible heat transfer from the oceans to the atmosphere depends on the difference in temperature between the relatively warm ocean surface and the overlying air. In high latitudes, especially in winter, the water is considerably warmer than the air above, and the convective transfer of heat is much more rapid than in the middle and low latitudes, where this temperature difference is not as great. In extreme circumstances, such as when very cold, dry air moves from a continent to an ocean in winter, the fluxes of both latent and sensible heat are greatly magnified. Off the northeast coast of the United States, for example, over 150,000 and 30,000 ly yr$^{-1}$ are expended for latent and sensible heat transfer, respectively, and most of this occurs under the conditions just described. (Compare these values with values in Table 2.1.)

The figures for the earth as a whole in Table 2.1 combine both land and water values, but, because water comprises almost three-quarters of the earth's surface, the mean figures are heavily weighted toward ocean conditions. At 40° to 50° S, for example, the amount of land is so limited that the figures for oceans and the earth are nearly identical. It also should be noted that the figures following the last row heading ("Earth as a whole") are not simple averages of those above, but are weighted according to the area of the earth's surface in each latitudinal span (for an example, see the illustrations in Section 8.3).

Table 2.2 shows the components of the surface heat balance for the continents. The $R$ values reflect the general position of the continents with respect to latitude; thus Africa, Australia, and South America have higher values than the remaining continents. Maximum values of $E$ and the highest ratio of $E$ to $R$ occur in South America, which is the most humid continent. The other large landmasses have lower values of $E$, but the ratio of $E$ to $R$ is also high in Europe and

**TABLE 2.2**
Annual energy balance of the continents, in kilolangleys per year.

| Continent | R | E | % of R | H | % of R | H/E |
|---|---|---|---|---|---|---|
| Europe | 39 | 24 | 62 | 15 | 38 | 0.62 |
| Asia | 47 | 22 | 47 | 25 | 53 | 1.14 |
| North America | 40 | 23 | 58 | 17 | 42 | 0.74 |
| South America | 70 | 45 | 64 | 25 | 36 | 0.56 |
| Africa | 68 | 26 | 38 | 42 | 62 | 1.61 |
| Australia | 70 | 22 | 31 | 48 | 69 | 2.18 |
| Antarctica | −11 | 0 | 0 | −11 | 100 | — |
| All land | 49 | 25 | 51 | 24 | 49 | 0.96 |

SOURCE: Adapted from *Physical Climatology* by William D. Sellers by permission of The University of Chicago Press. Copyright © 1965.

North America as well. In Africa and Australia, both of which have extensive desert areas, the proportion of R used to evaporate and transpire water is less, and the H values are correspondingly higher.

The H/E ratio, called the **Bowen ratio,** is a useful measure of the aridity, or dryness, of an area. It gives an indication of the disposition of R by E and H. Over oceans this value is very nearly 0.10. For the most arid continent, Australia, it is 2.18; and for the most humid (least arid), South America, it is 0.56, as Table 2.2 indicates.

One more change of perspective is needed in our attempt to understand the space-time variations of the energy balance components. We began by specifying the energy fluxes across the interface and then gave examples of the diurnal and annual variation of these components. The horizontal scope also was broadened in this process, from individual locations to entire latitudinal spans. Now we will once again examine energy fluxes in the vertical, but this time for the entire earth-atmosphere system in an average year.

Figure 2.15 shows the energy fluxes that occur within the three parts of this system: space, atmosphere, and the earth's surface. The amount of solar radiation arriving at the outer limits of the atmosphere, averaged over a year, is 256,000 langleys (ly) or 256 kilolangleys (kly), which is equivalent to 0.5 ly min$^{-1}$. Of this amount 49 kly are absorbed in the atmosphere (8 by clouds and 41 by noncloud constituents of the atmosphere). Reflection from these amounts to

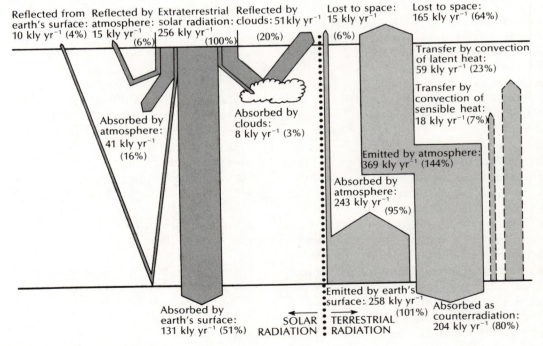

**FIGURE 2.15**
Disposition of solar and terrestrial radiation, and the energy balance, during an average year, in kilolangleys per year and percentage of extraterrestrial radiation.

51 and 15 kly, respectively. Thus, of the 256 incoming units, a total of 115 are either reflected by or absorbed in the atmosphere. The remaining 141 kly arrive at the surface, where 10 are reflected. This proportion, 10/141 (or 7.1 percent), indicates that the earth's surface as a whole has a low albedo with respect to short-wave radiation, as is expected because most of its surface is water. The 131 kly available at the interface is slightly more than half (51 percent) of the amount at the outer limits of the atmosphere, and includes direct beam, scattered, and diffuse radiation.

The value for infrared radiation emitted by the earth's surface may be obtained through an equation which relates this radiative flux to the mean surface temperature. (See page 6, where the earth-atmosphere system's mean planetary temperature is calculated.) Assuming that the mean surface temperature is about 285K, an infrared flux ($I\uparrow$) of 258 kly yr$^{-1}$ results. Most of this flux (94 percent) is absorbed by the atmosphere, especially by clouds, and by water vapor and carbon dioxide, which are highly absorptive in the infrared. Also, most of this absorption takes place in the lowest layers of the atmosphere, where clouds and water vapor are concentrated.

This differential transparency of the earth's atmosphere to short- and long-wave radiation is of special interest. The atmosphere is essentially transparent to solar radiation; of the 190 kly yr$^{-1}$ available to it after reflection, only 49 are absorbed and the remaining 141 reach the surface. Of the 258 kly of infrared radiation emitted by the earth's surface, 243 are absorbed by the atmosphere. The net effect is that surface, and near-surface, temperatures in the atmosphere are higher by about 33C° than they would be without this differential transparency. This net heating effect has been called, somewhat erroneously, the *greenhouse effect* because of a presumed similarity to the energy exchanges that take place within a greenhouse. The preferred term is now the *atmospheric effect*.

The latent and sensible heat fluxes to the atmosphere in an average year for the entire globe amount to 59 and 18 kly, respectively. You will recall that the transfer of latent heat from the interface requires both evapotranspiration of water at the interface and condensation in the atmosphere—that is, a loss and gain, respectively. The atmosphere also gains energy, as noted, from both short-wave (49) and infrared (243) radiation, for a total of 369 kly yr$^{-1}$. This is radiated back to the earth (204) and out to space (165). It may seem surprising that a part of the system, in this case the atmosphere, can radiate more heat energy than that which arrives from the sun. This is possible because of the atmospheric effect. The atmosphere almost completely blocks terrestrial radiation, requiring the surface temperature to increase by some 33C° over what would be necessary in the absence of an atmosphere to produce the needed flux to space and thus achieve a balance.

An accounting of both gains and losses for all three parts of the earth-atmosphere system will show a balance, as of course it must. This has already been shown for the atmosphere; for space, notice that of the 256 kly yr$^{-1}$ "lost" by space, a total of 76 kly (30 percent) is reflected back to it. This percentage, 30, is the earth's planetary albedo. The remaining units are represented by terrestrial radiation escaping to space (15) and radiation from the atmosphere to space (165). At the interface, from Equations (2–2) and (2–4), we get

$$(Q + q)(1 - a) + I\downarrow - I\uparrow = R = (E + H)$$
$$(141)(1 - 0.071) + 204 - 258 = 77 = 59 + 18$$

Remember that G can be omitted because, over the year, it is zero.

Thus a balance also is attained at the interface. These figures should be compared with those in the lower right of Table 2.1. Notice that the 59 (E, for the earth as a whole) agrees, but that R (77 vs. 72) and H (18 vs. 13) are somewhat different. The figures in Figure 2.15 are probably more accurate because they incorporate the most recently available satellite data; the figures in Table 2.1 do not.

There are significant differences in the values in Figure 2.15 from place to place, and even from one time of the year and day to another at one location. Many of these differences already have been shown, as in Table 2.1, and in the figures for individual locations. Still, we will reemphasize some of the most obvious departures from the global figures for an average year. In hot, dry areas or times, E would be negligible and H much larger. $I\uparrow$ would increase because of high surface temperatures, and the proportion of it absorbed in the atmosphere would be less because of the lack of water vapor and clouds. The $(Q + q)$ term would be higher, the albedo higher (assuming sparse vegetation and light-colored soils), and the proportion of solar radiation absorbed by clouds considerably less. Over a snow-covered surface in high latitudes all of the values would be reduced, proportionately, from the figures in Figure 2.15, and their ratios would be different. Solar radiation absorbed by the atmosphere and reflected from the surface would be proportionately greater. Heat loss by terrestrial radiation ($I\uparrow$) would exceed that gained by counterradiation from the atmosphere ($I\downarrow$) and, even in the presence of insolation, which would be greatly reduced, a negative value of R would result.

This would be compensated by negative values (fluxes toward the interface) of the net effect of $H$, $E$, and $G$.

Figure 2.15 shows that both atmosphere and surface gain and lose equal amounts of heat energy during a year. This must be so, because if it were not, one or more parts of the system would have a net surplus or deficit and would be warming or cooling. With respect only to radiation balance, however, the atmosphere has a deficit and the surface a surplus. It is customary to speak, then, of the atmosphere as a sink and the surface as a source of radiation. This disparity is compensated by the upward transfer of latent and sensible heat at just such a rate, over long periods of time, to keep the entire earth-atmosphere system in energy balance.

This energy balance for the earth's surface and its atmosphere, the earth-atmosphere system, does not apply among individual latitudinal spans, as shown in the upper part of Figure 2.16. There is a surplus of energy from 30° N to 30° S and a deficit poleward of these latitudes. This part of the diagram appears to show an overall deficit of heat, but it should be remembered that the units are thousands of calories per square centimeter per year. If the values indicated by the bar lines were multiplied by the area within each latitudinal span, the total length of the lines above and below zero would be the same; that is, the earth-atmosphere system, across all latitudes, would be in balance.

This interlatitude imbalance must be compensated, otherwise the middle and high latitudes would progressively cool and the low latitudes would warm. The mechanism which redresses this imbalance is atmospheric and ocean circulation, which must transport heat energy poleward in both hemispheres. This constraint on the way the atmosphere and oceans circulate has very important implications, as we will see in Chapter 4.

The manner in which circulation achieves this compensation is shown in the lower part of Figure 2.16. Here the vertical length of the bars is proportional to the difference between the energy balances of adjacent latitudes. The bars directed upward represent northward transport, or flux, and vice versa. This difference is greatest between 40° and 50° N and between 50° and 60° S, so it is in these latitudinal spans that the poleward energy transport is greatest. From these latitudes transport decreases both poleward and equatorward in both hemispheres. In the Northern Hemisphere all fluxes are everywhere northward and in the Southern Hemisphere everywhere southward, except for the latent heat flux in low latitudes, as is explained later.

There are three means by which this poleward transport of heat is accomplished: through sensible and latent heat by atmospheric transport, and by ocean currents. The relative importance of each is indicated in the lower part of Figure 2.16. Overall, the poleward flux of sensible heat is the greatest of these

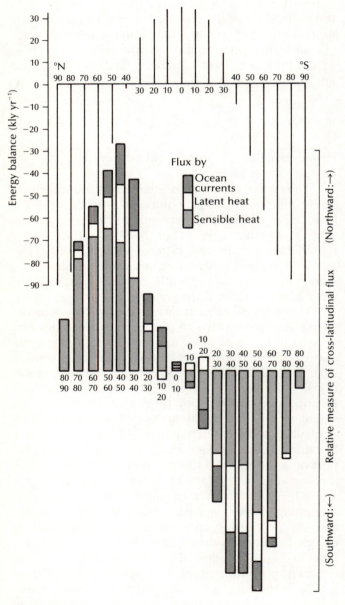

**FIGURE 2.16**
Energy balance by latitude (top), and the means by which interlatitude energy imbalances are compensated by cross-latitudinal transport of sensible and latent heat and ocean currents (bottom).

(about 60 percent of the total transport) and is evident in all latitudinal spans. Poleward transport by ocean currents, which accounts for about 20 percent of the total, is significant in the middle latitudes, somewhat less so in the low latitudes, and is negligible near the poles.

The remaining 20 percent is accomplished by the transfer of latent heat. Overall, evaporation exceeds condensation in the low latitudes, while the reverse is true in the middle and high latitudes. An interesting exception to the poleward transport of heat occurs in the spans 10° to 20° N and 0° to 20° S. Here the atmosphere circulates so as to carry latent heat equatorward. So in these latitudes, while the net transport—considering all three means—is poleward, the latent heat flux is directed toward the equator.

We may conclude, then, that atmospheric circulation must compensate energy imbalances in both the vertical and horizontal. The upward flux of sensible and latent heat, by convection, compensates the radiation deficit in the atmosphere, which is most apparent in the tropics. At the same time, latitudinal imbalances are compensated by poleward circulations of sensible and latent heat and by ocean currents. This transport is greatest in the middle latitudes, so we may infer that circulation is most vigorous here. We will explore circulation further in Chapter 4.

## SUGGESTED READING

BARRY, R. G. and R. J. CHORLEY. *Atmosphere, Weather, and Climate.* New York: Methuen, 1977.

HERMAN, J. R. and R. A. GOLDBERG. *Sun, Weather, and Climate.* Washington, D.C.: Superintendent of Documents, 1978.

NEUBERGER, H. and J. CAHIR. *Principles of Climatology: A Manual in Earth Science.* New York: Holt, Rinehart and Winston, 1969.

# 3
## TEMPERATURE AS A CLIMATIC ELEMENT

INTRODUCTION
THE PHYSICAL BASIS OF MEASUREMENT AND TEMPERATURE
TEMPERATURE IN TERMS OF THE ENERGY BALANCE EQUATIONS
TIME VARIATIONS OF TEMPERATURE
THE SPATIAL VARIATION OF TEMPERATURE
CONTINENTALITY
OTHER CLIMATIC ASPECTS OF TEMPERATURE

## 3.1
### INTRODUCTION

Temperature is the most important of the elements that comprise weather and climate. No other aspect of our atmospheric environment is as pervasive, and there is hardly any natural or artificial substance that is unaffected by the temperature of the air surrounding it. The patterns of our daily lives, the clothes we wear, the composition and structure of the earth's soils, the yields of crops and the patterns of natural vegetation, the rates of biochemical reactions within organisms, the energy expended in heating and cooling buildings—all are consequences of temperature.

Meteorologically, temperature is decisive. The phenomenon we call weather, and in particular the motions of the atmosphere which produce weather, is a consequence of the place-to-place variations of temperature, which in turn result from similar variations in the components of the energy balance equations. For some purposes, the climate of a location can be described adequately by temperature and by some indication of moisture, such as precipitation.

The layperson can grasp the concept of temperature easily. We readily understand that the temperature of a substance—the air around us, a pan of boiling water, or an ice cube—simply reflects the degree of hotness or coldness of that substance, or, in other words, the amount of heat that it contains. Various scales have been devised to quantify temperature. Here, at the bottom of the earth's atmosphere, the substances we ordinarily encounter have temperatures ranging from somewhat below water's freezing point to its boiling point. The **Fahrenheit** scale sets this freezing point at 32° and the boiling point at 212° (a range of 180°), while on the **Celsius** scale these points are 0° and 100°. On the **absolute,** or **Kelvin,** scale these points are at 273° and 373°. Figure 3.1 shows these comparisons.

The temperature climatologists focus upon is that of the earth's atmosphere and its surface substances. When we speak of temperature in everyday terms, as when we hear that the weather forecast is for a high of 90°F (32°C), we should understand that this figure refers to the temperature of the freely circulating air at a height of about 5 or 6 ft (nearly 2 m). In this chapter, then, we will explore the reasons for specifying "freely circulating air," and a particular height. In addition, we will consider the physical basis of temperature, and how this free-air temperature relates to the fluxes of energy expressed by the energy balance equations. Most of this chapter will deal with observa-

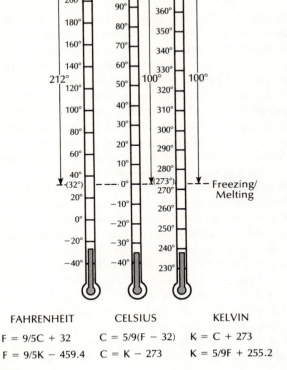

**FIGURE 3.1**
Comparisons and conversions among the three important temperature scales.

tional aspects and with spatial and temporal patterns of temperatures averaged over specified time periods.

## 3.2
### THE PHYSICAL BASIS AND MEASUREMENT OF TEMPERATURE

The temperature of a gas such as the earth's atmosphere is a measure of the internal energy of the individual molecules comprising the gas. Air with a relatively high temperature consists of molecules that move faster (have greater kinetic energy) than do molecules of air at a lower temperature. When air temperature is measured with a conventional mercury-in-glass thermometer, the molecular energy of the air surrounding the bulb is transferred to it; this in turn affects the height of the mercury column, and thus the temperature we read. In this application, we are defining temperature according to the *kinetic theory of gases*.

At the same time, because the thermometer itself is a solid substance, it is much more likely than the air around it to be warmed (or cooled) by objects with which it is in radiative exchange. The most obvious example of this occurs when the thermometer is held in full sun. In this case the temperature recorded by the thermometer will reflect not only the internal energy of the gas (atmosphere) surrounding it, but also an increment of heat due to the absorption of solar radiation. The reading obtained is thus higher than the true air temperature. How often have we heard statements like "The temperature is 95° (35°C) in the shade," with its implication that the air temperature in the sun is different from that in the shade? It should be clear now that there is no difference between the air temperatures in sun and shade. We will feel warmer in the sun, but this is because our bodies act much like a thermometer. The difference in temperatures results from the additional solar heat absorbed by the sensing device.

This difference between the temperature of exterior surfaces exposed to the sun and that of air with which they are in contact suggests that the temperature of the former should be defined in some other way than according to the kinetic theory of gases. Any solid substance immersed in our atmosphere will tend to have the temperature of the air around it in the absence of any radiation sources or sinks, that is, if it is not being heated or cooled by objects with which it is in radiative exchange. In the presence of solar radiation that substance will attain a higher temperature, such that it radiates to its environment as much heat energy as it receives both from the air around it and from the absorption of solar radiation. We say that this substance is in radiation equilibrium with its surroundings, and its temperature is a **radiation equilibrium temperature**. The earth's mean planetary temperature was calculated (page 6) according to the assumption that the earth-atmosphere system is in radiation equilibrium with the sun.

To a lesser extent this temperature difference between solid objects and the air around them arises when the substance is in radiative exchange with nearby objects such as walls, which can have a temperature much different from that of the air (see, for example, Figure 14.5).

At high elevations, where solar radiation is more intense because there is less of the atmosphere to absorb it, there may be a considerable difference between the temperature of the free air and that of the surfaces of solid substances exposed to the sun. This helps to explain why we prefer sun to shade when the air temperature is low, particularly in the low temperatures associated with high elevations.

In Chapter 2 we learned that the fundamental heat transfer processes at the interface can be expressed by the energy balance equations. Through these equations a substance's temperature can be related to the flux density of radiation emitted by it (see page 6). Unfortunately, however, the radiating substances in these equations do not include air at the height at which its temperature is customarily measured—about 2 m. At this point, therefore, we do not have a direct means for relating the fluxes of heat at the interface to the temperature of the air above it.

You might then ask, why do we not report the temperatures of earth surface substances? Why do we describe climate in terms of a characteristic—the temperature of the air—which is not readily deducible from the equations that describe the processes underlying it?

To answer this question we must recognize that there can be large simultaneous differences in temperature from place to place at the surface. When the surface is being warmed by the absorption of solar radiation, as on a hot summer day, an asphalt parking lot may be tens of degrees warmer than an adjacent grassy field. At night, under conditions of marked radiational cooling by the earth's surface, similar but less marked differences may develop. At levels of a few tens of feet, provided there is some air circulation, these differences are reduced, and air temperatures are reasonably uniform from place to place at that level. In general, the farther air is from the surface, the less local differences in the thermal characteristics of surface substances will affect air temperature, and the more nearly the same the air temperature becomes at a given level.

It is not practical, however, to measure temperature routinely at such a height that the air temperature no longer reflects the influence of surface substances. But we are most concerned with the level in which we live, that is, within the lowest 2 m of the atmosphere. Thus, a compromise is reached. Air temperatures are measured in an instrument shelter at the so-called "screen level," or at about 2 m. The shelter is painted white and is double roofed to minimize the absorption of solar radiation. It is vented to permit air to circulate through it, and, at least where possible, is placed on a grassy surface. Figure 3.2 shows the standard instrument shelter used by the U.S. National Weather Service.

## 3.3

**TEMPERATURE IN TERMS OF THE ENERGY BALANCE EQUATIONS**

Although the temperature of freely circulating air at 2 m is not given explicitly by the energy balance equations, it is reasonably well approximated by the temperature at the interface, especially for long time periods. In fact, the mean annual temperature from the depth to which soil is heated to a height of a few meters above the surface is essentially the same. The surface temperature can be determined from $I\uparrow$, the flux density of infrared radiation emitted by the interface to the atmosphere. By rearranging Equation (2-3), we can write

$$T \propto I\uparrow = (Q + q)(1 - a) + I\downarrow - H - E - G$$

where $T$ is temperature of the air at screen level and $\propto$ means "is proportional to." (The other terms are defined in Chapter 2.) This expression says that air temperature will be higher with more intense solar radiation, a low surface albedo, and a greater flux of counterradiation from the atmosphere ($I\downarrow$). Air temperature will decrease as greater proportions of the available energy are expended for upward fluxes of sensible ($H$) and latent heat ($E$) by convection and for heat conduction into the soil ($G$.) From a long-term, or climatological, point of view the solar radiation term is predominant, and air temperatures follow its variations rather closely. There are exceptions to this, especially for short time periods. These will be considered later.

**FIGURE 3.2**
An instrument shelter. The shelter is 2 m (6.5 ft) high and contains maximum and minimum thermometers, a hygrothermograph, and a barograph. In the background are an evaporation pan and a rain gage (see Figure 5.10). (Photo by Dennis M. Driscoll.)

## 3.4 TIME VARIATIONS OF TEMPERATURE

Temperature closely follows periodic fluctuations, or cycles, of insolation. There are two such cycles—the annual and the diurnal. They refer to changes that occur over a year and a day, respectively. We need to examine each of these cycles and consider the place-to-place variations of temperature that result from them.

Figures 2.4 and 2.6 show that latitudes poleward of 23.5° experience a single maximum, and a single minimum, of solar radiation at the respective solstices. They also show that seasonal, or month-to-month, variations in solar radiation are more pronounced with increasing latitude. Seasonal temperature variations are similar, but with two important exceptions. First, the range of temperature, from the coldest to the warmest months, varies around the earth at the same latitude. Second, the times (months) at which the maxima and minima are experienced do not coincide with the solstices. Both of these exceptions result from the differential thermal responses of water and the various land surface substances.

We learned in Chapter 2 that, when the same amounts of net radiation are available at the interfaces of land and water, land heats and cools more rapidly than water. From the heat transfer theory that underlies these processes we learned that two thermal properties—heat capacity and thermal conductivity—are chiefly responsible for this differential land-water response. These two properties can be combined into one by multiplying the heat capacity by the square root of the thermal conductivity, and this is called **conductive capacity**. It is measured in units of ly s$^{-1}$ degree$^{-1/2}$ (langley per second per degree to the one-half). Representative values are water, 7.0; air, 0.1; and land, 0.05. Keep in mind that these are rough approximations. The value for air will vary by orders of magnitude, depending on the extent of air movement. For example, for still, moderately stirred, and strongly stirred air, the conductive capacities are $10^{-4}$, $10^{-2}$, and $10^{-1}$, respectively.

The conductive capacity is a very useful characteristic because it implies a number of important consequences for climate. First, the amount of heat generated at, or lost from, the interface is shared in proportion to the conductive capacities of the substances in contact there. Thus, of the available heat (not including expenditures for evaporation), air over land gains about twice as much heat as the land does, while air over water gains only a negligible amount, provided there is strongly stirred air in both cases.

Second, a consideration of the different conductive capacities suggests that the range of temperature over heating and cooling cycles will be greatest at the interface and decrease with distance away from it. The temperature range will also be proportional to the sum of the conductive capacities there. Working with simple proportions, we can compare the range of temperature over land with that over water. For air over land, this sum is about 0.15; for air over water, about 7.1. Thus, the range of temperature for the land-air interface is almost 50 times that of the water-air interface (7.1/0.15). Diurnal ranges do, in general, reflect this difference. Representative values are 0.2 to 0.5C° (0.4 to 1.0F°) over water, and 10 to 25C° (18 to 45F°) over land.

Annual temperature ranges are not quite so disparate. Since air is constantly being exchanged between oceans and continents, this ratio of 50 times is approached only in the deep interiors of the largest continents. Elsewhere, and especially in coastal areas, annual ranges are more maritime (less extreme), particularly if the prevailing atmospheric circulation is from ocean to land. (We will consider the effects of atmospheric circulation in Chapter 4.)

There are two other factors which influence temperature ranges. Both involve the presence of water or water vapor, either at the interface or in the atmosphere. We learned that when equal amounts of net radiation are available to both a land and a water surface, the temperature response would be more extreme over land. But remember that over water as much as 90 percent of this net radiation is used in evaporation, which is another reason for the greatly reduced temperature ranges there. Water available for evapotranspiration over land varies considerably from place to place and from time to time at any place. Where water is abundant, as in swamps, marshes, and in heavily vegetated areas such as tropical rainforests, the temperature response tends to be similar to that over water. In very dry areas, however, with negligible expenditures for evapotranspiration, the diurnal and annual extremes are much more pronounced.

Not only the surface water but water vapor in the atmosphere influences extremes of temperature. In Section 2.3.5 we examined mean annual values, for the earth as a whole, of the energy balance components. We stated that wide variations in these values could be expected as the climate changed from very cold, to hot and dry, and to warm and humid. Without clouds and water vapor, terrestrial radiation ($I\uparrow$) escapes to space largely unabsorbed by the atmosphere, and the atmospheric effect ("greenhouse effect") is minimized. Surface temperature, and that of the air just above, drops rapidly, and the cool (winter and night) values are much lower than they would be with water present. At the same time, solar radiation reaches the surface largely unabsorbed, and warm (summer and day) temperatures are higher than in a humid atmosphere.

When clouds and water vapor are present they act not only to block both terrestrial and solar radiation but to increase atmospheric counterradiation ($I\downarrow$). This is not of great consequence during the day or in the summer when clouds obstruct surface absorption of solar radiation. At night or in winter, however, this counterradiation enhances the atmospheric effect and keeps temperatures from dropping as low as they would in the absence of the augmented downward flux.

Diurnal ranges of temperatures are influenced by two additional factors. Temperatures over a 24-hour period reflect the variation of the height of the sun in the sky, or its *elevation angle*. This is the diurnal counterpart of the annual variation of insolation, except that the latitudinal association is reversed. In low latitudes the sun's elevation angle varies by 90° over the day at the equinoxes and by almost 70° at the solstices. In high latitudes this diurnal difference is less. At 60°, for example, the diurnal elevation angle variation is from 6.5° to 53.5° at the winter and summer solstices, respectively. Thus, other influences being equal, low latitudes experience wider diurnal temperature ranges than high latitudes, and the latter experience their most pronounced day-night swings in summer. A world map of diurnal temperature ranges is shown in Figure 7.4. Figure 3.3 shows land areas where the average diurnal range of temperature exceeds the annual range.

Finally, on windy days vertical mixing of air promotes heat exchange through a relatively deep layer. Therefore, the range of temperature at the surface is less than on calm days.

Annual temperature ranges can be summarized as follows:

1. The variation in insolation over the year is latitudinal, with annual range increasing with latitude.
2. Since land heats and cools more rapidly than water, land areas show more extremes than oceans at the same latitude. In general, therefore, the temperature range increases with distance from a large water body.
3. The circulation of the atmosphere, particularly the exchange of air between continents and oceans, tends to minimize land-water differences, especially for land areas strongly affected by onshore winds.
4. The presence of water or water vapor, either at the surface or in the atmosphere, enhances the atmospheric effect and reduces seasonal temperature extremes.

Diurnal ranges are influenced to a lesser extent than is annual range by land-water differences and atmospheric circulation, but they are strongly affected by water. Also, they tend to be somewhat greater in low latitudes, where the daily variation of the sun's elevation is greater than in high latitudes, and by the extent to which mixing occurs in the layer of air near the surface. With greater mixing, diurnal ranges are reduced. These remarks apply principally to land areas; diurnal changes over water are very small.

The second exception to the similarity between insolation and temperature mentioned earlier involves the lag between the respective maxima and minima. To understand why this occurs, consider the temperature response of a substance with negligible heat capacity subjected to a periodic variation of radiation—for example, a very thin plate heated by a sinusoidal change of insolation, but removed from any air temperature or other radiative influences. The flux density of radiation emitted by the plate to its environment, and thus its temperature, would rise and fall in correspondence with insolation. That is, the temperature of the plate would be at all times proportional to the absorbed radiation.

The difference between the flat plate and the various earth-atmosphere interfaces is that the latter have considerable heat capacities. Increasing insolation received at a land-air or water-air interface results in positive net radiation; this heat is used to accomplish the processes represented by the right-hand side of Equation (2–3), thereby raising the interface temperature. Since these processes act as a drain on the available heat, temperature lags insolation. Also, when insolation decreases, the stored heat (and remember that storage is not possible in the flat plate) is returned to the interface, principally by upward soil heat flux. Temperature continues to lag radiation, although after sunset (diurnal cycle) there is no appreciable insolation.

Earth-air interfaces not only have considerable heat capacities, but these vary as well. A dry land surface, especially one with a dry atmosphere above it, has a moderate heat capacity; that of a water-air interface is much higher. It follows, then, that the annual and diurnal time lags over water will be greater than those over land.

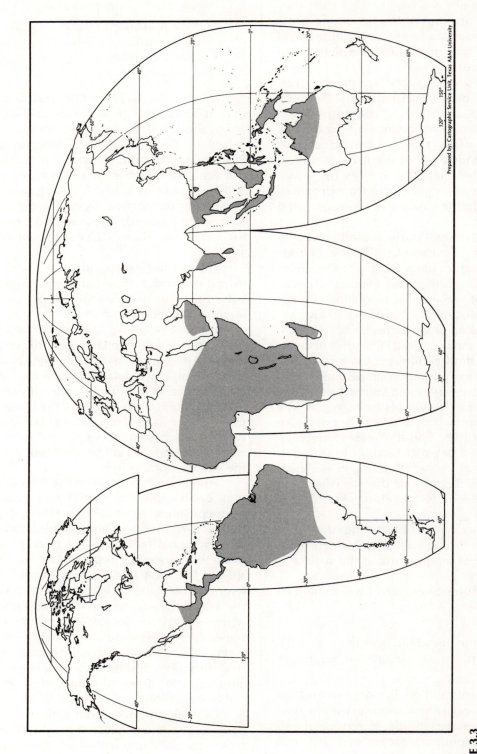

**FIGURE 3.3**
Land areas where the mean diurnal range exceeds the mean annual range.

## SEC. 3.4 | TIME VARIATIONS OF TEMPERATURE

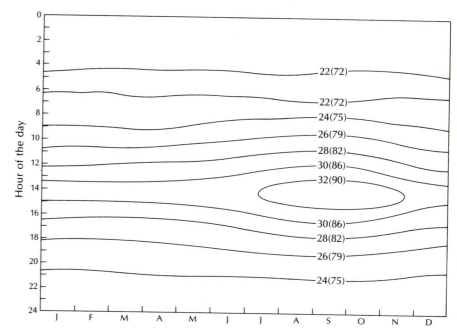

**FIGURE 3.4**
Thermoisopleth diagram for a representative humid tropical climate.

Thus we see that land-water differences account not only for increased temperature ranges over land, but also for longer time lags between insolation and temperature over water than over land. Over diurnal cycles this difference is not appreciable. Maximum temperatures in the instrument shelter are reached at about 2 to 3 P.M. local time, and minimum temperatures around sunrise, over both land and water. On an annual basis, however, continental temperatures lag the solstices by about one month, while over the oceans, and in ocean-dominated land climates, the lag increases to as much as ten weeks. In most marine climates this lag is more apparent in summer than in winter.

We need to examine one final aspect of temperature before looking at temperature-versus-time curves for various climates. Earlier we learned that the annual temperature variation for latitudes poleward of 23.5° is essentially sinusoidal, with maximum and minimum values reached from one to two months after the solstices. Diurnal variations are not sinusoidal, but are asymmetrical. For all but the highest latitudes in summer, the time from minimum to maximum is less than that from maximum to minimum. This is illustrated later (Figure 3.9), when the influence of weather conditions on diurnal variations of temperature is considered. This asymmetry arises because insolation is interrupted over the diurnal cycle by sunrise and sunset, whereas it varies continually over the annual cycle. Of the 24 hours in a day, nonzero amounts of insolation occur during roughly one-half, except in very high latitudes (see Figure 2.3).

Figures 3.4, 3.5, 3.6, and 3.7 are generalized representations of the diurnal and annual variation of temperature for four distinctly different climates. These **thermoisopleth diagrams** show mean temperatures for all hours of the year. Such diagrams are of course constructed only after a long period of record is available. By a careful study of these figures, we can verify several features.

Figure 3.4 represents a tropical humid climate, one at or very near the equator, with year-round, plentiful rainfall. The isotherm orientation is almost entirely horizontal, indicating that the major temperature changes occur diurnally, as would be expected. Seasonal changes of insolation in very low latitudes suggest that somewhat higher and lower temperatures will occur at the equinoxes and solstices, respectively (see Figure 2.6). These differences are quite small, however, and are generally not reflected in temperature. Instead, the maximum that occurs in August through September results from somewhat less rainfall in these months. This is not a dry or drought period, but simply one in which rainfall decreases to only about 76 to 127 mm (3–5 in.) a month instead of the 203 to 305 mm (8–12 in.) or more that is characteristic of the remaining months. The decreased surface water and atmospheric water vapor result in a somewhat expanded diurnal range to almost 11C° (20F°), whereas in the more rainy months a diurnal range of about 9C° (16F°) is typical.

A similar diagram for a wet-dry climate is shown in Figure 3.5. This area is located at about 15° latitude and receives abundant rains during the warm season

**FIGURE 3.5**
Thermoisopleth diagram for a representative tropical wet-dry climate.

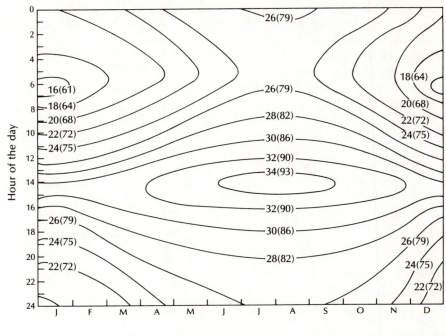

**FIGURE 3.6**
Thermoisopleth diagram for a representative continental climate.

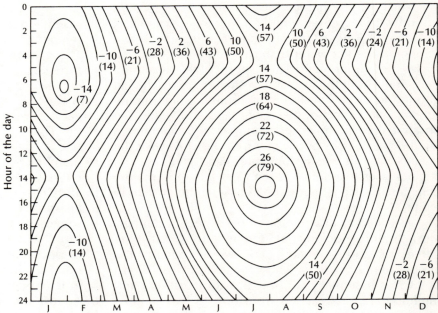

(April–September), but is much drier during the remaining half-year. The isotherm orientation still indicates that diurnal changes are predominant, but some annual variation is apparent. The higher temperatures in the warm six months reflect greater insolation receipts. Notice also the greater diurnal range in December, January, and February than in June, July, and August. Sun-angle variation is not of great consequence at this latitude, but the change from wet to dry seasons is. During the dry (cool) season the lack of water results in a diurnal range of about 19C° (34F°).

During the wet (warm) season the presence of clouds, water vapor, and surface water, depresses the diurnal range to about 10C° (18F°). This change, together with the month-to-month warming and cooling, results in a much greater difference between cool and warm season minima [about 9C° (16F°)] than between cool and warm season maxima, which are different by only about 4C° (8F°). Figure 3.5 also shows that the time of minimum temperature exhibits a small seasonal variation, that is, it occurs sooner in the warm than in the cool season because of earlier sunrise. The temperature experience of a Southern Hemisphere tropical wet-dry climate can be represented by shifting the months forward or backward by six months (by exchanging January for July, July for January, etc.).

A thermoisopleth diagram for a constantly dry climate at 15° latitude would be somewhat the same. The principal difference would appear during the warm season, when the diurnal range would be greatly amplified. Overall temperatures would be somewhat higher, because almost all of the net radiation would be used to heat the air and ground.

A high-latitude continental climate is represented in Figure 3.6. This area is well inland and removed from the moderating influence of large water bodies, at about 55° to 60° N. (Notice that this figure must exemplify the Northern Hemisphere, since there are no such areas in the Southern Hemisphere.) In this climate the annual changes exceed the diurnal; this is indicated by the predominantly vertical orientation of the isotherms. The greater number of isotherms here than in Figures 3.4 and 3.5 (the isopleth interval is the same in all four diagrams) indicates that this area experiences wide temperature variations. Overall, lowest temperatures occur just before sunrise in mid- to late January, and the highest at about 3 P.M. in mid- to late July. Water and water vapor are not of great consequence for temperature, but it is assumed that precipitation is more or less uniform throughout the year.

The diurnal range is greater in summer than in winter because the sun-angle variation is greater in summer. Also, the difference between the time of minimum temperatures is more pronounced than in Figure 3.5—reflecting the much higher latitude—and the most abrupt changes from month to month occur in spring and fall.

Figure 3.7 represents an area at about 60° latitude with a pronounced marine climate and year-round precipitation. Consequently, both diurnal and annual ranges are reduced compared with the continental climate in Figure 3.6, although the greater diurnal range in summer than in winter is still evident. Again, spring and fall are the times of most rapid change, and the time of minimum daily temperature moves backward from winter to summer. Of special interest in Figure 3.7 are the months of highest and lowest temperatures. These occur in August and February, respectively, because of the moderating effects of oceans.

More detail on the annual march of mean monthly temperature is shown in Figure 3.8. Included are the four representative climates just discussed, as are curves for two additional locations. The mean temperature for any month for the four climates may be obtained from Figures 3.4 to 3.7 by interpolating and averaging the 24 hourly values at mid-month.

The (e) curve in Figure 3.8 shows a humid subtropical climate at about 30° with year-round precipitation. Here summer temperatures are higher than in the equatorial climate (a). This should not be surprising, since for all latitudes maximum amounts of insolation (see Figure 2.6) are received at about 30° in the summer months.

Curve (f) represents the most extreme continental climate. This occurs in northeastern Siberia, where

**FIGURE 3.7**
Thermoisopleth diagram for a representative middle-high latitude marine climate.

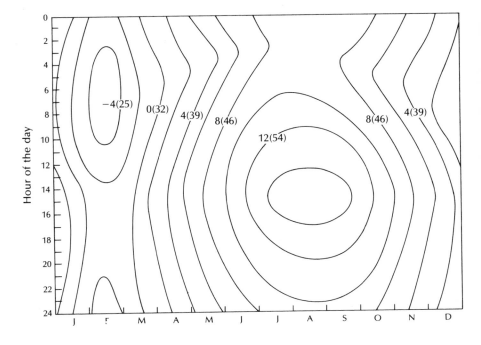

high latitude and distance from a large body of water combine to maximize annual range. Notice that summer temperatures are higher than in the middle-to-high latitude maritime climate (d), but winters are extremely cold and the annual range is the largest of any climate.

Figures 3.4 to 3.8 show the *temperature climate*, or long-term temperature means. As such they smooth out the sometimes wide variations that can occur from one year to the next at the same hour. Similarly, mean monthly temperatures vary from one year to the next. What is needed to complete our description, then, is some indication of how weather—as contrasted with climate—shows up in temperature-time curves. Specifically, we now discuss the *variability* of the diurnal and annual courses of temperature.

Although we will explore the causes of variability in later chapters, we can state here that in general extratropical climates are more variable than those of the tropics, at least in terms of temperature. Temperature also tends to be more variable in continental than in marine climates. Thus, in any year, the actual temperature experience of the humid tropical climate represented in Figure 3.4 will be quite close to the long-term mean values shown. On the other hand, the actual temperatures for any year for the continental climate of Figure 3.6 will show wide variability.

An example of how the diurnal progression of temperature varies according to the weather conditions is shown in Figure 3.9, which depicts a day in July for a mid-latitude continental climate. The mean values for each of the 24 hours are taken directly from Figure 3.6 for the middle of July to calculate the overall mean. The curve with greater amplitude represents clear sky conditions, with very little water at the surface or in the atmosphere. The flatter curve shows how temperature varies on an overcast day with rain. Notice the asymmetry of the curves.

How much the mean temperature for a month varies over a long period of record is shown in Figure 3.10. Again we are using the continental climate of Figure 3.6, since such a climate experiences great temperature variability. The height of the vertical lines (only every other month is shown) indicates the temperature span that includes 95 percent of the mean monthly temperatures. For example, 95 percent of the mean January temperatures have varied from −18°C (0°F) to −2°C (28°F). Thus, in only 5 percent of the years were mean January temperatures warmer or colder. Another interesting aspect of this figure is that winter temperatures display more variability than those of summer, while spring and fall are intermediate (compare the heights of the lines). This diagram also shows that the 95 percent marks are equidistant from the mean. This is not always the case, however, and in some climates the lower mark will be farther from the mean than the upper, or vice versa. We will

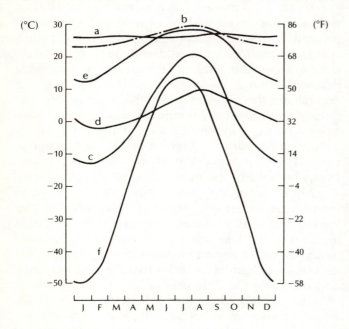

**FIGURE 3.8**
The annual march of mean monthly temperature for six diverse climates. Climates (a) through (d) are those shown in Figures 3.4 to 3.7, respectively. Curves (e) and (f) show a humid subtropical and an extreme continental climate.

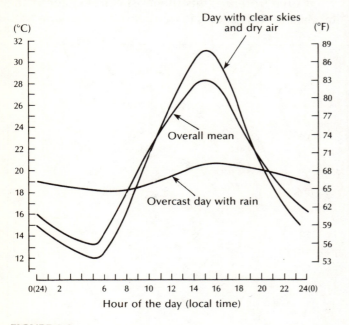

**FIGURE 3.9**
The diurnal progression of temperature as influenced by various weather conditions.

## SEC. 3.5 | THE SPATIAL VARIATION OF TEMPERATURE

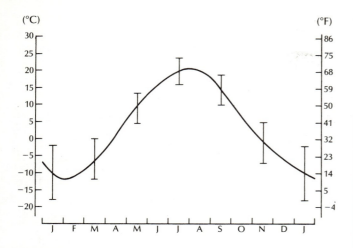

**FIGURE 3.10**
The variability of mean monthly temperature in a continental climate.

consider this subject further in the discussion of frequency distributions in Chapter 6.

## 3.5
### THE SPATIAL VARIATION OF TEMPERATURE

The variation of temperature with time, discussed in Section 3.4, dealt with temperature at shelter level, or at about 2 m above the interface, and at specific locations. The change of temperature in the vertical and horizontal dimensions also is of concern in climatology. The global perspective is the first to be examined.

Figures 3.11 and 3.12 show the mean temperature of the atmosphere from the surface (where mean sea level pressure is 1013 millibars) to where pressure is 100 millibars (mb), or about 16 km (10 mi), for 10° S to 75° N for January and July. These values are zonally averaged, that is, they represent averages for all longitudes. January and July are shown because these are the coldest and warmest months (warmest and coldest for the Southern Hemisphere) for extratropical latitudes. Within the tropics this may not be the case, but here there is very little difference among the months.

Many features of these cross sections are of interest. In both, monthly temperature decreases with height, at least to about 200 mb, except at 70° N in January, where there is an inversion. In the lowest latitudes (10° S to 20° N), temperatures are virtually the same in both months. Poleward of 20°, however, the differences between the months become apparent. Here the atmosphere is colder in January than in July at all but the highest levels.

Of great importance to atmospheric circulation is the latitudinal gradient of temperature, or how much temperature changes with latitude. Especially at the surface, but evident at heights up to about 200 mb, the poleward decrease of temperature north of about 20° N is greater in January than in July. At the surface, in the span from 20° N to 75° N, temperature decreases by almost 50C° (90F°) in January, but only by about 20C° (36F°) in July. At 300 mb this difference is about 20C° (36F°) and 10C° (18F°) in January and July, respectively.

A comparison of Figures 3.11 and 3.12 with Figure 2.6 reveals the reason for the seasonal disparity in latitudinal temperature gradients. The change in insolation with latitude is more pronounced in the winter than in the summer hemisphere, which is reflected in temperature gradients in these seasons. In Figure 2.6

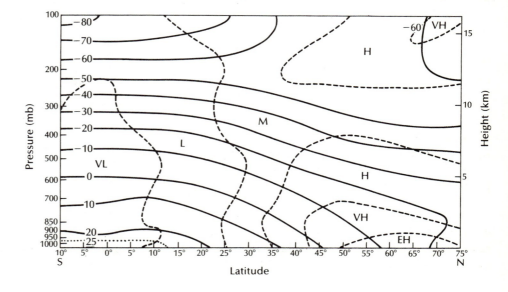

**FIGURE 3.11**
Mean temperature of the atmosphere (to 100 mb) and its annual variability, for 10° S to 75° N in January. Temperature (solid lines) is in °C. Variability (dashed lines) is scaled from standard deviation as very low (VL), low (L), moderate (M), high (H), very high (VH), and extremely high (EH). (From *Atmospheric Circulation Statistics*, NOAA Professional Paper No. 5, 1971.)

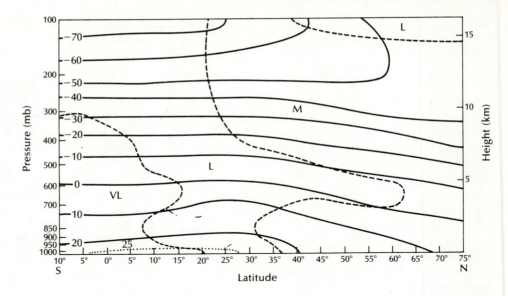

**FIGURE 3.12**
Mean temperature of the atmosphere (to 100 mb) and its annual variability, for 10° S to 70° N in July. (From *Atmospheric Circulation Statistics*, NOAA Professional Paper No. 5, 1971.)

compare the variation in insolation in June and December from the equator to the poles. In addition, because continents display greater ranges of temperature from summer to winter than do oceans, this latitudinal gradient is augmented in the Northern Hemisphere winter, when the continents are very cold. This factor is not present in the Southern Hemisphere because there are no continents in high latitudes except for Antarctica.

The temperature structure for that part of the atmosphere not shown in Figures 3.11 and 3.12, the southern latitudes south of 10° S, is generally the "mirror image" of the same latitudes in the Northern Hemisphere. An exception is that the latitudinal gradient in the lower atmosphere is not as pronounced in the middle-to-high latitudes in winter because of the absence of continents, although at the latitude of Antarctica it increases somewhat.

The dashed lines in Figures 3.11 and 3.12 illustrate the variability of temperature. That is, how different one January (or July) is from another is indicated by letters that represent variabilities from very low to extremely high. The difference between winter and summer is striking. Through the depth of the atmosphere shown, temperature variability changes from very low to moderate in July. But in January this variation is from very low to extremely high. The latter occurs near the surface in the highest latitudes. Here again the continental influence is apparent, since temperature variability is higher over continents than over oceans.

Additional detail on the change of temperature with height is given in Figure 3.13, which shows lapse rates for a tropical area, a mid-latitude continent in summer, and a high-latitude continent in winter. These lapse rates are taken directly from Figures 3.11 and 3.12, but extend somewhat higher to 50 mb (21 km, or 13 mi). The tropical lapse rate shows somewhat higher temperatures than those for the mid-latitude continent in summer up to about 200 mb. In this same depth temperatures over a high-latitude continent in winter are coldest up to the same level. Above 200 mb temperatures in the tropical atmosphere continue to decrease, while those in the other two locations are isothermal, or show a slight

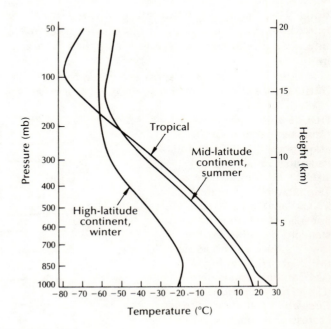

**FIGURE 3.13**
Lapse rates through the troposphere and lower stratosphere for various conditions.

increase. The result is that temperatures at about 200 mb in the tropical area actually are lower than in higher latitudes. This can be seen also in Figures 3.11 and 3.12.

The level at which temperature either changes from decreasing to increasing, or becomes isothermal, is called the **tropopause.** This marks the upper boundary of the lowest atmospheric layer, the **troposphere.** The average lapse rate in the troposphere is 6.5C° per km (3.6F° per 1000 ft). Immediately above the troposphere is the **stratosphere.** Characteristically, the tropopause is highest in low latitudes. Representative heights are 14 to 18 km for the tropics and 8 to 10 km over the poles.

The reason for the difference in height of the tropopause can be understood if we consider the prefix *tropo-*, which means mixing. Thus the troposphere is a layer in which mixing takes place, both in the horizontal and the vertical. The horizontal exchange is essentially synonymous with *advection,* and the vertical with *convection,* although radiative processes also are important, especially in the vertical. Since the earth's atmosphere is heated most strongly in low latitudes, it is here that the greatest mixing, and thus the greatest vertical transport, takes place. The mixed layer is therefore higher in low than in high latitudes.

The change of temperature closer to the interface also is of interest. We learned earlier that the conductive capacities of interface substances suggests that over cycles of heating and cooling the range will be greatest at the interface and will decrease with both height and depth. This is most conveniently shown by diurnal changes, as in Figure 3.14. Both scales are somewhat arbitrary, that is, temperature could be fixed at any reasonable range of interface values and the greatest height can be approximated at tens of meters. Clear sky conditions, and at least negligible advection, also are assumed.

Since the range of temperature at the interface is inversely proportional to the sum of the conductive capacities of the substances there, day-night differences should be something like 50 times as great over land as over water (see Section 3.4). Figure 3.14 shows an inversion over land at night, although such a reversal of temperature with height in the lowest layer is not always present. The penetration of the diurnal wave of temperature is more extensive in the atmosphere than in the soil, for a dry land interface, but this difference is much less over water.

Figure 3.14 also shows that the lapse rate becomes very steep, that is, temperature decreases very rapidly with height in the air next to the land surface during the day. In this situation vertical overturnings of air, or convection, are promoted. This has implications for the formation of precipitation, as we will see in Chapter 5.

Temperature inversions are a more conspicuous feature of the atmosphere than has yet been indicated. There are two general kinds of such inversions: ground and upper-air (Figure 3.15). Upper-air inversions, which commonly occur at heights of about 50 to 1500 m (150 to 4500 ft), develop as the result of subsidence, or sinking, of air. We will see in Chapter 5 that compression accompanies subsidence, and this leads to warming. Such air thus becomes warmer than

**FIGURE 3.14**
Generalized diurnal temperature profiles over land and water under clear sky conditions with neglible advection.

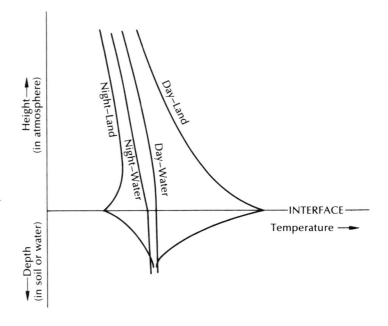

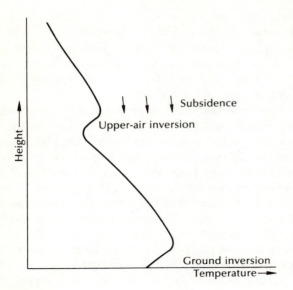

**FIGURE 3.15**
Ground and upper-air inversions.

the air both above and below it. These upper-air inversions are an important feature of some climates (see Section 4.6.2 on the subtropical highs).

Ground inversions develop principally from three causes. The most common inversion occurs when the ground and the layer of air immediately above it cool more rapidly, by radiation, than the air at higher levels. This is the primary cause of the surface inversion shown in Figures 3.11 and 3.13. Another ground inversion of temperature results from the drainage of cold, relatively dense air into topographic depressions. Third, inversions in the air adjacent to the surface may form as the result of the advection of cold air over a relatively warm surface. Given sufficient water vapor in the surface air layer, fog may form in the first two of these conditions; in fact, fogs are classified according to their origin.

# 3.6
## CONTINENTALITY

In the preceding sections we have mentioned differences in temperature between marine and continental climates. We have noted the larger amplitude of the diurnal and annual temperature cycles, and the great variability of temperature, especially for continental climates in high latitudes. A quantitative measure of the difference between continental and marine climates is called **continentality**.

A map of continentality shows the extent to which temperature is controlled by continental influences. We recognize that the mean annual range of temperature—that is, the difference between the mean temperatures of the coldest and warmest months—is such a measure in part. However, mean annual temperature range also is influenced by latitude. An index of continentality, then, is a measure that uses mean annual range but incorporates a correction based on latitude so that an attempt is made to nullify the influence of latitude.

Figures 3.16 and 3.17 show continentality in North America and western Europe. The units of the values shown and the way in which they are derived are not important. The numbers should be regarded simply as scalar quantities that indicate the degree of continentality. As expected, values are lowest closest to the oceans, and highest in or toward continental interiors. There are some important exceptions, however. The most apparent of these is in North America, where the lowest values occur along the west coast and east coast values are moderately high. This is because the prevailing circulation in the middle and middle-high latitudes (40° to 70° N and S) is westerly, or from west to east. Marine influences thus are advected to the east, and west coasts experience marine characteristics. The east coasts of continents, on the other hand, have more continental characteristics because the air brought to them is largely from the interiors of continents.

Another exception involves the influence of topographic barriers. The mountains of western North America, which run north-south, act as a barrier to the inland penetration of marine air. Thus we see a marked gradient, or sharp transition, from coastal California inland. In Europe, on the other hand, the gradient is more diffuse, and marine influences penetrate far inland (or conversely, continental influences are restricted to areas farther inland). This is due to the absence of north-south topographic barriers in Europe, although somewhat of an exception does occur in Norway.

Two other relatively minor influences can be discerned in Figures 3.16 and 3.17. The Great Lakes of North America are large enough to exert a moderating influence, as shown by the trough (an area of relatively low values) there. The other influence is the dryness of the climate. This leads to increased annual range, which in turn increases continentality. Thus, the intermontane area of the United States and the Prairie Provinces of Canada have somewhat higher values of continentality than they would if they were more humid. Similarly, the rather high values in Spain are in part due to dryness.

# SEC. 3.6 | CONTINENTALITY

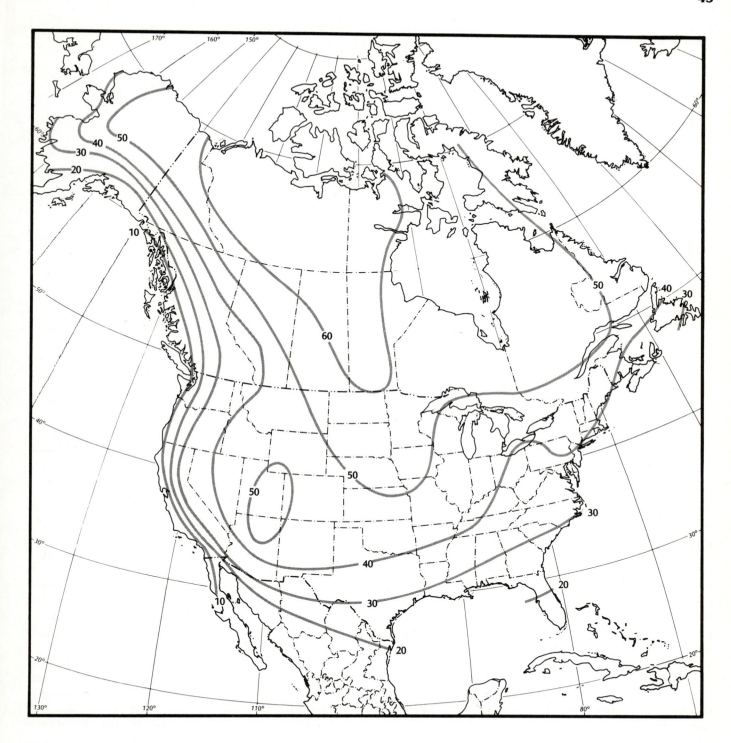

**FIGURE 3.16**
Continentality in North America.

Continentality is a useful climatic indicator, not only of the continental-versus-oceanic nature of temperature, but also of the strength and direction of atmospheric circulation, the influence of topographic barriers, and aridity. We will return to these influences in Chapter 4.

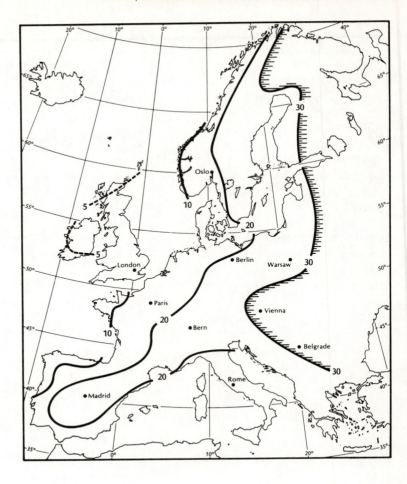

**FIGURE 3.17**
Continentality in western Europe.

## 3.7

### OTHER CLIMATIC ASPECTS OF TEMPERATURE

Portraying and analyzing the aspects of temperature considered in this chapter have helped to explain and describe climate. Not every aspect has been shown, however. In particular, maps of mean sea level temperature in January and July and maps of the mean annual range of temperature are instructive, but these are found later in Chapter 7.

In applied climatology there are many other important aspects of temperature. Very often the duration of temperatures above or below certain values, or the first or last occurrence of a specified temperature over the annual cycle, is of greater significance than mean values for a day, month, or year. The probability of the occurrence of temperature above or below certain values also is significant.

Suppose, as an example, you were to seek a climate in which temperatures never become excessively hot, that is, over 32°C (90°F). Figure 3.18, a map that shows the average number of days each year this temperature will be exceeded, would be helpful. As expected, the number of days with temperatures over 32°C decreases to the north, but exceptions occur along the Gulf coast, the east coast, and around the Great Lakes, where adjacent water bodies moderate temperature extremes. The influence of elevation also is apparent in the Appalachians and Adirondacks.

The mean seasonal temperature for a location correlates closely with the amount of fuel consumed for heating. An even better measure, however, is **heating degree-days.** We assume that most homes and places of work will be heated when the mean daily temperature drops below 65°F (18°C). For any day, the fuel consumed is proportional to the difference between the mean daily temperature (the average of the high and the low for the day) and 65°F. The sum of all such values during the interval when mean daily temperatures are less than 65°F is calculated as heating degree-days (Figure 3.19). The reverse of this procedure is used to calculate **cooling degree-days,** although 75°F (24°C) is a more appropriate threshold

## SEC. 3.7 | OTHER CLIMATIC ASPECTS OF TEMPERATURE

temperature for this measure. Heating and cooling degree-days will be considered in more detail in Chapter 15.

Of particular interest to agriculture is the mean date of the last temperature of 0°C (32°F) or below in spring, and the first such occurrence in fall. The difference between the two, in days, is the mean length of the **growing season** (Figure 3.20). Notice the moderating effect of the Great Lakes on the immediately adjacent land areas. How much the growing season varies from year to year in a particular location is shown in Figure 3.21.

Still another temperature-related aspect of interest to agriculture, and to transportation and communications as well, is the likelihood of the occurrence of some threshold temperature, or below or above, either within a specified time span or before or after specified dates. Figure 3.22 shows the probability of a "freeze" [roughly defined as the occurrence of a temperature of 0°C (32°F) or below] before October 1 in Wisconsin. Notice the expected overall increase in probability from south to north, and the moderating influence of Lake Superior to the north and Lake Michigan to the east.

Keep in mind that in Figures 3.19 through 3.22 local, or small-scale, effects may produce values somewhat different from those shown for particular locations. The first and last occurrence of freezes, for example, may vary appreciably within a few kilometers depending on topographic conditions, especially where cold air drainage is likely.

In this section we have been concerned with mean, or long-term, values. However, we must stress that the variability associated with mean values is as important to climate as the means themselves. A point that will be made in Chapter 6, when we discuss statistical applications in climatology, is that the representativeness of a mean decreases as the variability associated with it increases.

Most of us at one time or another have heard some long-time resident of a particular area say something like, "If you don't like the weather here, wait a bit, it'll change." Where is the most variable weather? What can climatologists offer as an answer? The answer depends, of course, on what constitutes weather. That is, which element or elements are regarded as most important? At least a partial answer involves examining temperature and its day-to-day fluctuations.

Figures 3.23 and 3.24 show the variability of temperature over short time periods, and at least partially answer our question about the most variable weather in the conterminous United States. The values isoplethed are the mean day-to-day changes in daily maximum temperature, regardless of whether they are positive or negative. In general, January is the month in which the greatest day-to-day changes occur, although the values then are only marginally greater than those of spring and fall. In summer the variability decreases appreciably. To answer our question, it appears that residents of the Midwest and northern Great Plains have the most valid claim on the changeability of weather, at least as represented by temperature. The smallest changes occur in coastal California and southern Florida. We will return to this subject in Section 7.2.1.

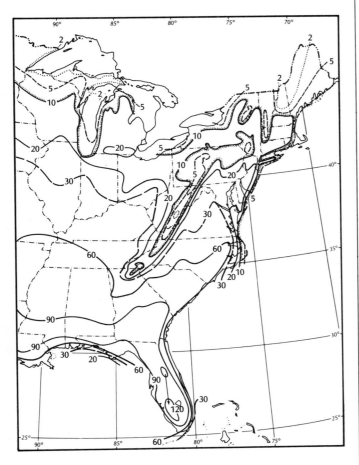

**FIGURE 3.18**
Mean annual number of days with temperatures equal to or above 32°C (90°F). (From *Climatic Atlas of the United States*, Environmental Data Service, Environmental Science Services Administration, U.S. Department of Commerce, 1968.)

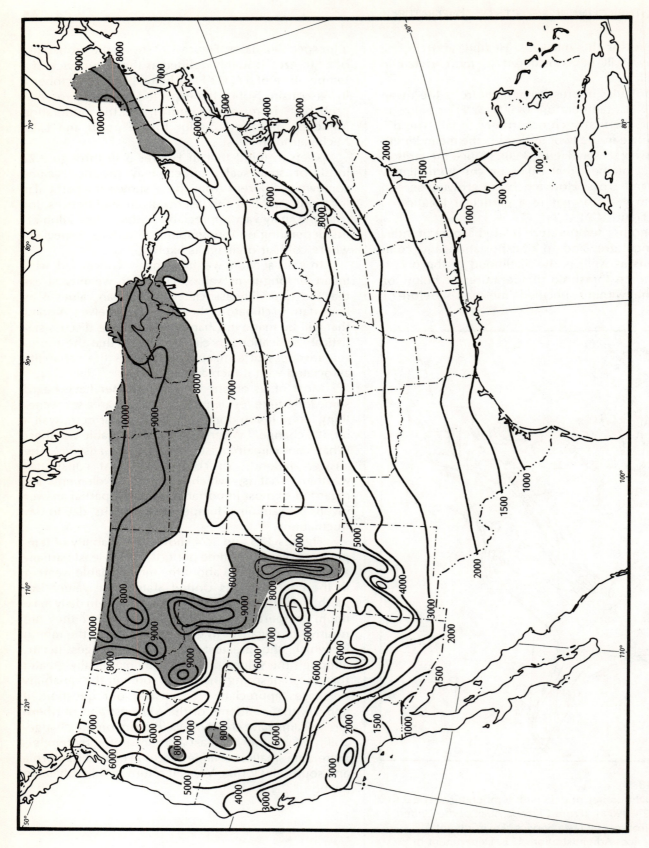

**FIGURE 3.19**
Average seasonal heating degree-days from 1941 to 1970. The base temperature is 18°C (65°F). (From U.S. National Weather Service.)

**FIGURE 3.20**
Length of the growing season, in days. (From *Climatic Atlas of the United States*, Environmental Data Service, Environmental Science Services Administration, U.S. Department of Commerce, 1968.)

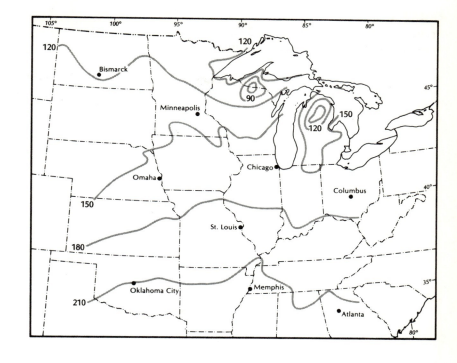

**FIGURE 3.21**
Dates of the first and last temperature of 0°C (32°F) or below, and length of the growing season (in days) for Lincoln, Nebraska, from 1941 to 1970. Earliest and last dates for this 30-year period also are shown.

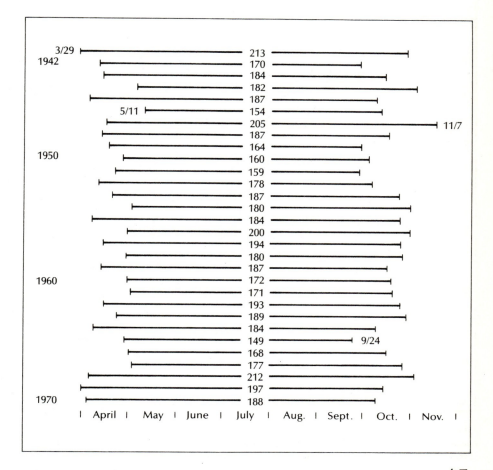

**FIGURE 3.22**
Probability of a temperature of 0°C (32°F) or below before October 1 for Wisconsin.

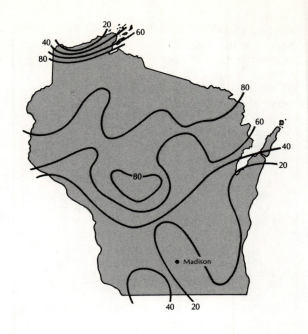

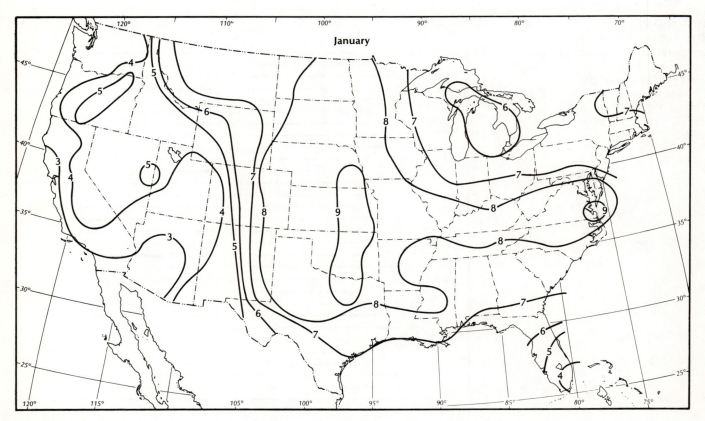

**FIGURE 3.23**
Mean interdiurnal variability of maximum temperature (°F) in January in the United States. (From H. E. Landsberg, *Interdiurnal Variability of Pressure and Temperature in the Conterminous United States,* Technical Paper No. 56, Environmental Science Services Administration, 1966.)

## SEC. 3.7 | OTHER CLIMATIC ASPECTS OF TEMPERATURE

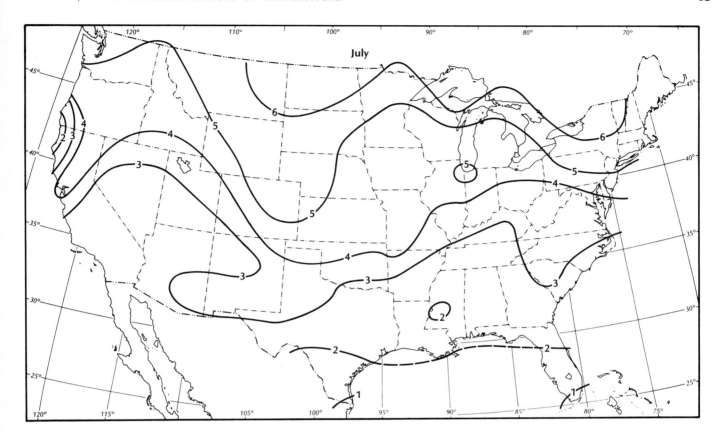

**FIGURE 3.24**
Mean interdiurnal variability of maximum temperature (°F) in July in the United States. (From H. E. Landsberg, *Interdiurnal Variability of Pressure and Temperature in the Conterminous United States,* Technical Paper No. 56, Environmental Science Services Administration, 1966.)

## SUGGESTED READING

BRYSON, R.A. and K. HARE. *World Survey of Climatology: North America.* Vol. 11. Amsterdam: Elsevier Scientific Publishing Co., 1974.

CALDER, N. *The Weather Machine.* New York: Viking Press, 1975.

*Climatic Atlas of the United States.* Superintendent of Documents, Washington, D.C., 1968.

# 4
# CIRCULATION OF THE ATMOSPHERE AND OCEANS

**INTRODUCTION**
**REQUIREMENTS FOR ATMOSPHERIC AND OCEANIC CIRCULATION**
Conservation of Mass, Heat, and Moisture
Redistribution of Heat and Moisture
Conservation of Angular Momentum
**THE BASIS OF AIR MOVEMENT**
Forces Affecting Air Movement
Rules for Wind Flow
The Geostrophic Approximation
The Effects of Friction
**CONVERGENCE AND DIVERGENCE**
**SCALES OF ATMOSPHERIC MOTION**
**SURFACE CIRCULATION**
Fundamental Considerations
Centers of Action
Transient Disturbances on the Synoptic Scale
Climatological Characteristics of Transient Disturbances
**UPPER-AIR CIRCULATION**
The Hadley Cells
The Zonal Circulations of Extratropical Latitudes
Summary of Surface and Upper-Air Circulation
**CLIMATIC INFLUENCES OF THE OCEANS**
The Transfer of Heat by Ocean Currents
Surface Circulation and Its Climatic Influences

## 4.1
### INTRODUCTION

The earth's atmosphere is a fluid medium in which various amounts of heat and moisture are constantly being moved, both horizontally and vertically. These atmospheric motions vary in size, and in their significance for climate, from the smallest gusts and eddies to the enormous "rivers" of air that circle the globe in the middle latitudes. The oceans, virtually limitless reservoirs of heat, also circulate and have a profound influence not only on the climate of the air over them, but also on the climates of the landmasses, especially their coastal areas.

Essentially all atmospheric motion can be regarded as the consequence of the conversion of thermal energy to kinetic energy, the energy of motion. Because solar radiation is unevenly distributed over the earth's surface and the reflectivities of both surface substances and the atmosphere vary, there are place-to-place differences in heating and cooling, both at the interface and in the atmosphere. These induce pressure differences, which in turn promote an exchange of air. In a continuous process the atmosphere strives to mitigate these differences, even as they are changing.

In the preceding chapters we emphasized the vertical exchanges of heat and moisture. We have seen that many of the temporal and spatial variations in climate can be explained in these terms. To appreciate how atmospheric motions can influence climate, consider these two extreme cases. First, consider the climate of a low-latitude, very dry desert. Temporal variations of temperature here can be explained simply by incoming solar radiation (insolation). On an annual basis, temperatures are highest shortly after the high sun period, and vice versa. On a diurnal basis, maxima occur shortly after the sun has attained its highest elevation angle, and minima about sunrise. Even though winds may be from any point of the compass, the lack of any horizontal gradient of temperature precludes any appreciable change in this element. Because rain is a rare event in this area, we do not need to take it into account. In short, the vertical exchanges considered earlier are generally sufficient to explain the climate here.

Now consider the influences on climate in a mid-latitude, humid continental area. The same vertical exchanges take place, but they are greatly modified by the air arriving from different sources and directions. Although there is some seasonal variation, horizontal gradients of temperature exist at all times of the year, and air brought to this location will be at least moderately different from that which was there originally. At any time of the year cold (or cool), dry air may move equatorward, or warm (or hot), moist air may move poleward, past this location. The annual and diurnal variations of temperature often are interrupted by such cold, or warm, advection. Also, the state of the sky and the presence of clouds and rain are dependent on the way the atmosphere circulates. Indeed, without atmospheric motion there would be no precipitation!

Clearly, then, an examination of air movement is indispensable to an understanding of the genesis of climate. In this chapter we consider the implication of various scales of atmospheric and oceanic circulation for weather, and hence for climate.

## 4.2 REQUIREMENTS FOR ATMOSPHERIC AND OCEANIC CIRCULATION

Throughout Chapters 2 and 3 certain requirements of atmospheric—and to a lesser extent, oceanic—circulation were described. There are other requirements that we must specify before we can understand fully why the atmosphere and oceans circulate as they do.

### 4.2.1 CONSERVATION OF MASS, HEAT, AND MOISTURE

Compared with the changes it has undergone in the hundreds of millions of years the earth has had an atmosphere, climate in this century has been comparatively stable. One year at a particular location is very much like other years. There will of course be occasional droughts and periods of below or above average temperatures, but these are minor when compared with changes that have occurred in the distant past. It follows, then, that the mechanisms governing climate also are reasonably stable over periods of a year or more. The fluxes of energy represented by the energy balance equations (shown in Figure 2.13) may show some year-to-year variation, but in the long run can be regarded as almost constant.

Similarly, the role that advection plays in influencing the components of the energy balance equations, and hence the more common elements of temperature and precipitation, must be much the same from one fundamental period of change—a year—to another. This imposes a constraint on the atmosphere. Over many years there can be no appreciable surplus or deficit in mass, heat, and moisture at any location. In other words the atmosphere cannot circulate in such a way as to accumulate or deplete these quantities. The mass of the atmosphere above a point must be nearly constant over long periods of time, and a given location cannot become progressively warmer or colder, or drier or more humid.

It is not apparent how these constraints on the atmosphere determine how it circulates. Actually, the causes of atmospheric circulation are very complex and not very well understood. But we do know that mass, heat, and moisture are conserved; that is, they must be very nearly constant in the long run at all locations.

### 4.2.2 REDISTRIBUTION OF HEAT AND MOISTURE

Figure 2.16 shows that the low latitudes have a surplus of heat, and the high latitudes a deficit. Figure 2.15 shows that a convective transfer of sensible and latent heat from the earth's surface to the atmosphere is required to keep the surface and the atmosphere in energy balance. Thus, we see that the atmosphere and oceans must circulate in a way that transports heat poleward and upward.

This does not mean that the atmosphere must everywhere have a net motion upward and poleward. The remarks in Section 4.2.1 obviously preclude this. But a net poleward transfer of heat can be accomplished by warm air or water moving poleward and cooling, or by cool air or water moving equatorward and warming. This is an especially effective mechanism when moisture condenses from the poleward moving air and evaporates into air moving equatorward. Similarly, warm air or water rising is accompanied by sinking cooler air or water. In air, evaporation occurs predominantly at the surface, and condensation in the atmosphere.

Thus, the redistribution of heat and moisture requires mainly convective transport, and *meridional* (from north to south or south to north) motion of the atmosphere and oceans. This is surprising, since atmospheric motion is predominantly *zonal* (east to west or west to east). Still, the necessary transports are fulfilled, both by the mean state of the atmosphere and oceans and by transient (eddy) motions not apparent in long-term averages.

### 4.2.3 CONSERVATION OF ANGULAR MOMENTUM

Any wind with an east-west component exerts a force, through frictional drag, on the earth's surface and thus tends to alter its rotational rate. If winds everywhere were easterly, or had a component of motion from east to west, the effect would be to reduce the rate at which the earth rotates from west to east. Conversely, a westerly wind would increase the rotation rate. This frictional force varies with latitude: a wind of a given speed at low latitudes exerts more force than the same wind at high latitudes because it acts farther from the axis of rotation.

The earth's rotation is essentially constant, although there has been a very gradual slowing since the formation of our planet. For example, some 400 million years ago day length was 21.5 hours. Thus, the net frictional force exerted by zonal surface winds must be zero; that is, easterly and westerly winds must balance when weighted according to their area, strength, and distance from the earth's axis.

# 4.3
## THE BASIS OF AIR MOVEMENT

Air moves in response to a combination of a few forces. Forces are **vector quantities,** and thus have both magnitude and direction. Wind (air movement) also is a vector, since it has both magnitude (speed) and direction. In contrast, **scalar quantities** have only magnitude; examples are distance, mass, and time.

These few forces may combine in such a way as to produce a net force, causing an imbalance. The wind, in response, accelerates (speeds or slows), changes its direction, or both. If the forces balance, no net force results and there is no air movement (no wind), or, the wind may be in **uniform motion.**

To understand this concept we must consider Newton's first law of motion and its implications for the atmosphere. Uniform motion is motion along a straight line that is neither positively nor negatively accelerated (neither speeded nor slowed). Newton was the first to recognize that *a body in uniform motion will continue in uniform motion unless acted upon by an unbalanced force.* Thus, if at any time air is moving in uniform motion, we know that a balance of forces must exist. In some circumstances the wind blows in just this way, and by specifying the particular manner in which this balance of forces is achieved, we can deduce both its speed and direction from the way the forces are represented.

### 4.3.1 FORCES AFFECTING AIR MOVEMENT

Before we attempt to balance the forces affecting air movement, however, we must specify the characteristics of each. The first force is **gravity,** which acts downward or toward the earth's center at every point. The second is the **pressure gradient force,** which results from place-to-place differences in atmospheric pressure, or the weight per unit area of overlying air. Both the magnitude and direction of the pressure gradient force change. In an atmosphere in which pressure changes only in the vertical, this force is directed upward (or exactly opposite to gravity) from high to low pressure. Figure 4.1(a) illustrates this. In this cross section the isobars (lines connecting points of equal pressure) are parallel to the earth's surface, and the entire pressure gradient force (**P**) is directed upward at right angles to the isobars. A balance attained in this way, between gravity and pressure, is called **hydrostatic balance.**

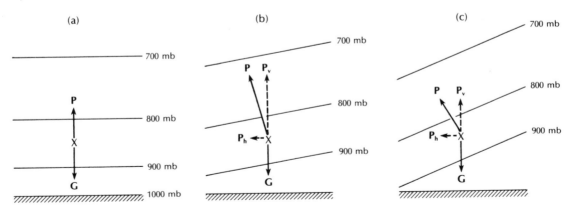

**FIGURE 4.1**
The effect of varying inclinations of pressure surfaces on the variation of the horizontal ($P_h$) and vertical ($P_v$) components of the total pressure gradient force (**P**). **G** represents gravity. As inclination increases, $P_h$ increases. The numbered values are the pressure surfaces measured in millibars.

In Figure 4.1(b) the isobars are moderately inclined. For a mass of air at point X the pressure gradient is again directed at right angles to the isobars toward lower values. Now, however, the vector representing the pressure gradient force can be broken down into two components, the vertical ($P_v$) and the horizontal ($P_h$). Notice that the vector addition of $P_v$ and $P_h$ equals **P**. Again $P_v$ balances gravity (or very nearly so), but there is as yet an unbalanced force, $P_h$, acting from high to low pressure at the level of X.

In Figure 4.1(c) the isobaric surfaces are even more inclined. Again gravity is compensated by $P_v$, but $P_h$ is larger in magnitude, although its direction is the same. We see, then, that except for the situation in which pressure surfaces parallel the earth's surface, $P_h$—the as yet unbalanced force—increases as the inclination of the pressure surfaces increases.

It should be pointed out that the slope of the surfaces in Figure 4.1 is greatly exaggerated. That is, in cross-sectional representations of the atmosphere the vertical scale is almost always increased relative to the horizontal in order to show the desired features better. In the case of Figure 4.1, an exaggeration of the inclination, or slope, of the pressure surfaces results. In the real atmosphere this inclination is very slight.

The next force acting on air is not really a force! To understand why this factor (we will call it a factor for the moment rather than a force) must be accounted for, we will have to stretch our imaginations. Indeed, some element of faith is needed to appreciate the following descriptions.

Imagine that a missile is fired from the North Pole toward New York City. The path intended is straight "down" the 74th meridian west, the longitude of this city. The perpetrators of this dastardly act are not knowledgeable about ballistics, however, because rather than arriving at its target, the missile lands in a cornfield in Illinois! They failed to realize that while the missile was in flight, the earth was turning under it. Their weapon appeared to describe a curved path such that while it initially sped directly southward, its direction changed gradually to southwestward.

We say that the missile "appeared" to follow a curved path because the path was not curved—it was straight. It only appears curved to us because we are constrained to view its motion from a rotating earth. If we could view this motion from a point outside the earth (from the moon, for example) the path would appear correctly to be a straight line.

This observation can be generalized. Picture the same event in the Southern Hemisphere. Now the deflection, or turning, is to the left of the intended path. Furthermore—and here is where the element of faith is required—any object, such as a mass of air moving over the earth's surface, is subject to this deflective factor, or "force." It acts to the right of the direction of motion in the Northern Hemisphere and to the left in the Southern Hemisphere. Whether the object is moving to the east or west or to the north or south, and regardless of latitude (except at the equator), it undergoes a turning that is apparent to any observer who rotates with the earth, but not to one who views the motion from outside it.

Imagine the flight of an object (e.g., missile, airplane, or a mass of air) that begins its flight from any location other than the poles to verify that this deflecting factor exists. This may prove frustrating, however, since it is difficult to conceptualize such motion when the object has an initial speed due to its original velocity vector (i.e., the earth is already turning at that location). Notice that the earth is not turning at the poles. It should be possible, however, for you to verify that if the motion begins at the equator and stays at that latitude, there will be no deflection of the motion.

This deflecting factor is therefore not really a force, by the physical definition of the term, but it must be regarded as having the same consequences as if a force acted at a right angle (90°) to the right of the motion in the Northern Hemisphere and to the left of the motion in the Southern Hemisphere. We will therefore need to account for it when we specify the forces that act on air in motion. This deflecting "force" is called the **Coriolis force** after G. G. Coriolis, the French mathematician who first quantitatively demonstrated its effects.

The direction of this force has already been specified. What determines its magnitude? For our purposes two factors are significant. The first is the speed with which an object, or a mass of air, moves over the earth's surface. The higher the speed, the greater is the force. The Coriolis force also is proportional to the sine of the latitude. Latitude varies from 0° to 90° in both hemispheres, and thus the sine of the latitude varies from 0 to 1, respectively. In other words, the Coriolis force vanishes at the equator and is at a maximum at the poles.

In addition to the horizontal component of the pressure gradient force and the Coriolis force, we need to consider **frictional force**. Friction acts opposite to the wind direction, and thus slows the wind. Its effects are greatest near the surface and decrease upwards. Above about 1 km (0.6 mi) frictional drag on the wind vanishes, although the depth of the **friction layer** varies considerably, depending on the stability of the atmosphere. Friction also affects wind direction, as we will see. There are other forces that affect wind speed and direction, but these are relatively minor and can be ignored.

Above the friction layer we frequently observe the wind blowing in uniform motion. That is, its speed

## SEC. 4.3 | THE BASIS OF AIR MOVEMENT

is constant or nearly so, and its movement is in a straight line or nearly so. According to Newton's first law of motion, then, a balance of forces must exist. We can use this deduction to infer both the speed and direction of the wind.

### 4.3.2 RULES FOR WIND FLOW

If the wind is to move in uniform motion without friction, the horizontal component of the pressure gradient force ($P_h$) must be balanced by the Coriolis force so that there is no net force, and hence no acceleration [see Figure 4.1(c)]. This requires that the Coriolis force be equal in magnitude and opposite in direction to $P_h$. This is shown in cross section in Figure 4.2(a), where the Coriolis force is labeled C. In this diagram the Coriolis force is a vector of the same length as $P_h$, but directed to the right.

Now that the direction of the Coriolis force has been specified, recall that it always acts at right angles to the right of the direction of motion (in the Northern Hemisphere). We are then left with only one direction the wind can blow—into the page. If Figure 4.2(a) were applied to the Southern Hemisphere, where the Coriolis force acts at right angles to the left of the direction of motion, the movement of air (the wind) would be out of the page.

Since it is customary to view atmospheric motion in plane view, or as shown on a map or chart, we must change our perspective. If the left side of Figure 4.2(a) is north and the right side south, and if we assume that the pressure surfaces extend as flat surfaces (like flat sheets of paper) both into and out of the page so that they have an east-west extent, we can represent pressure change in map view. This is done in Figure 4.2(b). Now we are looking down, instead of sideways, at the variation of pressure at the level which runs through X, shown by the dotted line in Figure 4.2(a). Low pressure is to the north, and thus $P_h$ is a vector at right angles to the isobars—that is, along the horizontal gradient of pressure and directed from high to low pressure. Since the forces are balanced, the Coriolis force must be equal in magnitude and opposite in direction. Balanced flow results, and the wind direction is from left to right, or from the west. Because winds are named for the direction from which they blow, this is a west, or westerly, wind.

Under conditions of uniform motion and with the assumption that $P_h$ and C are the only appreciable forces at work, we can specify wind direction. How do we determine the other aspect of this vector quantity, the wind speed? Imagine that the pressure surfaces of Figure 4.2(a) have been inclined at an even greater angle with respect to the surface. Now the level at X intersects more isobars. It should be easy to imagine that the pressure change from north to south, which was from 650 to 800 mb originally, now might be from 600 to 900 mb. In the same north-south distance we now have a greater pressure change, and in the map view the isobars would be closer together. The horizontal component of the pressure gradient force has thus increased. Accordingly, the $P_h$ vector would have to be longer; and if balanced flow is again to be attained, C would have to increase by an equal amount. In this case the wind direction is unchanged, but its speed is increased, since the Coriolis effect and the wind increase together. Conversely, if the isobaric surface were less inclined (more nearly parallel to the earth's surface), the pressure change in the horizontal would be less, the isobars farther apart, and the wind speed slower.

With the assumptions just specified, we can conclude that the wind blows parallel to straight isobars, and in such a direction that if we stand with the wind at our back, low pressure will be on the left in the Northern Hemisphere and on the right in the Southern Hemisphere. Also, wind speed will be inversely proportional to the spacing of the isobars; that is, the closer the isobars, the faster is the wind. In other words, wind speed increases as the horizontal pressure gradient increases.

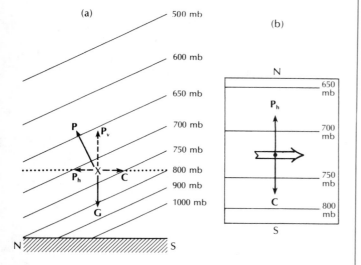

**FIGURE 4.2**
The balance of forces in the geostrophic approximation to wind speed and direction, in cross section (a) and in map view (b). C represents the Coriolis force, and the broad arrow indicates the direction of the geostrophic wind.

### 4.3.3 THE GEOSTROPHIC APPROXIMATION

It may seem that this approach to the determination of wind speed and direction is contrived, and that the conclusions drawn are reasonable only within a set of very special circumstances. Does the atmosphere really behave in this way? Surprisingly, it does. At levels above the influence of friction the wind is observed to flow along (parallel to) the isobars, accelerating where they converge (become more closely spaced), and vice versa. There is a name for the wind that develops under these conditions, where the horizontal component of the pressure gradient force is balanced by the Coriolis force and the flow is not affected by friction. It is called the **geostrophic approximation** to the wind, or simply the **geostrophic wind**.

Of course, this balance of forces (the geostrophic balance) is not always attained, even at high levels. When isobars curve, the motion is no longer uniform, which implies an imbalance of forces. This same imbalance applies, at least momentarily, when the wind speeds or slows. Nevertheless, it seems that the atmosphere is constantly striving to attain the balance of forces required by the geostrophic approximation, and the wind flow rule derived from it applies, at least to some degree, to all above-surface isobaric configurations.

With the wind flow rules derived from the geostrophic approximation, we can characterize the motion of air with respect to curved as well as straight isobars. Actually, when the flow is curved another approximation—the *gradient wind approximation*—applies, but the wind flow rules just developed are essentially unchanged. Figure 4.3, a map view, shows air movement around atmospheric pressure features, at levels above about one kilometer, that is in accordance with the geostrophic approximation. These features are **cyclones** (centers of low pressure, or lows, indicated by L), **anticyclones** (centers of high pressure, or highs, indicated by H), and **troughs** and **ridges** (areas of relatively low and high pressure, respectively, but without closed isobars). The axis, or center line, of the trough in Figure 4.3 is shown as a dashed line, and the jagged line shows the axis of the ridge.

Notice that in this Northern Hemisphere representation the flow around lows is counterclockwise and the flow around highs is clockwise. Because the deflection due to the earth's rotation is to the left in the Southern Hemisphere, the opposite relationships apply there. Regardless of the hemisphere, the flow around lows and troughs is called *cyclonic* and that around highs and ridges is *anticyclonic*.

It also is apparent in Figure 4.3 that wind speed, which is proportional to the length of the arrows, is inversely related to isobar spacing. Where the isobars are close together, as around the low and at the axis of the trough, wind speeds are high. Where the gradient is weaker (i.e., where the isobars are farther apart), as around the high, wind speeds are slower.

Although the relationships between wind velocity (speed and direction) and pressure configuration are idealized in Figure 4.3, they do apply to atmospheric motion in general above the friction layer. There are, of course, occasional departures from the geostrophic and gradient relationships. The wind is *ageostrophic*, for example, when it moves in a tightly curved path and centripetal forces become dominant (a centripetal force produces a centrifugal reaction). In this case there may be strong cross-isobaric flow. And, since the Coriolis force is weak in low latitudes (and nonexistent at the equator), air again tends to flow across, rather than parallel to, the isobars in these latitudes. Finally, it should be stressed that the geostrophic and gradient approximations apply mainly to large-scale motion (large macroscale and macroscale in Table 4.1).

### 4.3.4 THE EFFECTS OF FRICTION

It has been stressed that the geostrophic approximation to the wind applies only above the friction layer. When frictional drag is present, a different relationship exists between pressure and wind. To complete

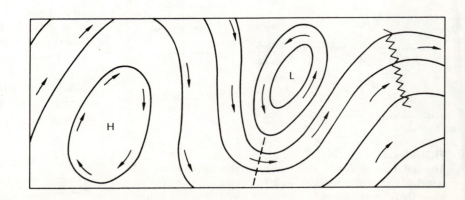

**FIGURE 4.3**
Air movement around pressure features above the friction layer, in accordance with the geostrophic approximation. H is at the center of an anticyclone, or high; L is at the center of a cyclone, or low. The dashed line follows the axis of a trough, and the jagged line the axis of a ridge. The length of the wind arrows is proportional to wind speed.

our discussion of the basis of air movement we need to examine the effect of this third, and final, force.

If friction slows the wind, the Coriolis force decreases because this force is proportional to wind speed. Momentarily, the forces no longer balance as in Figure 4.2(b), that is, $P_h$ exceeds $C$. Now the wind will tend to flow somewhat to the left of its direction in geostrophic balance and will also be slower. It is thus possible to construct a new balance of forces, now involving all three, as shown in Figure 4.4. The sum of the vector forces is zero, and uniform motion results. There is now a component of the wind from high to low pressure, and the wind speed is less than that expected from the geostrophic approximation.

Frictional effects on wind speed and direction vary, depending largely on the nature of the surface and on the extent of convection. Over extremely irregular terrain, frictional drag is great, wind speed is considerably less than that expected from the geostrophic approximation, the angle between the isobars and wind direction may be as much as 45°, and frictional effects extend farther into the atmosphere. Over water and level terrain, the angle between isobars and wind direction is less and the friction layer is thinner. With pronounced surface heating and developing convective currents in the lowest few hundred meters of the atmosphere, there is a vertical exchange of horizontal momentum. Vertical mixing tends to decrease the *vertical wind shear,* which is the change in both speed and direction of the wind with height.

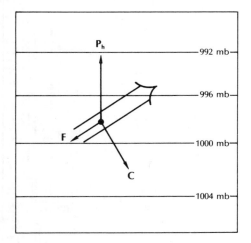

**FIGURE 4.4**
Frictional effects on the wind. The vector sum of the three forces (**F** is the frictional force) is zero, and balanced motion results. The wind blows with a component of motion across the isobars, and the wind speed is less than that derived from the geostrophic approximation.

Instead of a rapid increase in wind speed with height, such as might be expected on a day with little surface heating, faster horizontal winds aloft are mixed with slower, near-surface winds. The result is that the friction layer is deeper, that is, frictional effects imposed at the surface are propagated higher into the atmosphere. When frictional effects occur, the wind is called the **friction wind** or the **subgeostrophic wind**.

Wind and pressure relationships at the surface are illustrated in Figure 4.5. This is a map of pressure distribution and observed winds and weather for an instant of time (1 A.M., Eastern Standard Time, March 23, 1966). The isobaric analysis, or the drawing of lines of constant pressure, is made from pressure readings taken at the stations shown.

The values plotted for this analysis are not, however, the actual barometric pressures observed at the stations. Because pressure changes in the vertical are many times greater than those over a similar distance in the horizontal, a map of station pressure would show little more than differences in elevation. The United States from the Rocky Mountains west would thus show low pressure, while areas at and near sea level (e.g., the southeast United States) would show relatively high pressure. Since it is the horizontal variation of pressure at the same level (shown in Figures 4.1 and 4.2 as $P_h$) that is of interest in weather analysis, an attempt is made to construct the pressure pattern that would apply if all stations were at sea level. This is accomplished, for stations above sea level, by adding an increment of pressure to the observed station reading in order to approximate the value that would be obtained if that station were at sea level. An isobaric map drawn from such sea level values (actual or approximated) is a more meaningful representation of the horizontal variation of pressure and is more useful for weather analysis and interpretation. Still, this procedure does have inherent inaccuracies, and in areas where station elevation changes markedly over short distances, isobaric maps may be difficult to interpret.

Figure 4.5 shows much more than the relationship between isobaric configuration and wind speed and direction at the surface, so we will return to this map later in the chapter to discuss other features. But for now, notice that the wind arrows "fly" with the wind. At Jackson, Mississippi, for example, the wind is southerly; at Casper, Wyoming, it is northwesterly. The number of barbs on the wind arrows increases as the wind speed increases.

We can make the following generalizations about wind speed and direction from Figure 4.5. Wind speeds are highest where the pressure gradient is

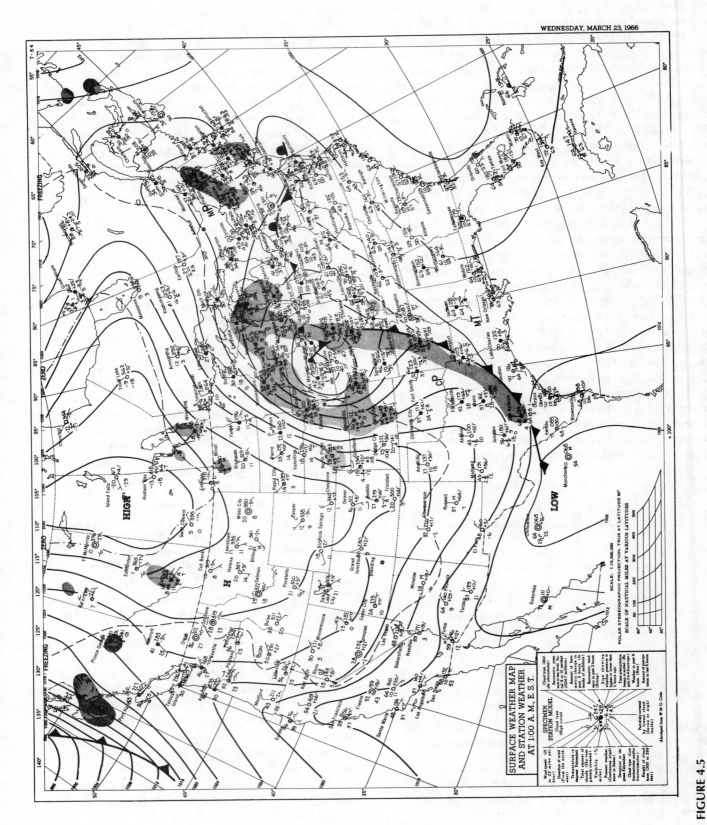

**FIGURE 4.5**
Surface weather map and station weather at 1 A.M. Eastern Standard Time, March 23, 1966. Shaded areas show precipitation.

greatest, as occurs to the west of the low center in western Iowa and eastern Nebraska. Where this gradient is weaker, as in Wyoming and Montana, wind speeds are less. Except where the pressure gradient is weak, flow is generally parallel to the isobars, but with a component of motion from high to low pressure, such as at Jackson, Memphis, and Oklahoma City. This correspondence is not as good in parts of the western United States. Here the gradient is weaker and the isobaric analysis is not exact because sea level pressure must be approximated.

There is an exception to the statement that wind speeds are highest where the pressure gradient is greatest. This relationship is reasonable only over a limited latitudinal span. Across the 35 or 40 degrees of latitude in Figure 4.5, and the even wider span of Figures 4.27 and 4.28, this relationship must be qualified.

Recall that the geostrophic approximation requires a balance between the pressure gradient and Coriolis forces, and that the friction wind requires a balance of the pressure gradient, Coriolis, and frictional forces. We have seen that the Coriolis force varies with latitude, from zero at the equator to a maximum at the poles. The speed of the geostrophic and friction winds thus varies with the sine of the latitude. This affects wind speed in such a way that, for a constant pressure gradient, wind speed is high in low latitudes and vice versa (just the opposite of the variation of the Coriolis force itself). For example, since the sine of 90° (1.0) is twice the sine of 30° (0.5), wind speed will be twice as fast at 30°, given the same pressure gradient at both latitudes. To put this another way, if a given pressure gradient, $X$, produces a wind speed of 10 m s$^{-1}$ at 30°, the pressure gradient must be 1.7$X$ and 2.0$X$ (isobars twice as close together), at 60° and 90°, respectively, to produce the same wind speed. This is illustrated in Figure 4.6.

This is an important consideration because pressure maps covering wide latitudinal spans invariably show more pronounced gradients in high latitudes. We see now that this does not necessarily mean faster winds.

## 4.4
### CONVERGENCE AND DIVERGENCE

We have seen that frictional effects promote a cross-isobaric flow. This flow is inward around lows and troughs and outward around highs and ridges (Figure 4.5). In such large-scale motion the atmosphere is

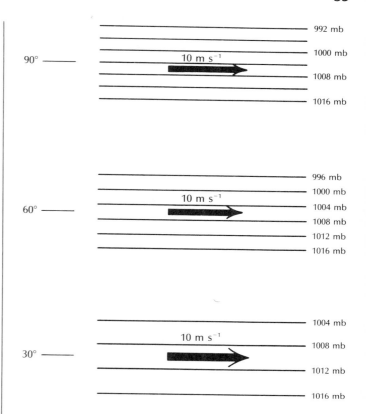

**FIGURE 4.6**
Because of the variation of the Coriolis force with latitude, a larger pressure gradient is required in high latitudes to produce the same wind speed.

essentially incompressible, that is, there can be no net accumulation or deficit of mass. As air moves into a low or trough, it must ascend. Air moving outward from a high or ridge must, by the same reasoning, descend. These motions are called **convergence** and **divergence**, respectively. This is illustrated, for both map and cross-sectional views, in Figure 4.7.

The extent of convergence and of divergence depends not only on the angle the wind makes with the isobars (both motions becoming more pronounced as this angle increases), but also on conditions in the atmosphere above the surface. Low-level, or surface, convergence often is accompanied by upper-level divergence, which maintains and may even accentuate this surface convergence. Surface divergence may also be affected in the same way by upper-level convergence.

In Figure 4.7 only the vertical component of the motion is shown for above-surface flow. Actually, the motion here is still overwhelmingly horizontal, with only small upward (convergent) and downward (divergent) displacements. This has been shown by mak-

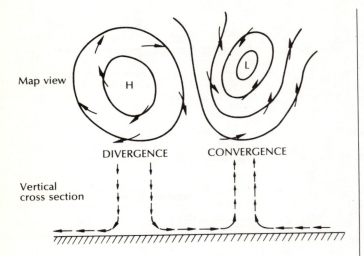

**FIGURE 4.7**
The flow around lows is inward and upward, and flow around highs is outward and downward. These motions are referred to as convergence and divergence, respectively. The speed of the wind is proportional to arrow length.

ing the arrows very small. We must also remember that for all but small-scale disturbances horizontal motions are one to two orders of magnitude larger than vertical motions. Typically, it may take days for large-scale vertical motions of a kilometer or two to occur; during the same time air may be displaced hundreds of kilometers in the horizontal.

Although vertical displacements are comparatively small, they are extremely important to weather. In Chapter 5 we will see how vertical motions are accompanied by cooling or warming of parcels, or of entire layers, of air. Condensation and evaporation resulting from such motions are thus closely linked to "bad" weather (clouds and rain) and "good" weather (clear skies). This can be verified by a closer examination of Figure 4.5. Notice that rain is occurring around the low pressure center and along the trough which extends to the south and southwest from it. To the west and north of the low, where divergence is occurring around the high, skies are generally clear, as shown by the open circles.

# 4.5
## SCALES OF ATMOSPHERIC MOTION

We have seen that horizontal variations in temperature lead to pressure differences, and that air moves in response to the latter in the process of transforming thermal to kinetic energy. However, the actual flow of air is complicated by the earth's rotation and by the irregular nature of its surface, which varies in both composition and topography. Frictional effects vary considerably from place to place, and convection, including turbulence, promotes vertical motions which lead to additional irregularities in the horizontal flow. These irregularities are apparent at all scales, but especially so with small-scale atmospheric motions.

Thus, if we were to observe the motions of the atmosphere at an instant of time, a broad spectrum of scales of motion would be apparent. Earlier in this chapter we mentioned these scales of motion, with gusts and eddies at one extreme and the "rivers" of air that circle the globe in middle latitudes at the other. They are categorized with respect to both size and duration in Table 4.1.

Although the phenomena associated with each of these four scales are important to climate, some are more significant than others. The large-scale variations in climate, which we will discuss in Chapter 9, are associated with the large macroscale phenomena. We will show in Section 4.6 that we can make some reasonable deductions about the nature of global climate from a study of the phenomena included in this category.

Macroscale, or synoptic scale, phenomena also are important, especially in the middle and high latitudes. The weather here, and thus the climate, is the result of transient storm systems (cyclones, anticyclones, and their associated fronts). Mesoscale phenomena are significant, too, but of course only at smaller space-time scales. They help to explain the climate of localized areas, and at particular times of the year. The smallest scales of motion (microscale) are significant only in very restricted locations and for short periods of time, and their influence on motions of a larger scale is probably not appreciable.

*Duration* as used in Table 4.1 needs some explanation. Generally, this refers to the length of time the indicated phenomenon lasts. Some of these motions are periodic (that is, they occur at regular time intervals), and some are not. In general, in the progression from the largest to the smallest scales, the motions tend to become less well ordered, or more irregular in time, and thus less periodic. For example, the semipermanent features of surface circulation have periods of a year. Typically, they are best developed, and most extensive, at one time of the year, and least developed and extensive six months later. In other words, these features ebb and flow with a period of a year.

Macroscale features have less predictability in this sense. Travelling storm systems, and waves in the mid-troposphere westerlies, last a few days on the average, but the time interval between successive beginnings of such features may vary widely. As with the large macroscale features there are seasonal variations, both in the frequency of occurrence and in intensity. Mesoscale features contradict somewhat the generalization about the progression from the largest to the smallest scales of motion, because some of the phenomena on this scale have diurnal periodicities. Land-sea breezes and mountain-valley winds tend to ebb and flow over a 24-hour period, and in many areas thunderstorm occurrence clusters around a particular time, or times, of the day. These phenomena also may show seasonal variations; for example, land-sea breezes and thunderstorms are best developed and more frequent in the summer months in the mid-latitudes.

The concept of **filtering** also is important in characterizing and illustrating the various scales of atmospheric motion. If pressure patterns—and thus the circulations associated with them—are averaged over successively longer periods, the phenomena in the various scales, from the shortest to the longest, are progressively filtered out and are no longer apparent. For example, Figures 4.12 and 4.13 show mean January and July surface pressure (reduced to sea level) for the entire globe. Only the large macroscale surface features are shown. Lesser scales of motion, and in particular the transient storm systems of the middle and high latitudes, disappear as a result of this averaging process. Therefore, to illustrate the presence of travelling lows and highs with a map of surface pressure, we would have to show surface pressure configuration for an instant of time, as is done in Figure 4.5.

Scales of motion below the macroscale are not evident in most pressure maps for yet another reason. Stations that observe the weather and routinely report their readings to a central agency at which weather maps are prepared are typically spaced at distances of 100 to 200 km. The circulation features of the meso-

**TABLE 4.1**
Scales of atmospheric motion.

| Scale of Motion | Typical Phenomena | Approximate Scale | |
|---|---|---|---|
| | | Horizontal Dimension | Duration |
| Large macroscale (also called *planetary scale*, and general circulation) | Semipermanent features of surface circulation (also called centers of action), mid-latitude westerlies (troposphere), Hadley cell | $10^4$ km | 1 year |
| Macroscale (also called *synoptic scale*) | Transient cyclones and anticyclones and associated fronts, easterly waves (lower troposphere), waves, or trough and ridge patterns (middle and upper troposphere) | $10^3$ km | A few days |
| Mesoscale | Squall lines, severe storms (including thunderstorms), land-sea breezes, mountain-valley winds, urban effects on climate, tropical storms and hurricanes (typhoons) | 10–100 km | A few hours to a day |
| Microscale | Convection and turbulence (including gusts, eddies); local wind regimes around obstacles and in and around crops, forests, etc. | Tens of meters | A few minutes to an hour |

and microscale thus fall within this grid, or network, of observing stations, and do not show up on maps of global and continental scale. Mesoscale analyses, which are indispensable to the study of severe storms and urban effects on climate, require a special network of more closely spaced stations.

Still another feature distinguishes the various scales of motion: the relative magnitude of vertical motions. Upward vertical motion is required for cloud formation and precipitation, while downward vertical motion characterizes fair weather (excepting haze and other obstructions to vision). In large macroscale and macroscale motions vertical speeds are from one to two orders of magnitude less than the horizontal motions. As the size of the circulation features decreases, the vertical motion relative to the horizontal increases. For microscale features vertical and horizontal motions are roughly comparable.

## 4.6 SURFACE CIRCULATION

In this section we will describe the near-surface circulation features of the atmosphere, emphasizing the locations and seasonal variations of global (large macroscale) circulation systems. Associations between these systems and the climatic elements will be mentioned, but this correspondence is treated in detail in Chapter 7 and in the chapters on regional and small-scale climates (Chapters 9 and 10).

### 4.6.1 FUNDAMENTAL CONSIDERATIONS

To simplify our description of surface circulation, we will begin by making an assumption about the nature of the earth's surface, constructing representative surface pressure features, and then applying the friction wind approximation to these features. Thus, it seems likely that if the earth were of a uniform substance—water, for example—pressure would vary only latitudinally, that is, in a north-south, but not east-west, direction.

Figure 4.8 shows, in cross section, the latitudinal variation of pressure in these simplified circumstances. High pressure prevails at 30° N and S and at the poles. Around the equator, and at latitudes 60° N and S, the pressure is low.

As was done in Figure 4.2, we now change our perspective by showing this same pressure distribution in map view in Figure 4.9, with isobars delineating the various pressure "belts." By applying the friction wind approximation (Section 4.3.4), we may deduce the direction of surface winds as shown. With the geostrophic approximation, the wind must flow parallel to the isobars, with low pressure on the left when viewed in the direction of motion (or when we stand with our back to the wind). With modification due to friction, there will be a component of motion from high to low pressure—that is, a turning of the wind from along the isobars toward low pressure. With this stipulation we deduce that the friction wind at latitude

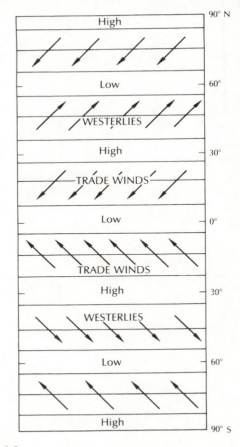

**FIGURE 4.9**
Generalized schematic representation of the surface pressure belts and corresponding winds that would develop on an earth of uniform substance.

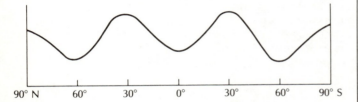

**FIGURE 4.8**
Generalized schematic representation of surface pressure, which will vary only latitudinally on an earth of uniform substance.

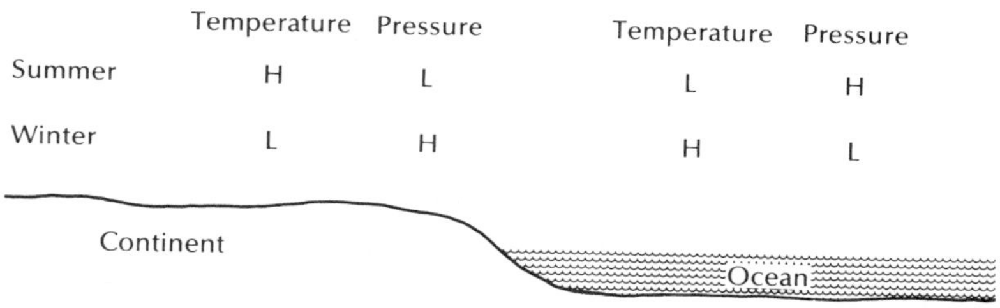

## SEC. 4.6 | SURFACE CIRCULATION

**FIGURE 4.10**
Representative seasonal variations of surface temperature and pressure over continents and oceans. In summer, when temperatures are high over continents (H) and low over oceans (L), surface pressure tends to be low and high, respectively.

|  | Temperature | Pressure | Temperature | Pressure |
|---|---|---|---|---|
| Summer | H | L | L | H |
| Winter | L | H | H | L |

Continent / Ocean

15° N (see Figure 4.9) must be approximately northeasterly. At a similar latitude in the Southern Hemisphere, and because the Coriolis force acts in a direction opposite to its influence in the Northern Hemisphere, the wind must be southeasterly. The mid-latitude westerlies and the northeast winds of the very high latitudes in the Northern Hemisphere (or the southeast winds of the very high latitudes in the Southern Hemisphere) can be deduced in a similar fashion.

Even though Figure 4.9 is a highly generalized representation of actual surface wind patterns, there is sufficient similarity to permit the designation of two of these wind belts. Remember that nearly three-quarters of the earth's surface is water, so some correspondence is to be expected. The northeasterly and southeasterly winds that flow from about 30° N and S toward the equator are called the **trade winds**. The latitudinal zones between about 30° and 60° in both hemispheres contain the **westerlies**. Belts of convergence, both at the equator and at 60° N and S, and a zone of divergence at 30° N and S will be named in the diagrams to follow.

The next step requires accounting for the effect of the varying nature of the earth's surface on pressure, and thus on circulation. Having learned that horizontal variations in temperature lead to variations in pressure, we can use this correspondence to deduce how continents and oceans have different influences on circulation systems. Recall that continents experience greater annual temperature extremes than oceans. They are warmer during summer, and colder during winter, than oceans at the same latitude. In general, air density decreases with temperature, and nontransient cyclones and anticyclones tend to be associated with relatively low and high pressure, respectively. Within the same latitude zone, then, continents in winter are colder and have higher pressure than oceans, while the reverse holds for summer. These associations are shown in Figure 4.10.

It must be stressed that this correspondence applies only to the large macroscale (general) circulation features at the surface—those that are present, to some degree, throughout the year. Transient low and high pressure systems (macroscale) have a different origin and are largely dynamically, rather than thermally, based. The dynamics of circulation systems also play a role in the development of surface large macroscale systems, but not to the extent that the correspondence in Figure 4.10 is invalidated.

In the absence of latitudinally aligned wind and pressure belts, as would develop on an earth of uniform substance, and considering only the land-water temperature differences and their effect on pressure, we can presume a cell-like pattern of highs and lows. For example, in the winter hemisphere there would

**FIGURE 4.11**
Pressure patterns likely to develop on an earth with greatly simplified continental locations and configurations.

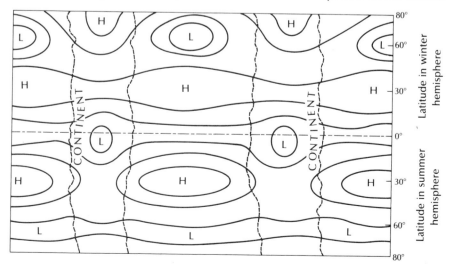

be cells of low pressure over the oceans and cells of high pressure over the continents. When both aspects are combined, the result comes reasonably close to depicting actual surface circulation features. This combination is shown in Figure 4.11. Here continental location and configuration are greatly simplified to show their influence, and pressure patterns are shown for months of extreme temperature (January and July). Above the equator the winter hemisphere is shown; below the equator are conditions in the summer hemisphere. This means that the map could represent January, if imagined as extending from 80° N (top) to 80° S (bottom), or as July if turned upside down.

In the lowest latitudes the prevailing low pressure is deepened over the continents, with the lowest pressure generally corresponding to the area of highest temperature, and thus displaced somewhat into the summer hemisphere. At about 30° N and S high pressure prevails; and, where oceans are cool relative to land at this latitude, as in the summer hemisphere, well-developed anticyclones dominate the surface circulation. In the winter hemisphere at this latitude there is less temperature contrast from land to water, and thus the ridge of high pressure is not strengthened to the point that cells develop.

From about 40° to 80° in the Northern Hemisphere winter there are marked temperature contrasts from land to water, which is reflected in pressure patterns. The circumpolar belt of low pressure (Figure 4.9) is deepened over the relatively warm oceans, while high pressure develops over the continents and becomes co-extensive with the polar highs. In the summer hemisphere there is little land-water contrast, and the circumpolar trough retains its identity.

The circulation features just described are not static throughout the year but tend to migrate latitudinally, northward from January to July, and southward from July to January. There is some latitudinal variation in the extent of this displacement, but a representative average is 5° to 10° of latitude.

Actual sea level pressure patterns are shown in Figures 4.12 and 4.13, which are very similar to the idealized configurations of Figure 4.11. The differences in the patterns can be accounted for principally by the actual location, size, and configuration of the continents. For example, at latitudes 40° to 70° N in January (the Northern Hemisphere winter), the oceanic lows and continental highs are much more pronounced than in Figure 4.11 because land occupies a much larger proportion of the surface area than it

**FIGURE 4.12**
Mean sea level pressure in January. (From C. Schutz and W. L. Gates, *Global Climatic Data for Surface, 800 mb, 400 mb: January.* Santa Monica, Calif.: Rand Corporation, 1971.)

does in the schematic diagram. Eurasia is such a large continent, in fact, that winter cooling and summer warming promote a well-developed high and a very deep low, respectively. These features eliminate the largely zonal pattern of Figure 4.11.

At the same time, the small proportion of land between the northern tip of South America and about 30° N in the western hemisphere precludes the development of a closed cyclonic circulation, although there is a trough along the west coast of Mexico in January. In July a closed low is displaced far northward into the southwest United States; this might be regarded as an extension of the low pressure trough around the equator. The last exception to the idealized arrangement of Figure 4.11 is that from about 40° S to the latitude of Antarctica there is virtually no land, and a strongly zonal configuration results.

## 4.6.2 CENTERS OF ACTION

Some of the features of Figures 4.12 and 4.13 are sufficiently extensive and persistent—that is, they exist to some degree throughout the year—to justify additional description. These features are closely associated with the other climatic elements, especially wind flow and convergence/divergence. These are called the semipermanent features of surface circulation, or **centers of action.**

### The ITCZ

The first of these centers of action is the **intertropical convergence zone,** or **ITCZ.** Here, for the most part, the northeasterly trades of the Northern Hemisphere meet the southeast trades of the Southern Hemisphere along a zone of convergence (see Figure 4.9). In Figures 4.12 and 4.13 the ITCZ is shown by a dashed line. Notice the latitudinal migration: north in July, south in January.

Additional detail in the ITCZ is shown in Figures 4.14 and 4.15. The August satellite photograph in Figure 4.16 delineates the clouds resulting from the convergence of the trades. Typically, this convergence zone is better developed over the oceans than over land, where variations in the continents' width, terrain, and surface composition disrupt the zonal configuration of the isobars in Figures 4.12 and 4.13. Again, notice the seasonal displacement of the ITCZ.

Also illustrated in Figures 4.14 and 4.15 is the relationship of wind speed and direction to isobaric con-

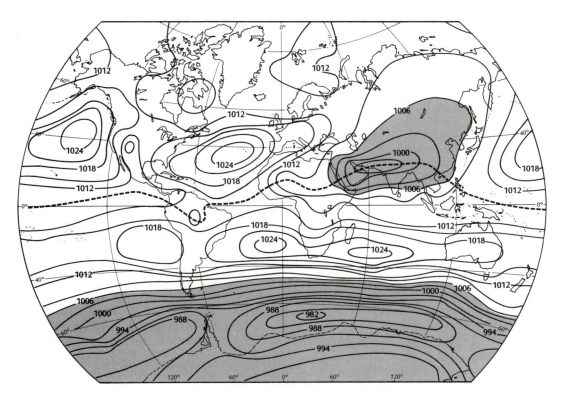

**FIGURE 4.13**
Mean sea level pressure in July. (From C. Schutz and W. L. Gates, *Global Climatic Data for Surface, 800 mb, 400 mb: July.* Santa Monica, Calif.: Rand Corporation, 1971.)

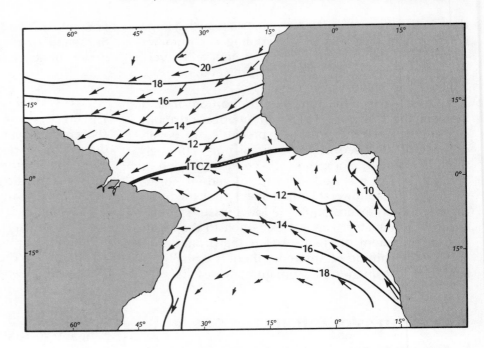

**FIGURE 4.14**
ITCZ location, pressure (1000 mb+), and winds in the tropical Atlantic in January. The mean resultant wind (the vector average of wind velocities over a long period of record) is proportional to arrow length. (Adapted from S. Hastenrath and P. J. Lamb, *Climatic Atlas of the Tropical Atlantic and Eastern Pacific Oceans*, Madison: University of Wisconsin Press, 1977.)

figuration. In general, the friction-wind relationship is validated, and the circulation around the equatorward sides of the subtropical cells of high pressure is clockwise in the Northern Hemisphere and counterclockwise in the Southern Hemisphere, with a component of motion from high to low pressure. Wind speeds generally are highest where isobars are closely spaced, and vice versa.

Notice, however, the weakening influence of the Coriolis force and the concomitant inapplicability of the geostrophic wind approximation as the equator is approached. The flow toward the equator becomes increasingly cross-isobaric as the Coriolis force decreases to zero; at the same time wind speed diminishes appreciably. Where the ITCZ is well removed from the equator, and the flow is across the equator toward the ITCZ, the trades recurve in response to the changing direction of the Coriolis force. The result is an area of westerly winds, the **equatorial westerlies.** This is apparent west of the

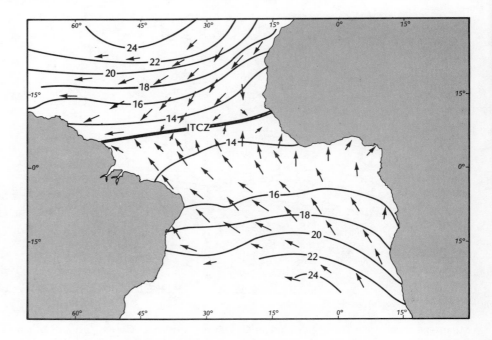

**FIGURE 4.15**
ITCZ location, pressure (1000 mb+), and winds in the tropical Atlantic in July. The mean resultant wind (the vector average of wind velocities over a long period of record) is proportional to arrow length. (Adapted from S. Hastenrath and P. J. Lamb, *Climatic Atlas of the Tropical Atlantic and Eastern Pacific Oceans*, Madison: University of Wisconsin Press, 1977.)

**FIGURE 4.16**
Satellite photograph showing the ITCZ. (Courtesy of NOAA.)

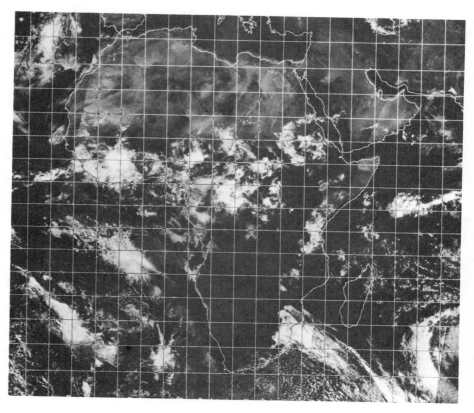

"bulge" of Africa, south of the ITCZ, in July. Equatorial westerlies are fairly widespread, as shown in Figure 4.17.

### The Subtropical Highs

The latitudes 15° to 35° or 40° in both hemispheres are dominated by belts of high pressure, the **subtropical highs** (see Figures 4.12 and 4.13). Over the oceans these belts are accentuated by cells centered somewhat to the east of mid-ocean. Because of the limited amount of land in this latitude span in the Southern Hemisphere, cells are evident throughout the year in the eastern Pacific, eastern Atlantic, and Indian Ocean and undergo only slight changes in latitudinal position and intensity. In the Northern Hemisphere the Pacific and Atlantic cells are well developed in summer, when land-water temperature contrasts are most pronounced, and have a mean position of about 34° to 38° N. In winter they are less pronounced and are at somewhat lower latitudes.

In Section 4.4 divergence was shown to be characteristic of high pressure areas. This promotes subsidence, or sinking, of air. There is, however, differential subsidence around the subtropical highs; that is, areas to the east of the meridional axis through them experience marked subsidence, while areas to the west of this axis experience only moderate subsidence that decreases to the west (Figure 4.18).

Also shown in Figure 4.18 is the outflow from the southern limits of the high, the generally northeasterly trades, and that from the northern limits, the (south) westerlies. With the former pressure decreases to the ITCZ; with the latter, to the subpolar lows. At and near the center of the subtropical highs circulation is poorly developed, as suggested by the widely spaced isobars, and winds are calm or light. Over the continents this flow pattern is altered, especially in the Northern Hemisphere summer, when lows in northern Mexico and northwest India (which may, according to some definitions, be regarded as extensions of the ITCZ) constitute marked exceptions.

### The Subpolar Lows

Figure 4.11 shows a circumpolar trough at about latitudes 60° N and S, which is apparent to varying degrees in each hemisphere in actual mean surface pressure (Figures 4.12 and 4.13). In the Southern Hemisphere the absence of land at this latitude permits a well-developed circumpolar trough that is centered at a slightly higher latitude (the opposite of what would be expected), and is deeper, in July than it is in

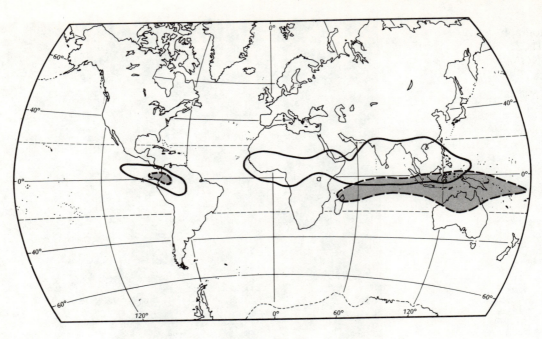

**FIGURE 4.17**
Areas of equatorial westerlies in January and July. (From Jen-hu Chang, *Atmospheric Circulation Systems and Climates*. Honolulu: Oriental Publishing Company, 1972.)

January. As the very "tight" pressure gradient here suggests, westerly winds are well developed and very persistent throughout the year, but especially in July. As will be explained in Section 4.6.4, this is an area of very frequent, transient disturbances.

The large amount of land in the Northern Hemisphere leads both to a breakdown of the circumpolar pressure configuration and to large seasonal differences in pressure patterns. In winter (Figure 4.12) pronounced temperature contrasts between the cold continents and relatively warm oceans promote the development of deep closed lows over the northern North Pacific and northern North Atlantic (the Aleutian and Icelandic lows, respectively). At the southern limits of these lows, where they merge with the subtropical highs, a generally westerly circulation prevails. At their northern limits the circulation is mostly easterly, and it is here that the **polar easterlies** (of the Northern Hemisphere) are best developed. In summer, when there is considerably less land-water temperature contrast, the Aleutian and Icelandic lows lose their identity. This latitudinal span (45° or 50° to about 70° N) shows poorly defined pressure systems (Figure 4.13).

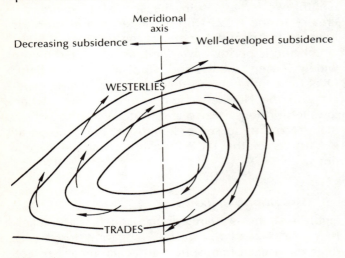

**FIGURE 4.18**
Representative subtropical anticyclone in the Northern Hemisphere. Subsidence is well developed to the east of the meridional axis, but decreases to the west of it. Arrows give the sense of the circulation.

### The Continental Highs

According to the schematic representation in Figure 4.10, high pressure develops over the continents in winter. The absence of land in the middle-high latitudes of the Southern Hemisphere precludes such a development there, but over Antarctica—an ice-covered landmass—relatively high pressure (reduced to sea level) prevails throughout the year. The great landmasses of North America and Eurasia display the expected anticyclonic circulations, but to different degrees. Eurasia is not only larger, but its southern and eastern margins are ringed by mountains which act as a barrier to warmer oceanic (maritime) air from the adjacent oceans and thereby intensify both the cold and the resultant high pressure. In North America, the absence of topographic barriers east of the Rocky Mountains and the smaller size of this continent promote only moderately high pressure.

The Eurasian winter anticyclone (more often called the Siberian high), a permanent feature of this season, acts as a barrier to travelling storm (synoptic scale) systems. In North America, on the other hand, such disturbances are quite frequent in winter. This is especially true in southern Canada and the continental United States, which are shown in Figure 4.12 as having moderately high pressure. These frequent disturbances are due in part to the readily facilitated north-south exchange of warm and cold air masses east of the Rockies.

Finally, notice that in winter the continental anticyclonic circulations of the Northern Hemisphere, in combination with the subpolar lows at about the same latitude, promote a circulation which is predominantly westerly. The presence of topographic barriers hinders the eastward movement of both continental air over the oceans and maritime air over the continents, and the absence of such barriers facilitates such movement. Thus in North America the north-south mountain chains (the Sierras, the Cascades, and the Rockies) act as a barrier to the inland penetration of maritime air which accompanies travelling disturbances. Along the eastern margins of this continent there is an essentially unhindered movement of continental air eastward over the Atlantic. In contrast, along the western margins of Eurasia, from the British Isles to the Ural Mountains on the western margins of the Siberian high, there are no such north-south barriers. Thus maritime air is carried far inland. Although summer inflow to southern and eastern Eurasia is hindered by mountains, in winter there are periodic outbreaks of cold continental air eastward to adjacent ocean areas.

### Monsoonal Circulations

In the preceding section we noted that continents are warmer in summer and colder in winter than oceans at the same latitude, and that this promotes a seasonal reversal of wind flow. The term **monsoon** describes this circulation pattern. The monsoon is more pronounced the larger the continent and the higher its latitude (thus maximizing seasonal temperature extremes), and it is for this reason that the monsoon is most pronounced in Asia. Africa and North America also have a monsoonal circulation, but the seasonal differences are not nearly as extreme as in Asia. For North America this is apparent in maps of mean surface flow (see Figures 4.12 and 4.13).

The term *monsoon* often is used incorrectly, or at least loosely. It is derived from the Arabic word *mausim*, which means season, so the expression "monsoon season" should be avoided. According to Colin Ramage, author of *Monsoon Meteorology*, the best definition of a monsoon area involves four criteria:

1. The prevailing wind direction shifts by at least 120° between January and July.

2. The average frequency of the prevailing wind direction (using an eight-point compass) exceeds 40 percent in both January and July.

3. The mean resultant wind (explained in Section 12.3.2 and Figure 12.16) in at least one of the two months of January and July exceeds $3 \text{ m s}^{-1}$.

4. Fewer than one cyclone/anticyclone alteration occurs every two years in either month.

When these criteria are applied, the main monsoon area is between 35° N and 25° S latitude, and between 30° W and 170° E longitude (see Figure 4.12). More particularly, areas which receive abundant monsoon rains in summer, or the high-sun period, are northwestern India and southern China.

Circulation reversals also occur diurnally as well as seasonally. When forced by land-sea temperature differences, they are referred to as sea (or land and sea) breezes. This is discussed in Section 10.2.1.

### 4.6.3 TRANSIENT DISTURBANCES ON THE SYNOPTIC SCALE

#### Extratropical Transient Disturbances

We have used a number of terms for the extratropical (middle and high latitude) surface circulation features identified as macroscale, or synoptic scale, in Table 4.1. Because of the weather presentations in the newspaper and on television, the layperson knows these features as storms, or storm systems. Although we have used also the terms *cyclonic storm* and *wave cyclone*, we will now apply the term *transient disturbance* to this very important aspect of the weather—and hence of the climate—of extratropical areas. Since this term is sufficiently general to apply to all

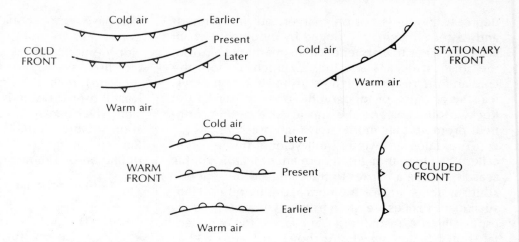

**FIGURE 4.19**
Symbolic representations of fronts.

areas, it will be used to indicate transient tropical features as well.

Figure 4.5 illustrates a mid-latitude transient disturbance. Included in it are the low pressure center and the troughs extending from it, and the areas of relatively high pressure which are generally centered to the northwest and southeast of the low (Northern Hemisphere). Also included are fronts (Figure 4.19). A **front** is a transition zone, or discontinuity, between bodies of air of different density, temperature, and humidity (air masses). It is produced, in part, by horizontal convergence—that is, when these contrasting bodies, or masses, of air are brought together. It follows, then, that fronts will be located in low pressure centers and troughs, around which the flow is convergent. This is clearly the case in Figure 4.5.

The questions we want to answer now are, How did the transient disturbance begin? What are the circulation patterns at the different stages of its growth and dissipation? How does the system move with respect to the surface? What are the consequences for weather elements in the areas affected by the disturbance? Why are there such disturbances?

It is not easy to generalize the formation, growth, and dissipation of transient disturbances. Nature seldom duplicates the atmospheric models we construct, and the variations between model and reality are often so great as to make the relationship between the two very tenuous. Some disturbances lack clearly defined fronts. Topographical features may obscure existing fronts or may be partly responsible for generating new ones. The disturbance's life cycle may last anywhere from a few hours to a week or longer; on the other hand, the development of a disturbance may be arrested at any point.

Furthermore, we do not know the dynamics of extratropical disturbances well enough to explain their causes and effects. It is easy to imagine that a well-developed anticyclone may "push" the cold front ahead of it, or that the cold front, or low, may "pull" the high in behind it; but neither is necessarily true. Also, although the strength of the circulation depends on temperature contrast across the frontal zones, the circulation itself helps to intensify this contrast, at least during the formative stages. In this instance, as in many meteorological problems, we are faced with "the chicken or the egg" quandary.

Nevertheless, the principles involved in transient disturbances can be illustrated by the sequence of events shown in Figure 4.20. The starting point of this sequence is somewhat arbitrary, but it is convenient to begin with the formation of the low center. Such centers form in troughs, which are areas of horizontal convergence at the surface. In part (a) typically cold, dry air meets warm, moist air in a belt, or zone, of convergence. Because the front is not moving with respect to the surface, it is shown as stationary. Notice that cold, dry air underlies the warm, moist, air—a necessary condition since the former is more dense.

The formation of a low, called **cyclogenesis**, begins when pressure falls at some point along the front, and a counterclockwise circulation (Northern Hemisphere) begins. This promotes undercutting of the warm air by the cold to the left (west) of the low center, while to the east of it the warm air begins to push back the cold [Figure 4.20(b)]. The frontal boundaries, shown in the left-hand portion of Figure 4.20, reflect these motions; they are displaced southward to the west of the low and northward to the east of it, forming a wave (hence the use of the term *wave cyclone*). The cold and warm fronts now mark the advance of cold and warm air, respectively.

The initial formation of the low and its continued deepening depend on many factors. Conditions in the upper air, topographic effects, and the extent of tem-

## SEC. 4.6 | SURFACE CIRCULATION

**FIGURE 4.20**
Stages in the development of a mid-latitude transient disturbance (Northern Hemisphere). (Right, from A. N. Strahler, *Introduction to Physical Geography.* New York: John Wiley & Sons, 1965.)

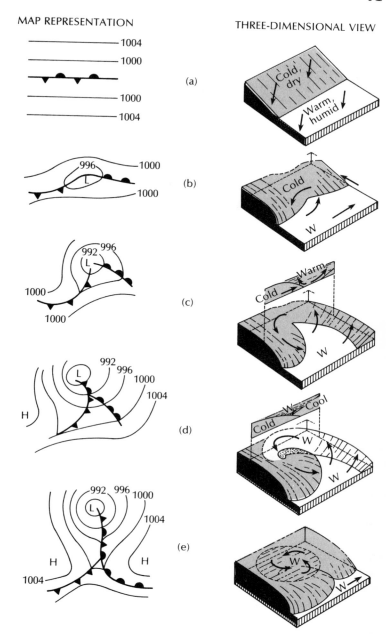

perature contrast along the front are instrumental. How surface lows relate to upper-air conditions will be considered in Section 4.7.

With continued deepening, and thus progressively strengthened cyclonic circulation, the advection of cold air to the west of the low, and of warm air to the east of it, intensifies the temperature contrast, which in turn helps to strengthen the circulation [Figure 4.20(c)]. Cold surface air invariably advances more rapidly along the cold front than warm air advances along the warm front. Usually about the time of lowest central pressure the cold front overtakes the warm, beginning at the apex of the frontal wave, and lifts the warm air off the ground [Figure 4.20(d)]. This is called the **occlusion** process, and northward of the new apex the front is said to be occluded.

In the final, or dissipating stage [Figure 4.20(e)], the cold front has completely overtaken the warm, and a mass of warm air still rotating counterclockwise has been lifted from the surface. Subsequently, the occluded front vanishes in the absence of temperature and moisture contrast, and the low eventually fills. The disturbance has now exhausted the energy available to it, the warm and cold fronts in part (e) become linked as a stationary front, and a situation much like part (a) is resumed, although the position of

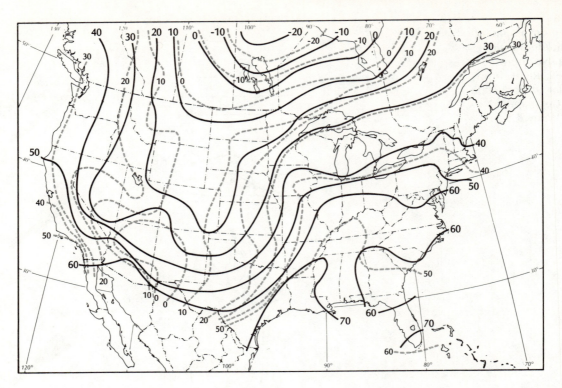

**FIGURE 4.21**
Temperature (solid lines) and dewpoint fields (dashed lines) corresponding to Figure 4.5.

this new front is usually somewhat southward of its position at (a).

The cross section above the block diagram in Figure 4.20(c) shows that the cold front intersects the surface at a steeper angle than does the warm front. This is typical, with the slope of the cold front being about 1 in 60 and that of the warm front 1 in 150 (one distance unit of height in 150 distance units in the horizontal).

These developments do not occur in place, but tend to have a generally eastward motion in keeping with the predominantly westerly flow in the upper air. Lows move generally to the east, northeast, or southeast (in both hemispheres), while following highs are displaced toward the south or southeast (toward the north or northeast in the Southern Hemisphere).

We can now visualize the weather associated with the passage of a transient disturbance. Surface air, from the cold front counterclockwise around to the warm front, is relatively warm, and, more often than not, moist as well. From the warm front around to the north and west, the surface air is cool or cold and relatively dry. These associations apply best to North America east of the Rocky Mountains, with some variations in other parts of the extratropical latitudes.

Figure 4.21 shows the temperature and dewpoint fields corresponding to Figure 4.5 and to the generalized representation of a transient disturbance in Figure 4.20(c). **Dewpoint**, the temperature to which air must be cooled to produce saturation at constant pressure, is a measure of the moisture content of the air. Notice the tongue of warm air (warm sector) extending northward and eastward from southeast Texas and Louisiana through the middle Mississippi valley and into Virginia and North Carolina. Typically, the decrease of temperature is quite marked west and north of the cold front; but also typical is the more gradual decrease of temperature across the warm front (notice that the front in Figure 4.5 is stationary). The moisture content of the air, shown by the dewpoint temperature (or, simply, dewpoint), shows similar patterns. Air in the warm sector is moist, with dewpoints of about 10°C (50°F), while the decrease of moisture in back of the cold front is even more pronounced than the decrease in temperature. The Southwest, Rockies, western plains, and interior Canada are dry to very dry, with dewpoints of −18° to −29°C (0° to −20°F), while the western states are moderately warm and moist.

The precipitation associated with transient disturbances is for the most part restricted to areas of relatively low pressure, that is, around the low center and within troughs. Here convergence occurs, with an accompanying slow upward motion of air. The mechanical lifting of warm air over cold, which takes place at cold, warm, and stationary fronts, is another means by which precipitation may be formed. Figure 4.5, in which areas experiencing precipitation are shaded, is reasonably typical. There are many exceptions to this pattern, however, depending largely on the amount of moisture available and the extent of convergence at the frontal boundaries.

Synoptic weather representations, such as Figures 4.5 and 4.20, unfortunately impart a static quality to circulation patterns, and we need to be reminded that our atmosphere is actually a dynamic, constantly changing system. Synoptic meteorology takes "snapshots" of a part of a system at regular intervals, but we must put these snapshots together, in sequence, to better appreciate the dynamics of the atmosphere.

Consider Figure 4.22 (pp. 74-75), which is a synoptic representation of the weather over the United States and southern Canada from March 20 to March 24, 1966 (including the time depicted in Figure 4.5). The important things to notice are the formation, deepening, and dissipation of the low; its track to the northeast; the intense anticyclone that follows the cold front and whose center moves southeastward, then southward; and the areas of precipitation, especially those associated with the disturbance.

The weather for a particular location (southeast Texas) during this period is shown in Figure 4.23 (p. 76). Prior to the cold front passage winds are southeasterly, but they shift abruptly to northwesterly and northerly after its passage. Sky cover changes from mostly overcast to clear within about 24 hours after the front goes through. The arrival of increasingly higher pressure after a cold front usually leads to clearing because of the dry air and subsidence associated with divergence, but this is not always the case. Temperature shows the expected diurnal variation both before and after the front, but of course postfrontal temperatures average, in this case, about 14C° (25F°) cooler. The diurnal range, then, in the drier air is greater. The moisture content of the prefrontal (warm sector) air is rather high for this time of the year, but air arriving from the northwest is much drier, as indicated by the dewpoint.

Notice in Figure 4.22 that southeast Texas—and the entire Gulf Coast area, for that matter—are affected only by the cold front and its associated pre- and postfrontal weather. This is often the case in this area. Farther north the sequence of weather events would be more complex, with an area such as the Midwest experiencing pre– and post–warm-front and pre– and post–cold-front conditions, as well as the weather in and around the low center.

Why are there such disturbances? Recall from Chapter 2, Section 3.5, and from Section 4.2 that the atmosphere must circulate in such a way as to transport heat both poleward and upward. Any meridional motion tends to accomplish the first of these because cold air is moved equatorward and warm air poleward. Around lows the circulation brings warm air poleward to the east of the low center and cold air equatorward to the west of it. Around highs the opposite relationships apply. Also, the ascent of warm air over cold fronts and in and around the low center promotes an upward transport of heat.

The intensity and frequency of transient disturbances is not uniform at all extratropical latitudes, but tends to maximize, as Figure 2.16 suggests, at about 40° in both hemispheres. Also, it is not just the transport of heat, as sensible heat, that is responsible; latent heat transfer also is involved. We have seen that precipitation is a frequent result of the slow upward motions associated with transient disturbances. For precipitation to occur there must be condensation of water vapor, that is, water must change from gas to liquid. This is a heat-releasing process, and thus the latent heat of vaporization—first obtained by evaporation from surface water—is released in the upper air when condensation occurs. Furthermore, the advection of water vapor, such as occurs to the east of lows and on the western sides of highs (Northern Hemisphere), provides a poleward transport of heat. Moisture evaporated from the Gulf of Mexico, the Mediterranean, and the South China Sea may condense and subsequently fall as precipitation over Canada, Scandinavia, and Siberia.

### Tropical Transient Disturbances

The horizontal temperature differences that are a conspicuous feature of middle and high latitudes are largely absent in the tropics. Place-to-place temperature differences that do occur are largely the result of the unequal heating of land and water. Thus, throughout the year and especially during the high-sun season, the landmasses are somewhat warmer than the adjacent oceans. As shown in Figure 4.10, surface flow tends to be from ocean to continent in these latitudes, although most of these motions are minor compared with those of the dominant features of tropical circulation—the trade winds and the ITCZ.

There are, however, some tropical circulation features that are sufficiently organized to qualify as transient disturbances. One of these, **easterly waves,** is large enough to be discerned in routine isobaric analyses. Tropical depressions, tropical storms, and hurricanes (typhoons in the Pacific Ocean) are smaller (mesoscale) features, although a few of the latter are large enough to be classified as macroscale. Nonetheless, these features are monitored closely by other than the usual observing methods and can be represented in synoptic weather analyses. Still smaller scales of motion in the tropics include squall lines, which are also a mid-latitude phenomenon; convective cloud clusters; shearlines; and disturbances occurring in association with monsoonal circulations, the seasonal reversal of flow over the largest continents (see Figure 4.10 and Section 4.6.2), and with land-sea breezes, a similar reversal that occurs over diurnal cycles.

Two characteristics of tropical circulation features distinguish them from those of higher latitudes. The first is the diminished effect of the Coriolis force near the equator. As we learned in our discussion of the ITCZ (Section 4.6.2), the flow is strongly ageostrophic within a few degrees of the equator, and the correspondence between wind and pressure suggested by the friction wind approximation no longer holds. The variation with latitude of the Coriolis force also results in a tendency toward closed lows, or **vortices,** to be largely absent in low latitudes, although an exception is well-developed systems such as hurricanes. In low latitudes even slight perturbations on otherwise stable flow patterns can produce marked convergence, divergence, or both.

The second distinguishing characteristic of tropical circulation features is that the primary source of energy for tropical transient disturbances is not horizontal temperature differences, but the latent heat

**FIGURE 4.22**
Synoptic weather patterns over the United States at 1 P.M. Eastern Standard Time, March 20–24, 1966.

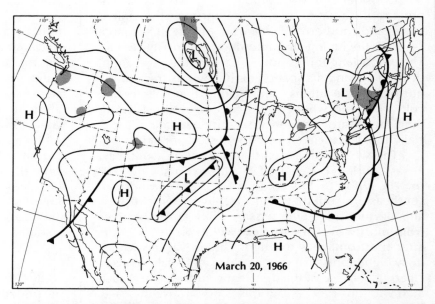

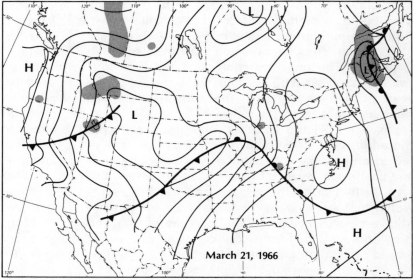

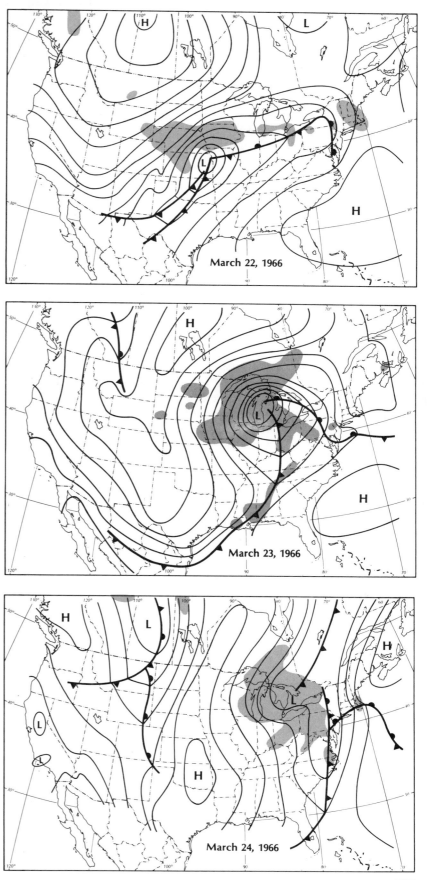

**FIGURE 4.23**
The sequence of weather events at College Station, Texas, before and after a cold front passage, March 20–24, 1966.

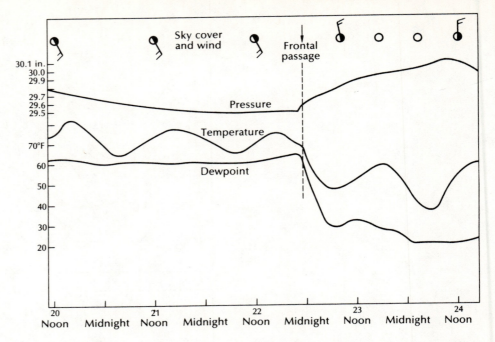

released when water, previously evaporated from warm tropical oceans, condenses in the upper air. This mechanism also is at work in extratropical circulation features, but becomes of paramount importance in the tropics.

Easterly waves form as troughs between the subtropical highs and the ITCZ, or within the trade winds, and move toward the west (Figure 4.24). Convergence, with accompanying clouds and rain, is found to the east of the trough axis, while to the west of it divergent motion tends to bring fair weather. Typically, easterly waves have a wavelength of 15° of longitude and move toward the west at 5 to 8 m s$^{-1}$ (10 to 15 mi h$^{-1}$).

Hurricanes are nature's most awesome display of meteorological might. Tornadoes pack more concentrated power but are quite short-lived and very localized. In contrast, hurricanes may have lifetimes up to three weeks and diameters of as much as 800 km (500 mi) They are not as extensive, however, as a mature extratropical disturbance, as Table 4.1 indicates.

Hurricanes are distinguished from other tropical disturbances by both a pronounced vortex with very low central pressure and sustained winds in excess of

**FIGURE 4.24**
Surface chart showing a model of a wave in the tropical easterlies. Pressure is in millibars. (From *Climate and Weather*, by H. Flohn. Copyright © 1969 by McGraw-Hill Book Company. Used with the permission of McGraw-Hill Book Company.)

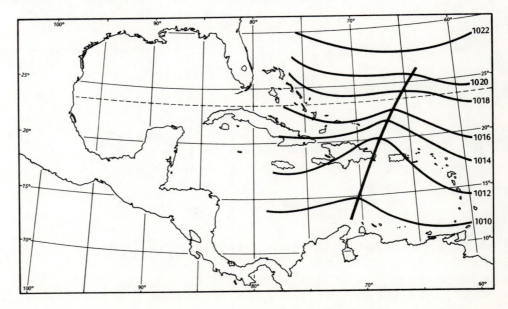

**FIGURE 4.25**
Vertical section of a hurricane and the associated patterns of wind, pressure, and rain. (From A. Miller and J. C. Thompson, *Elements of Meteorology*, 3rd ed., Columbus, Ohio: Charles E. Merrill, 1979)

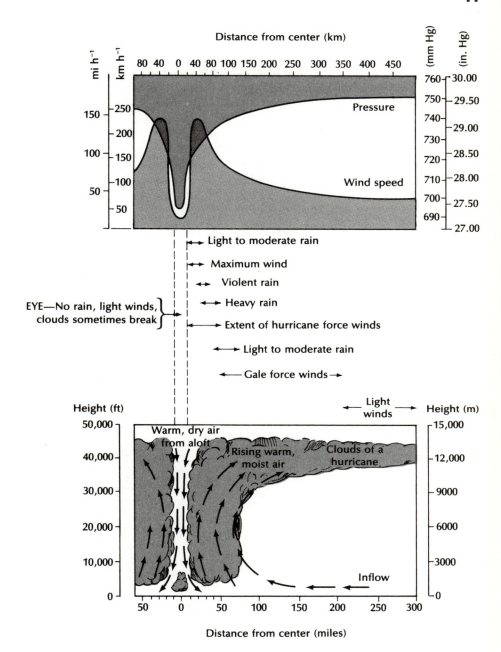

33 m s$^{-1}$ (74 mi h$^{-1}$). Another feature of these storms is the approximately symmetrical distribution of clouds, wind speed, and rain (Figure 4.25), although less well-developed disturbances—tropical depressions, with sustained winds of less than 17 m s$^{-1}$ (39 mi h$^{-1}$), and tropical storms, with winds of 17 to 33 m s$^{-1}$ (39 to 74 mi h$^{-1}$)—may also display symmetry. Hurricanes are associated in the layperson's mind with enormous amounts of rainfall, but this is not always the case. Although few in number, there have been "dry" hurricanes in which only nominal amounts fell.

At least three necessary conditions for hurricane development have been singled out. First, the Coriolis force must exceed some minimum value, and thus hurricanes almost never form within 5° of latitude from the equator. Second, sea surface temperature must be 26° or 27°C (79° or 81°F) or above. Third, the change of the horizontal wind velocity in the vertical, or the *vertical wind shear*, must be weak.

Prior to its designation as a hurricane, a disturbance is recognized as one of three forms. It may begin as a tropical depression and then grow to tropical storm stature. Easterly waves also may intensify to the point that a closed center of low pressure emerges. Of course, not all tropical storms and depressions grow to hurricane strength; many lack sufficient organization, or, for other reasons not well understood, do not develop to the hurricane stage.

Tropical disturbances play a very necessary role in atmospheric circulation. They help to promote a flux of heat upwards, and, since their general motion is poleward (Figure 4.26), toward higher latitudes as well. The energy required for evaporation at the surface in low latitudes is released, by condensation, in the upper atmosphere and/or in higher latitudes.

Areas where hurricanes originate are shown in Figure 4.26. Principal tracks, or the general direction of storm movement, also are indicated. Notice the generally westward movement in both hemispheres, with a tendency to recurve, or move poleward and then eastward, as higher latitudes are reached. Also, hurricanes—or tropical cyclones in general—originate a few degrees from the equator, again reflecting the greatly diminished Coriolis force.

### 4.6.4 CLIMATOLOGICAL CHARACTERISTICS OF TRANSIENT DISTURBANCES

Transient disturbances, especially those of the middle and high latitudes, produce relatively short-term variations of weather. Now we will discuss the effect of such disturbances on climate, particularly on long-term means of temperature and precipitation. To do so we will examine seasonal and geographic variations of these circulation systems.

It is likely that long-term means of temperature of a month or longer are not appreciably affected by transient disturbances. The warmer weather associated with pre–cold-front conditions is more or less balanced by the colder weather that follows. But climate, as we have stressed, is more than just long-term averages. The day-to-day, and even hour-to-hour, changes of weather are as integral a part of climate as averages and enable us to differentiate climate, both among the seasons and from one geographic location to another.

On the other hand, climatic means of precipitation are greatly affected by transient disturbances. The precipitation that results from the relatively slow ascent of air associated with convergence around lows and fronts contributes significantly to long-term means, especially in extratropical latitudes in winter. In the tropics and subtropics transient disturbances can easily account for one-quarter to one-half of the annual rainfall, although this proportion varies depending upon which other precipitation-producing mechanisms are present.

The influence of transient disturbances is not apparent on maps of mean pressure for about a month or longer because systems of length-time

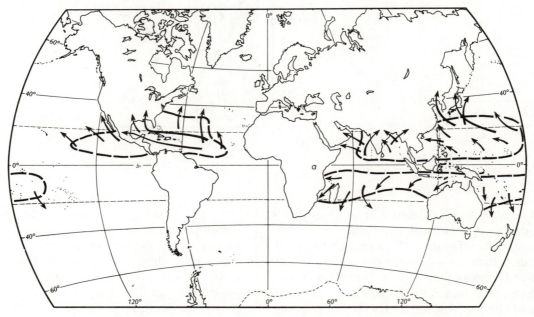

**FIGURE 4.26**
Hurricanes: where they originate and their principal tracks.

scales less than the large macroscale (Table 4.1) are masked by the averaging process. Or, in other words, transient systems have been filtered out in the averaging process, and virtually no trace of them is left in maps such as Figures 4.12 and 4.13.

Let us approach this idea from another point of view. Suppose you were to study a number of maps showing instantaneous values of pressure over a major portion of the globe—the Northern Hemisphere, for example—thereby delineating circulation features such as highs, lows, troughs, and ridges. Would the centers of action be obvious? Would their seasonal differences in intensity and location be apparent? What other circulation features would you find?

These questions can be answered by studying Figures 4.27 and 4.28. These maps show surface pressure and fronts for instants of time in January, 1969, and July, 1968. Try to visualize the probable positions of pressure patterns and fronts a few hours before and after these times to provide some continuity. In Figure 4.27, for example, a cold front stretches in a circular fashion from western Florida, through the Gulf of Mexico, and into central and northern Mexico. Since this is a cold front, we know that relatively cold air has pushed out over the Gulf and northeastern Mexico and will undoubtedly continue its advance in a generally southward direction. In the Rocky Mountains the front is stationary, and recently there has been no push of cold air toward the west or of warm air to the east. We also can speculate that the low at the southern tip of Hudson Bay was situated somewhat to the southwest a few hours earlier and should continue its northeastward movement in the next few hours. The continuity for other transient disturbances can be visualized in a similar way.

### Centers of Action

The areal coverage of the maps in Figures 4.27 and 4.28 does not extend far enough south for the ITCZ to be apparent. The subtropical highs, however, are obvious in July, when high pressure dominates the southern parts of the North Atlantic and North Pacific oceans. Such a pattern, with slight variations, could be expected at any time during the summer months of June, July, and August. Also on the July map are troughs, with closed lows, occupying northern Mexico and northwestern India. Note the similarity of these patterns to those in Figure 4.13. The pattern north of about 40° is much more complicated than is suggested by Figure 4.13, although there is low pressure over northwest Africa and the interiors of North America and Asia. There is no evidence of a circumpolar low at about 60°.

In January the subtropical highs are no longer apparent; instead, these areas contain travelling (transient) disturbances. We can expect, according to Figure 4.12, that in the mean moderately high pressure prevails over these oceans in latitudes of about 25° to 35°, and that the subtropical highs have been displaced southward and reduced in intensity. The continental highs in Asia and North America, however, are easily seen. If we could see many more such maps of winter, we would observe that there is virtually continuous high pressure from central to northeastern Asia. In North America, however, the tendency is for highs to form, intensify over central Canada, and then move southward or southeastward into the United States, as shown in Figure 4.27. The subpolar lows on this map are not as clearly delineated as on the map of mean conditions, but in both the North Atlantic and the North Pacific north of 60° there is generally low pressure. Transient lows are characteristic of this area.

### Transient Disturbances

Around the Northern Hemisphere in the middle and high latitudes transient disturbances are evident, especially in January (see Figure 4.27). In general their movement is westerly, or from west to east. In January the eastern half of North America is occupied by the high centered over Missouri and a low south of Hudson Bay. In western North America the pattern is much more complex, but a low pressure center is moving into the northwest United States and extreme southwestern Canada.

In the mid-Pacific an almost classical synoptic wave is developing, with a closed low and accompanying warm and cold fronts, and with high pressure from the northwest replacing the eastward-moving low. Farther west in the Pacific fronts have formed around a general area of low pressure.

In the western Atlantic a transient high "pushes" a cold front eastward, while east of this front, and the trough that it fills, another high moves eastward into Europe. All of central and northern Asia is dominated by the Siberian high. Except for the Siberian high, and the tendency for low pressure to occupy more or less continuously the northern oceans around Iceland and the Aleutians, the area of the Northern Hemisphere shown in Figure 4.27 is being affected by transient disturbances.

**FIGURE 4.27**
Sea level pressure and fronts, January 1, 1969, 1200 GMT. (Courtesy of NOAA.)

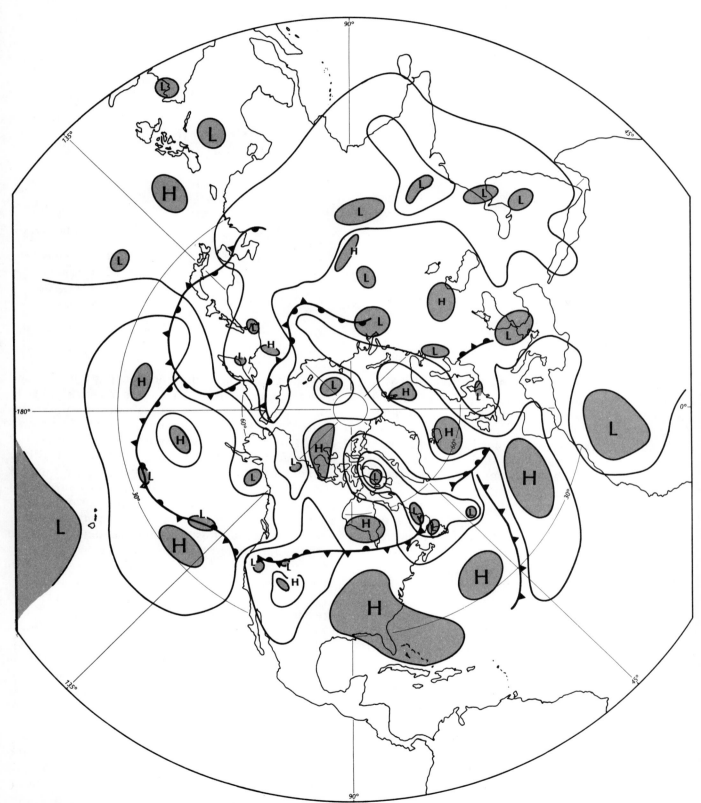

**FIGURE 4.28**
Sea level pressure and fronts, July 16, 1968, 1200 GMT. (Courtesy of NOAA.)

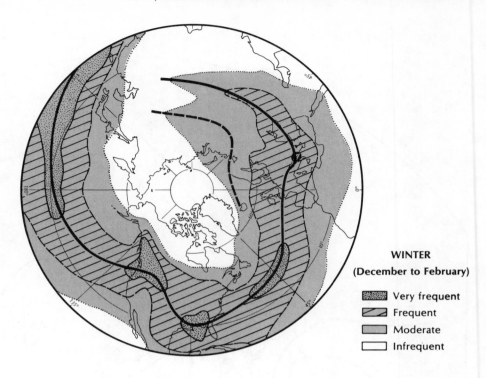

**FIGURE 4.29**
Frequency of frontal occurrence in winter (December, January, February) for the Northern Hemisphere north of 20°. Solid lines locate axes of maximum frequency; bold lines indicate the polar front; and dashed lines, the arctic front. (From R. J. Reed, "Principal Frontal Zones of the Northern Hemisphere in Winter and Summer," *Bulletin of the American Meteorological Society* 41 (1960): pp. 591–98.)

### Seasonal Differences and Frontal Zones

The most striking difference between the winter and summer maps (Figures 4.27 and 4.28), evident even to a casual observer, is the density of lines. There are many more isobars on the winter map, indicating a tighter pressure gradient, and thus a stronger (or more intense) surface circulation. The atmosphere can be likened to a rubber band, one that is wound tightly in winter but left slack in summer. In winter highs are higher, lows lower, and their movement is faster. In general, spring and fall are intermediate in this respect.

Earlier in this chapter we learned that the atmosphere circulates in response to, or is driven by, temperature differences. On a long-term, or climatic, basis these differences occur predominantly in a north-south direction. Figures 3.11 and 3.12, as well as Figures 7.1 and 7.2, show that this gradient is greater in winter than in summer. We can associate this seasonal difference with an even more fundamental cause—the seasonal variation of the north-south gradient of insolation. In Figures 2.4, 2.6, 2.7, and 2.8, we see that the latitudinal change of surface radiation receipts is greater in winter than in summer. Also, meridional circulations are better developed in winter, when east-west temperature differences, from cold continents to warm oceans, are most pronounced.

Winter-summer differences in the intensity and speed of transient disturbances are thus a response to the seasonally varying horizontal gradient of temperature. These disturbances constitute one of the mechanisms by which the required poleward transport of heat is accomplished. The more pronounced winter gradient requires that the atmosphere respond by accelerating the rate at which circulation systems transport cold air equatorward and warm air poleward. Of course, it is horizontal temperature differences throughout the atmosphere, and not just at the surface, that govern these motions; we will return to this subject in the discussion of upper-air circulation in Section 4.7.

The gradient of temperature from the equator to the poles is not uniform, either at any instant of time or in the climatic mean. We now recognize the fronts in Figures 4.27 and 4.28 as locally intensified temperature gradients. These are restricted, in general, to latitudes above about 30°, but extend farther toward the equator in winter than in summer in response to the equatorward shift of pronounced temperature gradients. At most times it is possible to assign names to the bodies, or masses, of air separated by these fronts. Thus, equatorward of 30° tropical air dominates. Northward of 30° to about 60° or 70° polar air holds sway, while the very highest latitudes are occupied by Arctic air (Antarctic in the Southern Hemi-

## SEC. 4.6 | SURFACE CIRCULATION

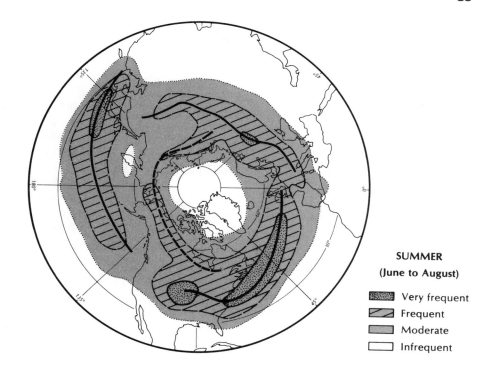

**FIGURE 4.30**
Frequency of frontal occurrence in summer (June, July, August) for the Northern Hemisphere north of 20°. Solid lines locate axes of maximum frequency; bold lines indicate the polar front; and dashed lines, the arctic front. (From R. J. Reed, "Principal Frontal Zones of the Northern Hemisphere in Winter and Summer," *Bulletin of the American Meteorological Society* 41 (1960): pp. 591–98.)

sphere). These terms not only indicate something of the origin of the air so described, but also give some qualitative indication of their temperature.

By studying many maps like Figures 4.27 and 4.28, we can deduce the areas of greatest and least frontal frequency and the mean position of the various frontal zones. Figures 4.29 and 4.30 show the variation of frontal frequency in the Northern Hemisphere; the thick lines running along the axes of highest frequencies show the average positions of the boundaries between polar and tropical air, the **polar front,** and between polar and arctic air, the **arctic front.** Notice that

1. During both seasons the middle latitudes experience the highest frontal frequency, with decreases both equatorward and poleward, but in winter this area of highest frequency is farther south. For example, the mean position of the polar front is at about 30° to 40° in winter and at about 40° to 50° in summer.

2. Areas of highest frequency occur from the east coasts of North America and Asia into the adjacent oceans and in the center of North America.

3. The zone of infrequent fronts comprises the lowest latitudes and extends northward to the pole in winter over central and east central Asia. This zone reflects the dominance of the Siberian high, with its tendency to act as a barrier to eastward-moving transient disturbances.

These maps and the way they are constructed are discussed further in Chapter 12.

Two additional remarks about Figures 4.29 and 4.30 are necessary. First, frontal frequency and the mean position of the polar front are influenced by topography. High mountains act as barriers both to the inland penetration of maritime air and to the outward movement of continental air to the oceans. This is evident in North America, where the western mountains, especially the Rockies, keep cold or cool continental air from much of the southwestern and western United States and western Canada in all seasons. Thus, the polar front in winter swings northward to the west of the central plains. In summer mountains bar the inland movement of moist air from the Pacific; thus the interior of the United States and Canada is drier and experiences somewhat higher temperatures than it would without these topographic barriers. In contrast, western and central Europe, without such barriers, are occupied by maritime air most of the year.

Second, although the frequency of fronts is shown in these maps, the extent of temperature contrast is not. Because of the more conservative thermal properties of water, air temperatures over the oceans are more uniform over horizontal distances, and frontal contrasts in temperature are less than over the continents. Somewhat of an exception occurs off the east coasts of the Northern Hemisphere continents in winter, when very cold continental air spills out over the oceans, but here such air quickly acquires maritime characteristics (see Section 2.3.4). Over land, on the other hand, inland-moving maritime air is modified relatively slowly. Thus cold continental air meets warm maritime air in winter, and cool continental air meets very warm to hot and moist maritime air in summer, producing greater temperature contrast across fronts.

Another consequence of the differences in thermal properties of land and water relates to the distances, generally in a north-south direction, over which fronts move. Over oceans fronts tend to be "anchored" very near the position shown by the climatological averages. But cold continental air in winter may sweep as far south as Central and even South America and into North Africa. Similarly, maritime air in summer may penetrate far northward into central Canada and Scandinavia.

It is clear, then, that even in the climatic mean there are significant discontinuities, or breaks, in the decrease of temperature from the equator to the poles. These frontal zones not only separate air masses of differing temperatures and humidities, but also often act as breeding grounds for the formation of transient disturbances. In addition, the location and intensity of wind speed maxima in the upper air are governed largely by the location of these frontal zones.

## 4.7 UPPER-AIR CIRCULATION

Our discussion of upper-air circulation begins with Figure 4.31, a generalized, schematic representation of surface pressure and winds which eliminates east-west differences. From it we can infer something about the circulation above the surface. If surface convergence (see Figure 4.7) occurs at the equator and at about 60° N and S, and surface divergence at 30° N and S, we know that air is rising and sinking, respectively, at these latitudes. Continuity for these motions is provided by circulation cells within the tropics and subtropics in which heated, moist air rises over the ITCZ, spreads poleward to about 30°, and then sinks.

### 4.7.1 THE HADLEY CELLS

The thermally direct circulations (warm air rising, cool air sinking) providing continuity of motion are called the **Hadley cells** (Figure 4.31). They are symmetrical about the ITCZ, and thus the midpoint of the two cells has a mean position near the equator. Although there are exceptions to this circulation pattern that are imposed by land-water differences, these cells are the dominant features of half of the earth's surface. The Hadley cells are not maintained by surface horizontal temperature differences, which are largely lacking in the low latitudes.

The Hadley cells accomplish a part of the necessary heat transfer poleward and upward. Diverging and subsiding air at 30° warms and picks up moisture as it moves equatorward. Zonally averaged rates of evaporation for 70° N to 60° S (Figure 4.32) are highest in the latitude of the trade winds, with a relative minimum in the lowest latitudes. The higher rates in the Southern Hemisphere are due to the large amount of water there. Equatorward-moving air thus acquires latent heat, and transport is toward lower latitudes, as Figure 2.16 indicates. The energy represented by latent heat is then released by moist convection (discussed in detail in Chapter 5) in the upper air, predominantly in and around **the ITCZ** but not restricted to it.

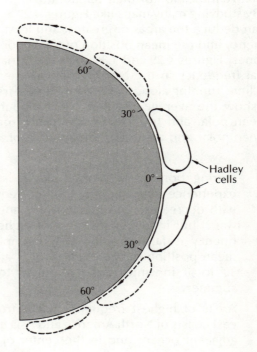

**FIGURE 4.31**
Generalized cross section of the meridional and vertical components of the general circulation. Weaker and more variable mean motions are dashed.

## SEC. 4.7 | UPPER AIR CIRCULATION

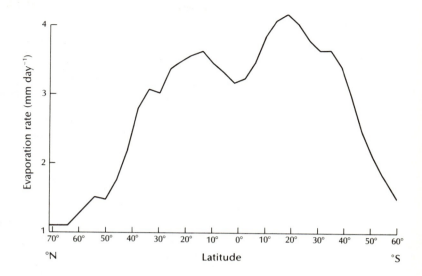

**FIGURE 4.32**
Zonally averaged evaporation rate, in millimeters per day, for 70° N to 60° S, averaged for the months of January, April, July, and October. (From C. Schutz and W. L. Gates, *Global Climatic Data for Surface, 800 mb, 400 mb*. Santa Monica, Calif.: Rand Corporation, 1971.)

It is through this cycle, then, that the Hadley cells accomplish the necessary transport of heat upward and poleward. The predominant transfer mechanism in the vertical is the acquisition of latent heat at the surface and its release in the upper air. The meridional transport is achieved mostly through poleward-moving air at upper levels, which carries both sensible and latent heat to higher latitudes.

### 4.7.2 THE ZONAL CIRCULATIONS OF EXTRATROPICAL LATITUDES

Dashed lines are used in Figure 4.31 for latitudes poleward of 30° N because the meridional component of these motions is weak. Instead, the flow in these latitudes is strongly zonal and also highly variable. This point must be stressed because cross sections such as Figure 4.31 can be misleading. Actually, the zonal circulation in the middle and upper troposphere overwhelms the meridional motion. This is especially true in extratropical latitudes, but also applies to a lesser extent to the Hadley cells. In addition, the vertical motions, such as occur in the ascending and descending parts of the Hadley cells, are quite small compared with the meridional and zonal components. It is the vertical exaggeration that suggests otherwise.

#### Mean Flow Patterns

Temperature gradients, which occur mostly in a north-south direction, are the forcing mechanism for tropospheric circulation in the extratropical latitudes. The result is a strongly zonal (in this case, west to east) flow in both hemispheres. There are, however, significant latitudinal and height differences that are apparent in zonally averaged cross sections. Meridional (longitudinal) differences must be shown on hemispheric maps of the mean flow at various levels.

To understand the reason for this strongly zonal flow, and to substantiate the repeated theme that horizontal temperature differences are the forcing mechanism, we need to consider the association between the temperature of a column of air and the rate at which pressure changes in the vertical through that column. If we combine the *general gas law*, which specifies the pressure exerted by a gas as a function of its temperature and density, and the *hydrostatic equation*, which gives the rate at which pressure changes in the vertical as a function of density and expresses the balance between gravity and the vertical component of the pressure gradient (Section 4.3.1), we can derive the **hypsometric relationship.** This relationship shows that pressure change in a column of air is inversely related to the mean temperature of that column. In other words, pressure decreases more rapidly upward in cold air than in (relatively) warm air. A corollary is that the vertical distance between adjacent pressure levels, called *thickness* (for example, between 500 and 400 mb), increases as the temperature increases.

The hypsometric relationship is illustrated in Figure 4.33, which is a schematic cross section of the atmosphere from 90° to 30° N. It is reasonable to expect that at all times the air temperature decreases with higher latitude across this span. We know now that pressure will change less rapidly in the vertical in low latitudes than in high latitudes, and, if we assume a uniform decrease of temperature with latitude, the isobaric surfaces will be as shown in Figure 4.33(a). These surfaces therefore tilt downward toward the cold air. The more these surfaces are inclined (which corresponds in map view to more east-west aligned

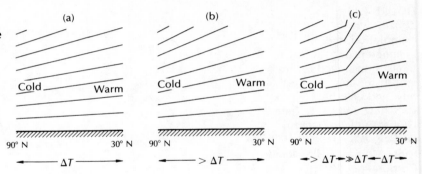

**FIGURE 4.33**
The influence of the south-to-north temperature decrease on upper-level winds, which is expressed through the hypsometric relationship. The lines are isobaric surfaces. Parts (a) and (b) simulate summer and winter conditions, respectively, and part (c) presumes a locally intensified temperature gradient. The west wind is proportional to the inclination of the isobaric surfaces.

isobars, or a greater gradient), the faster is the wind.

Notice the similarities between Figures 4.33 and 4.2. In the latter we saw an unbalanced component of the pressure gradient force ($P_h$), which, when combined with the Coriolis force, resulted in the geostrophic wind. In both illustrations, then, the direction of the geostrophic wind is into the page, or, in the Northern Hemisphere, from west to east—a west wind. In the Southern Hemisphere the tilt of the isobars would still be toward the cold air, which would be to the south. Since the Coriolis force acts at right angles to the left of the direction of motion, the result is the same as in the Northern Hemisphere—a west wind.

We conclude that in the presence of a poleward decrease of temperature in the atmosphere winds above the surface will be westerly. Regardless of the pressure distribution at and near the surface—and here winds may be from any point on the compass—with sufficient height westerly winds will prevail.

Other aspects of extratropical upper-air circulation can be deduced from the hypsometric relationship. These are seasonal differences in wind speed, an increase of wind speed with height to the tropopause (the top of the troposphere), and localized increases in wind speed. Figures 3.11 and 3.12 show zonally averaged temperatures for January and July. North-south temperature differences in the depth of the atmosphere shown become apparent northward of about 35° or 40° in July and northward of about 25° or 30° in January, with considerably more poleward decrease, at levels up to about 200 millibars, in January. The seasonal variations in wind speed which result from these differences are shown schematically in Figures 4.33(a) and (b). Since temperature decreases poleward more rapidly in winter ($\Delta T$ in January $> \Delta T$ in July), isobaric surfaces are more inclined in this season and wind speeds are correspondingly higher. Also, since this gradient extends to lower latitudes in winter, the zone of westerly winds is farther south then, although this is not shown in Figures 4.33(a) and (b).

In addition, it is clear from these cross sections that with increasing height at any latitude (again, assuming that the mean temperature of successive columns of air decreases uniformly with latitude through this depth of the atmosphere) there is an increase in the inclination of isobaric surfaces. The result is higher wind speeds with height. This relationship holds up to the tropopause, but there the poleward decrease of temperature weakens, and so also does the westerly wind regime.

The final variation of the association between temperature gradient, the inclination of pressure surfaces, and resultant winds is shown in Figure 4.33(c). To this point we have assumed a uniform north-south temperature gradient. A locally intensified gradient ($\gg \Delta T$ compared with $> \Delta T$ and $\Delta T$) produces a relatively narrow zone of winds that are faster than those of other latitudes at the same height. We have already discussed (Section 4.6.4) locally intensified surface temperature gradients at the polar front. Provided that these gradients exist as well at higher levels—and for the most part they do—we should expect a core of high wind speeds approximately above the polar front.

With this background Figures 4.34 and 4.35 can be readily understood. These diagrams show zonally averaged wind speed in January and July for the same latitudinal span shown in Figures 3.11 and 3.12. Remember that only the east-west, or zonal, component of the wind is shown; these diagrams do not indicate the meridional component. However, zonal components, especially in the extratropical latitudes, are considerably higher than the meridional. The latter have mean values that show only a small variation with height and latitude, ranging generally from near zero to $-2$ (mean northerly component) and $+2$ m s$^{-1}$ (mean southerly component).

## SEC. 4.7 | UPPER AIR CIRCULATION

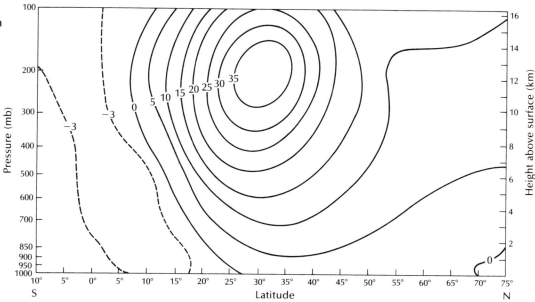

**FIGURE 4.34**
Mean zonal wind speed, in meters per second, from 10° S to 75° N, in January. Positive values are west winds; negative values are east winds. (From *Atmospheric Circulation Statistics*, NOAA Professional Paper No. 5, 1971.)

In January (Figure 4.34) a core of strong west winds is centered at 30° N and 200 mb (about 12 km, or 7.5 mi, in the U.S. Standard Atmosphere). This is called the **climatological jet stream.** Speeds decrease with latitude away from this core, especially toward the equator, where the wind shear (the place-to-place variation in wind speed and/or direction) is very pronounced. Weak east winds are found in low latitudes through the entire depth of the atmosphere shown, but extend farther poleward at the lowest levels. A small sector of easterly winds—the polar easterlies— also is apparent at and near the surface northward of 70° N (see Figure 4.9).

In July (Figure 4.35) there is also a core of westerly winds (the climatological jet stream), but two important differences from the January cross section are apparent. First, the considerably reduced north-south temperature gradient results in slower winds. Second, the jet core is centered at about 40° N, reflecting the fact that the locally intensified gradient is farther north (compare with Figures 4.29 and 4.30). East winds now occupy a much larger latitudinal extent than in

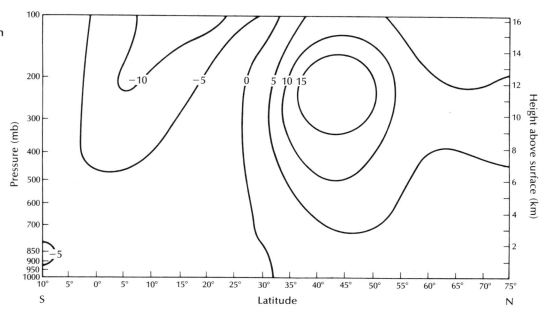

**FIGURE 4.35**
Mean zonal wind speed, in meters per second, from 10° S to 75° N in July. Positive values are west winds; negative values are east winds. (From *Atmospheric Circulation Statistics*, NOAA Professional Paper No. 5, 1971.)

winter (recall that the surface trades show a cross-latitude variation with the seasons as the subtropical highs move), and these winds at high levels are faster. The small sector of surface easterlies has moved northward to latitudes higher than 75° N.

Mean circulation characteristics of the Southern Hemisphere are very similar to those of the Northern Hemisphere. The different locations and configurations of the continents, however, do produce small differences. The jet stream core has slightly higher wind speeds in both winter and summer—the result of a somewhat stronger temperature gradient—and the latitudinal extent of the low-latitude easterlies is somewhat greater in the Southern Hemisphere.

This description of upper-air circulation has, because of zonal averaging, eliminated any east-west differences. Although mean meridional components of the motion are quantitatively much smaller than the zonal, they are significant in terms of climate. There are preferred positions of troughs and ridges, and these appear to be influenced by continental size and configuration and the presence of mountains. Ocean surface temperatures also may play a role.

Before we study hemispheric maps of the mean flow in the mid-troposphere, we must understand how these maps are presented. Perhaps the most straightforward way to illustrate such circulation would be to plot the mean wind speed for a specified level and thus show its geographic variation. The manner of presentation preferred by meteorologists, however, is different.

To this point all of the surface circulation maps, including the upper-air flow shown in Figure 4.3, have shown the variation of pressure at constant height. This height has been the surface, with pressure reduced to sea level where necessary, and, for Figure 4.3, it was an unspecified height in the upper atmosphere. Variation of pressure can also be shown for levels above the surface—at 6095 m (20,000 ft), for example. The speed and direction of the wind could then be inferred from the geostrophic approximation. Instead, the meteorologist prefers to show, with contour lines, the geographic variation of the height of a constant pressure surface. Notice in Figure 4.33 that there is a correspondence between the pressure at any level and the height of an isobaric surface passing through this level. Thus, maps of the height of a constant pressure surface and maps of pressure at a constant height show the same features. All that is required to associate contour configuration with wind speed and direction is to modify the wind flow rule by saying that the wind flows parallel to the contours, with low heights on the left (when looking downwind) and with a speed that is inversely proportional to the contour spacing. The direction of flow is again reversed for the Southern Hemisphere.

Figures 4.36 and 4.37 show the mean height of the 500-mb surface in January and July. These contours in extratropical latitudes have a circumpolar orientation, with heights decreasing poleward. When the modified wind flow rule is applied, this means that the flow is westerly in both months. It is not possible to associate contour spacing with the mean wind speed, however (except around a limited latitudinal span), for the reasons given in Section 4.3.4.

Significant interseasonal differences are apparent in these maps. In January the height gradient is greater, and the zone of westerlies extends farther equatorward, than in July. Typically, winter wind speeds are twice those of summer. These differences are, of course, a response to seasonal variations in the extratropical temperature gradient; since temperature decreases more rapidly with increasing latitude in winter than in summer, heights also decrease more rapidly [compare with Figures 4.33(a) and (b)]. The Northern Hemisphere subtropical highs are evident at 500 mb in July and to a lesser extent in January. Also, the Icelandic and Aleutian lows of Figure 4.12 are evident at 500 mb (the former as a closed low, the latter as a trough), although their centers have shifted northwestward relative to their surface positions.

The major climatological troughs and ridges at 500 mb (Figures 4.36 and 4.37) produce undulations, and thus meridional components, in the mean circumpolar flow which are called **long waves** or **planetary waves**. A trough and an adjacent ridge, or vice versa, comprise a wave. Long waves vary with time in both position and number. In the Northern Hemisphere five such waves usually predominate, although the mean 500-mb map shows only three in winter and four in summer.

### Short-Term Variations

As was the case with surface pressure representation, mean flow patterns are of limited usefulness when we try to relate long waves to surface weather. The short-term variations in the waves' location and amplitude are more relevant. Figure 4.38 shows schematically how the amplitude of long waves increases to the point that closed centers of low and high pressure (or heights)—representing pools of cold and warm air, respectively—are cut off. After the point where meridional motions are best developed [part (d)], the atmosphere reverts back to the initial, strongly zonal pattern. The entire sequence, lasting a few days to a few weeks, is called the **index cycle**.

## SEC. 4.7 | UPPER AIR CIRCULATION

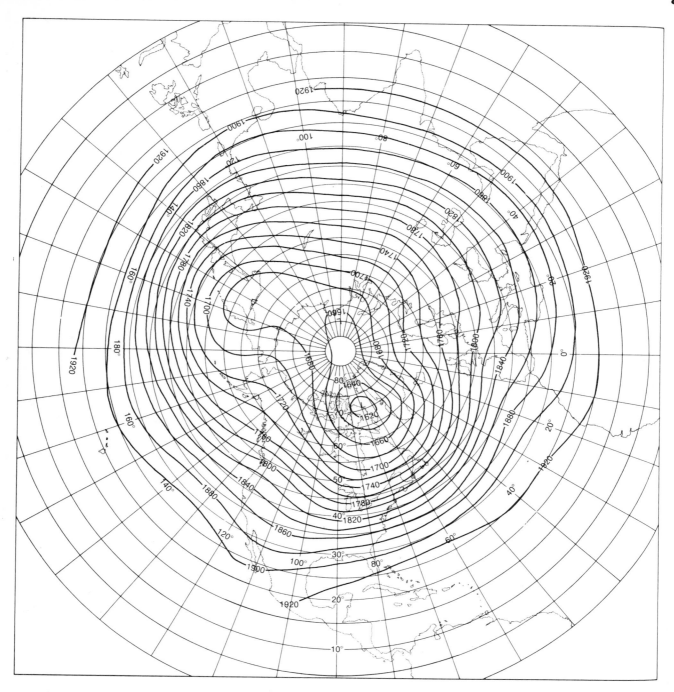

**FIGURE 4.36**
Mean height of the 500-mb surface, in tens of feet, in January. (From U.S. Department of Commerce, Weather Bureau.)

Also shown in Figure 4.38 is a core of fastest winds, the jet stream (indicated by the longest arrows). The climatological jet stream is very apparent in Figures 4.34 and 4.35, but remember that only its zonal component is shown. Notice that in the instantaneous (as opposed to climatological) representations in Figure 4.38 the jet stream meanders around troughs and ridges, and thus its meridional motions are pronounced during periods of large amplitude of the long waves.

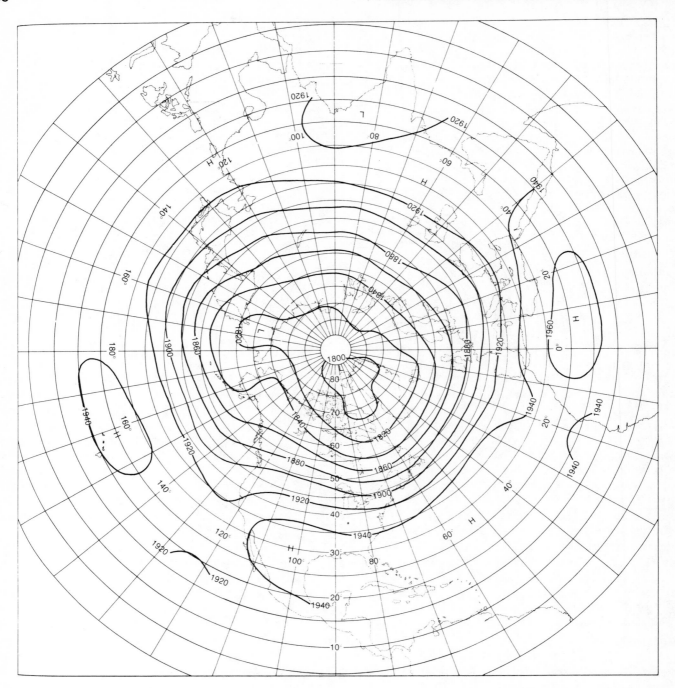

**FIGURE 4.37**
Mean height of the 500-mb surface, in tens of feet, in July. (From U.S. Department of Commerce, Weather Bureau.)

Actually, the fastest winds in the troposphere are found at somewhat higher levels than those shown in Figure 4.38, which illustrates conditions at about 500 mb. Figure 4.39 shows the major axes of the jet stream at a height of 10,700 to 12,200 m (35,000 to 40,000 ft) in January and July. These heights correspond to pressures of about 250 and 200 mb, respectively. In both months there are cores of maximum speed. In winter these cores are off the east coasts of North America and Asia and south of the Mediterranean Sea. In sum-

## SEC. 4.7 | UPPER AIR CIRCULATION

**FIGURE 4.38**
The index cycle of mid-tropospheric flow. Patterns at or near 500 mb characteristically evolve through stages. In (a) the flow is strongly zonal (west to east), but in (b) the wave amplitude has increased. By stage (c) the flow is strongly meridional—the opposite of zonal—and polar air is being carried into the middle and low latitudes, and tropical air to the middle and high latitudes. At stage (d) the greatly amplified waves are cut off, leaving "pools" of cold air in the south and warm air in the north. After stage (d) the atmosphere reverts to stage (a). (From "The Jet Stream" by J. Namias. Copyright © 1952 by Scientific American, Inc. All rights reserved.)

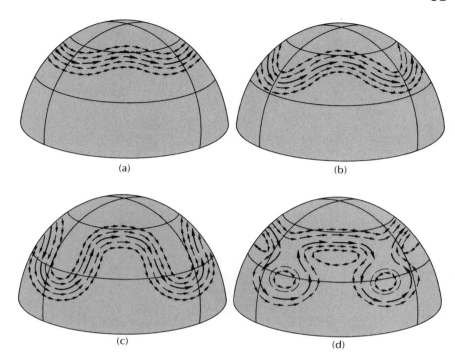

mer the longitudinal position of these cores stays the same, but they are farther poleward.

Compare these locations with those of the principal frontal zones shown in Figures 4.29 and 4.30. Just as Figure 4.33 suggests, wind speeds are highest above the polar front zone, where horizontal temperature contrasts are most pronounced. The jet stream cores, therefore, are approximately above the surface polar front zones. The term **thermal wind** is used to refer to such high-level mid-latitude flow. A thermal wind flows with cold air to the left and warm to the right when viewed downwind; the opposite applies to the Southern Hemisphere. Notice that the wind depicted as blowing into the page in Figure 4.1, or from the west, also is a thermal wind. This category is not distinct from geostrophic and gradient winds; for example, a wind can be both thermal and geostrophic.

Upper-air circulation patterns are not, of course, independent of those at the surface. At the same time it is not always possible to infer definitively the position of mid-tropospheric troughs and ridges on the basis of surface features. Some associations that can be made include the vertical continuity of surface lows, the position of surface lows and highs relative to upper-air troughs and ridges, and the vertically limited extent of cold surface highs. We will examine these associations further in Section 12.3.2.

Figure 4.20 is a generalized, or idealized, diagram of the stages in the development of an extratropical transient disturbance, or wave cyclone. In the same way we can generalize upper-level flow patterns, illustrated in Figure 4.40, for a Northern Hemisphere situation. Here the stages (a), (b), (c), and (d) correspond with stages (a), (b), (c), and (e), respectively, of Figure 4.20, but only the surface frontal positions are shown. The thin lines indicate the height of a mid-tropospheric pressure surface, and the arrows have been added to indicate the direction of flow. As explained previously, the closer the height contours, the faster is the wind.

In the initial state [Figure 4.40(a)] an east-west aligned stationary front is at the surface. Above the front the wind is strongly zonal, with fastest winds (the jet stream), located above the zone of strongest north-south temperature contrast. As the wave begins forming and surface pressure falls, a wave also forms in mid-troposphere, with the beginnings of a trough to the west and a ridge to the east [Figure 4.40(b)]. In part (c), when the surface temperature contrast is strongest, the upper-air trough and ridge are still better developed. At this level cold air is brought southward to the west, and warm air northward to the east, of the surface low. In the occluded stage at the surface [Figure 4.40(d)], the upper-air wave amplitude increases still further. If the temperature contrast is sufficiently large, an upper-level closed low forms to the west or northwest of the surface low, although this may at times be only an upper-level trough (an area of low heights without a closed low center).

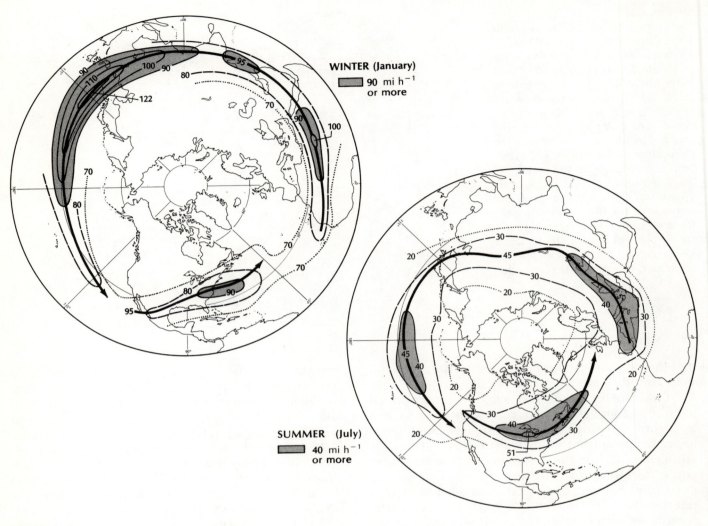

**FIGURE 4.39**
The mean jet stream in winter (upper left) and summer (lower right). (After J. Namias and P.F. Clapp. From *Introduction to Meteorology*, 3d ed., by S. Petterssen. Copyright © 1969 by McGraw-Hill Book Company. Used with the permission of McGraw-Hill Book Company.)

In terms of vertical continuity from the surface to mid-troposphere, the surface low tilts upward toward the cold air. Because of the cyclonic circulation, the cold air invariably is found (in the Northern Hemisphere) in the northwestern quadrant (see Figure 4.21). The surface high, which follows the cold front, is replaced at upper levels by a trough. Both relationships are consequences of the hypsometric relationship—namely, that cold lows tend to incline upward toward the coldest air, while above the "pools" of cold or cool air encompassed by anticyclones, the pressure eventually reverses. Thus, cold surface highs tend to be relatively shallow features, while cold lows are likely to retain their identify aloft (see Figure 12.21).

A consideration of convergence/divergence also suggests that the surface–upper-air coupling should be as described. To maintain vertical continuity, surface convergence should be accompanied by upper-air divergence, and vice versa. This is shown in Figure 4.41, which should be compared with Figure 4.7. When we consider the consequences for weather—particularly the occurrence of clouds and precipitation, which result from upward vertical motion—it is clear that "good" and "bad" weather occur in and around surface highs and lows, respectively.

The circulation patterns of Figure 4.40 are not fixed with respect to the surface, but tend to move to the east. Thus, the upper-level trough and the wave that follows the surface low also move to the east.

### SEC. 4.7 | UPPER AIR CIRCULATION

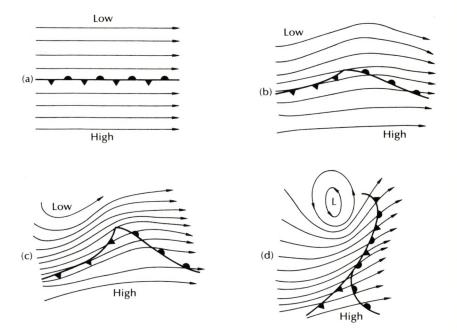

**FIGURE 4.40**
Upper-air contours (thin lines) associated with the various stages of a surface transient disturbance, which is shown by frontal positions. (Adapted from *Introduction to Meteorology*, 3d ed., by S. Petterssen. Copyright © 1969 by McGraw-Hill Book Company. Used with the permission of McGraw-Hill Book Company.)

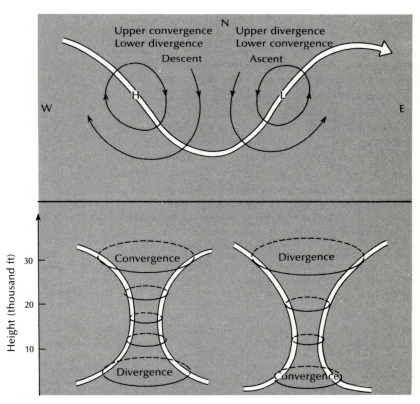

**FIGURE 4.41**
Coupling between waves in upper westerlies and high- and low-pressure centers near the surface. At the top is a horizontal (map) view of upper jet stream core in relation to surface systems. The bottom diagram is a vertical cross section showing the convergent and divergent parts of the motion in the horizontal plane, and the vertical motion. (From *Introduction to the Atmosphere* by H. Riehl. Copyright © 1965 by McGraw-Hill Book Company. Used with the permission of McGraw-Hill Book Company.)

#### 4.7.3 SUMMARY OF SURFACE AND UPPER-AIR CIRCULATION

This section concludes with a summary of the principal features of both surface and upper-air circulation as seen in cross section. Figure 4.31, our first attempt at depicting mean meridional and vertical motions, used dashed lines to illustrate where the motions were weak and/or highly variable. The details, principally of extratropical features, are now more completely illustrated in Figure 4.42.

In the tropics the Hadley cell dominates the circulation, with upward transport of sensible and latent heat by convection near the equator. Poleward-moving air cools by radiation and sinks at about 30° (variable with the seasons), arriving at the surface quite warm and dry. The cell is completed by the trade winds, in which areas large quantities of water are evaporated during the equatorward trajectory.

The zone of transient disturbances is indicated by the polar front zone, where mixing by surface cyclones and anticyclones and upper-air troughs and ridges accomplishes the required poleward transports of heat and water vapor. Here warm air rises over the polar front, and cold air sinks beneath it. Because the intent is to show mean conditions, the arrows in Figure 4.42 are shown as cutting through the polar front zone. At any particular time, however, this is not the case, because this zone separates these contrasting motions. At higher latitudes a secondary front, the arctic front, is another zone of steepened horizontal temperature contrast. At the poles there is a slow descent of radiatively cooling air. This motion, and similar subsidence at the poleward margins of the Hadley cell, tends to produce surface or near-surface temperature inversions. The very high latitude inversion is shown in Figure 3.11; the subtropical inversion will be discussed further in Chapter 7.

Just poleward of the Hadley cell, at about 12 km, is the **subtropical jet (STJ)**. The **polar front jet (PFJ)** is located over the polar front, with a core of maximum winds at a somewhat lower level than the STJ. The PFJ results from horizontal temperature contrasts throughout the depth of the atmosphere shown, while the STJ results essentially from upper-level temperature contrasts. Interestingly, in terms of climatological averages, these two wind maxima merge as one central core: the climatological jet (see Figures 4.34 and 4.35). At most times, however, there are two reasonably distinct jet maxima. Their meanders, or large north-south displacements (especially those of the PFJ), cause them to be shown as one central core in long-term zonal averages.

Two tropopauses are shown in Figure 4.42. Recall that a tropopause is that level at which temperature stops decreasing with height and either becomes constant with height (isothermal) or increases slowly (see Figure 3.13). In the tropics and subtropics this level varies from about 14 to 18 km, and in high latitudes from about 8 to 10 km. At the polar front zone there is generally no clearly marked level at which this change in lapse rate occurs.

The conditions in Figure 4.42 represent winter in the Northern Hemisphere, when the principal features of atmospheric circulation are best developed and most obvious. With minor modifications this cross section also could be applied to the Southern Hemisphere winter. Figure 4.42 could represent summer conditions if it were modified to account for the

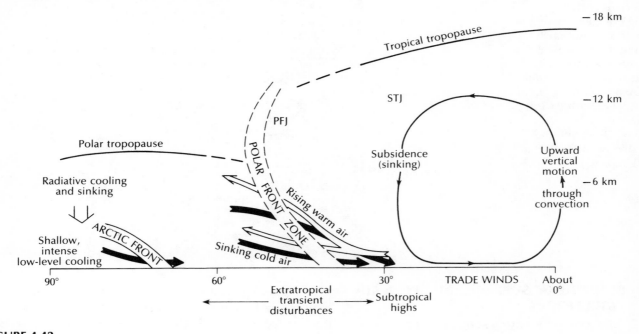

**FIGURE 4.42**
Principal features of the general circulation for the Northern Hemisphere winter. (From E. Palmen and C. W. Newton, *Atmospheric Circulation Systems*. New York: Academic Press, 1969.)

## 4.8
### CLIMATIC INFLUENCES OF THE OCEANS

The time and space distribution of insolation is a primary control of climate. Equally important is the nature of the surface receiving the insolation, the interface. The nature of the interface controls the disposition of the energy created there or directed toward it, and thus influences the magnitude and the rate of change with time of the energy balance components. The modification of these components by advection completes this process. In these few sentences we have summarized the most salient features of the physical basis of climate, and particularly its differentiation in time and space.

The fact that 72 percent of the earth's surface is covered by water thus has profound consequences for climate. Our climate would not be the same if the present land-water proportions were greatly different. With a greater amount of land the expanded continents would be much drier and would experience greatly amplified seasonal temperature variations. The climatic elements, especially temperature, would be considerably more variable than they are now. On an earth covered almost entirely by water we could expect the reverse of these conditions—in short, more stability and moderation of climate.

Three properties of the earth's surface substances help to explain why our climate would not be the same were the land-water proportion greatly different from what it is now. These characteristics were discussed earlier in Sections 2.3.4 and 3.6, but it will be helpful to review those for water. First, because water has a very large heat capacity, the oceans contain enormous amounts of heat. It has been estimated that as much heat is present in a depth of 3 m (10 ft) of ocean water as in an equivalent cross-sectional area of the atmosphere from the surface to its outer limit.

A second consequence of water's large heat capacity is that temperature changes in water, occurring from the loss or gain of heat, are much less over annual and diurnal cycles than those in land substances. The oceans, therefore, have a moderating effect on temperature. Horizontal circulation in the atmosphere advects maritime air over the continents and continental air over the oceans. However, because of the very conservative thermal properties of water as compared with those of land and air, this influence is virtually all one-way. That is, maritime influences can be carried far inland, but continental influences are restricted to a few tens of kilometers offshore. Where mountains bar the movement of air into the continents, this tempering effect may be limited to the immediate coastal area; this is the case in western North America, where the continentality index (see Figure 3.16) increases rapidly with distance inland from the coast. Conversely, maritime influences penetrate far inland in Europe (see Figure 3.17).

The second property of water is thermal conductivity. Although minimal for still water, thermal conductivity is enormously magnified for mixed water. As a fluid water is capable of motion within itself—as is air—and its properties, including heat content, can be mixed. This occurs principally in the vertical near the surface. Heat generated at the water-air interface is mixed downward or may be brought upward at times of surface cooling. The *conductive capacity* of water, a combination of its heat capacity and thermal conductivity (see Section 3.4), is thus very large, with the result that heat is rapidly mixed into water bodies and the accompanying temperature changes are quite small. Also, solar radiation can penetrate water, which diffuses the heat production through the uppermost few meters. Figure 4.43 indicates how the temperature of the uppermost 20 to 40 m changes seasonally through vertical mixing. The mixing, or mobility, of water also means that it can move horizontally. Surface—and to a lesser extent, underwater—currents move warm water from low to high latitudes, and vice

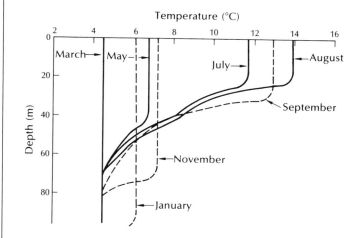

**FIGURE 4.43**
Temperature profiles at 50° N, 145° W in the North Pacific. (After John A. Knauss, *Introduction to Physical Oceanography,* © 1978, p. 48. Reprinted by permission of Prentice-Hall, Inc., Englewood Cliffs, N.J.)

versa. Obviously such heat transport is not possible in solid land substances.

The third aspect of the oceans' influence on climate involves phase changes of water and the hydrologic cycle. The earth's oceans constitute a vast evaporation surface, where heat from the absorption of solar radiation turns water to water vapor. Because evaporation is a heat-requiring process, it removes large quantities of heat from the water surface. The atmosphere transports this water vapor both upward and poleward and thus helps to compensate for latitudinal inequities in the heat balance (see the conclusion of Chapter 2).

The continuation of the hydrologic cycle involves the subsequent condensation of water vapor and the formation of clouds and rain. Virtually all (an estimated 80 to 90 percent) of the precipitation that falls on land was previously evaporated from the oceans; the remainder is derived from evaporation and transpiration from the land. We will discuss additional aspects of the hydrologic cycle in Chapter 5.

We will explore three major aspects of the oceans' influence on climate. The manner in which ocean circulation accomplishes the required poleward and upward heat transport will be considered first. Then we will examine the way sea surface temperatures help to differentiate the earth's climates. The role of the oceans in initiating, or responding to, climatic change and their role in numerical models of the climate system will be covered in Chapter 11.

### 4.8.1 THE TRANSFER OF HEAT BY OCEAN CURRENTS

Two components of heat transfer by the oceans will be considered here: vertical and lateral. In terms of the vertical transfer by the convection of both latent and sensible heat (excluding radiative transfer), the oceans act as a source and the atmosphere as a sink. This concept is illustrated in Figures 4.44 and 4.45, which are generalized maps of all oceans. The rate at which sensible heat is transferred upward depends on the quantity (sea surface temperature − air temperature), which is nearly always positive, and the extent of circulation (or mixing) of the air. Latent heat transfer depends, of course, on the evaporation rate, which is proportional to the difference between the vapor pressure of the air immediately adjacent to the

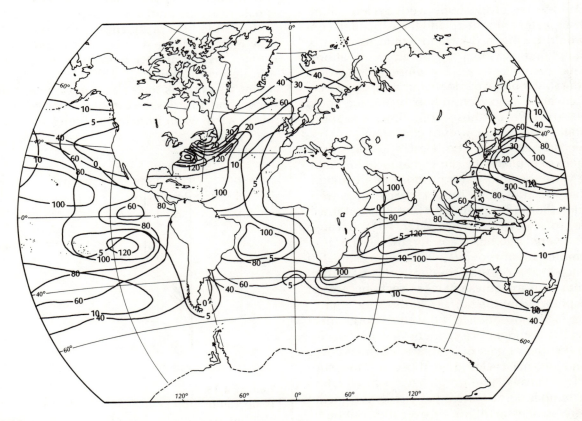

**FIGURE 4.44**
Latent heat flux from the oceans, in kilocalories per square centimeter per year. (From M. I. Budyko, *Climate and Life*. New York: Academic Press, 1974.)

### SEC. 4.8 | CLIMATIC INFLUENCES OF THE OCEANS

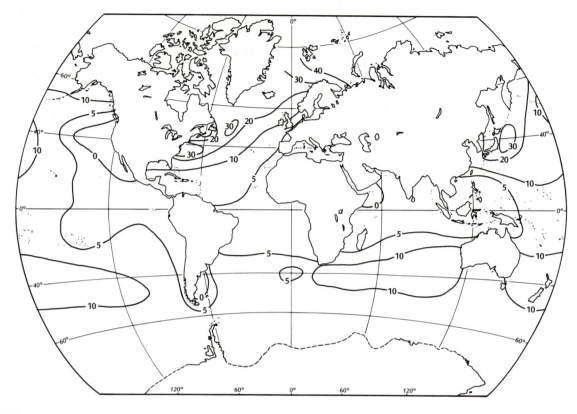

**FIGURE 4.45**
Sensible heat flux from the earth's surface into the atmosphere from the oceans, in kilocalories per square centimeter per year. (From M. I. Budyko, *Climate and Life*. New York: Academic Press, 1974.)

water surface and the vapor pressure of the air at higher levels, and on the amount of atmospheric mixing. More details of the evaporation process are given in Chapter 5.

Thus, where cold, dry air moves over a warm water surface the fluxes of both sensible and latent heat are very large. This is a frequent occurrence in winter off the east coasts of the continents in the middle latitudes. There outbreaks of cold continental air move eastward, with the prevailing westerly circulation. This air is rapidly modified—warmed and moistened—by contact with the warm waters of the Gulf Stream and Japanese Current (see Figure 4.49). Of the two fluxes latent heat is by far the larger. Another area, of much greater extent, in which the latent heat flux is large is that of the trade winds (see Figure 4.32 and Table 2.1).

The total energy balance of the ocean surface includes radiation exchange as well as the transfer of latent and sensible heat. When direct-beam and sky radiation and the balance of infrared radiation emitted by the water surface and of atmospheric counter-radiation ($I\uparrow - I\downarrow$) are considered, we have a complete accounting of the vertical components of the energy balance. These fluxes are represented in Figure 2.12, where it is understood that $\pm G$ refers to the vertical transfer of heat beneath the water surface.

To complete an accounting of the energy balance at the water-air interface, we must consider another process. In Section 2.3.5, with reference to Table 2.1, we learned that, unlike solid land substances, water can transport heat laterally. In general the low latitudes have a surplus of heat, and the high latitudes a deficit. This means that the vertical balance at a particular location, averaged over a year, need not be zero; in fact it rarely is. Lateral circulation is the process by which surpluses and deficits are compensated, and thus there is no appreciable net warming or cooling at particular locations over long periods of time.

Figure 4.46 shows the latitudinal variation of all vertical components of the oceans' energy balance for the Northern Hemisphere. (The symbols used in this diagram are defined in Figure 2.12.) The summation, by latitude, of all these components is indicated at the bottom. It is clear that up to about 30° there is a gain, and beyond 30° there is a loss. It appears that loss exceeds gain in this diagram, but, as has been

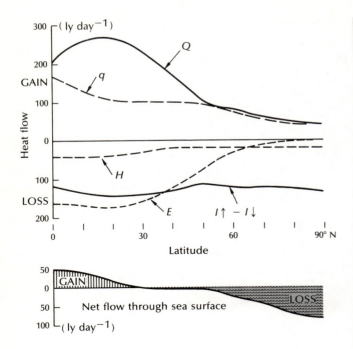

**FIGURE 4.46**
Latitudinal variation of the components of the energy balance for the Northern Hemisphere oceans. (After G. L. Pickard, *Descriptive Physical Oceanography*. Oxford, England: Permagon Press, 1963.)

explained in connection with Figure 2.16, these quantities will be identical (or nearly so, within errors of measurement) when the area between latitude spans is taken into account.

Although the principal geographic variations in the balance of the energy components (Figure 4.46) are latitudinal, there are significant east-west differences as well, as shown in Figure 4.47. In this map positive values indicate a net annual transfer of heat laterally into the area, and vice versa. Notice the large heat gain in the northern North Atlantic, and, to a lesser extent, in the western Pacific (Northern Hemisphere). Conversely, the west coasts of South America and Africa are areas where heat is lost laterally. Obviously, ocean currents are the mechanism that provides for this transfer. We should expect to see that the oceans circulate, just as the atmosphere does, so as to transport heat from areas of surplus (negative values in Figure 4.47) to areas of deficit.

## 4.8.2 SURFACE CIRCULATION AND ITS CLIMATIC INFLUENCES

From a climatic viewpoint the chief function of the oceans is the same as that of the atmosphere: to transfer heat upward and poleward. Perhaps the simplest way for the oceans to accomplish this would be for water to rise to the surface in low latitudes, absorb warmth from relatively intense surface heating, move poleward and release its heat to the atmosphere—increasingly so as the difference between sea surface temperature and air temperature increases—and then sink to lower levels and move equatorward to complete the cycle.

In fact, oceanic circulation is much more complicated than this, and there are several factors that combine to produce the patterns we observe. Atmospheric circulation is the most important of these factors. The combination of wind direction and the earth's rotation results in a tendency for surface currents to move at an angle somewhat less than 45° to the right of the wind direction in the Northern Hemisphere and to the left in the Southern Hemisphere; that is, the Coriolis force affects water movement just as it does air movement. Contrasts in density that arise between salt and fresh water, as where rivers empty into the oceans, and between water masses of differing temperature also influence circulation patterns. Finally, the location and volume of contributing rivers and the configuration of the continents help shape ocean circulation.

It is clear, then, that the oceans and atmosphere are a coupled system. Near-surface prevailing winds to a large extent control the movement of surface water; at the same time the energy transferred from the ocean surface to the atmosphere provides "fuel" for the engine that governs the way the earth's atmosphere circulates.

### Surface Circulation Patterns

Before examining the specific patterns of oceanic circulation it will be helpful to study a model, or a simplified schematic, of the dominant features (Figure 4.48). The circulation cells, here shown as one in each hemisphere, are called **gyres,** and their direction of rotation is opposite: clockwise in the Northern Hemisphere, counterclockwise in the Southern Hemisphere. Notice the similarity to the surface anticyclonic cells in Figures 4.12 and 4.13. In high latitudes in the Northern Hemisphere the eastward-moving current splits—one part moving northward, the other southward. In the Southern Hemisphere, because of the absence of continents southward of about 40° S, one branch flows equatorward and the other maintains the eastward movement.

The portions of these gyres indicated as warm or cool are of special interest. In equatorward-moving flow the absorption of solar radiation warms surface waters, which lose heat by long-wave emission as well as by the release of latent and sensible heat, as they curve around the western parts of the gyres. But because of the very conservative thermal properties of water, sea surface temperatures remain high compared with those at other longitudes in the same latitudinal span. Further cooling occurs as currents move eastward, but even here temperatures remain relatively warm for the latitude. In the Northern Hemisphere a portion of the surface water moves northward, maintaining comparatively warm temperatures, while the other moves southward. As the latter branch reaches lower latitudes, and thus increasingly absorbs solar radiation, the lag due to water's thermal conservativeness now results in cool surface waters, which remain so around the southeastern half of the gyre. In an upside-down sense, the same applies to the Southern Hemisphere, except where currents split, as just described.

The relative coolness of the southeastern quadrants (northeastern in the Southern Hemisphere) is accentuated by **upwelling.** In this process wind stress on the ocean tends to move surface water away from the coast; the surface water is then replaced by cooler water from below.

The average position and extent of the principal surface ocean currents is shown in Figure 4.49. Notice that the major gyres, those in the north and south Atlantic and Pacific oceans, conform rather closely to the schematic model in Figure 4.48. Thus, the California, Peruvian, Canaries, and Benguela currents flow equatorward and are anomalously cool, while the Kuro Siwo (or Japan)–North Pacific Drift, East Australian, Gulf Stream–North Atlantic Drift, and Brazilian currents have a poleward component of motion, and

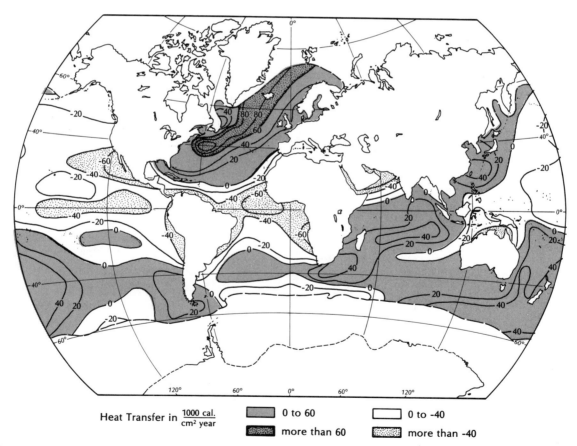

**FIGURE 4.47**
Lateral heat transport by the oceans. Positive values indicate a net transfer of heat laterally into the area, and vice versa. (After map by Rudolph Geiger, Justes Perthes Geographische Verlagsanstalt, 6100 Darmstadt, West Germany.)

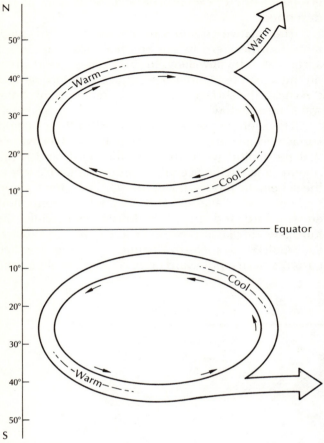

**FIGURE 4.48**
Schematic representation of ocean surface circulation. Sea surface temperature indicators mean that these areas are warm or cool relative to all sea surface temperatures at that latitude.

are anomalously warm. South of the Indian subcontinent, from Africa to Australia, the surface circulation is more complex than is indicated by the model. In the Southern Hemisphere the absence of continents from about 40° to 75° results in a circumpolar flow, the West Wind Drift, which of course cannot be considered anomalously warm or cool because it occupies essentially the same latitudinal span at all longitudes. Additional currents not indicated by the model are the relatively cold currents off the northeast coasts of North America and Asia (the Labrador and Okhotsk currents) and the Equatorial Countercurrent of the Pacific, which flows eastward between the westward-flowing portions of the Pacific gyres.

## Climatic Effects

The circulation patterns of the ocean surface and the areas of anomalously cool and warm ocean currents have been emphasized because of their pervasive effects on climate. They affect not only the atmosphere over the oceans, but, more important, the climate of continents in general, especially their coastal areas. We will use the term **littoral** to refer to those coastal or near-coastal areas where oceanic effects are appreciable. The **littoral** may extend far inland in the absence of topographic barriers, or it may be restricted to a few tens of kilometers inland where mountains block the inward penetration of maritime air.

It is difficult to generalize the oceans' effects on climate. There may be considerable seasonal variation in their effectiveness in moderating extremes in land climate, and much depends on whether the prevailing atmospheric circulation brings air from the oceans to the continents, or vice versa. However, it is possible to characterize, at least approximately, the variations of two very important climatic elements: air temperature and humidity. In most instances we can infer space and time variations of elements such as cloudiness, precipitation, and insolation from the variations of these two principal elements. Obviously, the advection of maritime air that has passed over a warm water surface will bring air with that characteristic to an adjacent land area. However, *warm* refers to sea surface temperature. In winter this produces a moderating effect, and littoral air temperatures will be raised because of the ocean-to-land circulation. But in summer, when continents are warmer than oceans in the middle and high latitudes, the effect will still be a moderation of the littoral air temperatures because, although still relatively warm, ocean surface temperatures are less than those over land. In short, warm ocean currents moderate the temperatures of littoral areas, but do so by producing appreciable increases in winter and only modest decreases in summer. The reverse reasoning applies to cool currents; summer temperatures are appreciably decreased, and those of winter only slightly increased.

The most continental climates are those which experience the least maritime influence. Thus, Figures 3.16 and 3.17 show not only continentality, but its inverse—*oceanicity*. The effect of topographic barriers is clear from these figures: maritime effects penetrate far into western and central Europe, but are greatly restricted in North America.

The water vapor content of the air, which in most locations is closely correlated with precipitation amounts, is influenced by the air's prior experience. After long residence over a continent, air is likely to contain little water vapor, while air over the oceans invariably is quite humid. But the rate of evaporation from a water surface, because it depends on the tem-

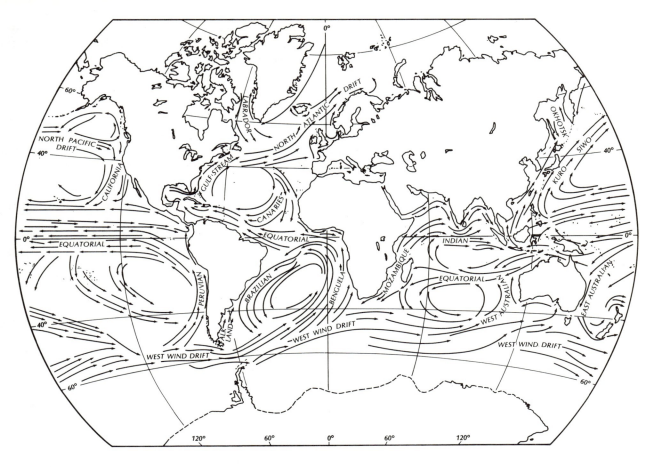

**FIGURE 4.49**
The average position and extent of the principal ocean currents. The length of the arrows gives some indication of the speed of the currents. (From *Weather and Climate* by C. E. Koeppe and G. C. DeLong. Copyright © 1958 by McGraw-Hill Book Company. Used with the permission of McGraw-Hill Book Company.)

perature of that surface, increases at an increasing rate with the temperature. Thus a fairly modest increase in water temperature, such as occurs across portions of gyres in the same latitude, can result in considerably more water vapor in the air over the warmer current.

Such elements as humidity and precipitation are substantially influenced by water temperature. Windward coasts adjacent to warm ocean currents have higher precipitation amounts and less insolation and are more humid and cloudy than windward coasts adjacent to cool water. Of course, climatic influences other than sea surface temperatures also play a role in global patterns of temperature and precipitation and will be considered in Chapter 7.

A few examples of climate in particular locations will illustrate these influences. The waters of the northwestern quadrant of the North Atlantic and North Pacific gyres, being anomalously warm, greatly modify (warm and humidify) the prevailing westerlies in the winter. The climate of the western coast of North America, from northern California to the Gulf of Alaska, and that of the British Isles and Scandinavia is considerably moderated and moistened. Winter temperatures are some 17C° to 22C° (30F° to 40F°) warmer than the average for that latitude (including continents and oceans), and annual ranges of temperature are greatly reduced. Although precipitation amounts are not large, these areas are overcast and rainy throughout much of the year, with relatively few sunny days. Because of the prevailing westerlies, however, the east coasts of continents in these latitudes are largely unaffected by the oceans' moderating effects.

In the equatorward/eastern quadrants of the gyres cool currents tend to parallel the coasts of areas such as southern and Baja California, Chile and Peru, and northwestern and southwestern Africa (Figure

4.49). Here the heat of summer is reduced, annual ranges of temperature are smaller, and aridity prevails. Perhaps surprisingly, fog is a frequent occurrence. With an onshore trajectory, winds blow from a warm to a cool water surface and are chilled from below, often to the point of condensation (Figure 7.9).

**SUGGESTED READING**

DUTTON, J. A. *The Ceaseless Wind: An Introduction to the Theory of Atmospheric Motion*. New York: McGraw-Hill Book Company, 1976.

"Oceans and Climate." *Oceanus* 21 (1978).

THURMAN, H. V. *Introductory Oceanography*. 3rd ed. Columbus, Ohio: Charles E. Merrill, 1981.

# 5
# MOISTURE IN THE ATMOSPHERE, CLOUDS, AND PRECIPITATION

**INTRODUCTION**
**PHASE CHANGES**
**MEASURES OF WATER VAPOR**
**THE HYDROLOGIC CYCLE**
**GLOBAL VARIATIONS**
**EVAPORATION**
Measurement
Potential Evapotranspiration
**WATER BUDGETS**
**CLOUDS AND PRECIPITATION**
Cloud Formation and Vertical Motion
Condensation
The Formation of Precipitation
Forms of Precipitation
Characteristics of Precipitation

## 5.1
### INTRODUCTION

Of all substances in the physical environment of planet Earth, water is probably the most important. In its various forms water is almost as significant to weather and climate as temperature. Together with oxygen it provides sustenance for virtually all living things; its absence can be detrimental, and at times even lethal.

In addition to providing a medium in which nutrients are transported and used in living organisms of all orders, from crops to the human body, water as an erosive agent shapes the contours of the earth's surface. Earth is unique among the planets in our solar system because it contains such an abundance of water in the liquid state. Recent fly-by missions by satellites have confirmed earlier indications that while the other planets have small amounts of water vapor in their atmospheres, only Mars has solid and liquid water, and there it exists in very limited amounts.

Water's role in weather and climate is twofold. First, its relative presence or absence provides a principal means by which climate can be differentiated, both spatially and temporally. On a short-term basis changes of state of water are a conspicuous indication of present weather—for example, increasing cloudiness, decreasing visibility, and the beginning of rain. Indeed, many weather changes result simply from water's changes of state.

Second, these same changes of state, as when water condenses or evaporates, are an integral part of the energy exchange processes at the interface and in the atmosphere. Explicit account must be taken of these processes in the mathematical formulas that describe atmospheric energetics. Furthermore, the presence of water in the atmosphere, either as clouds or as water vapor, modifies the transfer of both long- and shortwave radiation. In modeling atmospheric dynamical—and especially thermodynamical—processes, we have been hampered by our inability adequately to account for water's manifold influences.

## 5.2
### PHASE CHANGES

We will use the term *water* for both the chemical compound in all its phases and the liquid phase of this compound as well. When referring to the latter, we will add the word *liquid* to avoid ambiguity.

Within the earth-atmosphere system water exists in three phases: as a gas (water vapor), as a liquid (water), and as a solid (as ice, snow, hail, and other solid forms). Which of these forms water will assume at a given place and time depends on the environmental conditions of pressure and temperature, but temperature is by far the more important. Water is solid below 0°C (32°F) at the earth's surface, but is mostly vapor in the atmosphere at such temperatures, and is liquid from 0° to 100°C (212°F). It can exist as water vapor over virtually all temperatures encountered in the terrestrial environment.

The terms used for the phase changes of water are shown in Figure 5.1. Within any one of these phases the internal energy which that particular water phase possesses is approximately proportional to its absolute temperature (for temperature in degrees Kelvin, see Figure 3.1). However, this proportionality does not hold for comparisons of the internal energies of different phases. During phase changes additional amounts of heat, which are not indicated by temperature, are either acquired from or released to the environment. This is an extremely important consideration involving the concept of hidden, or *latent*, heat.

To understand this acquisition or release of latent heat, consider what happens when heat is added to a gram of ice at −2°C. Since the specific heat of ice is about 0.5, the addition of one calorie of heat will raise its temperature to 0°C. If we continue to add heat to this gram of ice, the temperature will not change until another 80 calories have been added. At this point the ice will melt to water, and, if we continue to add heat, the water temperature will increase. Therefore, every gram of liquid water contains hidden, or latent, heat beyond the sensible heat indicated by its temperature. In the reverse process, freezing, this same 80 calories per gram is liberated. The particular latent heat that applies here is the **latent heat of fusion,** or the **latent heat of melting.**

To convert liquid water to water vapor requires even larger amounts of latent heat, about 580 calories per gram at terrestrial temperatures. Conversely, this same amount is released when vapor condenses. This is the **latent heat of vaporization,** or the **latent heat of condensation,** respectively. An important difference from the latent heat of melting/freezing is that evaporation and condensation can occur across a broad range of temperature. Whether or not these phase changes occur depends on other factors, too, such as the water vapor content of the air (see Section 5.3).

The importance of latent heat acquisition and release for weather and climate cannot be overemphasized. Water continually undergoes phase changes in a never-ending cycle—the hydrologic cycle—that involves incredibly large amounts of energy. A typical hurricane, in condensing water to water vapor, produces $24 \times 10^{11}$ kilowatt-hours in a day. By comparison, the U.S. consumption of electric energy in 1971 was essentially the same, $15 \times 10^{11}$ kilowatt-hours. Since evaporation and condensation occur overwhelmingly at the surface and in the atmosphere, respectively, these phase changes constitute an extremely significant means of transporting heat upward. Figure 2.15 indicates that in an average year 59,000 langleys per square centimeter (59 kly cm$^{-2}$) are used in the evaporation process. This amount is 77 percent of the total upward transport of both latent and sensible heat by convection. In terms of the previous comparison, which was on a daily basis, this is equivalent to $10^{19}$ kilowatt-hours—a figure over 4 million times larger than that for the hurricane.

# 5.3
## MEASURES OF WATER VAPOR

There are a number of ways to express the quantity of water vapor in the atmosphere. Water vapor, nitrogen, and oxygen, with much smaller amounts of carbon dioxide and trace gases, make up the earth's atmosphere. Since the atmosphere's total pressure is the sum of the pressures exerted by each component gas, it is possible to specify the partial pressure of an individual gas. The partial pressure of oxygen, for example, is of concern in the study of respiration. Similarly, the quantity of water vapor in the atmosphere at any time and place can be measured by the partial pressure it exerts. In terms of percentages, water vapor comprises from near zero to as much as about 4 percent of total atmospheric pressure. Thus at sea level, where pressure averages 1013 mb, water vapor exerts a pressure varying from near zero (extremely dry) to 40+ mb (very humid).

Sometimes water vapor pressure is expressed as its equivalent in height of a mercury column, which is one way of measuring atmospheric pressure. Thus, if air at a temperature of 26°C (79°F) has all the water vapor it is capable of holding (i.e., is saturated), it will exert a partial pressure of 33.6 mb, which is equivalent to 2.5 cm (1 in.) of mercury.

Another way of expressing the water vapor content of air is to relate the mass of this vapor to either a standard volume or a standard mass. In the former the result may be in grams per cubic meter, or ounces per cubic foot in the English system. For the latter the ratio may be applied to either a mass of dry or a mass of humid air. Thus, we can speak of grams of water vapor in a kilogram of humid air (which includes both water

## SEC. 5.3 | MEASURES OF WATER VAPOR

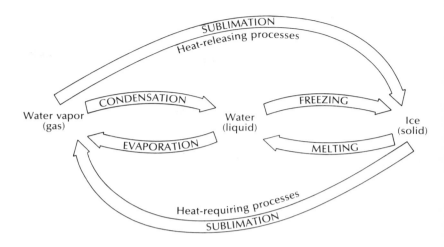

**FIGURE 5.1**
The phase changes of water.

vapor and the other gases), or as grams of water vapor in a kilogram of dry air (air with the mass of water vapor deleted). This second way, mass per mass of dry air, is preferred by meteorologists, and is called the **mixing ratio.** For saturated air at 26°C the mixing ratio is 21.3 grams per kilogram (g kg$^{-1}$); that is, in a volume of air with a total mass of 1021.3 g, 1000 g (1 kg) is dry air, 21.3 g is water vapor.

We stated earlier that air may contain all the water vapor it is capable of holding. This implies that there is an upper limit to the amount of water vapor that air can contain. This upper limit is determined almost entirely by temperature. The warmer air is, the more water vapor it *can* hold; notice we said *can*, not *does*. This is not a linear or a proportional relationship. Instead, the water vapor holding capacity of air increases at an accelerating rate with increasing temperature.

**Relative humidity** indicates how close air is to being saturated, that is, how close the *actual* amount of water vapor is to the *theoretical* amount the air can hold. This ratio of actual to theoretical (or capacity) amounts requires some absolute measure of water vapor as both numerator and denominator. For example, if air has a mixing ratio of 21.3 g kg$^{-1}$ (the previous example) and its temperature is 26°C, then the ratio of actual to capacity is 21.3/21.3 = 100 percent, or saturation. If the air temperature were 30°C, we would use 21.3 as the numerator and the mixing ratio that would apply to air saturated at 30°C, which is 27.2, as the denominator. Thus, 21.3/27.2 = 78 percent. Increasing the air temperature to 35°C, which requires 56.2 in the denominator, decreases the relative humidity to 38 percent. (Values for the mixing ratio have been taken from standard tables.)

As an element of weather, relative humidity is not as useful as is commonly supposed. It is important in the process of condensation, which occurs only at 100 percent (although there are some exceptions to this). It is unfortunate, however, that relative humidity as an indicator of the water vapor content of air is more familiar to the layperson than are the absolute measures we have discussed. Strictly, relative humidity is not such an indicator; it is instead a number that represents the combined influences of the actual water vapor content of air and the temperature, which determines the saturation amount. To emphasize this inappropriateness, consider the following analogy. You are asked "What's the humidity?" (that is, "What is the water vapor content of the air?"), and your reply is "60 percent." This is very much like responding to the question, "How far is it from New York to Boston?" with the reply "55 miles an hour." Remember that 60 percent relative humidity can apply to a very wide range of conditions, and, unless the temperature is specified as well, no indication is given of the actual water vapor content of the air.

Except when atmospheric circulation brings more humid or drier air to a locality, or in the presence of local sources or sinks of water, the water vapor content of air remains very much the same from day to day. When temperature fluctuates in its usual way (see Figure 3.9, the clear-skies day), relative humidity will be high when the temperature is low, and vice versa. Although the relative humidity is changing, the water vapor content is not, at least not appreciably. Figure 5.2 illustrates this.

More widely used measures of water vapor—largely because they are easier to measure than the absolute measures given earlier—are the dewpoint and the wet-bulb temperature. The **dewpoint,** or dewpoint temperature, is the temperature to which air must be cooled at constant pressure to produce saturation. In our example, with temperatures and relative humidities of 35°C and 38 percent, 30°C and 78 percent, and 26°C and 100 percent, the dewpoint is 26°C

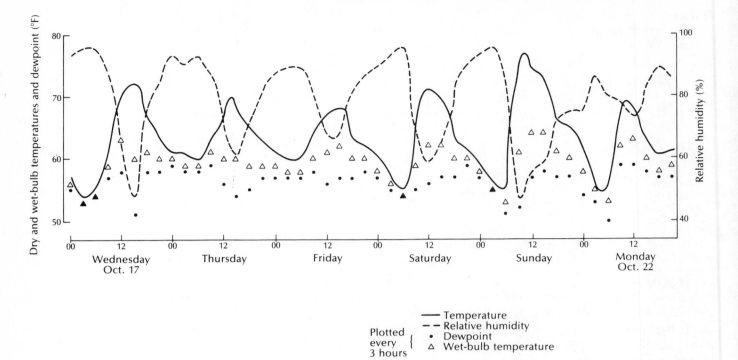

**FIGURE 5.2**
Six-day variation of temperature, relative humidity, dewpoint, and wet-bulb temperature at Los Angeles, California, October 17–22, 1962.

because this is the temperature at which saturation occurs. As an absolute measure of air humidity (i.e., its water vapor content), either the mixing ratio or vapor pressure is preferred in most meteorological applications. However, for most practical purposes dewpoint can be regarded as essentially proportional to these two measures.

The **wet-bulb temperature** is the lowest temperature to which air can be cooled by evaporating liquid water into it. The bulb of an ordinary mercury-in-glass thermometer is covered by a water-saturated wick and then thoroughly ventilated so that water is evaporated from the wick. Since evaporation is a heat-requiring process, the thermometer is cooled, and the lowest temperature attained at equilibrium (when no more water can be evaporated) depends on the initial amount of water vapor in the air surrounding the wet bulb. In very dry air a relatively large amount of water will be evaporated, requiring a relatively large amount of heat from the bulb, and the wet-bulb temperature will be less than if the air is relatively humid and cannot absorb much liquid water by evaporation. If the air around the wet bulb is motionless, evaporation cannot occur, and the reading will be erroneous. The wet-bulb temperature is always equal to or greater than the dewpoint, and the disparity increases as relative humidity decreases (see Figure 5.2)

All measures of the water vapor content of air can be obtained from the others. Given the temperature and any one measure, the others may be calculated, or, more commonly, obtained from tables.

# 5.4

## THE HYDROLOGIC CYCLE

Water in the earth-atmosphere system not only continually undergoes phase changes, but is constantly being transported from place to place as well. Horizontal transport of water vapor is accomplished by atmospheric circulation, which, for the most part, advects humid air masses from the oceans to the continents. Vertical transport is accomplished by both convection and the slow uplift of air associated with convergence and the overrunning of dry air masses by more humid ones. Vertical movement of water occurs during rainfall, and horizontal movement by the slower motions of streams and groundwater. The possible routes that water may follow, and the phase

## SEC. 5.4 | THE HYDROLOGIC CYCLE

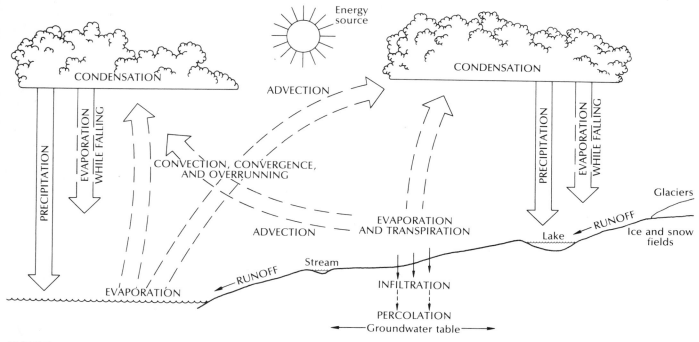

**FIGURE 5.3**
The hydrologic cycle.

**FIGURE 5.4**
Water storage in the earth-atmosphere system. (Adapted from R. J. More, "Hydrological Models and Geography" in *Models in Geography*, ed. R. J. Chorley and P. Haggett. London: Methuen, 1967.)

| Oceans: 97% of all water | Atmosphere: 0.035% of all fresh water | | Ice sheets and glaciers: 75% of all fresh water |
|---|---|---|---|
| | Rivers: 0.03% | Lakes: 0.3% | |
| | Soil moisture: 0.06% | | |
| | Groundwater (above 2500 ft): 11% | | |
| | Groundwater (2500–12,500 ft): 14% | | |
| Oceans (saline water) | Continents (fresh water) | | |

changes that often accompany its vertical fluxes, are shown in Figure 5.3. These processes and transports are collectively called the **hydrologic cycle**.

Since this is a continuous cycle, any starting point is arbitrary. We will start with evaporation from the ocean surface. Water has changed state from liquid to gas; as water vapor, it will eventually condense and form clouds in the atmosphere, either while remaining over the oceans or after horizontal transport to a land surface. At this point the cloud droplets may evaporate, merge (or grow by other means) into raindrops, or be transformed into snow crystals. Thus, some form of precipitation results: rain at temperatures generally above freezing, and snow or other solid forms at subfreezing temperatures. In falling the precipitation may evaporate, or melt and evaporate in the case of solid precipitation, but more often than not it reaches the ground.

At the ground liquid precipitation can either *run off* into successively larger streams until it reaches the ocean, evaporate from standing water, or *infiltrate* the soil and remain there for widely varying lengths of time depending on what happens to it after infiltration. Within the soil water may be absorbed by the roots of plants and subsequently transpired back into the atmosphere, or it may move back upward through the soil and be evaporated at the surface. Under widely different conditions, but principally when there is an excess of it in the soil, water may *percolate* downward to the water table, within which it moves horizontally into streams or lakes and hence eventually back to the oceans. When solid precipitation reaches the ground, the snow or ice pack deepens and eventually melts or sublimes (see Figure 5.1) with seasonal variations of temperature, or, at high latitudes and high elevations progressively accumulates in glaciers or more or less permanent ice and snow fields.

Having described the processes of the hydrologic cycle, we now need to describe the various quantities involved. Figure 5.4 shows how much water is stored in the various parts of the hydrologic cycle. Of all water, 97 percent is saline, and almost all of this is in the oceans. Of the remaining 3 percent (fresh water), about three quarters is contained in ice sheets and glaciers and almost all of the remaining is in ground storage. The percentage amounts of water and water vapor in the atmosphere and of water in rivers and lakes and in the soil are very small.

The quantities of water involved in the various processes of the hydrologic cycle are shown in Figure 5.5, where the width of each tube is proportional to the volume involved in that phase of the cycle. The first number indicates the annual quantity of water, in thousands of cubic kilometers; the second, in parentheses, is the equivalent in both centimeters and inches. The second number applies only to evaporation and precipitation and incorporates the areas of continents and oceans. Notice that by far the largest exchanges are at the surface of the oceans (evaporation) and above them (precipitation over the oceans). Also, the runoff is equal to precipitation minus evaporation over land; this assumes that in the mean, over several years, there is no net gain or loss of either ground or surface water. Finally, some 89 percent (94/106) of land precipitation comes from the oceans.

# 5.5

## GLOBAL VARIATIONS

The latitudinal variation of precipitation and evaporation, and of water surpluses and deficits, is illustrated in Figure 5.6. Notice that the abscissa (x-axis) of this graph is corrected for the varying amounts of land in latitude belts. As Figure 5.5 has already indicated, precipitation and evaporation over the oceans greatly exceeds that over land. Areas of water surplus are found from 40° to the poles and from 15° N to about 10° S. There is a deficit of water in the remaining lati-

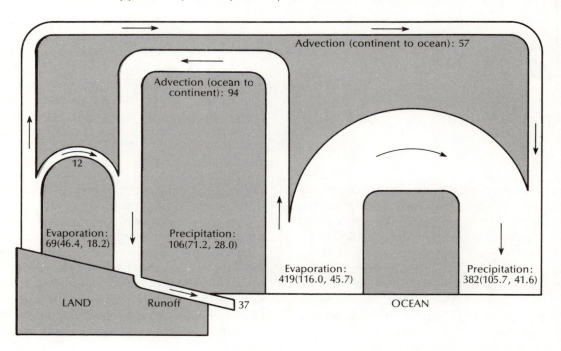

**FIGURE 5.5**
Schematic representation of the quantities of water involved in the hydrologic cycle. The width of each tube is proportional to the volume of water involved in that phase of the hydrologic cycle. The first figure is water volume in thousands of cubic kilometers. Figures in parentheses are the equivalent depths, corrected for areas of ocean and continents, in centimeters and inches. (After R. Geiger. From J. Mather, *Climatology: Fundamentals and Applications*. New York: McGraw-Hill, 1974.

**FIGURE 5.6**
Latitudinal variation of evaporation and precipitation for the earth as a whole, and for land and sea areas. (After R. Geiger. From J. Mather, *Climatology: Fundamentals and Applications.* New York: McGraw-Hill, 1974.)

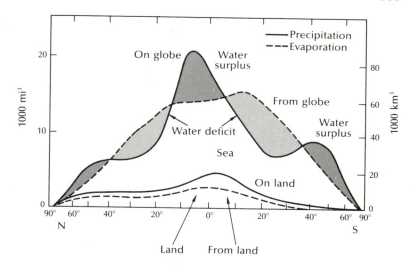

tudes which is greatest at 25° N and S. The humid areas are those associated with the ITCZ (low latitudes) and the transient disturbances of the middle and high latitudes. The subtropical belts of high pressure, with their subsidence and general lack of precipitation-producing mechanisms, are of course associated with areas of water deficit.

How absolute and relative humidity vary with latitude is shown in Figure 5.7. The water vapor content of air decreases uniformly away from the lowest latitudes. Relative humidity, as we have learned, is determined by both water vapor content and temperature. In the mean the poleward decrease of temperature to about 40° is less rapid than the decrease of water vapor over this span. Thus, from 30° to 40° N and from 20° to 30° S, while the capacity for vapor remains fairly high because of high temperatures, the moderate amounts of water vapor result in low values of relative humidity. Poleward of 40° the opposite relationship applies: Temperature decreases more rapidly, water vapor less so, and relative humidity increases.

If, as in Figure 5.5, we assume that there is no net accumulation or loss of water in streams and groundwater, then evaporation and runoff must equal precipitation on the continents. In Table 5.1 the water balance of the continents is shown. Clearly, South America is the wettest continent, and Australia the driest. Compare this table with Table 2.2, and notice the similarity between $E$ (latent heat energy) and evaporation in Table 5.1.

**TABLE 5.1**
Water balance of the continents (in millimeters per year).

| Continent | Precipitation | Evaporation | Runoff |
|---|---|---|---|
| Europe | 640 | 390 | 250 |
| Asia | 600 | 310 | 290 |
| North America | 660 | 320 | 340 |
| South America | 1630 | 700 | 930 |
| Africa | 690 | 430 | 260 |
| Australia | 470 | 420 | 50 |

SOURCE: Reprinted from *Physical Climatology* by William D. Sellers by permission of The University of Chicago Press. Copyright © 1965.

Absolute humidity, and its spatial and temporal variations, is a key to understanding the occurrence of related weather elements such as precipitation and cloudiness. Figures 5.8 and 5.9 show mean vapor pressure in January and July for the United States and southern Canada. This horizontal variation of water vapor is a reflection both of temperature, since warmer air is capable of holding more water vapor and the higher temperatures promote increased evaporation, and a tendency in North America for inflow (ocean to continent) in summer and outflow in winter.

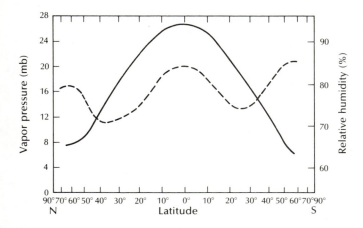

**FIGURE 5.7**
The latitudinal variation of absolute humidity, as water vapor pressure (solid line) and as relative humidity (dashed line).

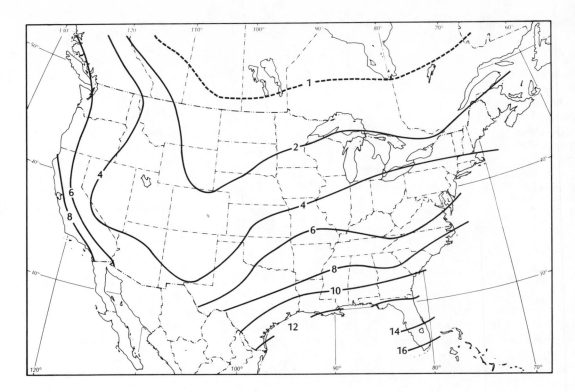

**FIGURE 5.8**
Mean vapor pressure (mb) in January.

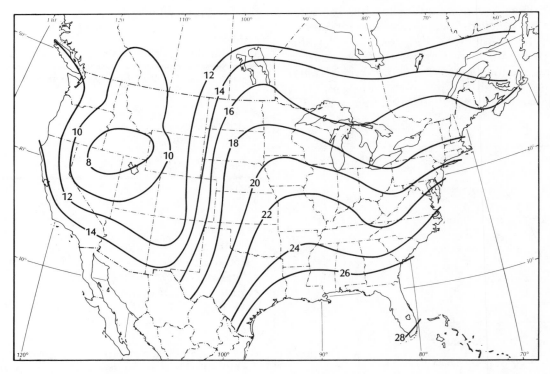

Figure 5.9
Mean vapor pressure (mb) in July.

The influence of elevation also is apparent in these two maps. The proportion of the air's total water vapor content in middle latitudes is 46 percent for the lowest 1 km and 95 percent for 5 km. Much of western North America is dry for this reason and because the penetration of humid air from the Pacific is blocked by mountains. The effect of elevation also is apparent, to a lesser degree, in the Appalachians, especially in summer.

Mean atmospheric flow patterns can be approximated by Figures 5.8 and 5.9. In July, the strong ridge of high water vapor extending from the lower Mississippi valley through the central and northern plains and into southcentral Canada results from the advection of water vapor northward on the western side of the North Atlantic subtropical high (see Figure 4.13). Water vapor from the Gulf of Mexico and the Atlantic thus extends far into the interior of the continent. In winter, when the flow is more zonal, the pattern tends also to be zonal except for the west coast and the mountainous west. Compare Figure 5.9 with Figure 4.21, in which the water vapor content is indicated by dewpoint, and notice the similarity of patterns.

## 5.6
### EVAPORATION

At the air-water or air–moist land interface, there is a constant flux of water both ways in the vertical. Water molecules move upward from the water surface and evaporate and move downward from the air and condense. What we have called evaporation and condensation to this point is what results from the balance of these two processes. If the former process prevails, there is (net) evaporation; if the latter prevails, there is (net) condensation.

For evaporation to occur two conditions must be met. Evaporation is a heat-requiring (and thus cooling) process, so there must be a source of energy. Both air and water supply this energy, but because of water's higher heat capacity it supplies virtually all of it. Thus the water surface is cooled during evaporation; but, because of water's high heat capacity, this temperature drop is minimal.

The second requirement is a gradient of water vapor above the water surface; that is, the water vapor concentration must decrease with distance from the water. We can approximate this gradient by taking the difference between the vapor pressure of the air in direct contact with the water—and here we assume that the air is saturated at the water surface temperature—and the vapor pressure in the air at some specified distance above, or in the so-called free air. In symbols, if $e_{sw}$ is the water vapor pressure exerted by the saturated air in contact with the water, and $e_{aa}$ is the actual vapor pressure in the free air, evaporation will occur at a rate proportional to $(e_{sw} - e_{aa})$.

Evaporation is a negative feedback (self-defeating) process. That is, it proceeds in such a way as to impede progressively the factors that initiate it. Consider a closed container half filled with water. If we make the air above the water very dry, the quantity $(e_{sw} - e_{aa})$ is initially large. As evaporation adds water vapor to the air, this quantity decreases (since $e_{aa}$ increases), and thus the evaporation rate decreases also. Finally the gradient vanishes, and as many water molecules are leaving the surface and evaporating as are leaving the air and condensing.

There is an important difference, however, between this closed container and evaporation from the earth's surface into the air above it. Since the atmosphere is almost always in motion, the free air moistened by the evaporative flux along the gradient can be replaced by air with a lower $e_{aa}$, either by advection or convection (horizontal or vertical import of drier air) or by both. The evaporation thus becomes more or less continuous. In the absence of any change in water temperature, $e_{sw}$ stays the same, and $e_{aa}$ is maintained at some lower value by the continuous import of relatively dry air.

Mathematical expressions for evaporation rate based on the $(e_{sw} - e_{aa})$ difference therefore require some measure of air movement, or wind. One commonly used formula is

$$E = Kw(e_{sw} - e_{aa})$$

where $E$ is the flux density (or, simply, flux) of evaporation (how much water vapor passes through a given area in a given time); $w$ is wind speed; and $K$ is a constant whose units make the expression dimensionally correct and which, presumably, incorporates factors unaccounted for.

Another way of calculating evaporation involves the energy budget. We know that the energy available through net radiation ($R$) is used to promote an upward transfer of both sensible and latent heat by convection and/or advection, and to conduct heat downward into the soil. From Equation (2–4), we get

$$R = H + E + G$$

assuming positive values for all four fluxes. Recall from Chapter 2 that if there is water to be evaporated, virtually all of $R$ will be used to do it. As a first approximation, then, assuming $H$ and $G$ are negligible, we get

$$E = R$$

If this assumption is not reasonable, as when the quantity of water available to be evaporated is not unlimited, we may write

$$E = \frac{R - G}{1 + H/E}$$

Notice that when $H$ and $G$ are very small this expression is nearly identical to the one above. When derived from the energy balance equation, $E$ has dimensions of energy flux density (e.g., cal cm$^{-2}$ min$^{-1}$ or watts m$^{-2}$), but these are easily converted to inches or centimeters.

The vapor gradient above a water or wet land surface and the quantities in the energy balance equation are not easily obtained and require rather elaborate instrumentation. For this reason approximations to the evaporative loss have been made by using the more or less routinely measured elements such as air temperature, solar radiation, and absolute or relative humidity. These elements do not directly influence evaporation, but rather are intermediate influences that affect the water vapor gradient, or the proportion of energy used by the processes represented by the right-hand side of Equation (2–4). For example, the absorption of solar radiation at a water surface raises the temperature of that surface, providing more energy for evaporation and increasing both $e_{sw}$ and the vapor gradient. This process promotes a more rapid upward flux of water vapor. Relative humidity, or some other measure of the capacity of the air to absorb or transmit water vapor, also is associated with the rate of evaporation because of the effects on evaporation of the import and mixing of dry air.

Mixing of the air above the evaporating surface is probably the most difficult aspect of this process to express quantitatively. **Eddy diffusivity,** the term used to characterize this mixing, essentially provides a numerical descriptor of the amount of convection, including turbulence, that occurs in the lowest few meters (see the general definition of convection in Section 2.3.2). The eddy diffusivity increases with height above the surface and varies also with wind speed, surface roughness, and the extent of surface heating.

To this point in our discussion of evaporation, we have focused on ways to calculate the flux from relatively restricted areas such as ponds, lakes, or even pans, or from saturated soil. Of equal, if not greater, climatological significance are approximations of the total surface water loss from large areas of the earth's surface. Over homogeneous surfaces such as the oceans there is an endless supply of water to be evaporated. Our estimates of evaporative loss in these areas are reasonably correct and representative (for example, see Figures 4.44 and 4.46). But over land the water available to be evaporated varies enormously from time to time and place to place. There is obviously a great difference between evaporation from a nearly dry soil and that from the same soil after a heavy rain.

In addition, water loss from land areas includes that from **transpiration,** the process by which water in plants is transferred as water vapor to the atmosphere, or, more simply put, evaporation from plants. Transpiration varies spatially according to the extent of plant cover, and temporally (in this case, seasonally) with the stage of plant growth. An acre of corn transpires 1,226,460 liters (324,000 gal) during a growing season—the equivalent of 30 cm (12 in.) of precipitation. If the corn field is in central Iowa, it transpires about two thirds of the water that would be evaporated from an acre-sized lake during the same period. Deciduous forests transpire only about 4 percent of total annual transpiration during the winter months, showing that seasonal variations also occur.

In view of such large space and time variations in evaporation and transpiration, or **evapotranspiration (ET)**, it can be very misleading to depict average annual actual water loss from land areas. Because of the extreme variability in these dimensions, an average annual figure would have less meaning than monthly means of atmospheric pressure (see Section 4.5). However, in many agricultural and hydrologic applications *the actual water loss is not as significant as the capacity of the atmosphere to evaporate and transpire water.* Remember that actual water loss from a land surface depends on two factors: the extent, or quantity, of water at or near the surface available to be evapotranspired, and the atmosphere's capacity for evapotranspiration. It is possible to standardize the former and thus arrive at a measure of ET which depends almost entirely on this capacity.

### 5.6.1 MEASUREMENT

Measuring evaporation is done in two ways. First, uniform evaporation surfaces, such as water pans (Figure 5.10), can be exposed to the elements and the water level measured. Lakes also may be similarly monitored, and evaporation measured, but this is a much more difficult undertaking. Figure 5.11 shows mean annual evaporation from pans for the United States east of the Rockies. Because of large elevation differences within short distances in the West, the pattern there is very complex and is not shown. For an area of this size associations can be made between

## SEC. 5.6 | EVAPORATION

**FIGURE 5.10**
An evaporation pan. The three-cup anemometer mounted on the platform measures wind speed, which is an important influence on evaporation. In the background is a rain gage. (Photo by Dennis M. Driscoll.)

evaporation, temperature, and air humidity. Thus, the Great Lakes region is relatively cool and fairly humid, while western Texas is both warm (and thus able to supply more energy for evaporation) and quite dry.

There is a general relationship between size of the evaporating surface, the relative humidity of the overlying air, and the rate of evaporation from that surface (Figure 5.12). At high relative humidities evaporation is independent of the size of the surface from which moisture is lost, but at low values it tends to increase as this size decreases.

### 5.6.2 POTENTIAL EVAPOTRANSPIRATION

The second way to approximate the evapotranspirating capacity of the atmosphere is to express it in terms of the elements that govern it. To simplify, we assume that for a given land area the soil and vegetation are saturated; that is, there is an unlimited water supply. There is therefore a potential for evapotranspiration that depends only on the capacity of the atmosphere to remove water from the surface, which is called **potential evapotranspiration (PET)**.

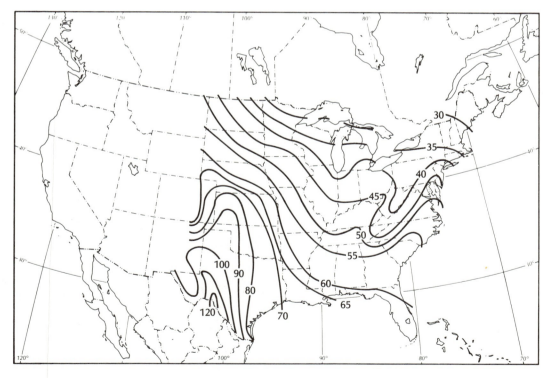

**FIGURE 5.11**
Mean annual pan evaporation (inches). (From *Climatic Atlas of the United States*, Environmental Data Service, Environmental Science Services Administration, U.S. Department of Commerce, 1968.)

**FIGURE 5.12**
Schematic relation between evaporation rates under different relative humidities and with varying surface areas. (After J. Mather, *Climatology: Fundamentals and Applications*. New York: McGraw-Hill, 1974.)

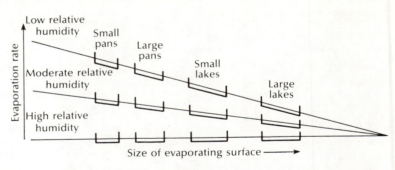

In application PET is used to approximate the atmosphere's capacity for evapotranspiration over large areas of the earth's surface, which is significant in evaluating climate's regional and temporal variations. C. W. Thornthwaite, a prominent climatologist, developed a measure of PET to use as a basis for classifying climate. When used for this purpose, estimates of PET based on temperature alone give fairly good results when compared with water loss from pans and lakes for particular regions. Thornthwaite's formula for total monthly PET is

$$PET = 1.6 \left[ \frac{10T}{\sum_{i=1}^{12} \left(\frac{T_i}{5}\right)^{1.514}} \right]^a$$

where $T$ is the mean temperature of the month of measurement in °C; $T_i$ are the 12 monthly means; and $a$ is a lengthy function of the quantity in the denominator. This formula requires adjustment for variations in day length (a simple function of latitude) and applies best to temperatures in the range of 0° to 26.5°C (32°–80°F). PET is easily computed from this formula because it uses only temperature.

Figure 5.13 shows Thornthwaite's map of annual PET for the eastern United States. Notice that the pattern is very similar to that of Figure 5.11. However, Thornthwaite's map underestimates water loss in arid areas, such as the Great Plains and western Texas. The highest value on his map of the 48 states is about 152 cm (60 in.) at the California-Arizona border (not shown), while mean annual lake evaporation there has been measured at 218 cm (86 in.). In much of the United States, the calculated seasonal pattern of PET has been shown to be very similar to that of observed evapotranspiration.

# 5.7
## WATER BUDGETS

In many applications of climatology, such as in agriculture, hydrology, and forestry, how much precipita-

**FIGURE 5.13**
Mean annual potential evapotranspiration. (Reprinted from C. W. Thornthwaite, "An Approach Toward a Rational Classification of Climate," *Geographical Review*, Vol. 38, 1948, with the permission of the American Geographical Society.)

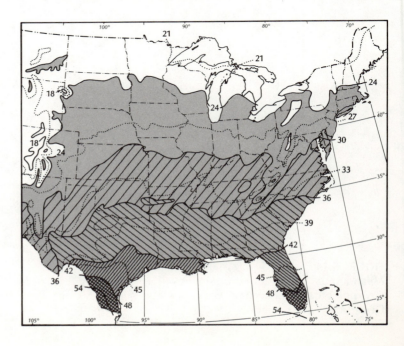

tion falls is not nearly as important as how much of it is available to the soil. In Section 5.4 we learned that precipitation may run off or be evaporated; in either case it is unavailable to the soil. Within the soil water may move downward to the groundwater table or upward to replace that lost from a dry topmost soil layer.

If we are concerned with only a specified depth of the soil, from the interface down, we can assume that vertical movement of water below the bottom of this layer is of no consequence. If it is possible to approximate the maximum amount of water that this layer holds (in most soils, about 10 to 15 cm), a budgeting procedure can be employed. Just as heat energy exchange at the interface in the energy balance equations can be assumed to equal zero when all factors are accounted for [see Equation (2–1)], it is possible to keep track over time of gains and losses of water within this soil layer, much as one would do with a bank account. Gains to this layer result from precipitation; losses, from evapotranspiration, which can be approximated by PET.

This budgeting procedure can be applied to climatic averages of the factors that influence soil water for individual areas, helping to differentiate place-to-place variations in what becomes the *water climate*. The same procedures can be applied continuously to recently measured factors—as opposed to long-term averages—to provide an up-to-date accounting of soil water. Both of these applications are described next.

Consider a very rainy climate, one in which precipitation is sufficient throughout the year to maintain a surplus of soil water. Even though large amounts of water are evapotranspired, mean monthly precipitation always exceeds PET, and all 12 months have a water surplus. This is illustrated by the example of Brevard, North Carolina, in Figure 5.14. At Bar Harbor, Maine, there also is a surplus of water in the period from November through April. But in May the increasing PET overtakes precipitation (*P*), and from then until September water is withdrawn from the soil (soil water use). The water storage in the soil is sufficient to compensate the excess of PET over *P*, and there is no period of water deficiency. From September through November *P* again exceeds PET, with the difference used to recharge the soil (soil water recharge). By November the soil has been completely recharged and thereafter there is a water surplus.

At Pullman, Washington, *P* is less than that at the previous two stations and PET is about the same. As at Bar Harbor there is a period—in this case, from December to April—when the soil has all the water it can hold, and then the reserve in the soil is used.

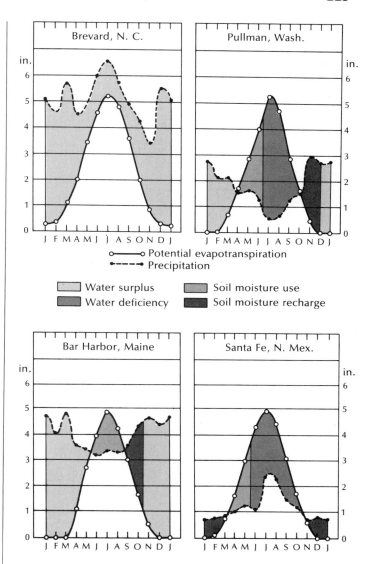

**FIGURE 5.14**
Annual variation of water budget factors at selected stations. (Adapted from C. W. Thornthwaite, "An Approach Toward a Rational Classification of Climate," *Geographical Review*, Vol. 38, 1948, with the permission of the American Geographical Society.)

Here, however, the cumulative effect of moderate PET and low *P* means that by late June soil water has been depleted, and from then until October there is a period of water deficiency. After this, when *P* minus PET is positive, the soil is first recharged so that by December it once again has all the water it can hold. Thereafter there is again a period of water surplus. At Santa Fe the sequence is much the same, except that *P* is not sufficient to produce a period of water surplus. At one time of the year (November to March) *P* exceeds PET. Although the soil is being recharged, there is insufficient water to reach the storage capacity; thus there is never a period of water surplus.

These budgeting procedures only approximate the soil water content, because the water budget is actually more complicated than we have indicated. Soils vary widely in their capacity for storing and transmitting water, depending largely on their physical properties (e.g., predominant particle size and organic content). Also, actual evaporation does not proceed at a rate independent of the amount of water, as is assumed in these simple water budgets. For water-saturated soils the loss will be reasonably well given by PET, but with less soil water the rate will be less. In general, then, this method overestimates the evaporative loss.

The **Palmer Drought Index (PDI)** is an example of a budgeting procedure designed to give current values of soil water. Antecedent conditions of precipitation, PET, and the stored water in two soil layers are incorporated into a formula that expresses the current soil water content by a number ranging from around +4.0 (very much wetter than usual) to −4.0 (extreme drought). This index has had wide application, especially to agriculture, and has the distinct advantage of defining drought as a significant reduction of available water below that required for the near-normal operation of a region's established economy. The U. S. National Weather Service routinely issues maps of the **Crop Moisture Index,** which is of the same form as the PDI but does not emphasize antecedent conditions. An example of the Crop Moisture Index is shown in Figure 5.15.

# 5.8
## CLOUDS AND PRECIPITATION

A **hydrometeor** is a product of condensation or sublimation of atmospheric water vapor, whether formed in the atmosphere or at the ground. Precipitation, a kind of hydrometeor, is distinguished from others because it falls from the atmosphere and reaches the ground. Other hydrometeors are cloud droplets, fog, and dew, while the forms of precipitation are drizzle, rain, snow, hail, and sleet. The amount of precipitation that falls in periods such as a month or year, or a long-term average of these quantities, is a principal distinguishing feature of climate. Traditionally, the two elements used to describe climate, and differentiate it over the earth's surface, have been monthly means of both temperature and precipitation.

The way clouds form, and how precipitation develops in them, is a fascinating—and yet not well understood—aspect of atmospheric science. Entire books have been devoted simply to the microphysical processes that lead to clouds and precipitation, or to the atmospheric motions, on practically all scales, that govern cloud formation and the growth of cloud droplets to raindrop size, and produce the ever-changing panorama of the sky. Here we restrict our discussion to the essentials of these processes, emphasizing those that help to differentiate space-time variations in climate. Details can be found in any standard text in meteorology.

### 5.8.1 CLOUD FORMATION AND VERTICAL MOTION

To understand the process which ends in precipitation, we need to consider first how and under what circumstances water vapor becomes cloud droplets. To form a cloud air must be brought to the point of saturation; as we have learned, this happens when the air has the maximum amount of water vapor it can hold at its temperature. Saturation is attained in two ways: by evaporating water into, or by cooling, the air. In either event the relative humidity is brought to 100 percent (there are exceptions to this, as we will see), and, ordinarily, condensation begins. In terms of changes in the ratio that expresses relative humidity, evaporation increases the numerator and cooling decreases the denominator; saturation is attained when these are equal (see the discussion of relative humidity in Section 5.3). Evaporating water into the air is fairly common, occurring when rain falls from the base of clouds and is evaporated into the drier air below them. Or, water may be evaporated from the earth's surface and then condense above it or at some distance away, forming fog, which is simply a low cloud.

The second way in which saturation is attained, however, is by far the more important. Before considering condensation further, then, we must examine the circumstances that promote the cooling of air. Air cools when it moves upward in the atmosphere—as individual air parcels, as entire layers, or as air masses. This upward movement is called *vertical motion.* Unless otherwise specified, this term refers only to rising air. Vertical motions are the result of three mechanisms: thermal convection, convergence, and mechanical uplift. These may occur singly, but more than one, and sometimes all three, may be present in a particular instance of cloud formation. We will now consider each of these separately.

**Thermal Convection**
*Convection* has been defined generally (Section 2.3.2) as the mass motions within air which result in trans-

## SEC. 5.8 | CLOUDS AND PRECIPITATION

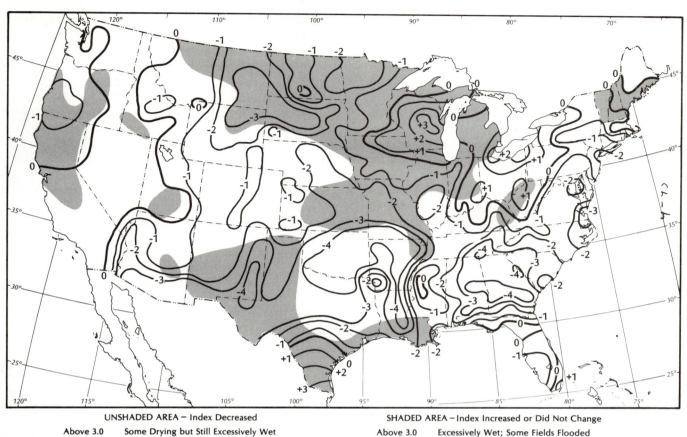

UNSHADED AREA – Index Decreased

| | |
|---|---|
| Above 3.0 | Some Drying but Still Excessively Wet |
| 2.0 to 3.0 | More Dry Weather Needed; Work Delayed |
| 1.0 to 2.0 | Favorable Except Still Too Wet in Spots |
| 0 to 1.0 | Favorable for Normal Growth and Field Work |
| 0 to -1.0 | Topsoil Moisture Short; Germination Slow |
| -1.0 to -2.0 | Abnormally Dry; Prospects Deteriorating |
| -2.0 to -3.0 | Too Dry; Yield Prospects Reduced |
| -3.0 to -4.0 | Potential Yields Severely Cut by Drought |
| Below -4.0 | Extremely Dry; Most Crops Ruined |

SHADED AREA – Index Increased or Did Not Change

| | |
|---|---|
| Above 3.0 | Excessively Wet; Some Fields Flooded |
| 2.0 to 3.0 | Too Wet; More Standing Water |
| 1.0 to 2.0 | Prospect Above Normal; Some Fields Too Wet |
| 0 to 1.0 | Moisture Adequate for Present Needs |
| 0 to -1.0 | Prospects Improved but Rain Still Needed |
| -1.0 to -2.0 | More Improvement but Still Too Dry |
| -2.0 to -3.0 | Drought Eased but Still Serious |
| -3.0 to -4.0 | Drought Continues; Rain Urgently Needed |
| Below -4.0 | Not Enough Rain; Still Extremely Dry |

**FIGURE 5.15**
Crop Moisture Index, August 9, 1980.

port and mixing of its properties. **Thermal convection** is more specific, because it refers to the vertical motions that originate when near-surface air is heated, and replaces the cooler, denser air above. This upward motion may be sustained, leading to cooling, saturation, condensation, and precipitation. To understand how thermal convection originates and why it occurs, we need to learn four new terms, or concepts: adiabatic processes, process rate and lapse rate comparisons, buoyancy, and instability.

In the turbulent motions of the atmosphere at and near the surface individual entities of air are constantly rising and falling as well as moving horizontally. These entities, or air *parcels*, are assumed to be insulated from ambient conditions (the conditions in the environment around them) so that there is no mixing of the parcel air with that of the environment and, more importantly, no heat exchange between the parcel and surrounding air. It is as if a barrier existed—like the skin of a balloon—between the parcel and the surrounding air. In this case density and temperature conditions within the parcel change only in response to its vertical displacement; such a process is called **adiabatic.** As the parcel rises, it moves from higher to lower pressure and expands. In expanding it cools (just as the air rushing out of a tire valve cools), and its density decreases. The opposite changes in temperature and density occur when the parcel is displaced downward.

This change of temperature in rising and falling parcels is constant with height in the atmosphere, as long as the parcel is not saturated, and is equal to 1C° per 100 m (5.5F° per 1000 ft). In other words, if a parcel moves downward 100 m it warms by a degree; if it moves upward through this distance it cools by the same amount. This temperature change, called the

**dry adiabatic process rate,** applies only to nonsaturated parcels. It must be stressed again that *adiabatic* means that parcel changes of temperature and density occur without heat transfer across the imaginary parcel boundary. For the moment we are concerned only with the dry adiabatic process rate.

Thus, the atmosphere's turbulent motions constantly displace air parcels up and down, and the parcels warm or cool at a constant rate with height as long as the parcel is not saturated. Although the pressures both inside and out of the parcel are equal, temperature and density need not be. To explain why this is so, we need to introduce two more terms: process rate and lapse rate comparisons, and buoyancy. Imagine a parcel initially at rest, so that its conditions of pressure, temperature, and density are the same as those of its environment. When displaced upward it will cool at the dry adiabatic process rate; at any level thereafter it can be colder (and thus more dense) or warmer (and thus less dense) than its environment. In the first case, because it is more dense, it will have *negative buoyancy* and will return to its original level. In the second case its buoyancy will be positive, and the parcel will continue to rise. This upward motion will be sustained as long as the parcel is warmer than the air around it. Recall that **buoyancy** means that an object of greater density than the fluid in which it is immersed will sink, and if it is less dense than the surrounding fluid, it will rise (this applies only in the presence of gravity). Thus, a parcel of air "nudged" upward by turbulent motion will continue to rise if its temperature is warmer, but will return to its initial level if colder, than the temperature of its environment. By similar reasoning, for downward displacement a parcel colder than its environment will continue to sink, whereas a warmer parcel will return to its original level.

A physical analogy may be helpful at this point. Imagine a marble placed at the bottom of a cereal bowl. Because the bowl curves upward away from the lowest point, after a slight nudge (the counterpart of turbulence) the marble will return to its starting point. Now turn the bowl upside down and place the marble at the top of it. A slight nudge now is enough to propel the marble down and off the bowl, that is, to sustain the motion.

The concept of **instability** is now easily understood. If, after displacement of an air parcel, its motion is sustained, the layer through which the displacement has taken place is *unstable.* If the parcel returns to its starting point, the layer is *stable.* When the parcel and ambient temperatures are equal, the layer is *neutral,* and the parcel will remain at the level to which it was moved by its initial displacement.

How do meteorologists determine if, after a parcel is displaced, it will be warmer or colder than the air around it and thus stable, unstable, or neutral? They compare the **process rate** with the **lapse rate** of temperature. It is very important that the difference between the two be understood. In Chapter 2 we learned that lapse rate is the measured change of temperature with height in the atmosphere (see Figure 3.13). Obviously, lapse rates can be very different from time to time and place to place. At times they are very steep (temperature decreases rapidly with height), and at other times temperature may increase with height (temperature inversion). A lapse rate in which the temperature does not change with height is isothermal. Lapse rates of temperature and humidity are determined from a *sounding* of the atmosphere over a location. This sounding is obtained by *radiosondes* or *rawinsondes* as they are carried aloft, usually by helium-filled balloons. These instruments contain radio transmitters which relay conditions of pressure, temperature, and humidity back to the location from which the launch was made.

The key to determinations of stability and instability lies in comparisons of process rate with lapse rate. Suppose that in the lowest 100 m the lapse rate is 1.5C° per 100 m; that is, temperature decreases by 1.5C° in this layer. A parcel is displaced upward and cools at the dry adiabatic process rate (1C° per 100 m). At any distance above the surface through this layer the parcel will have positive buoyancy and thus be unstable, because it will be warmer than the ambient air. Assuming the parcel and ambient air temperatures are 30°C at the surface, their respective temperatures (in °C) will be 29.90 and 29.85 at 10 m, 29.50 and 29.25 at 50 m, and 29.00 and 28.50 at the top of this layer. The higher the ascent, the greater is the difference in temperature. It is important to understand that a continued upward displacement of the parcel is possible because it is always warmer and less dense, and thus positively buoyant, with respect to the air around it.

Rather than make calculations as we have just done, meteorologists plot the temperature sounding on a **thermodynamic diagram,** also known variously as pseudoadiabatic charts, tephigrams, and skew-*t*/log-*p* diagrams. They can then infer stability conditions by comparing the slope of the measured lapse rate with that of the process rate (dry or moist) shown in the diagram. A much simplified thermodynamic diagram is shown in Figure 5.16. Pressure and height are on the ordinate, temperature on the abscissa. Thus, any point on this chart specifies these three factors. The solid lines are soundings (lapse rates of temperature), and the dashed lines are **dry adiabats,** which show the temperature change that takes place in a parcel of air

**FIGURE 5.16**
Stability, instability, and conditional instability in the lower atmosphere, as depicted on a thermodynamic diagram.

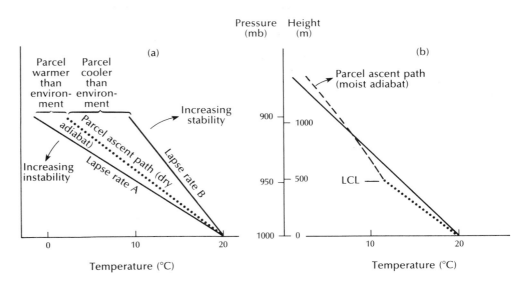

as it moves vertically (the dry adiabatic process rate). Notice that since the dry adiabatic process is constant with height, the lines are straight. For simplicity the lapse rates also are shown as straight lines. In part (a) a parcel is displaced upward from the surface (assumed to be at 1000 mb). At any height the parcel temperature, which is read vertically downward to the abscissa, is higher and lower than the ambient air temperature at the same height, as deduced from lapse rates A and B, respectively. At 900 mb, for example, the parcel is about 3C° warmer than the ambient air for lapse rate A, and 6C° colder than the ambient air for lapse rate B. The layers of air through which these lapse rates extend are thus unstable and stable, respectively.

We can now generalize these results. A layer of air is unstable when the decrease of temperature from bottom to top (i.e., the lapse rate) is more rapid than the decrease in temperature of an air parcel moving upward through the layer. Graphically, an air layer is unstable when the lapse rate is steeper than (rotated in a counterclockwise direction with respect to) the dry adiabatic process rate. A layer is stable when the lapse rate is less than the dry adiabatic process rate. If it happens that the lapse rate and dry adiabatic process rate are equal, a displaced parcel will be slowed by internal friction and will remain the same temperature as its surroundings. This is a neutral layer.

To determine the likelihood of sustained vertical motion through the mechanism of thermal convection, then, meteorologists have only to compare the lapse rate in an air layer with the dry adiabat that intersects the bottom of this layer. If the lapse rate is clockwise from the dry adiabat (that is, if there is a lower rate of temperature change with height or the slope is less steep), the layer is stable; if the lapse rate is counterclockwise (representing a higher rate and a steeper slope), the layer is unstable [Figure 5.16(a)]. Also, there are degrees of stability and instability. A layer of air in which the lapse rate exceeds the dry adiabatic process rate is called **superadiabatic** and is very unstable. An adiabatic layer (or atmosphere) is one in which the lapse rate equals the dry adiabatic process rate and is neutral with respect to stability. Proceeding farther clockwise, an isothermal layer is stable, and an inversion of temperature represents extreme stability. Both of these layers are **subadiabatic.**

The highest lapse rates, indicating unstable atmospheres, invariably occur next to the surface, since it is here that heat is added diabatically (the opposite of adiabatically) by the absorption of terrestrial radiation. For example, see the land-day lapse rate in Figure 3.14. There are times, however, when lapse rates can be steepened in higher layers, as when cold air is advected aloft. Temperature inversions, which of course suppress vertical motions, also may occur at upper levels. A subsidence inversion is shown in Figure 3.15.

One aspect of lapse rates, and its implication for stability, remains to be described. Until now we have considered a dry adiabatic process, one in which the parcel is not saturated. If the parcel cools to the point of saturation and continues rising and cooling, latent heat is liberated as water vapor condenses. The net effect of cooling by ascent and warming by condensation is a lower process rate than the dry adiabatic; that is, it is a less steep line on the thermodynamic diagram. Also, this rate is not constant; the decrease of temperature is slight at first, and then proceeds at a greater rate. Figure 5.16(b) shows the path of a parcel which is first displaced upward and cools at the dry rate. At the **lifting condensation level (LCL)** saturation is reached and the parcel cools moist adiabatically (less rapidly). In Figure 5.16(b) the moist adiabatic process rate is shown by a dashed line. The rate looks constant, but this is because it is shown through only a limited layer. Notice that in this figure the layer of air is stable with respect to a dry parcel, but unstable (parcel to the right of the lapse rate, hence warmer, with positive buoyancy) shortly after condensation begins. This is another stability condition, called **conditional instability;** that is, stability is conditional upon the parcel being saturated. The LCL (actually, the CCL, or cumulus condensation level) is the base of cumulus-type clouds; because moisture and temperature conditions are often reasonably homogeneous in the horizontal over great distances, we see why the bases of cumulus clouds are at about the same height.

In general the magnitude of turbulence decreases with distance from the surface, because it is at the surface that heat is added to the atmosphere and where the wind encounters obstacles such as topographic variations and buildings. We have seen that the vertical motions resulting from such turbulence, when coupled with lapse rates greater than the dry adiabatic (superadiabatic), give individual air parcels enough upward displacement to make them positively buoyant. This turbulence need not be extreme; in the case of lapse rate A in Figure 5.16 even the slightest "nudge" will propel the parcel upwards. Whether or not a cloud grows to the point that rain (or, rarely, snow, since such large lapse rates are not common in relatively cold temperatures) falls out of it, however, depends on other factors. Water vapor content of the air is the most important of these. Other conditions being equal, the more humid the air, the greater is the possibility of precipitation. Imagine that the parcel in Figure 5.16(b) becomes saturated (reaches the LCL) at a lower level than is shown, as it would if it were more humid. Its "free ride" upward then commences at a lower level, and the initial "nudge" by convection need not be as great to produce positive buoyancy. Furthermore, the liquid water content of the cloud so produced will be greater.

If the cloud exhausts its store of energy—the latent heat released during condensation—further upward development may stop. Determining the conditions that govern whether a cloud continues to grow or dissipates is the focus of cloud physics and atmospheric thermodynamics. For the most part, we do not know why some clouds grow sufficiently to produce precipitation while others do not, although generally the larger the cloud, the more likely it is to continue to grow to rain-producing size.

The space and time scales of vertical motions produced by atmospheric instability are quite different from those of the other mechanisms that produce rising air. Typical horizontal dimensions of a convective cloud are several tens of meters to a few kilometers, and such clouds are usually separated by at least several kilometers. Vertical motions range from a few to as much as 20 or 30 m s$^{-1}$, and there are accompanying downward motions, peripheral to the core of upward moving air, that tend to be more dispersed and therefore weaker. Typical durations of convective clouds are several tens of minutes to several hours.

With respect to geographical and seasonal variations, convective clouds produced by atmospheric instability are, in general, common throughout the year in the tropics—although in the wet-dry parts of the tropics, they are limited to the high-sun period—but are restricted to spring and summer in the middle and high latitudes. We would of course expect clouds to be most frequent in summer in climates that are at least somewhat humid, because this season exhibits not only the highest surface temperatures, but the highest absolute humidity as well. Spring tends to be a period of more frequent convective activity than fall because of the lag between surface and upper-air warming. Since mid-tropospheric temperatures take longer to increase in spring than do those at the surface, lapse rates are steepened, increasing the likelihood of instability.

### Convergence

In the broadest sense, convergence implies a coming together and divergence a spreading apart. Somewhat more rigorously, convergence occurs when the horizontal area occupied by a given mass of air decreases; divergence occurs when this horizontal area increases. Figure 4.7 shows that at the surface, in part because of the effects of friction, air is converging toward the low center and diverging away from the high. In such large-scale (synoptic-scale) motions, air is essentially incompressible, unlike the behavior of air parcels. For this reason surface convergence must lead to vertical motion. Convergence, therefore, is the second mechanism the atmosphere affords for cooling air by forcing it to rise.

This ascending motion is obvious around a low-pressure center, where air spirals into, and upward within, the cyclone. Thus the transient lows of the

### SEC. 5.8 | CLOUDS AND PRECIPITATION

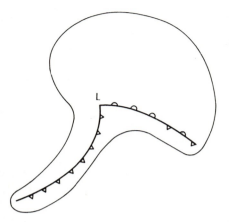

**FIGURE 5.17**
A mature wave cyclone and the area in which precipitation is likely.

middle and high latitudes are often associated with clouds and precipitation. Even better examples of marked convergence and uplift are tropical storms, including hurricanes. This association of surface lows with vertical motion can be generalized as follows: Any time the near-surface flow is cyclonic (counter-clockwise in the Northern Hemisphere, clockwise in the Southern Hemisphere), such as occurs in and around cyclones and troughs, there is accompanying vertical motion. We have learned that very often fronts fill troughs, and in this case the ascent of air may also be due to the mechanical lifting of warm air over cold. Within a trough, then, frictional effects tend to produce convergence and consequent vertical motion. Figure 5.17 shows a wave cyclone in the mature state (compare with Figure 4.20). The outer line represents an envelope that encloses areas in which clouds and precipitation are possible through the mechanism of convergence. It must be stressed that other mechanisms may be involved; frontal lifting is one, and thermal convection also is likely in relatively humid areas in which air is given an initial uplift by convergence or frontal lifting, or both. Also, keep in mind that this envelope encloses only those areas in which clouds and precipitation are possible. Whether or not they actually occur depends on other factors, chiefly moisture availability. In Figure 4.22, for example, although cloudiness is not indicated, precipitation is. Throughout this five-day sequence there is precipitation at or near the cold front only at 1 A.M. EST on March 23.

Convergence and vertical motion also can occur where frictional effects on the near-surface wind change abruptly in the horizontal. Recall that the frictional drag on the surface wind is least over flat surfaces and greatest over highly irregular terrain. In Figure 5.18 we see that the surface wind is faster, and more nearly parallel to the isobars, over water than over land. Convergence at and just landward of the coast is the result, which may be an important factor in the sea-land breeze phenomenon (discussed in Chapter 10). Notice also that coastal divergence would be promoted if land and water were reversed in Figure 5.18, or if pressure decreased rather than increased to the east (i.e., if the high were replaced by a low). Convergence and divergence also may develop when winds are parallel to the coastline.

The ascent of air due to convergence is relatively slow but quite extensive, and systems that produce convergence last for comparatively long periods. Typical ascent rates in mid-latitude transient storms are in the order of a few centimeters per second. But, as Figures 4.22 and 5.17 indicate, the area affected either by cloudiness or precipitation or both may be as large as several states, or even the greater part of a continent (compare the scales of atmospheric motion given in Table 4.1). Also, these transient storms may last from a day or so to as long as a week. We considered the geographical and seasonal variation of transient storms in Chapter 4 and will return to the subject in Chapter 12.

### Mechanical Uplift

It has been pointed out before that the motions of the earth's atmosphere are overwhelmingly horizontal. The movement of surface air away from the subtropi-

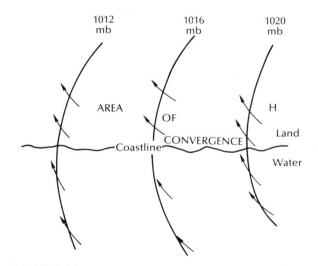

**FIGURE 5.18**
Convergence can occur at coastlines because frictional drag on the surface wind varies from water to land (Northern Hemisphere representation).

cal highs, both equatorward toward the ITCZ and poleward to the mid-latitudes; the surface circulation patterns accompanying cyclones and anticyclones; and the zonal, wavelike air patterns in the upper air all attest to this. At the same time it is the atmosphere's vertical motions, however small, that produce rising air and, in the presence of sufficient moisture, clouds and precipitation. We have seen that air may rise rapidly as discrete parcels, or slowly in fields of convergence. Air also rises when it is forced over barriers.

There are two barriers to wind flow. The first is density differences. The principle is a simple one: dense air sinks, less dense air rises. Density differences may result from differences of temperature or moisture, or both. The density of air increases both with decreasing temperature and decreasing moisture. In other words, the warmer and moister the air, the less dense it is. The influence of temperature is clearly seen, as in oil (less dense) floating on water. The effect of water vapor may not be as apparent. Because water vapor has a lower molecular weight than both nitrogen and oxygen, the density of a gas comprised of all three gases is less than that of dry air.

In the mid-latitudes, the area of transient disturbances, air masses of different characteristics are constantly being moved about. Especially when the flow is meridional rather than zonal, these air masses clash along frontal boundaries, where convergence tends to accentuate already existing differences. This clash is illustrated by Figure 4.20, particularly in its three-dimensional portions. Along the cold front cold air, usually from the west to north, undercuts the warm air. Usually, but not always, the cold air is dry and the warm air moist, both in an absolute sense. At the cold front the uplift is relatively abrupt and may be sufficient to instigate convective activity, which, with sufficient moisture, produces showers and thundershowers. Farther back, into the surface cold air, warm air may continue to rise over the cold air wedge, producing **overrunning.** At the warm front the uplift is more gradual, and the clouds formed by overrunning here are more likely to be of the stratus (layered) variety. Precipitation also may occur in the warm sector of the transient disturbance (the area clockwise from the warm to the cold fronts), but here the mechanism may be surface convergence accentuated by divergence aloft (see Figures 4.40 and 4.41), which is of course a feature of the entire transient disturbance.

The second barrier to horizontal wind flow, also of a mechanical nature, is topographic obstacles. The term *orography* has been used to describe this phenomenon, and precipitation so produced is called **orographic precipitation.** Clouds and precipitation are often the result of humid air being forced to rise over mountains or hills, and this process is of greatest significance when air has had a long trajectory over water.

Unlike precipitation produced by the other mechanisms we have discussed, orographic precipitation is site-specific. That is, particular locations have large mean annual amounts of precipitation because of their topographic circumstances. Among the most notable of these are the Cameroon highlands in Africa, the foothills of the Himalayas, the Coast Ranges and the Sierra Nevada in North America, and the island of Kauai in Hawaii. In areas where there is a considerable elevation difference, about 2500 m (8000 ft) or more, it is possible to specify a height interval on the windward side at which precipitation maximizes. Below this height the uplift of air is not sufficient to augment precipitation appreciably, while above it the reduced temperatures decrease the air's capacity to hold moisture, and the air may already have been depleted of it. The height of maximum precipitation is higher in mid-latitudes than in the tropics. This appears to be due to the concentration of moisture in a relatively shallow layer in the tropics.

A corollary of increases in precipitation on windward sides of orographic barriers is the decrease on their leeward sides. This is called the **rain shadow** effect. Areas that might otherwise be at least moderately rainy are dry and may even be deserts in some cases. Examples include the Patagonian Desert in southern Argentina (in the lee of the southern Andes); much of the interior of Asia, especially Mongolia and Tibet; and the intermontane area (roughly between the Sierra Nevada and the Rocky Mountains) of North America.

### *Layer Instability*

Instability, the tendency for air of lower density than the air around it to be accelerated upwards, occurs in conditions other than those we have already specified. Entire layers of air, rather than parcels, may be given an initial uplift by convergence or mechanical lifting, or both. Then, if this uplift has been sufficient to produce a superadiabatic lapse rate through it, the entire layer will become unstable.

When an air layer or mass is lifted, vertical stretching also occurs. In fact, this is a necessary condition of convergence. Since the density of air decreases upward, the vertical distances between adjacent isobaric layers (e.g., 1000 mb to 900 mb, 900 to 800, etc.) increase with height. In vertical stretching the upper part of a layer therefore ascends a greater distance than the lower part. The result is that the

## SEC. 5.8 | CLOUDS AND PRECIPITATION

**FIGURE 5.19**
Lifting an air mass or layer tends to steepen the lapse rate and make instability more likely. The reverse occurs when an air mass sinks.

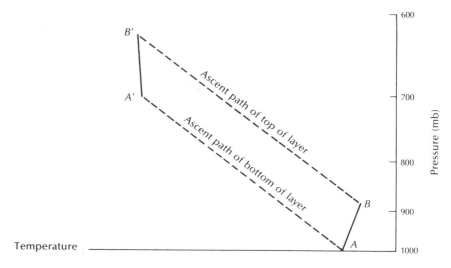

lapse rate of temperature through this layer is steepened, making instability more likely. In Figure 5.19 the layer of air between 1000 and 900 mb is lifted so that its pressure decreases by 300 mb. The initial lapse rate is $A - B$, an inversion. Because the top of this layer moves upward farther than the bottom (i.e., vertical stretching), it cools more, and the lapse rate after displacement ($A' - B'$) steepens. Notice that the dry adiabatic process rate is represented here by the ascent paths. The layer could have been made unstable (that is, the lapse rate would be steeper than the dry adiabatic process rate) if the layer had been lifted farther or if the initial lapse rate had been only slightly subadiabatic. In general, then, the steeper the initial lapse rate and the greater the lifting, the more likely is instability. The reverse of this process, such as occurs during the spreading out and sinking of an air layer or mass, tends to increase stability.

### Mixing

Clouds also form as the result of thorough mixing of the near-surface air. Mixing—the seemingly random motions of eddies which grow, mix with the surrounding air, and die out, only to be replaced by other eddies—tends to redistribute heat and water vapor, leaving the affected layer of air homogeneous with respect to these properties. The forcing mechanisms may be thermal or mechanical convection. If there is a strong vertical wind shear, wherein wind speed increases markedly with height, turbulent instead of laminar flow results, with pronounced vertical rather than horizontal components to the motion. This situation was considered in Section 4.3.4, where the effects of turbulence on friction were described.

Whatever forcing mechanism produces mixing, the result invariably is an altered stratification of heat and moisture. Interestingly, the lapse rate of temperature tends to become dry adiabatic (equal to the dry adiabatic process rate) or to approach a state of neutral stability. Absolute humidity, on the other hand, tends to become isohumic, or the same throughout the depth of the mixed layer. Since the usual conditions before mixing begins are a subadiabatic lapse rate and water vapor concentrated near the surface, the result of mixing is to warm, and thus decrease the water vapor content of, the surface air, and to promote the opposite changes at the top of the mixed layer. Thus, provided there is thorough mixing and sufficient water vapor, the relative humidity at the top will reach 100 percent and clouds will form. This is illustrated in Figure 5.20.

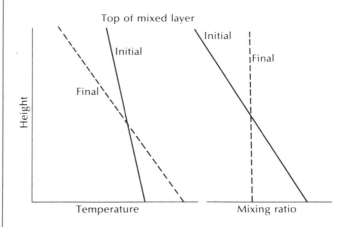

**FIGURE 5.20**
The effect of mixing on the vertical distribution of temperature and absolute humidity (as mixing ratio). The final lapse rate will be dry adiabatic, and the relative humidity at the top of the mixed layer may attain 100 percent, so that clouds form.

**FIGURE 5.21**
A very generalized vertical arrangement of cloud types. (From *Atmosphere and Weather Charts*, published by A. J. Nystrom and Company. Used by permission.)

A fairly typical daily sequence of events in warm, humid climates can now be detailed. The sun rises in a cloudless sky and begins heating the surface. Thermal convection ensues and, especially in the presence of horizontal winds and topographic or other irregularities, mixing results. Absolute humidity decreases at the surface and increases at the upper limit of mixing, at about 600 to 900 m (2000 to 3000 ft). At the same time the lapse rate, which characteristically is somewhere between isothermal and adiabatic (equal to the dry adiabatic process rate), becomes adiabatic. At the top of the mixed layer, clouds form when the combination of decreasing temperature and increasing absolute humidity produces a relative humidity of 100 percent. The presence of this relatively thin stratus layer decreases the surface heating, and thus the principal mechanism for producing mixing is curtailed—another example of negative feedback (see Section 5.6).

Although the cloud layer is highly reflective of solar radiation, enough is absorbed so that by about mid-morning this layer "burns off," and the surface is warmed once again. Now, since the lapse rate is adiabatic or nearly so, only a small increment of heat added to the surface air is required to produce a superadiabatic lapse rate. In this situation air parcels, given an initial displacement, will be unstable, and thermal convection results. By noon or early afternoon scattered cumulus clouds develop. Since moisture and temperature conditions are uniform for great distances in the horizontal, the base of these clouds will be at the same level—the lifting (or cumulus) condensation level. Solar radiation absorption at the surface will again be decreased, but since cumuli are typically widely scattered, heating will be sufficient to sustain growth and the clouds will continue to increase in number and size. By mid- to late afternoon, provided there is enough water vapor, the cumuli will merge as cumulus congestus, and perhaps even develop vertically to become cumulonimbus (Figure 5.21). Showers and thundershowers then result, and these may continue into early evening.

As the sun approaches the horizon, its heating effect is greatly reduced. Also, the other energy source—the liberation of latent heat during condensation—is curtailed as the available water, first transported from the surface both by evaporation and mixing, is condensed and returns to the surface as precipitation. The near-surface air temperature, which peaked in mid-afternoon, continues to fall until the sun comes up again the next morning. During the entire daily cycle water has served as a medium for the transfer of heat from the surface to the atmosphere, and of course sensible heat has been transferred upward as well. We do not mean to imply that every detail in the sequence of events is an everyday occurrence in warm, humid climates, but the sequence is fairly common and illustrates both the means by which heat is removed from the surface and the manner in which clouds and precipitation often form.

### 5.8.2 CONDENSATION

When we discussed the mechanisms that cause air to rise and thus to cool, we assumed that condensation would begin when the relative humidity increased to 100 percent, and that clouds would form and subsequently precipitate. The details of the condensation and precipitation processes now must be considered. You may already have some questions: What does water vapor condense on? How do the drops that form a cloud become raindrops or snow? What determines at what point these hydrometeors fall out of the cloud as rain or snow?

When the layer of air in contact with the earth's surface cools to the dewpoint, condensation begins. There is obviously an unlimited number of surfaces upon which condensation forms, although it begins on the coldest surfaces because they face upward, and thus lose heat rapidly by radiation to the relatively cold atmosphere. (Recall that dew first forms on the roof rather than on the sides of an automobile.) It might appear that there are no such surfaces in the atmosphere, but this is not the case. In even the "cleanest" air there are suspended solid particles, and to varying degrees these act as **condensation nuclei,** or objects upon which condensation occurs. These solid particles come from many sources, such as blowing soil, volcanoes, the oceans (sea spray), and even smokestacks, and have varying affinities for condensing water vapor. For **hygroscopic,** or water-attracting, particles, condensation begins at relative humidities considerably less than 100 percent. Salt particles are an example of hygroscopic particles. Other particles are **hydrophobic** (water-repelling), and air must be supersaturated (relative humidity slightly over 100 percent) for condensation to begin. Representative sizes of condensation nuclei are shown in Figure 5.22.

We see then that condensation in the atmosphere is not as straightforward as we might have expected. In the absence of hygroscopic particles condensation begins at relative humidities of 100 percent or a little greater, while in their presence the process may begin in air with a relative humidity as low as 80 percent. To complicate matters further, condensation on a curved surface is retarded relative to the rate at which it would occur on a flat surface (other conditions being

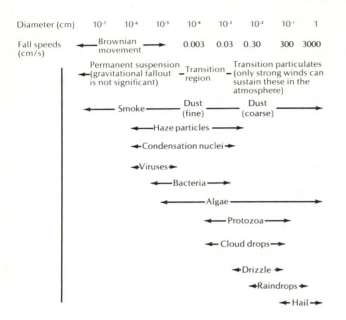

**FIGURE 5.22**
Typical sizes of particles and forms of precipitation in the atmosphere. (From *Patterns and Perspectives in Environmental Science,* National Science Foundation, 1973.)

the same). These two effects—the first of which enhances the cloud-forming process, while the latter retards it—are called the **solute effect** and the **curvature effect,** respectively. Both are significant chiefly at the beginning of condensation, when a thin film of water forms around a condensation nucleus. As the cloud drop grows by continued condensation or other means, the liquid becomes more dilute and the solute effect is weakened. Simultaneously, the curvature of the drop decreases as it grows, and the inhibiting effect is reduced.

At temperatures below freezing, condensation occurs as just described. That is, it is not only possible but very likely that the condensate will still be in liquid form. In the atmosphere liquid water can exist at temperatures down to about -40°C (−40°F); in this state it is referred to as **supercooled water,** and the drops are called **supercooled drops.** However, water vapor molecules may be converted directly to ice crystals by sublimation (see Figure 5.1). This process requires sublimation nuclei, which are not as numerous as condensation nuclei in the atmosphere. Once supercooled drops form, they will freeze rapidly if **freezing nuclei** are present. Freezing nuclei, more abundant than sublimation nuclei, may initiate freezing of a supercooled drop by contact with the drop or because the freezing nucleus is within the supercooled drop.

In a cloud with temperatures below freezing, then, there will be perhaps a thousand to a million more supercooled water drops than ice crystals. This has important implications for the growth of cloud constituents to a size such that they fall from the cloud as precipitation, and also for the artificial enhancement of precipitation ("cloud seeding").

We now have a cloud. The kind of cloud produced by the processes of condensation or sublimation, or both, depends on the manner of formation. In general, the mechanisms of convergence and mechanical uplift tend to produce layered, or **stratiform,** clouds; thermal convection and instability produce vertically developed, or **cumuliform,** clouds. Another way to distinguish stratiform clouds is according to the height at which they occur (see Figure 5.21).

Cloud drops are of course not raindrops. Typically, the former have sizes of $10^{-3}$ cm (one one-thousandth of a centimeter), while raindrops are two orders of magnitude or so larger (about 100 times larger, or $10^{-1}$ cm). Because the volume of a sphere is proportional to the cube of its radius, on the average there is a million times more water in a raindrop than in a cloud drop. Another important difference between cloud and rain drops is how rapidly they fall. The speed with which any solid or liquid particle falls in the atmosphere depends on its size; in general, the larger the particle, the faster it falls (see Figure 5.22). Thus, cloud drops fall at speeds relative to the air around them, but very slowly; a few tenths of to one centimeter per second is typical. Raindrops fall faster, about 200 to 1000 cm s$^{-1}$. Still, we have yet to explain how cloud drops grow to raindrop size so that the latter's terminal velocities are large enough to result in precipitation.

### 5.8.3 THE FORMATION OF PRECIPITATION

Atmospheric scientists have not been able to derive a satisfactory theory to explain fully how cloud drops grow to raindrop size. It is clear, though, that a continuation of the condensation process is not sufficient because the growth rate possible by this process is much too slow.

To describe our observations, we will categorize the precipitation-forming processes as either "warm" or "cold." In the tropics, where ambient temperatures in the cloud are always above freezing, it appears that **collision** and **coalescence** are the processes by which cloud drops grow to raindrop size. That is, in the turbulent motions inside the cloud there will be a number of collisions, causing the drops to coalesce, or merge. Coalescence seems to be favored when a larger drop, falling with a higher terminal velocity, overtakes and captures a smaller drop. It also appears, from laboratory experiments rather than by observation of nature, that drops may carry either positive or negative electrical charges. The fact that opposite charges attract therefore enhances coalescence.

In "cold" clouds—those that occur in the subtropics and mid-latitudes even in summer, where at least a small part of the cloud is above the freezing level—another mechanism comes into play. We have seen that in such a cloud there are both supercooled cloud drops and ice crystals. This is an unstable situation, however, because the two are unlikely to coexist for any length of time. Coexistence is not likely because the saturation vapor pressure over ice is less than that over water. This means that if the relative humidity is 100 percent with respect to water, the ambient air is supersaturated with respect to ice crystals; conversely, if the relative humidity is 100 percent with respect to ice crystals, there is a saturation deficit with respect to water. In a cloud with both forms the relative humidity is somewhere between these extremes, that is, supersaturated for ice crystals and subsaturated for water. The result, as should be anticipated, is that the water evaporates and sublimes on the ice crystals, and the latter grow at the expense of the former. This is called the **Bergeron-Findeisen process,** named for the scientists who first suggested that it plays an important role in cloud physics. In a continuous sequence of events, then, both supercooled cloud drops and ice crystals form in a cloud with temperatures below freezing. Because of the feedback processes involved, the ambient vapor pressure is kept at a value intermediate with respect to the saturation vapor pressure of these two forms, and the water evaporates and sublimes on the relatively very few ice crystals. Therefore, the crystals grow and fall faster than the drops. This three-phase process appears to be most effective at temperatures below about $-10°$ to $-15°C$ ($14°$ to $5°F$). The ice crystals aggregate, by collision and coalescence, and become snowflakes. If temperatures within and below the cloud remain below freezing, the result is snow; if not, the flakes melt and rain results. Thus, even in mid-latitudes in summer, it seems likely that rain begins as solid precipitation in clouds whose upper portions have below-freezing temperatures.

There is one drawback to the efficiency of the precipitation-producing process in cold clouds. Because of the scarcity of sublimation nuclei and other factors, there is a surplus of supercooled drops and a deficit of ice crystals. This suggests that if we could inject ice crystals into a supercooled cloud, we could accelerate the conversion of water to ice. Early attempts at **cloud seeding,** or *precipitation enhancement*, dating from the mid-1940s used dry ice, which was scattered into the cloud from above it. More recently silver iodide, a crystal with a structure very similar to that of ice, has been used. This seeding agent is injected into the cloud by aircraft flying just above the freezing level or is dropped into the cloud from above. Ground-based rockets also have been used.

### 5.8.4 FORMS OF PRECIPITATION

The manner of cloud formation influences the type of cloud produced and, if temperature is considered as well, the kind of precipitation that falls from it. In general, the thicker the cloud, the greater is its water content, both as liquid and vapor, and the greater the precipitation. **Rain** is the term applied to all liquid forms except drops of 0.5 mm or less, which are **drizzle.** Drizzle falls relatively slowly (see Figure 5.22) from thin stratus as a result of the "warm" cloud process. The upper limit to raindrop size is about 5 mm; if drops grow above this size, they tend to break up into smaller drops before reaching the ground. Steady, continuous rain is associated with stratus clouds formed as the result of convergence or mechanical uplift. **Rain showers** indicate rapid changes in intensity (amount of precipitation falling in a given time) and are usually the result of convective activity.

Solid forms of precipitation include **snow,** which can be individual ice crystals, or their aggregate; **snow flakes;** and **snow** and **ice pellets.** For solid precipitation to form, temperatures both within the cloud and between the base of the cloud and the ground must be below freezing, although the near-surface air may be slightly above freezing. Included in ice pellets is **sleet,** which occurs when raindrops freeze as they fall. Occasionally supercooled raindrops freeze as they strike the surface, resulting in **freezing rain. Hail** also is a solid form of precipitation, but unlike the others it is a consequence of vigorous convection. Hail forms when ice pellets remain in parts of a thunderstorm with supercooled water long enough to grow by collision with drops, which freeze on contact. To grow to an appreciable size, such stones must remain above the freezing level for several minutes. This is possible only in the very strong updrafts found in thunderstorms.

### 5.8.5 CHARACTERISTICS OF PRECIPITATION

In describing the climate of a location or area, the most frequently cited precipitation characteristic is the average amount accumulated during specified time periods, usually a month or a year. In the following chapters the various climates will be compared and contrasted in terms of the total mean annual precipitation and the variations throughout the year in average monthly amounts. The former characteristic is used to distinguish rainy, intermediate, and dry climates, while the latter indicates the seasonal regime of precipitation (e.g., summer wet–winter dry, winter wet–summer dry, or year-round precipitation). Global patterns of mean annual precipitation and precipitation regimes are considered in Chapter 7.

Another characteristic of precipitation, especially important in hydrologic applications, is intensity. Record intensities for short periods, such as five minutes or an hour, are of course greater than those for longer time intervals. This is because thunderstorm rainfall, the result of extremely well-developed convective activity, can produce copious amounts, but only over short periods. Intensity of precipitation is considered again in Section 7.3.2, and precipitation variability in Section 7.2.2.

The final aspect of precipitation to be described here is the **probability of occurrence** of a specified amount, or more, during a given time interval (Figure 5.23). This is of great concern to both agriculturists and hydrologists. Associated with such probabilities is **return period,** the reciprocal of the probability. Thus, if the probability of receiving 40 or more in. during a year for the northeastern Panhandle of Texas is 1 percent, the return period is 100 years; that is, this event can be expected to occur, on the average, once in 100 years.

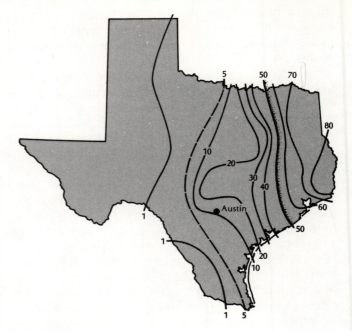

**FIGURE 5.23**
Probability (in percent) of receiving 40.0 inches or more annual precipitation in Texas. Data are from 1931 to 1960.

**SUGGESTED READING**

ANTHES, R. A., et al. *The Atmosphere*. 3rd ed. Columbus, Ohio: Charles E. Merrill, 1981.

BUDYKO, M. I. *Climate and Life*. New York: Academic Press, 1974.

FLOHN, H. *Climate and Weather*. New York: McGraw-Hill Book Company, 1969.

KRAUS, E. B. *Atmosphere-Ocean Interaction*. New York: Oxford University Press, 1972.

LEOPOLD, L. B. *Water: A Primer*. San Francisco: W. H. Freeman, 1974.

MATHER, J. R. *Climatology: Fundamentals and Applications*. New York: McGraw-Hill Book Company, 1974.

# 6
# ANALYTICAL METHODS IN CLIMATOLOGY

**INTRODUCTION**
**DEFINITIONS**
**DESCRIPTIVE STATISTICS**
Numerical Descriptors
Graphical Descriptors
**PROBABILITY**
The Binomial Distribution
The Normal Distribution
The Poisson Distribution
Other Theoretical Distributions
**SAMPLING AND TESTS OF STATISTICAL SIGNIFICANCE**
Nonparametric Tests
Parametric Tests
**REGRESSION AND CORRELATION**
Bivariate Linear Regression and Correlation
Multivariate Linear Regression and Correlation
**TIME SERIES ANALYSIS**

## 6.1
### INTRODUCTION

Climatology is a science of numbers. No other branch of the earth sciences, and perhaps even of the physical sciences, deals with numbers and their interpretation as much as climatology does. The storage facilities of the various national weather and climate services almost bulge with weather data accumulated since the advent of systematic, instrumented weather observations. Such data are available in a variety of forms, from handwritten copies, printed tabulations, and narrative climate summaries, to punched cards, microfilm, and magnetic tape.

Climatologists are responsible for extracting from these numbers essential information both to describe and analyze climate. To this end the science of statistics has proven to be an indispensable adjunct to the study of climate in general and to applications of climatology to other fields in particular.

Broadly speaking, the science of statistics is applied to climatic data in two ways. First, it can be used to condense and summarize observations and assign numerical descriptors (the descriptive approach). For example, it is common practice to describe a location's climate in terms of the arithmetic means of precipitation and temperature, where these means are calculated from a long period of record. This method is frequently employed in this book. Or, to go a step further, we can describe the variability of monthly values, or how much variation there has been in all the values over the period for which the means have been calculated. The descriptor used in this approach is the standard deviation (see Section 6.3.1), if a single number will suffice, or a graph of the frequency distribution of individual monthly values.

The second way in which the science of statistics can be applied to climatology is through the use of **statistical inference.** Climatological data sets can be regarded as samples drawn from a much larger, hypothetical set of numbers called a population (defined in Section 6.2). From this perspective it is possible to determine the statistical significance of climatological descriptors. The central question considered is, Could such a result have been obtained purely by chance? If it could have, we say the result is not statistically significant; if it could not, it is statistically significant. For example, at a certain station precipitation amounts have been increasing over the last several years. We ask, Is the mean precipitation for these years statistically significantly different from the mean for the prior years? In this case we can establish a mathematical-statistical model and test the claim, or hypothesis, that the increase is real in the statistical—as opposed to meteorological—sense. If, according to this model, the result could have occurred by chance, the result is not statistically significant, and vice versa. To use the statistician's terminology, we accept or reject the hypothesis that there has been a change. The methods of statistical inference, again involving the question of chance occurrence, also can be applied to determine if there is an association between two or among several variables. Are sunspots mathematically related to weather changes? is a question that might be answered this way.

In the actual decision-making process a hypothesis is accepted or rejected at a specified confidence, or probability, level. As a result, the investigator takes an $x$ percent chance of being wrong in his or her conclusion. Thus, probability theory is an essential part of statistics, forming the foundation on which statistical models are developed.

Two very important qualifications—indeed, warnings—must be made at this point. Statistical methods should be considered only as an adjunct, or supplementary means, of answering questions about weather and climate, not as a definitive solution in and of themselves. It is very easy for the novice to arrive at spurious conclusions using statistics if he or she is not familiar with the underlying physical mechanisms involved in meteorological phenomena. An investigator may find, for example, that one rain gage has a statistically significantly different catch than one that is spaced a few tens of meters distant. However, the second gage may be placed a meter or so higher than the first, and thus can be expected to measure less precipitation because higher wind speeds (expected with higher heights) blow more of the rain away from the gage.

It is important to know that statistical methods require certain assumptions to be fulfilled before tests are valid. In many cases meteorological data do not fulfill these tests. Unbiased, or random, samples must be used, and these are infrequently encountered in meteorology. Another disadvantage of the data with which the climatologist must work is that successive values of a weather element may not be unrelated, or independent, as is required by many statistical tests. Some tests require that data samples have a frequency distribution of a special kind (e.g., the normal distribution), and it is not often that climatologists enjoy the luxury of working with such an ideal case. This does not mean that the methods of statistics cannot be applied to weather data, but it does mean that interpretation of the results must be qualified in proportion to the extent to which these basic assumptions are not met.

In this chapter we will present some of the basic analytical tools used by climatologists, particularly those working in applications of climatology. We emphasize what the particular method accomplishes and how it has been used rather than review mathematical procedures. These are covered in any introductory level statistics text.

# 6.2
## DEFINITIONS

Like most other fields of study, statistics has its own terminology. It is important to know the following terms:

**Variable or variate:** The item—in our application, a meteorological element—being analyzed or under consideration.

**Observation:** A record of information relating to a variable.

**Set:** A group of observations possessing a common property.

**Frequency:** The number of occurrences of a specific observation; or, more often, the number occurring within a specified interval of the variable, all intervals of which comprise a *frequency distribution*.

**Continuous variable:** A variable that can assume any value between two given limits, given the precision of the measuring instrument. Examples are temperature and pressure.

**Discrete variable:** A variable that can assume only certain values between two given limits, given the precision of the measuring instrument. Examples are cloudiness in tenths, where any one observation must be in integer values of tenths (i.e., one tenth, two tenths, etc.); rain days; and wind direction given as 8 to 16 compass points.

**Domain:** The interval within which the values of the variable lie.

**Range:** The highest value in a set minus the lowest value.

**Population:** The complete set of observations of the variable. In meteorology, all of the observations of a particular variable (e.g., daily maximum temperature in July) that occurred in the past and will occur in the future.

**Sample:** A set of observations that is part of a population.

**Random sample:** A sample obtained in such a way that all observations in the population have an equal chance of being selected.

**Inference:** A deduction drawn from a sample, usually about the population.

**Hypothesis:** A contention or assumption, generally based on preliminary observation or experience and usually involving an inference from sample to population.

**Characteristic:** A feature of a set, or group of variates (e.g., the mean).

**Parameter:** A characteristic of a population.

**Statistic:** A characteristic of a sample. The singular form distinguishes a statistic from the field of study and from its general usage (statistics).

In meteorology there are four kinds of sets, which are differentiated by bounding conditions. As will be shown, this distinction becomes important when inferences are made from sample to population.

**Unbounded:** No practical limits at either end of the domain. For example, consider that all of the daily maximum air temperatures in July at a particular station constitute a population. In a sample from this population, there are no definite lower or upper limits. There may be lowest and highest values of record, but we know that there is a possibility, however small, that these could be exceeded. In this case, then, the domain is indefinite. Pressure is another example of an unbounded variable, as is precipitation in some circumstances.

**Bounded on upper end:** A fixed limit exists at the upper end of the domain. Variables so fixed are not frequent in meteorology, but examples are temperatures at the top of a snow layer, which cannot be above freezing, and relative humidities in a humid climate, which cannot be above 100 percent.

**Bounded on lower end:** A fixed limit exists at the lower end of the domain. Examples are frequent in meteorology: receipts of solar radiation, precipitation accumulated over short time intervals, and wind speeds.

**Bounded on both ends:** Fixed limits exist at both ends of the domain. This applies to discrete variables in general, and in particular to tenths of cloudiness (i.e., nothing below zero tenths or greater than ten tenths) and to number of rain days over relatively short periods such as a month or a week.

# 6.3
## DESCRIPTIVE STATISTICS

The characteristics of sets of numbers provide climatologists with a way of condensing and summarizing numerical information. Most of the salient features of such sets can be expressed either as single numbers or as graphical representations.

### 6.3.1 NUMERICAL DESCRIPTORS

There are three primary characteristics of a set of numbers. The first is that set's **central tendency,** which is expressed numerically by the following:

**Mean:** This is by far the most frequently used measure of central tendency. The mean is simply the sum of all observations in the set divided by the number of them.

**Median:** The middle observation in a set when they are ordered (arrayed) from lowest to highest. In the case of an even number of observations it is a value—not in the set—halfway between the two middle values.

**Mode:** Applies only to variables that are relatively discrete, so that duplications of a single observation are possible. The mode is the observation occurring with the greatest frequency.

**Mid-range:** The highest plus the lowest observation in the set divided by two.

Notice that among these measures of central tendency the mean and the median use all the observations in their calculation, but that the latter considers only their ranking and not their magnitude. The mid-range uses only two observations and thus is not as representative of the entire set as are the mean and the median.

The second primary characteristic of a set is the dispersion of the observations in it. Do they extend over a relatively wide range, or are they mostly clustered about a measure of central tendency such as the mean? Consider that the set $-50, -30, 0, 30,$ and $50$ has the same mean and median as the set $-5, -3, 0, 3,$ and $5$, but that the dispersion, or variability, of values is greater in the former. Measures of dispersion are described as follows:

**Standard deviation:** This is the square root of the mean of the squares of the deviations from the mean, or the root mean square error. We write this in symbols as

$$S = \left[\frac{1}{n}\sum_{i=1}^{n}(X_i - \bar{X})^2\right]^{1/2}$$

where $\bar{X}$ is the mean of the set, and $X_i$ the $n$ numbers in it. At times it is more convenient to find $S$ without first calculating the mean and finding the deviations from it. In this case we use

$$S = \left[\frac{\sum X_i^2}{n} - \left(\frac{\sum X_i}{n}\right)^2\right]^{1/2}$$

The square of the standard deviation, $S^2$, is the *variance*. The standard deviation is by far the most frequently used characteristic of dispersion. A slightly different formula is used when we want to estimate the standard deviation of the population rather than obtain a descriptive measure of dispersion. This is explained in Section 6.5.

**Mean deviation:** This is the average value of the absolute difference between each variable in the set and its mean:

$$|e| = \frac{1}{n}\sum_{i=1}^{n}|(X_i - \bar{X})|$$

where | | indicates the *absolute value* (value without regard to sign).

**Coefficient of variation:** In some meteorological data sets the variability increases with the mean. To obtain a comparable measure across a wide range of the mean, then, we divide the standard deviation by the mean:

$$V = \frac{S}{\bar{X}}$$

The third primary characteristic—**quantiles**—applies not to the entire set, but to the relative position of an observation in that set when all are arrayed. The most frequently used specific quantiles are quartiles, quintiles, deciles, and percentiles, which divide the set into 4, 5, 10, and 100 parts, respectively. The third quartile value is the variable which separates the lower 75 percent from the upper 25 percent; the second decile value separates the lower 20 percent from the upper 80 percent; and so on. Sometimes these values are not numbers in the set and must be interpolated between two adjacent variables. Some test scores are expressed in percentiles. If a student scores in the 84th percentile it means that he or she made a score higher than 84 percent but lower than 16 percent of other students tested similarly.

Quantiles also afford a means of measuring dispersion. For example, the interquartile deviation is expressed as

$$\frac{Q_3 - Q_1}{Q_2}$$

where $Q_3$, $Q_1$, and $Q_2$ are the third, first, and second quartile values, respectively. Notice that the second quartile value (the 50th percentile) is also the median.

### 6.3.2 GRAPHICAL DESCRIPTORS

Single number characteristics of a set are useful in many applications, but sometimes it is helpful, or even necessary, to visualize all of the variables as they are represented in a graph. In this form the frequency distribution can be visualized. To prepare a set for such representation the variables must first be combined into groups. The number of groups ($N_G$) chosen depends on the number of variables; a rule of thumb for relating the two is that $N_G$ should be approximately 5 log$_{10}n$, where $n$ is the number of variables in the set. For $n$ = 25, 50, and 100, $N_G$ = 7, 8, and 10, respectively (to the nearest whole number).

When the number of groups has been determined, the group limits must be fixed. A general guideline is to find a group interval of such size that when it is multiplied by the number of groups ($N_G$) the result is slightly greater than the range of the set (the highest value minus the lowest). This permits a little overlap at both ends (see the example in Table 6.1). After the group limits are fixed, each variable is placed in its respective group, and a frequency distribution results. This is then shown graphically, usually by a **histogram**, but sometimes by an **ogive**. A histogram is simply a graphical representation of the frequency of occurrence in each group. An ogive shows the cumulative frequency of occurrence expressed as a percentage of the total number of variables.

We can now illustrate most of the previous characteristics. In Table 6.1 are arrayed 100 mean January temperatures at Blue Hill, Massachusetts (near Boston), from 1856 to 1955. The mean for this 100-year period is 24.72°F, and the median is 25.0°F. Notice that because there is an even number of observations we must find the number halfway between the 50th and 51st observations; however, because they are both 25.0, no interpolation is necessary. Strictly, the mode is 25.6 because there are more occurrences of this

## SEC. 6.3 | DESCRIPTIVE STATISTICS

variable than of any other. But, and as we have learned, there is no point in assigning a mode to a continuous variable such as temperature. On the other hand, if all the observations were rounded to the nearest whole number, making the variables less continuous (or more discrete), there would be two modes—25 and 27—with eleven observations each.

Some measures of dispersion and the quartiles for this set also are shown in Table 6.1. To construct a frequency distribution, we use the rule of thumb to find that $N_G = 10$, and establish the group intervals so that there is a slight overlap at each end. The ogive is constructed by finding the cumulative percentage of variables for each successive group, starting with the lowest. The lowest group has 2 percent (2/100) of the total; thus the first mark on the ogive is at the upper limit of the first group and at 2 percent. Notice that from an ogive presentation it is possible to find any quantile. First, find the quantile value on the vertical axis, then find its intersection with the curve and read downward to the horizontal axis.

When presented in the form of frequency distri-

**TABLE 6.1**
Some characteristics of sets of variables, illustrated by 100 January mean temperatures (°F) at Blue Hill, Massachusetts (1856–1955).

| | | | | | | | | | |
|---|---|---|---|---|---|---|---|---|---|
| 1 | 14.6 | 21 | 20.8 | 41 | 23.9 | 61 | 25.7 | 81 | 28.2 |
| 2 | 15.5 | 22 | 20.8 | 42 | 24.0 | 62 | 25.7 | 82 | 28.2 |
| 3 | 16.3 | 23 | 21.0 | 43 | 24.0 | 63 | 26.0 | 83 | 28.6 |
| 4 | 16.7 | 24 | 21.1 | 44 | 24.2 | 64 | 26.2 | 84 | 29.1 |
| 5 | 17.2 | 25 | 21.4 ⟩1 | 45 | 24.4 | 65 | 26.6 | 85 | 29.5 |
| 6 | 17.2 | 26 | 21.6 | 46 | 24.6 | 66 | 26.8 | 86 | 29.7 |
| 7 | 17.2 | 27 | 21.8 | 47 | 24.6 | 67 | 26.8 | 87 | 29.9 |
| 8 | 17.4 | 28 | 21.8 | 48 | 24.7 | 68 | 26.8 | 88 | 30.3 |
| 9 | 17.4 | 29 | 22.0 | 49 | 24.8 | 69 | 27.1 | 89 | 30.8 |
| 10 | 18.2 | 30 | 22.1 | 50 | 25.0 ⟩2 | 70 | 27.1 | 90 | 31.2 |
| 11 | 18.5 | 31 | 22.3 | 51 | 25.0 | 71 | 27.1 | 91 | 31.3 |
| 12 | 18.6 | 32 | 22.3 | 52 | 25.1 | 72 | 27.3 | 92 | 31.4 |
| 13 | 19.2 | 33 | 22.6 | 53 | 25.1 | 73 | 27.4 | 93 | 31.7 |
| 14 | 19.2 | 34 | 22.8 | 54 | 25.1 | 74 | 27.4 | 94 | 31.9 |
| 15 | 19.4 | 35 | 22.8 | 55 | 25.3 | 75 | 27.4 ⟩3 | 95 | 32.0 |
| 16 | 19.9 | 36 | 23.0 | 56 | 25.4 | 76 | 27.5 | 96 | 33.8 |
| 17 | 20.0 | 37 | 23.0 | 57 | 25.6 | 77 | 27.7 | 97 | 33.8 |
| 18 | 20.2 | 38 | 23.1 | 58 | 25.6 | 78 | 28.0 | 98 | 34.6 |
| 19 | 20.4 | 39 | 23.6 | 59 | 25.6 | 79 | 28.0 | 99 | 35.0 |
| 20 | 20.7 | 40 | 23.8 | 60 | 25.6 | 80 | 28.0 | 100 | 35.6 |

The mean January temperature for these 100 years is 24.72°F, the median 25.0°F. The standard deviation is 4.63F°, the mean deviation 3.70F°.
1: First quartile value is 21.5 (interpolated between the 25th and 26th variables).
2: Second quartile value is 25.0.
3: Third quartile value is 27.45 (interpolated between the 75th and 76th variables).

**Frequency Distribution**

| | |
|---|---|
| 14.1/16.2 | 2 |
| 16.3/18.4 | 8 |
| 18.5/20.6 | 9 |
| 20.7/22.8 | 16 |
| 22.9/25.0 | 16 |
| 25.1/27.2 | 20 |
| 27.3/29.4 | 13 |
| 29.5/31.6 | 8 |
| 31.7/33.8 | 5 |
| 33.9/36.0 | 3 |
| | 100 |

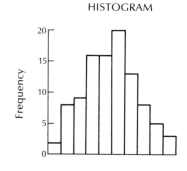

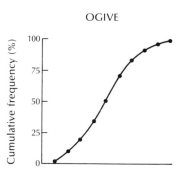

Horizontal axes of both the histogram and the ogive are scaled in either group limits or the midpoints of groups.

butions, many meteorological elements (variables) display distinctive shapes. A common type is the bell-shaped, or symmetrical, distribution shown in Table 6.1. Most of the variables are at or near the center, and the frequencies decrease toward both higher and lower values. The shapes of other frequency distributions depend on the meteorological variable and the basic interval of time used to form the average or accumulation.

Figure 6.1 shows some of the frequency distributions most frequently encountered in meteorology. Bell-shaped distributions (a) describe many iterations (repetitions) of temperature (maximum, minimum, mean) averaged for periods such as a day, week, month, or year; examples would be many years of the daily maximum temperature on January 15, of mean daily temperature during the first week of April, or of mean January temperature (as in Table 6.1), or many years of mean annual temperature. Precipitation amounts accumulated over a year, in all but very dry climates, and monthly amounts, at least in very rainy climates, also tend to be distributed almost symmetrically. One of the conditions necessary for an element to be so distributed is that it be unbounded or at least have no practical upper or lower limit.

As the likelihood of reaching an upper or lower limit increases, the frequency distribution tends to become skewed, or asymmetrical (b, c, and d). For example, in a humid climate with a mean annual precipitation of 1270 mm (50 in.), where the lowest annual accumulation ever observed was perhaps 500 mm (20 in.), the likelihood of no precipitation at all in a year is negligibly small. A frequency distribution of these amounts would be reasonably symmetrical. However, if the mean annual precipitation is 250 mm, even though the smallest accumulation in any year is only a few tens of millimeters, the likelihood of 0 mm is no longer negligible. Thus there is now a lower bound at zero. The result is usually that the frequency distribution is positively skewed (b), that is, it "tails off" toward higher values. With monthly precipitation amounts this skewness becomes even more apparent because in most climates there is at least a small chance of no precipitation falling in a month. In this respect the chance that no precipitation will fall in a day is even greater, and pronounced positive skewness results (c). In general, then, when there is an effective lower limit to the values that can occur (in our precipitation example, this is zero), frequency distributions are positively skewed and become increasingly so as it becomes more likely that the lower bound will occur. Another common example of positive skewness occurs with frequency distributions of wind speeds, which of course are bounded at the lower end by zero. Negative skewness ["tailed off" toward lower values, as in part (d)] is not common in meteorology, but can occur, for example, when there is an effective upper limit or bound.

We also have shown in Figure 6.1 that mean and median are identical, or nearly so, in a symmetrical distribution, but become more disparate as skewness increases. With positive skewness the median is less than the mean; the reverse occurs with negative skewness.

When a discrete variable is equally likely over its domain and is clearly bounded at both ends, a linear

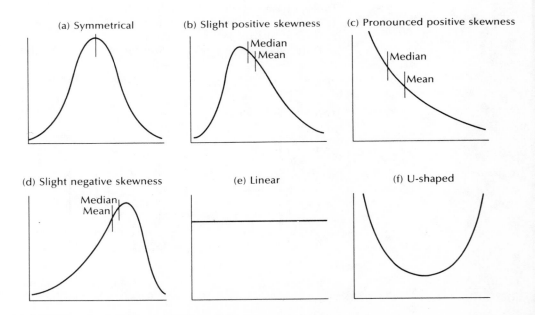

**FIGURE 6.1** Some frequently encountered frequency distributions of meteorological variables. The vertical axis shows frequency of occurence; the horizontal axis shows values of the variable, increasing to the right.

## SEC. 6.4 | PROBABILITY

distribution results [part (e) in Figure 6.1]. An example is cloudiness, in tenths, for some locations at particular times of the year. In mid-latitude climates in winter the weather often is produced by transient disturbances that feature prefrontal cloudiness and postfrontal clear skies. This tends to produce many observations at the low and high ends (0, 1, 2, and 8, 9, and 10-tenths cloudiness) and relatively few in the middle range. The result is a U-shaped distribution (f). In addition, some frequency distributions are bi- or multimodal, but one must question whether these more complex shapes are real—in the sense that they would be duplicated by additional samples—or simply a result of sampling variations.

The frequency distributions just described are derived by observation and thus are empirical. It also is possible to derive frequency distributions by theoretical, or entirely mathematical, means. Before we do this, however, we must discuss probability, because it is the foundation for both theoretical frequency distributions and tests that enable us to infer something about the parameters of a population from the statistics of samples.

## 6.4 PROBABILITY

**Probability** is the degree of expectation of the occurrence of an event. This expectation may be expressed qualitatively, using words such as slight, moderate, likely, or nearly certain. In statistics probability is more often expressed quantitatively by a number between 0 and 1, corresponding to 0 and 100 percent, respectively. Zero indicates no possibility of the event's occurrence and 1 (unity) is certainty, with a continuum of values between these two extremes. If the probability of an event is 50 percent (0.5), that event is as likely to occur as not to occur.

In climatology there are two ways of determining probability. The first is by observing what has happened in the past and using these observations to estimate what will occur in the future. This is called the *empirical method*. For example, if it is known that it has rained on 20 of the last 50 June 10ths, then the probability of rain occurring on any June 10 in the future is 20/50, or 40 percent.

The second way of determining probability, called the *theoretical method*, is by mathematical derivations often based on a physical analogy. We know that if a properly balanced coin is tossed it will certainly fall as either a head or a tail and that both outcomes are equally probable. If $p$ is the probability of a head and $q$ that of a tail, then $p = q = 0.5$, or 50 percent. In rolling a die, any one of six possibilities is equally probable. If $p$ is the probability of any one face being uppermost, then $p = 1/6$ and $q$, the probability of any other side, is 5/6. Of course, coins and dice have no direct application to determining climatological probabilities, so we need to show how frequency distributions can be derived from physical analogies.

### 6.4.1 THE BINOMIAL DISTRIBUTION

To construct a frequency distribution, consider first the outcome of a flip of one coin: either a head or a tail, and each is equally likely. If the coin is tossed twice, there are four possible outcomes, assuming for the moment that the order of occurrence makes a difference: HH, TH, HT, and TT. The two middle events can be considered the same in this development because both consist of a head and a tail. With this stipulation there are three possible outcomes, with corresponding probabilities of 25, 50, and 25 percent, respectively. The probabilities for the different results of each of these circumstances, as well as those for all other positive whole number values of $n$, may be obtained from the quantity

$$(p + q)^n$$

where $p$ and $q$ are defined as before and $n$ is the number of tosses.

Table 6.2 shows how the results vary for different values of $n$ and presents the general formula for $(p + q)^n$, the binomial formula. The **binomial distribution**, a special kind of frequency distribution, results because the frequency of occurrence of any particular combination of heads and tails is comparable to that occurring in groups (see Table 6.1). Each of the terms in this formula is a number between 0 and 1 and represents the probability that that particular combination of heads and tails will occur in $n$ tosses. All of the separate terms add up to one. Thus, in a toss of one coin ten times (or ten coins tossed simultaneously) the probability of 0 heads and 10 tails—which intuition tells us is very small—is $q^{10}$, or 0.001. This is also the probability of 10 heads and 0 tails. The probabilities of obtaining any other combination of heads and tails in ten tosses also is shown in Table 6.2. For convenience the $C_{n,r}$ terms (the number of combinations of $n$ things taken $r$ at a time) can be obtained from standard mathematical tables as well as calculated from the formula.

**TABLE 6.2**
Development of the binomial distribution.

| No. of Coin Tosses | Possible Results | Corresponding Probabilities | Expansion Whose Terms Correspond to These Probabilities |
|---|---|---|---|
| 1 | H, T | $p, q$ | $(p + q)^1 = p + q$ |
| 2 | HH, HT, TH, TT | $p^2, pq, qp, q^2$ | $(p + q)^2 = p^2 + 2pq + q^2$ |
| 3 | HHH, HHT, HTH, THH, HTT, THT, TTH, TTT | $p^3, p^2q, p^2q, qp^2, pq^2, q^2p, q^2p, q^3$ | $(p + q)^3 = q^3 + 3p^2q + 3q^2p + p^3$ |
| . | | | |
| . | | | |
| . | | | |
| n | ... | ... | $(p + q)^n =$ (see below) |

Thus, in $n$ tosses of a coin, the probabilities of the various possible events (i.e., of obtaining 0, 1, 2, . . . . $n$ heads) are given by the successive terms of the binomial expansion of $(p + q)^n$, which is

$$(p + q)^n = p^n + C_{n,1}p^{n-1}q + C_{n,2}p^{n-2}q^2 + \ldots + C_{n,r}p^{n-r}q^r + \ldots + q^n$$

where $p = q = 0.5$. This expression sums to unity.

The coefficient $C$ is the number of ways in which $n$ tosses of a coin can produce exactly $r$ tails and $n - r$ heads, which is the same as the number of combinations of $n$ things taken $r$ at a time. That is, we can write

$$C_{n,r} = \frac{n!}{r!\,(n - r)!} \qquad \text{(Note: } 4! = 4 \times 3 \times 2 \times 1 = 24\text{)}$$

For $n = 10$, the 11 terms of this expansion are shown in the following table:

| No. of Heads | Probability | No. of Heads | Probability |
|---|---|---|---|
| 0 | 0.001 | 6 | 0.205 |
| 1 | 0.010 | 7 | 0.117 |
| 2 | 0.044 | 8 | 0.044 |
| 3 | 0.117 | 9 | 0.010 |
| 4 | 0.205 | 10 | 0.001 |
| 5 | 0.246 | | 1.000 |

Theoretical and observed frequencies in tossing 10 coins 1000 times:

| No. of Heads: | 0 | 1 | 2 | 3 | 4 | 5 | 6 | 7 | 8 | 9 | 10 |
|---|---|---|---|---|---|---|---|---|---|---|---|
| Theoretical | 1 | 10 | 44 | 117 | 205 | 246 | 205 | 117 | 44 | 10 | 1 |
| Observed | 0 | 12 | 47 | 110 | 208 | 248 | 190 | 121 | 43 | 11 | 2 |

It is very important to realize that reality seldom agrees exactly with theory. Suppose ten coins are tossed simultaneously and the number of heads are recorded. This process is then repeated until there is a total of 1000 trials. The frequency distribution of the number of heads expected, according to the probabilities in Table 6.2, is given in the first line at the bottom of this table, or 1, 10, 44, and so on. The frequencies obtained from the actual experiment could be as shown in the second line. Of course, this outcome does not invalidate the binomial expression; it simply demonstrates **randomness.** We cannot expect exact correspondence to theoretical probability, but we do have the assurance that if the entire process just explained were repeated over and over, the frequencies in each of the eleven categories would average out to those shown in the first line. A mathematician would say that such frequencies *converge* to the values in the first line. However, it is not necessary to perform this experiment in order to obtain the probabilities; we know *a priori* (based on theory instead of experience or observation) what they are.

When $p$, $q$, and $n$ are known, the mean and standard deviation of a binomial distribution can be obtained. They are

$$\overline{X} = np \quad \text{(mean)}$$
$$S = \sqrt{npq} \quad \text{(standard deviation)}$$

Notice that in our example (the first line at the bottom of Table 6.2) $\overline{X}$ is

$$[1(0) + 10(1) + \ldots + 10(9) + 1(10)]/1000 = 5.00$$

### 6.4.2 THE NORMAL DISTRIBUTION

The binomial distribution, as developed to this point, has only limited application in climatological statistics. To make it more applicable, we must change it from discrete to continuous. Notice that all probabilities refer to discrete events; it would be unthinkable to try to determine the probability of 3.789 heads, so only whole numbers are possible. Remember also that we are still assuming $p = q = 0.5$. Now consider what happens when both the number of coins tossed and the number of trials (10 and 1000 in our example) increase to extremely large values. The histogram would expand to include a wider range of values along the X-axis, and the individual rectangles comprising the histogram would get narrower. At the limit the range of values (X-axis) would be from $-\infty$ ($\infty$ is infinity) to $+\infty$ (an unbounded distribution), the rectangles would decrease to zero, and a smooth curve would result.

The frequency distribution thus obtained is the **normal distribution** (also called the Gaussian distribution). It is the most frequently used theoretical distribution, not only in climatology, but in most other applications as well. Figure 6.2 shows the normal distribution. Certain of its characteristics are important:

1. The area under the curve, from $-\infty$ to $+\infty$, is unity (one).
2. About two thirds (68 percent) of the area under the curve is between ±1 standard deviation; 95 percent is between ±1.96 standard deviations (or approximately 2).
3. The curve is symmetrical, and thus the probability of the occurrence of a value greater than $X$ (on the X-axis) is equal to the probability of value less than $-X$ (see standard units in Figure 6.2).

The normal curve has been called an *error curve* because errors of observation often are distributed this way. That is, most observations will be at or near

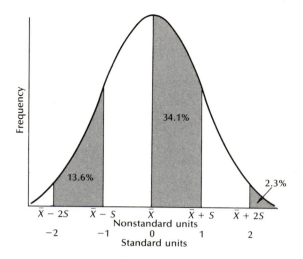

**FIGURE 6.2**
The normal distribution.

the mean; the larger the deviation, the less is the probability of its occurrence. As climatologists, we might say that nature "aims" for the mean (e.g., mean daily temperature) and comes close to it relatively often, but "misses" in such a way that moderate misses are more frequent than those far from the value aimed for. Meteorological variables for which no practical limits exist (unbounded distributions) are often distributed this way. It is important to remember that the normal curve is more than just a symmetrical, or bell-shaped, curve; the three characteristics and others not given make it unique.

The use of the normal curve is illustrated in the following example. To proceed from its representation in nonstandard units in Figure 6.2, we must *standardize* the variables. That is, the variables in their original (nonstandard) units (°F or °C, or inches or centimeters of precipitation, for example) must be converted to standard units. If $X$ is any original variable, then its counterpart in standardized units ($z$) is given by

$$z = \frac{X - \overline{X}}{S}$$

Only when the original variable has been standardized can the normal curve be employed to obtain probabilities.

Consider the 100 values of January mean temperature given in Table 6.1. Suppose we wish to determine the probability of a January mean less than 18°F by using the normal curve approximation. First, we find the standardized value:

$$z_{18} = \frac{18 - 24.72}{4.632} = -1.45$$

If we refer to a table of areas under the normal curve (not included here, but available in almost any introductory statistics textbook), we find that the area to the left of −1.45 is 0.07. We therefore expect 7 percent of the January mean temperatures to be below 18°F. Suppose we want to determine in how many of the next 50 years this temperature will be between 19.7° and 29.7°F (a deviation of 5° from the mean; the second decimal place has been dropped for convenience). Standardizing gives

$$z_{19.7} = \frac{19.7 - 24.7}{4.632} = -1.08$$

$$z_{29.7} = \frac{29.7 - 24.7}{4.632} = 1.08$$

The values are of course the same in absolute magnitude because we have gone an equal distance on either side of the mean. The area between these values is 0.72, or 72 percent (again, this is determined from a table of areas under the normal curve). In the next 50 years, therefore, we expect that 36 years (50 × 0.72) will have temperatures within ±5F° of the mean.

Notice that since the normal distribution is continuous there is no probability associated with a single variable. We cannot say what the probability of a temperature of $X$°F is, only that of greater or less than $X$ or between $X_1$ and $X_2$.

Why do we use the normal curve approximation to probabilities when they can be determined empirically by examining the record? For example, on an empirical basis the probability of a temperature less than 18°F is 9/100, or 9 percent (see Table 6.1). Similarly, the number of temperatures between 19.7° and 29.7°F is 71 (including the value 29.7), which gives a 71 percent probability and an expected number of years as 36, with rounding. The normal curve approximation is preferred because obtaining probabilities with this method is much easier. Particularly with modern computing facilities it is more expedient to find the desired probabilities with only the mean and standard deviation and a table of normal curve areas than to store all available data and determine the probabilities empirically. It also can be argued that if the particular variable can be assumed, or shown, to be normally distributed (or distributed according to some other theoretical distribution), future occurrences of that variable are more likely to conform to theory than to past data, which incorporate sampling variations. Of course, before the normal curve or any other theoretical approximation can be applied, it must be known beforehand that the variable fits such a curve. A frequent problem in applying statistical methods to climatology is finding the most appropriate theoretical curve (distribution) for a given variable.

### 6.4.3 THE POISSON DISTRIBUTION

In developing the normal distribution from the binomial, $p = q = 0.5$. If we change these probabilities, the event is still *dichotomous* (either one possibility or the other), but now the probabilities may be $p = 0.6$, $q = 0.4$, or 0.1 and 0.9, so long as $p + q = 1$. As $p$ decreases from 0.5, the distribution becomes slightly, then moderately, and finally markedly positively skewed (Figure 6.3). If $p$ were to increase from 0.5, the skewness would be negative.

When $p$ is very small (less than 0.1) and the number of times the event could occur $(n)$ is very large, an approximation to the binomial may be used to determine probabilities and hence form a frequency distribution. This approximation, which applies only to dichotomous, and thus discrete, events, is the **Poisson distribution**. The Poisson, a special case of the binomial, applies only when $p$ is approximately less than

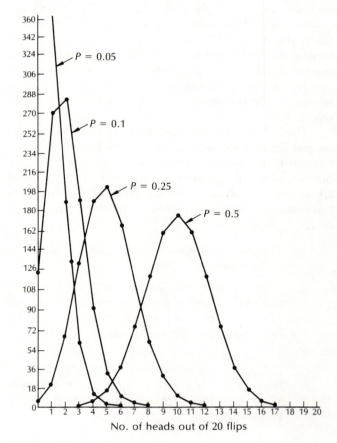

**FIGURE 6.3**
Theoretical binomial distribution for varying probability (1000 trials).

0.1 and $n$ is very large. If the average number of times an event occurs in a long period of time (a year, for example) is known, and if this average number is between somewhat less than 1 and about 8 or 10, then the probability of the occurrence of $X$ events in this long period may be approximated by

$$P(X) = \frac{\bar{X}^{-X} e^{-\bar{X}}}{X!}$$

where $\bar{X}$ is the mean number of occurrences in the period, $e$ is the base of the natural logarithms (2.718 . . .), and the ! indicates factorial (see Table 6.2 for an explanation of factorial).

As an example, suppose that at a particular station 76 mm (3 in.) or more of precipitation in a 24-hour period has occurred on the average twice a year. The mean for this period is thus 2, and the probabilities of 0, 2, and 5 such occurrences in a year are

$$P(0) = \frac{2^0}{0! e^2} = 0.135$$

$$P(2) = \frac{2^2}{2! e^2} = 0.271$$

$$P(5) = \frac{2^5}{5! e^2} = 0.036$$

A frequency distribution results when the probabilities of all whole number occurrences are similarly determined. In this example the mean is a whole number, but the Poisson distribution can be used when the mean has decimal places.

For the Poisson distribution, $\bar{X} = np = S^2$; that is, the mean equals the variance. Poisson distributions for different values of $\bar{X}$ are shown in Figure 6.4. As $\bar{X}$ increases, the frequency distribution becomes less skewed. However, because it is bounded by zero on the low end and essentially unbounded on the high end, this distribution will not be exactly symmetrical.

The probabilities in the preceding example could have been obtained by expanding the binomial—in this case $(p + q)^{365}$, where $p$, the probability of 76 mm (3 in.) or more in a day, is 2/365, and $q$ is 363/365. The last term in this expansion is the probability of zero occurrences, the next to the last term that of one occurrence, and so on. Obviously, this would be a very cumbersome calculation; thus, the Poisson is a "shortcut" method. Since the Poisson distribution requires that $p$ be less than about 0.1, it can be applied to determine the probability of rare events. If an event occurs on the average once in $n$ occasions (e.g., once in $n$ days, or, in the preceding example, 2/365 equals once in 182.5 days), we may wish to know the probability that it will or will not occur in $k$ occasions. The

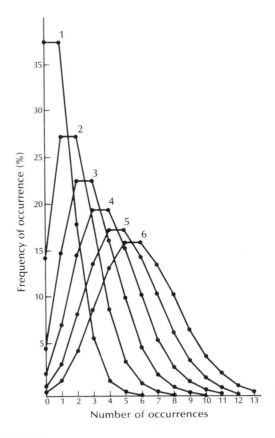

**FIGURE 6.4**
Frequency of occurrence of rare events by the Poisson distribution for the indicated means.

mean frequency of occurrence in $k$ occasions is thus $k/n$, and the probability of no occurrence during this time is

$$P(\text{no occurrence}) = e^{-k/n}$$

Notice that if we let $k = 365$ and $n = 182.5$, then this probability is $e^{-2} = 0.135$, which agrees with our example. The probability of at least one occurrence in these $k$ days is thus $1 - 0.135 = 0.865$.

As an example, consider that a dam has been designed so that its capacity can be expected to be exceeded once in 100 years. Its **return period**, the reciprocal of the probability of the event in a year, is thus 100 years. We want to know the probability of the dam's capacity being exceeded in the next 10, 20, and 100 years. The computations for these probabilities are as follows:

$$P(\text{no occurrence in 10 years}) = e^{-k/n}$$
$$= e^{-10/100} = 0.905$$
$$P(\text{no occurrence in 20 years}) = e^{-20/100} = 0.819$$
$$P(\text{no occurrence in 100 years}) = e^{-100/100} = 0.368$$

## 6.4.4 OTHER THEORETICAL DISTRIBUTIONS

A number of other mathematically derived frequency distributions are applicable to meteorological variables. Although we will not show their derivations or formulas, we will explain the circumstances under which they are used.

For variables whose frequency distributions are moderately positively skewed—a frequent occurrence because many are bounded at the low end by zero—it may be possible to modify the shape of the frequency distribution by transforming the original variables. This is most frequently done by taking the square root or the natural logarithm (ln) of the variable. In Figure 6.5, the square root transformation has the effect of spacing the frequencies at varying intervals and makes the resulting distribution more normal. The ln $X$ transformation changes the shape excessively and results in negative skewness. Of course, after probability estimates are made for the transformed variables from the table of areas under the normal curve, the variables must be transformed in reverse.

Notice that in Figure 6.5 we made a subjective judgement (an eye approximation) as to which transformation made the curve appear more normal. There are objective ways to make this analysis, and in Section 6.5.1 we will show how to determine if a given frequency distribution differs significantly, in a statistical sense, from the normal.

Some other important distributions in meteorology will be briefly described here, and you can find further details in the references cited in the suggested reading list. The *negative binomial distribution* can be used to model discrete events whose mean number of occurrences in a long time period (e.g., a year) is too high for the Poisson to be applicable. It has been used for mean annual hail days and mean annual number of tropical storms. The *gamma distribution* is well adapted to meteorological variables for which the occurrence of zero is frequent, such as precipitation accumulations over short periods (from a few days to 2 or 3 weeks). *Extreme value distributions* are used when the variable of interest is the most extreme in a period (e.g., a year). These distributions are especially important to consider in structural design, where less extreme occurrences are not as important.

# 6.5

## SAMPLING AND TESTS OF STATISTICAL SIGNIFICANCE

To this point we have been concerned with the use of various statistical measures to describe the salient features of sets of numbers. Now we consider tests of statistical significance. These tests are of two types: those in which inferences about the population are made on the basis of samples (**parametric tests**), and those that do not involve such inferences (**nonparametric tests**). In particular, such inferences pertain to population parameters—mean and standard deviation, and the distribution of the population.

### 6.5.1 NONPARAMETRIC TESTS

Nonparametric tests are in general simpler than parametric tests and require no assumptions about the population from which samples are drawn. A disadvantage is that nonparametric tests may not always make use of all available information.

#### *The Sign Test*

The **sign test** is one nonparametric test. To illustrate its use, we will use a climatological example. Suppose we want to determine if the annual frequency of hail days at two stations is statistically significantly different. That is, we will use the methods of statistics to determine if one station experiences significantly more hail days than the other. Two outcomes of the sign test are possible. We may find that whatever difference exists between the stations is likely to have occurred by chance alone, that is, due to random variations. Or,

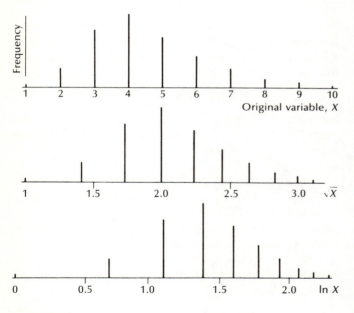

**FIGURE 6.5**
Transforming the original variable by taking its square root or natural logarithm can make a skewed distribution approximately normal. In the case of this moderately positively skewed distribution, the square root transformation works best.

we may find that the probability of the chance occurrence of the difference between the stations is so small as to be negligible. Of course, we must define what is meant by *negligible*, which is why significance levels must be specified.

Here is the climatic record for hail days per year for 11 years at the two stations:

| Year | Station A | Station B | Sign of A − B |
|------|-----------|-----------|---------------|
| 1    | 3         | 4         | −             |
| 2    | 3         | 5         | −             |
| 3    | 2         | 4         | −             |
| 4    | 4         | 5         | −             |
| 5    | 5         | 6         | −             |
| 6    | 3         | 3         |               |
| 7    | 1         | 4         | −             |
| 8    | 5         | 8         | −             |
| 9    | 4         | 6         | −             |
| 10   | 4         | 3         | +             |
| 11   | 4         | 5         | −             |

Just a glance at these figures reveals that station B has more hail days per year overall than does station A. But is this difference statistically significant? To answer this question, we hypothesize that A and B are equal; in other words, the probability that A has more hail days than B equals the reverse, and both equal 0.5, or 50 percent. Now we can rely on the binomial distribution and determine the probability of obtaining such a disparate result. If we omit the tie in the sixth year, this is analogous to finding the probability that 9 heads occur in 10 tosses of a coin, or, more exactly, the probability of 9 *or* 10 heads in 10 tosses.

A significance level must be specified before performing the test. Then, if the calculated probability is less than the significance level, the hypothesis that A and B are equal is rejected; if the reverse applies, the hypothesis is accepted. Suppose we use the 5 percent significance level, which is very common. In this case the calculated probability, that of obtaining 9 or 10 heads in 10 tosses of a coin, is the sum of the last two terms in the binomial expansion of $(p + q)^{10}$, where $p = q = 0.5$. Notice that the first two terms could be used as well, since the distribution is symmetrical. We compute the probability to be

$$\frac{1}{1024} + \frac{10}{1024} = \frac{11}{1024} = 1.1 \text{ percent}$$

Since the calculated probability is less than the significance level, we reject the hypothesis that A and B are equal and conclude that the hail day experience at station B is statistically significantly greater than that at station A.

Why is it necessary to specify a significance level? Significance levels are used because in significance testing there is no absolute certainty. It must be acknowledged that, however unlikely, a result such as 9 of 10 or 10 of 10 could occur by chance alone, and our conclusion could be wrong. By choosing the 5 percent level we acknowledge that 5 percent of the time—in the long run, over many repetitions of experiments at the 5 percent level—our conclusion will be in error. If we want to reduce the chance of making an incorrect conclusion, a lower significance level must be chosen (e.g., 1 percent), but even at this level we will be in error by this amount in the long run. Thus, specifying a significance level is a compromise between our unwillingness to make an incorrect conclusion and our attempt to account for the fact that chance plays a role in all significance testing.

What would the result have been if, in the preceding example, there had been 8 negative and 2 positive signs (again omitting the tie)? The last three terms total 5.5 percent. If we had again chosen the 5 percent significance level, we would accept the hypothesis that A and B are equal since the calculated probability now exceeds the significance level. We would be saying, in effect, that the probability that a result such as 8 or more of 10 could occur by chance is large enough to result in accepting the hypothesis that A and B are equal. And, of course, any less disparate result (e.g., 7 or more of 10, or 6 or more of 10) would produce the same conclusion.

There is yet another way to assess the statistical significance of the difference in the number of hail days between stations A and B. Suppose we want to determine the probability (chance) of a deviation as large or larger than *X* out of 10. In this case the calculated probability is doubled; that is, we would use 2.2 percent for 9 or more of 10, and 11 percent for 8 or more of 10. This is a *two-tailed test* since both ends (tails) of the frequency distribution are used, which is the customary procedure in the sign test. If we are interested in only one sign, however, a one-tailed test is used.

### The Run Test

Just as we can determine the probability of obtaining a given number of pluses and minuses, as in the sign test, it also is possible to determine the probability of a specified number of runs, where a **run** is one or more consecutive values of like sign. The median of a series of values ordered in time is first determined, and then the individual values are designated as plus or minus (above or below the median). The number of runs is then compared with the number expected if the series were random. A physical analogy would be getting 1, 2, 3, . . . *n* heads in the same number of tosses of a coin. If the calculated number of runs exceeds the number expected by chance alone, oscillations, or periodicities, are indicated. If too few runs occur compared with those expected by chance, a trend or shift may have occurred. In a series with the trend amplified to the point that each value is greater than the one before it, it is easy to see that there would be only two runs.

This test, the **run test,** often is used to determine the homogeneity of a climatological data series. A change of location of weather instruments might be indicated by this test, in which case the data series would be heterogeneous. Once the date(s) of break(s) in the record are known, a more powerful parametric test, the *t*-test for the significance of the difference of means (Section 6.5.2), can be used.

### The Chi-Square Test

The final nonparametric test we will consider is the **chi-square** ($\chi^2$) **test.** Again, a physical analogy will illustrate how this test quantifies the extent of chance occurrence. This time a die is used instead of a coin. The probability of occurrence of any one of the 6 sides of a die is of course 1/6. Suppose we roll a die 120 times. It would not be reasonable to expect that each face would turn up exactly one sixth of the times (20), because we expect some random variation. But how much random variation should we expect? When does the departure between the ideal (20 times for each face) and the actual frequencies cross the divide between random variations and whatever the alternative is (a "dishonest" die)?

To quantify the extent of chance occurrence, this experiment is repeated a very large number of times. A frequency distribution is then made of the differences between expected ($E$) and actual, or observed ($O$), frequencies. Expected frequency for each side, in our example, is simply the product of the probability of that (or any) side and the number of rolls, or 1/6 (120) = 20. More explicitly, $\chi^2$ is defined as

$$\chi^2 = \sum_{i=1}^{k} \frac{(O_i - E_i)^2}{E_i}$$

where the summation extends over $k$ classes (in our example, $k = 6$). From this formula it is clear that when $O$ and $E$ are equal for all $k$ classes, $\chi^2 = 0$. Conversely, the greater the differences between $O$ and $E$, the greater is $\chi^2$. When $E_i$ is small (less than about 5), it is necessary to combine classes.

Suppose that in this example of 120 tosses of a die $\chi^2$ is calculated as 20.5. A table of chi-square distribution (available in introductory statistics texts) shows that a value this large or larger occurs only 0.1 percent of the time with an honest die (for entry to such a table one must know the number of degrees of freedom that applies). Since this figure is well below the 5 percent level usually used, the hypothesis that the die is honest would have to be rejected. On the other hand, if $\chi^2$ was 8.0, the same table shows that this value or larger could occur roughly 15 percent of the time. The hypothesis would therefore be accepted.

Thus, $\chi^2$ is used to test whether the observed number of occurrences in each of $k$ classes or categories is statistically significantly different from an hypothesized number of occurrences. One particular example frequently applied to climatological data is testing for normality. Given a bell-shaped frequency distribution, we ask, Could the differences between the observed distribution and an exactly normal distribution with the same mean and standard deviation occur by chance alone? If not, we say that the two frequency distributions (observed and hypothesized) are significantly different.

Refer to the histogram at the bottom of Table 6.1. Is there a statistically significant difference between this distribution and an exactly normal distribution with the same mean and standard deviation? First, construct the normal distribution that has these statistics. (Details of this procedure can be found in other sources.) Then calculate $\chi^2$ and compare it with tabular values at specified significance levels. In the example given in Table 6.1, we obtain the following frequencies:

| | | | | | | | | | |
|---|---|---|---|---|---|---|---|---|---|
| Observed: | 2 | 8 | 9 | 16 | 16 | 20 | 13 | 8 | 5 | 3 |
| Expected: | 3.3 | 5.3 | 10.1 | 15.2 | 18.5 | 18.0 | 13.7 | 9.0 | 4.4 | 2.5 |

Are the differences statistically significant? We find that $\chi^2$ is 1.27, and a table of chi-square distribution shows that this value, or one larger, could easily have occurred by chance. Therefore, the differences between observed and expected frequencies are not statistically significant.

There are details of the chi-square procedure that we have omitted, but these are available in any standard statistics text. Also, in testing for normality the chi-square test is of very limited usefulness for all but very large $n$ (100 or more).

### 6.5.2 PARAMETRIC TESTS

Parametric tests require knowing or inferring the parameters of populations (e.g., the mean and standard deviation), and the distribution of the population (e.g., normal, linear, U-shaped). It then is possible to determine the probability that a given sample came from a population whose parameters have been so specified. Parametric tests rely on the **central limit theorem,** which is comparable to the physical analogies of coin and die tossing in nonparametric tests because it enables us to quantify the extent of chance occurrence.

The central limit theorem says that the variability of the means of samples varies inversely with sample size. This variability is measured by the **standard error** (*SE*), which is simply the standard deviation of sample means. Thus, the standard deviation (standard error) of the means of samples of large *n*—where *n* is sample size and *not* number of samples—will be smaller than that for samples of small *n*. The formula is

$$E(SE_{\bar{X}}) = \frac{\sigma}{\sqrt{n}}$$

where $\sigma$ is the standard deviation of the population from which samples of size *n* are drawn, and *E* indicates that this is an expected value. Unlike other mathematical formulas, the relationship is seldom exact because of sampling variations. In the long run, however, as the number of samples gets very large, $SE_{\bar{X}}$ will converge to the value on the right-hand side.

It is helpful to know how the dispersion (variability) of means depends on sample size, but probability estimates of the occurrence of a mean of size *X* or larger (or smaller) are not possible unless we also know the distribution of these sample means. The central limit theorem also says that, provided the sample size is large enough (roughly, larger than 30), *the means of samples will be distributed in a manner not significantly different from normal*. With this very important information in mind, we can now determine the probability that a sample whose mean is known was drawn from (came from) a population whose parameters are known or can be estimated. For example, suppose the long-term mean temperature for winter (December–February) at a station is 0°C and the standard deviation 5C°. We will regard these as parameters of the population. In the last 30 years the mean temperature was 5°C. Is this a statistically significant increase in temperature, or is it likely to have occurred by chance alone?

To answer this question, we find the probability of obtaining a sample mean equal to or greater than 5 from a population with a mean of 0 and $\sigma$ of 5. If this probability is smaller than a preestablished significance level such as 5 percent, we conclude that the sample is not from the population and thus that the two means (population and sample) are statistically significantly different. If this probability is larger than 5 percent, we accept the hypothesis that the sample came from the population.

The standard error of the mean in this case is $5/\sqrt{30} = 0.913$. Because sample means can be assumed to be distributed normally, we know that ±1.96 standard errors will include 95 percent of these sample means. The difference between population mean and sample mean is 5C°, which is 5.5 standard errors ($5/\sqrt{0.913} = 5.5$). This difference in means is thus statistically highly significant; this could have occurred by chance alone with only a negligibly small probability. Notice that a difference in means of about 1.8C° would be the "break-even" point at 5 percent, assuming a two-tailed test (i.e., a difference of 1.8C° or more on either side of the population mean).

This example is rather artificial, since in most applications to climatological data the population parameters are not known. The most frequently used test for the significance of the difference of means is the ***t*-test**. In one such application the population mean, $\mu$, is known (or can be hypothesized), but the population standard deviation, $\sigma$, is unknown and must be estimated from the sample standard deviation (*S*). The best estimate of $\sigma$ from *S*, written $\sigma_{\hat{p}}$, is given by $S\sqrt{n/(n-1)}$. In this case the formula for *t* is written as

$$t = \frac{(\bar{X} - \mu)(\sqrt{n-1})}{S}$$

The calculated *t* value can be compared with tabular values, as was done with $\chi^2$ to determine the statistical significance of the difference in means.

More frequently, the population mean is not known and cannot be hypothesized. In this second case, we are not testing the hypothesis that the sample came from the population, but rather that two samples came from the same population. Now, the formula for *t* is written as

$$t = \frac{\bar{X}_1 - \bar{X}_2}{\sqrt{\frac{n_1 S_1^2 + n_2 S_2^2}{n_1 + n_2 - 2}\left(\frac{1}{n_1} + \frac{1}{n_2}\right)}}$$

where the subscripts 1 and 2 distinguish the two samples.

A climatological example will illustrate how the *t*-test is used to determine the significance of the difference of means. The mean winter temperature from 1911 to 1940 at a station was 2.4°C, and the standard deviation was 2.1C°. From 1941 to 1970 these same statistics were 3.7°C and 2.3C°. Is this a significant difference, or could such a result have occurred by chance? If the former, there would be some statistical basis supporting a claim of climatic change. When the preceding formula is solved with these numbers, $t = 2.25$. At the 5 percent significance level (the tables for significance levels are not shown here), the hypothesis that these means are from the same population would be rejected; that is, at 5 percent the difference is significant.

A third situation in which the *t*-test may be used involves matching observations. Results are obtained under identical conditions except for the effect being tested, and the differences between matched observations are then tested for their significance. For example, we select two adjacent areas for an experiment to determine the efficacy of cloud seeding. On meteorologically suitable days we flip a coin to determine over which area a seeding agent will be disseminated into the clouds in hopes of converting water vapor to liquid water. Two nearby weather radars calculate the total liquid water content in the atmosphere above each of the two areas. Table 6.3 shows the results obtained during 15 seeding missions (not a real example).

Are the differences shown in Table 6.3 statistically significant? That is, if we hypothesize that the mean difference in the population is zero, what are the chances of obtaining a result as disparate as (or more disparate than) 3.88? If this probability is less than 5 percent, we will reject the hypothesis, which of course implies that seeding significantly increases liquid water content. The significance ratio *t* is now given as

$$t = \frac{\overline{D}\sqrt{n-1}}{S_D}$$

where $\overline{D}$ is the mean, and $S_D$ the standard deviation, of the differences. With the figures given above, $t = 2.87$. Referring to a table of the *t*-distribution, we find that a value at least this large could occur with a probability of between 1 and 2 percent. With the customary 5 percent significance level, the hypothesis would be rejected. However, notice that if the 1 percent significance level had been chosen the hypothesis would be accepted.

The sign test (Section 6.5.1) could also have been used in this example. If seeding makes no difference, we would hypothesize, what are the chances of obtaining three or fewer negative signs? This probability is 1.8 percent, and at the customary 5 percent significance level this hypothesis would be rejected. But the *t*-test is preferred here because it makes use of more information; that is, it uses scores, or counts, and not just rankings.

**Analysis of variance** is yet another statistical technique that has been helpful in climatological analysis. A particular variable may be influenced by a number of others. (Finding the mathematical relationship

**TABLE 6.3**
An application of significance of means testing to the determination of the efficacy of cloud seeding.

| Mission | Liquid Water Content (arbitrary units) | | |
|---|---|---|---|
| | Seed Area | Nonseed Area | Difference |
| 1 | 14.5 | 13.0 | +1.5 |
| 2 | 10.7 | 6.0 | +4.7 |
| 3 | 22.6 | 20.4 | +2.2 |
| 4 | 19.5 | 21.0 | −1.5 |
| 5 | 32.7 | 14.9 | +17.8 |
| 6 | 10.9 | 11.2 | −0.3 |
| 7 | 14.8 | 6.7 | +8.1 |
| 8 | 25.8 | 19.2 | +6.6 |
| 9 | 7.6 | 7.1 | +0.5 |
| 10 | 19.4 | 10.0 | +9.4 |
| 11 | 23.0 | 22.7 | +0.3 |
| 12 | 14.6 | 9.4 | +5.2 |
| 13 | 24.0 | 19.7 | +4.3 |
| 14 | 4.7 | 8.0 | −3.3 |
| 15 | 37.9 | 35.2 | +2.7 |
| $\overline{X}$: | 18.85 | 14.97 | 3.88 |
| $S$: | 8.84 | 7.78 | 5.06 |

among the variables is the subject of the next section.) In analysis of variance we first estimate how much of the total variance of the variable of interest is contributed to by different sources. Then we test the significance of these separate contributions to determine if each is real, in the statistical sense, or could have occurred by chance.

# 6.6
## REGRESSION AND CORRELATION

It often is of interest in climatological studies to estimate one variable from one or more other variables. Natural events are rarely independent, and weather variables in particular are strongly interrelated. Positive and negative feedback, so much a part of climatology, are certainly a case in point. The quantitative determination of the degree of the relationship of one variable to another, and the determination of the mathematical form of that relationship, are called **correlation** and **regression**, respectively.

These procedures answer two questions: Is there a relationship between variables, and if so, how strong? What is the mathematical formula that best relates the predictor(s) and the predictand? (The predictand is the variable being predicted, or estimated.) We will consider two specific kinds of regression and correlation: bivariate linear and multivariate linear.

## 6.6.1 BIVARIATE LINEAR REGRESSION AND CORRELATION

In bivariate linear regression and correlation there are only two variables: the predictand (dependent variable) and the predictor (independent variable). The procedure determines the linear (straight-line) relationship between the two. The bivariate linear procedure is justified if there is no reason to believe a nonlinear relationship exists; this is usually the case because the functional form of the relationship (linear or nonlinear) is not known and may not even be suspected when regression and correlation are applied.

The manner in which a linear or nonlinear relationship is determined is best shown graphically. Figure 6.6 is a **scatter diagram,** on which are plotted joint occurrences of the predictor (X-axis) and predictand (Y-axis). A climatological example would be solar radiation receipts in 24 hours and maximum air temperature at instrument shelter level (about 2 m, or 6 ft). It is clear from the graph that the two variables are directly (as opposed to inversely) related; that is, as the predictor increases, so does the predictand. The scatter of points at least suggests a linear relationship.

Of the two questions raised earlier, we first consider the second—finding the mathematical relationship between the two variables. Assuming that this relationship is linear, we know it will be of the form

$$Y = a + bX$$

The solution of this equation is a straight line, where $a$ is the intercept (the value of Y at which this line intersects the Y-axis) and $b$ is the slope of this line. Knowing these two **regression coefficients** enables us to predict Y from X. The regression coefficients are found by **least squares theory,** which says that the line of best fit is the one that minimizes the sum of squared deviations, point to line (shown as $d$ in Figure 6.6). In other words, no line other than that whose regression coefficients we wish to find has as low a sum of squared deviations. The values of $b$ and $a$ are given by

$$b = \frac{N\Sigma XY - \Sigma X \Sigma Y}{N\Sigma X^2 - (\Sigma X)^2}$$

$$a = \overline{Y} - b\overline{X}$$

The line of best fit goes through the point $(\overline{X}, \overline{Y})$.

The values of $a$ and $b$ found in this way are obtained only when Y is regressed on X, as we have done. If the regression is reversed (X on Y), $a$ and $b$ would not be the same. Which variable to regress on which is not a problem in climatological applications, since the causative variable (predictor) can be distinguished from the variable being caused (predictand).

The actual prediction equation for Y is as shown above, except that we need to use either a subscript or superscript to indicate that Y values are estimated (or predicted) from the relationship:

$$Y' = Y_e = a + bX$$

Now we need to show how closely the two variables are related. It should be clear that if every point in Figure 6.6 fell on the line, the relationship would be perfect; conversely, the more the scatter, the poorer is the relationship. We can express this degree of association by the sum of the differences (line to point) squared—the quantity which we just learned had to be minimized to find the line of best fit. If the $Y_e$ are the values estimated from the regression relationship and Y the actual (observed) values, then we have

$$SEE = \left[\frac{1}{n}\Sigma(Y - Y_e)^2\right]^{1/2}$$

where SEE is the standard error of estimate. This gives a measure of how closely the two variables, X and Y, are related.

But notice that SEE includes units of the dependent variable. It is highly desirable to have a measure of association that is independent of these units so that comparisons could be made for all kinds of associations. The **correlation coefficient,** which expresses the degree of association as a number between $-1$ (perfect inverse relationship) and $+1$ (perfect direct relationship), is such a measure. Zero indicates that

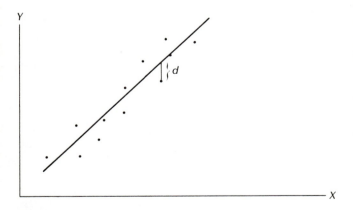

**FIGURE 6.6**
A scatter diagram, on which individual observations are plotted as points. The line of best fit is obtained by minimizing the sum of the squares of all vertical distances, point to line, one of which is shown as $d$.

there is no relationship. The correlation coefficient, $r$, is obtained directly from the data by the following equation:

$$r = \frac{N\Sigma XY - (\Sigma X)(\Sigma Y)}{\sqrt{\left[N\Sigma X^2 - (\Sigma X)^2\right]\left[N\Sigma Y^2 - (\Sigma Y)^2\right]}}$$

If *SEE* and the variance of the dependent variable are known, we use

$$r = \left[\frac{S_Y^2 - (SEE)^2}{S_Y^2}\right]^{1/2}$$

where the correct sign of $r$ (plus or minus) can be determined by inspecting the scatter diagram or from $b$ (the slope), which has the same sign.

The correlation coefficient must be tested for its statistical significance. That is, we must ask, What are the chances that a given $r$, or larger in absolute value, could have occurred by chance? Even in two populations known not to be correlated, nonzero coefficients will occur simply by random variations or chance. If many pairs of size $n$ ($n$ = number of pairs) are taken from two populations known not to be correlated (have a zero coefficient), the $r$'s obtained will in general be normally distributed with a standard error of

$$SE_r = \frac{1}{\sqrt{n-2}}$$

That is, the larger the number of pairs, the smaller is the standard error. For $n$ = 30 and 100, 1.96 standard errors (which includes 95 percent of the values) are 0.37 and 0.20, respectively. Suppose, as an example, that an $r$ of 0.35 is determined for 30 pairs. Is this statistically significant, or, alternatively, could such a value (or larger in absolute sign) have occurred by chance at the 5 percent level? As just stated, 1.96 standard errors includes the interval $-0.37$ to $+0.37$. Thus, since the calculated value is 0.35—indicating a moderate degree of correlation—this $r$ is not statistically significantly different from zero. Suppose the same $r$ (0.35) had been calculated from 100 pairs. Now it is significantly different from zero because it is beyond two standard errors.

### 6.6.2 MULTIVARIATE LINEAR REGRESSION AND CORRELATION

Multivariate linear correlation and regression is frequently used not only in climatological analysis, but in weather analysis and forecasting as well. In this case it is of interest to know the extent, or degree, of correlation between the predictand and two or more predictors, and the functional form of the relationship among them. In addition, there is the problem of determining which predictors to use. Very often there are several potential predictors (i.e., weather variables that might be correlated with the predictand), and the investigator must determine which of these to use. Various screening (selective) methods have been developed to facilitate this process.

The mathematical relationship for multivariate linear regression and correlation is

$$Y' = a + b_1 X_1 + b_2 X_2 + \ldots + b_n X_n$$

where $X_1, X_2, \ldots X_n$ are values of the predictors 1 through $n$, and the $b$'s are their respective regression coefficients. The latter are determined by solving equations similar to those given for bivariate linear regression. The measure of how successful such an equation is in predicting or estimating $Y$ is the unexplained (or explained) variance of $Y$. A perfect prediction would be one which explains 100 percent of the variance of $Y$. The more predictors, the lower is the unexplained variance, but a point of diminishing returns invariably is reached. The selection of which and how many predictors ($X$'s) should be used is a compromise between accuracy and economy.

Just as with bivariate correlation, it is possible to quantify the degree of relationship; in multivariate correlation we examine the degree of relationship between the predictand and the predictors. The measure is the multiple correlation coefficient, $R$, and $R^2$ is the proportion of the variance of the predictand that is accounted for by the regression formula. In a screening process potential predictor variables may be added until a desired value of $R^2$ is obtained. Although adding a predictor will increase $R^2$, this increase for a particular predictor may be negligible, and a compromise must be reached.

The predictors in climatological regression invariably are correlated among themselves, which leads to the problem of redundancy. Sometimes predictor variables can be transformed mathematically into variables which are **orthogonal** (essentially unrelated). In this case there is no redundancy and thus no duplication of explained variance, and each additional predictor now contributes a unique proportion to $R^2$. The method of **principal components** is a particular kind of regression and correlation which uses orthogonal predictors.

# 6.7
## TIME SERIES ANALYSIS

The behavior of meteorological variables with time is of great concern in climatological analysis. Some of these variables have definite periodicities, or cycles; temperature, for example, has both diurnal and annual periodicities and appears to have much longer cycles as well, due to changes in the earth's orbital parameters (see Section 11.7). Although they are much less definite, there may be periodicities in precipitation (a 20- or 25-year cycle in the Great Plains? cycles tied to periods of the moon?) and in small-scale fluctuations of the wind.

The techniques of **Fourier analysis,** also called *harmonic analysis,* have been used to accomplish two general objectives. First, the time behavior of a variable may be mathematically described (modeled) by the use of sine and cosine functions. For example, consider the following temperatures (°F), which are monthly means for a mid-latitude station:

| J | F | M | A | M | J | J | A | S | O | N | D | Mean |
|---|---|---|---|---|---|---|---|---|---|---|---|------|
| 27 | 27 | 36 | 48 | 59 | 67 | 71 | 69 | 62 | 51 | 40 | 30 | 48.9 |

The general formula which reproduces these temperatures is

$$T_t = \overline{T} + A_1 \sin\left(\frac{360}{P}t\right) + B_1 \cos\left(\frac{360}{P}t\right)$$

where $T_t$ is the mean temperature of month $t$ ($t = 1$ for January, $t = 2$ for February, etc.); $\overline{T}$ is the mean annual temperature; $P$ is the fundamental period (in this case, 12 months); and $A_1$ and $B_1$ are coefficients determined by the procedures of Fourier analysis. For the listed temperatures, $A_1$, $B_1$, and $P$ are $-12.91$, $-18.45$, and 12, respectively, and insertion of these into the formula reproduces the temperatures almost exactly.

The subscripts on $A$ and $B$ indicate the *first harmonic* (roughly, a time series with one maximum and one minimum in the fundamental period). One harmonic is all that is required to reproduce these temperatures since, following receipts of solar radiation, they are almost exactly sinusoidal. In Asian monsoon climates the situation is somewhat different. With the onset of rain around June, temperatures no longer increase but tend to level off. Consider mean monthly temperatures (°F) for New Delhi, India:

| J | F | M | A | M | J | J | A | S | O | N | D | Mean |
|---|---|---|---|---|---|---|---|---|---|---|---|------|
| 57 | 62 | 73 | 82 | 92 | 94 | 88 | 86 | 84 | 79 | 68 | 59 | 77.0 |

These temperatures obviously do not exhibit a simple sinusoidal shape, so additional terms (more harmonics) must be added to reproduce them. The formula that accomplishes this is

$$\begin{aligned}T_t = 77.0 &- 5.45 \sin\left(\frac{360}{P}t\right) - 16.02 \cos\left(\frac{360}{P}t\right) \\ &- 4.04 \sin\left(\frac{360}{P}2t\right) - 1.00 \cos\left(\frac{360}{P}2t\right) \\ &+ 0.67 \sin\left(\frac{360}{P}3t\right) - 1.33 \cos\left(\frac{360}{P}3t\right)\end{aligned}$$

where the second and third, fourth and fifth, and sixth and seventh terms on the right-hand side represent the first, second, and third harmonics, respectively.

Strictly, any time series (or a variable distributed in space rather than time) can be mathematically represented by a finite number of sine and cosine terms (harmonics). In general, the simpler the time series, the fewer harmonics are necessary to reproduce it.

Suppose we wanted to reproduce the climatological mean hourly temperature at a station. The formula would have to include terms for the first harmonic, which has a period of 8760 hours, and terms for the 365th harmonic (the daily cycle), which has a period of 24 hours. Such a formula for a station in western Texas is

$$\begin{aligned}T_t = \overline{T} &- 5.40 \sin\left(\frac{360}{P}t\right) - 19.86 \cos\left(\frac{360}{P}t\right) \\ &- 10.10 \sin\left(\frac{360}{P}365t\right) - 4.88 \cos\left(\frac{360}{P}365t\right)\end{aligned}$$

where $T_t$ is the temperature (°F) at hour $t$ (1 to 8760); $\overline{T}$ is the mean annual hourly temperature; and $P$ is the fundamental period of 8760 hourly values (the number of hours in a year). You can verify that, with a $\overline{T}$ of 59.7, inserting $t = 1901$ and $4695$ (which correspond to 0500 on March 21 and 1500 on July 15, respectively) gives a $T_t$ of 39.4° and 90.8°. Here the second and third terms on the right-hand side comprise the first harmonic (annual periodicity), and the fourth and fifth terms comprise the 365th harmonic (diurnal periodicity).

Notice that this aspect of time series analysis is similar to regression. With the latter, regression coefficients obtained from observed data—a scatter of points—give a line of best fit. Here the time series coefficients (the numbers preceding the sine and cosine terms) also are obtained from the data, and, as shown by the example, can be used to reproduce (model) the meteorological time series. Further manipulation of these coefficients gives the amplitude (one-half the range) and the phase (the time of maximum and minimum) of each harmonic.

The second objective of time series analysis is to find periodicities. **Spectrum analysis** is the term used for this search. If periodicities are found, time series analysis may lead to a better understanding of the fundamental physics underlying the time variation of meteorological variables. For example, it has been found that short-term fluctuations of the surface wind's vertical component can be differentiated according to whether turbulence is of thermal or mechanical origin.

In short, the methods of time series analysis (1) enable a long series of observations of a variable to be expressed in a single, relatively simple equation; (2) may lead to a better understanding of the physical processes that underly the time variation of meteorological variables; and (3) make it possible to predict the value of a meteorological variable at some time in the future.

**SUGGESTED READING**

BROOKS, C.E.P., and N. Carruthers. *Handbook of Statistical Methods in Meteorology.* London: Her Majesty's Stationery Office, 1953.

MUNN, R.E. *Biometeorological Methods.* New York: Academic Press, 1970.

PANOFSKY, H.A., and G.W. BRIER. *Some Applications of Statistics to Meteorology.* University Park: The Pennsylvania State University, 1958.

SIEGEL, S. *Nonparametric Statistics,* New York: McGraw-Hill Book Company, 1956.

U.S. DEPARTMENT OF COMMERCE, WEATHER BUREAU. *Climatology at Work: Measurements, Methods, and Machines,* ed. G.L. Barger. Washington, D.C., 1960.

# 7
# PATTERNS OF THE CLIMATIC ELEMENTS

**INTRODUCTION**

**TEMPERATURE**
Spatial Patterns
Temporal Patterns

**PRECIPITATION**
Spatial Patterns
Temporal Patterns
Other Aspects of Hydrometeors

**AIR MASSES**

**CLIMATIC ITERATIONS**
Temperature
Precipitation
Climatic Types

**CLIMATIC EXTREMES**
Temperature
Precipitation
Other Climatic Extremes

## 7.1 INTRODUCTION

The study of climate focuses primarily on the place-to-place (spatial) variations of the elements that comprise it. Indeed, the very term *climate* implies not only an average state of the atmosphere—with the recognition that actual conditions vary from one time to the next—but also that this average state is different from one area of the globe to another. Studying the spatial variations of the most important climatic elements is therefore indispensable to an understanding of climate classification systems considered in Chapter 8 and the geographical distribution of climatic types delineated by these systems.

Rather than attempt detailed analyses of each of the many climatic elements, we will emphasize just two: temperature and precipitation. For each, we are most concerned with mean monthly and mean annual values. For example, for temperature we are most interested in the months showing extremes, which for most of the earth are January and July. Of course, the nature of other elements may be deduced from temperature and precipitation. A very rainy month at a particular location implies relatively low insolation, extensive cloudiness, and high humidity. A month with little or no precipitation and high temperatures usually implies the reverse of such conditions.

We have already examined the reasons for the spatial patterns of temperature and precipitation in Chapters 2 through 5. Because Section 3.5 covers the spatial variation of temperature, in this chapter we will emphasize only the horizontal change of mean monthly and mean annual temperature at instrument shelter level. Remember also that smaller scale (meso- and microscale) influences cannot be shown on global maps, and thus the fields of temperature and, especially, precipitation are generalized.

## 7.2 TEMPERATURE

Global temperature patterns result from five controls: (1) insolation; (2) thermal contrasts arising from land-water differences; (3) elevation and topography; (4) ocean surface temperature; and (5) atmospheric circulation.

Insolation is the most dominant control of global temperature patterns. Solar radiation received at the earth's surface varies principally with latitude, with the largest receipts in low latitudes (see Figures 2.6–2.9). The variation of insolation throughout the year also varies with latitude, but in the opposite way; monthly receipts in low latitudes are very similar, while those at the poles vary markedly. These patterns are illustrated in Figure 2.10, which shows mean annual temperature and mean annual temperature range (difference between the warmest and coldest months) on an earth of uniform substance or in a highly generalized actual depiction.

The varying thermal properties of water and land surface substances, combined with much greater evaporation over water and the partial transparency of water to solar radiation, lead to higher temperature ranges over continents than over oceans. Continents, particularly their interiors, are colder in winter and warmer in summer than oceans at the same latitude. In the tropics these differences are not highly significant, but they become increasingly so poleward. This is such a dominant control of temperature that it has been quantified by the parameter *continentality*, which is explained and illustrated in Section 3.6. The differences between marine and continental climates are discussed further in Section 8.5.

The average lapse rate in the troposphere is 6.5C° per km (3.5F° per 1000 ft). Temperatures therefore decrease with elevation, a feature that is shown only on maps on which temperature is not reduced to sea level. Topography has the effect of blocking and channeling maritime influences. Mountains may act as a barrier to the inland penetration of mild air and, at the same time, channel this flow through relatively low passes. To a lesser extent the same is true for air moving outward from a continent. Such effects may be discernible on global maps of mean monthly temperature (see Figures 7.1 and 7.2).

The temperature of maritime air that is advected onto the continents is strongly influenced by ocean surface temperature. There are locations where the surface water, and thus the air temperature on the adjoining littoral, is anomalously cool or warm. As explained in Section 4.8.2, anomalously warm waters

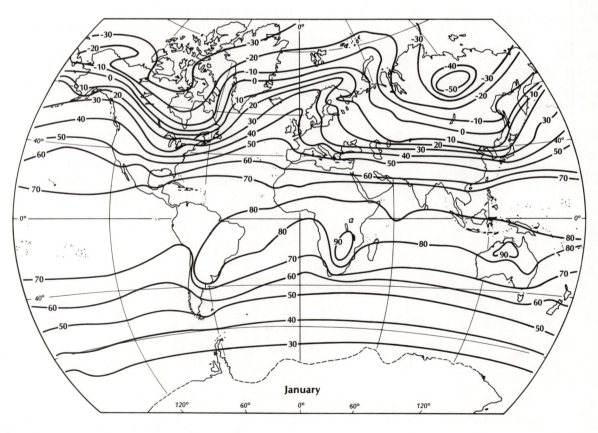

**FIGURE 7.1**
Mean January temperature at sea level. (From *Weather and Climate* by C. E. Koeppe and G. C. DeLong. Copyright © 1958 by McGraw-Hill Book Company. Used with the permission of McGraw-Hill Book Company.)

# SEC. 7.2 | TEMPERATURE

lead to appreciable warming in winter and to modest cooling in summer; anomalously cool waters lead to rather marked cooling in summer and slight warming in winter.

The preceding four controls are reasonably independent of one another. The circulation of the atmosphere, the fifth control, acts in concert with the first four. It is through a continuous exchange of air that latitudinal differences in insolation are compensated. Atmospheric circulation tends also to minimize land-water differences, and is the mechanism by which the influence of ocean surface temperature is propagated inland and modified by topographic barriers.

## 7.2.1 SPATIAL PATTERNS

Before we can describe spatial patterns of temperature, we must be familiar with the various measures of temperature. In any 24-hour period (midnight to midnight) there is a highest and a lowest temperature. The average of these is often taken as the mean daily temperature. (This is the practice in the United States; in some other countries the 24 on-the-hour temperatures are averaged.) All such means for a month, averaged over all the days in it, constitute the mean monthly temperature. Similarly, by averaging the daily maxima and minima for a month we obtain the mean maximum and mean minimum temperatures, respectively, for that month. This process is repeated for the same month of many years (usually 20 to 40), and long-term means of these three parameters are the result. However, these long-term means are—perhaps inconsistently—still called mean monthly, mean maximum, and mean minimum temperatures for a particular month, even though they are actually means of means.

Mean monthly sea level temperatures in January and July are shown in Figures 7.1 and 7.2. These are reduced to sea level, so the effects of elevation are not apparent. As explained previously, January and July are chosen because for most areas of the earth outside of the tropics these are either the warmest or

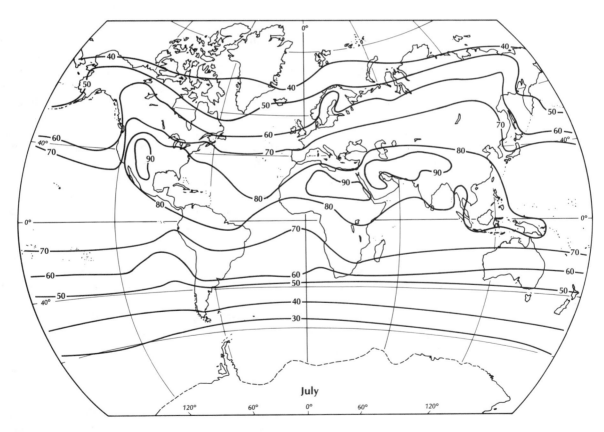

**FIGURE 7.2**
Mean July temperature at sea level. (From *Weather and Climate* by C. E. Koeppe and G. C. DeLong. Copyright © 1958 by McGraw-Hill Book Company. Used with the permission of McGraw-Hill Book Company.)

coldest months. In both maps the patterns are predominantly zonal, but are less so in the tropics, where the latitudinal variation in insolation is much less pronounced than in higher latitudes. Land-water differences are indicated by equatorward and poleward deflection of isotherms in the winter and summer hemisphere, respectively. This feature is of course much more apparent in the Northern Hemisphere, which has a much higher proportion of land to water.

The combined influences of atmospheric circulation and sea surface temperatures are most obvious in northwest Europe, and coastal Alaska and British Columbia in January, when these areas are quite mild for their latitudes. Topographic influences also are apparent in these areas, since maritime air is advected far inland in Europe but is restricted to coastal areas of both northern North America and southern South America. Relatively cool ocean currents result in summer temperatures somewhat lower than would otherwise be the case; this is especially apparent along the west coasts of both South America and South Africa in January, and off the west coast of North America and northwest coast of Africa in July. Finally, notice the overall latitudinal shift of temperature: northward in July, southward in January. Monthly mean temperatures, such as those shown in Figures 7.1 and 7.2 indicate nothing of the diurnal range. Two stations may have mean daily maximum and minimum temperatures for a month of 30° and 10°C, and 25° and 15°C, respectively. Both have a mean monthly temperature of 20°C, but they are otherwise different. To show how warm (hot) a particular location is, we use the mean monthly maximum temperature for the month in which this is the highest. We will call this parameter $H$. The mean monthly minimum temperature for the month in which this is the lowest is the parameter $L$, which indicates how cool (cold) a particular location is.

Maps of $H$ and $L$ are shown in Figures 7.3 and 7.4. Temperatures are not reduced to sea level. Notice that within the very lowest latitudes (±15°), $H$ is less than 38°C (100°F). Higher temperatures occur in somewhat higher latitudes, in what will be recognized as arid and semiarid areas. The zonal pattern is now less apparent on both maps than it was in Figures 7.1 and 7.2, but the effects of land-water differences (continentality) are even more pronounced. On the map of $L$, areas exceeding 21°C (70°F) experience little relief from the heat, and the effect of a warm sea surface combined with prevailing westerly winds is very noticeable, especially in northwestern Europe. Remember that these figures are considerably simplified. Some mountainous areas, such as the Tibetan massif and the Andes, are not well represented for this reason and because of the paucity of stations.

Figure 7.5 depicts the mean annual temperature variation by the difference between $H$ and $L$. This is not the same value that would be obtained by finding the difference between the mean temperatures of the coldest and warmest months; the latter is smaller because $(H - L)$ incorporates both annual and diurnal ranges. Values of $(H - L)$ below 11C° (20F°) are within the tropics in all except a few locations, such as those areas in the Southern Hemisphere with an ocean-dominated climate and parts of the west coast of the United States. In general the mid- and subtropical latitudes have values ranging from 22C° (40F°) to 44C° (80F°). A value above 44C° indicates extreme continentality. $(H - L)$ reaches a maximum of 72C° (130F°) at Verkhoyansk, Siberia. At Quito, Ecuador, a high elevation station almost exactly on the equator, $(H - L)$ is 16C° (29F°). This compares with a value for San Francisco of only 13C° (24F°).

In Figures 7.1 and 7.2 diurnal ranges of temperature are filtered out, while Figures 7.3, 7.4, and 7.5 incorporate both diurnal and seasonal ranges. In Figure 7.6 we show only the mean diurnal range of temperature. This is obtained by finding the difference between the mean maximum and mean minimum temperatures for each month and then averaging these twelve values. Diurnal range is fairly conservative in that it generally changes little from month to month at most locations. It increases with continentality and decreases with the moisture content of the air; that is, other things being equal, a dry location will have a larger diurnal range than one that is humid. There is also an effect related to latitude, but this effect is opposite to that of seasonal temperature range. In general, high latitudes have small diurnal ranges because the variation of the sun's altitude over a 24-hour period is greater in low than in high latitudes, especially in winter. See Section 3.4 for a more detailed discussion of the factors influencing diurnal range.

The mean diurnal range is therefore a maximum in low or mid-latitude desert areas that are continental. Values of 11 to 17C° (20 to 30F°) are typical of these areas. Lower values are found in high latitudes and in marine climates. Diurnal ranges in excess of 17C° occur in the arid and high elevation areas where, because of the reduced air density and lack of water vapor, the atmosphere's moderating effect on diurnal swings is greatly reduced (see the discussion of the greenhouse, or atmospheric, effect in Section 2.3.5).

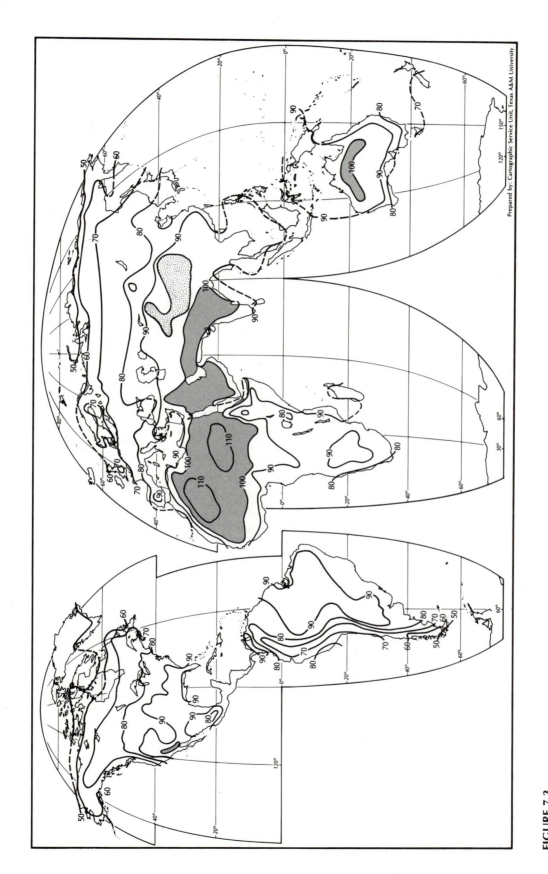

**FIGURE 7.3**
The highest mean monthly maximum temperature (°F).

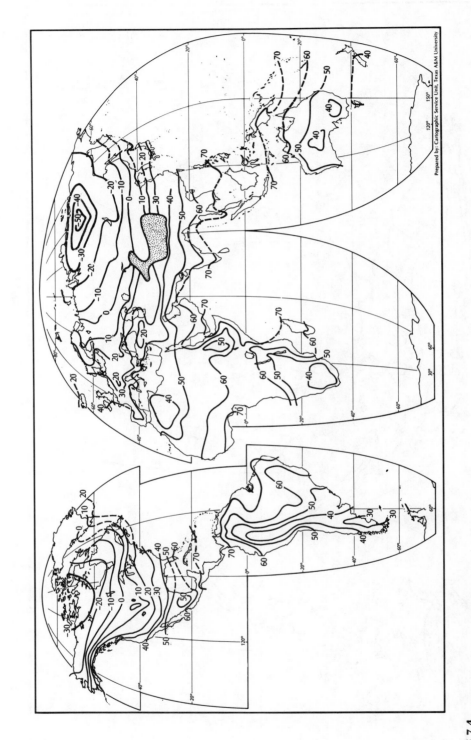

**FIGURE 7.4**
The lowest mean monthly minimum temperature (°F).

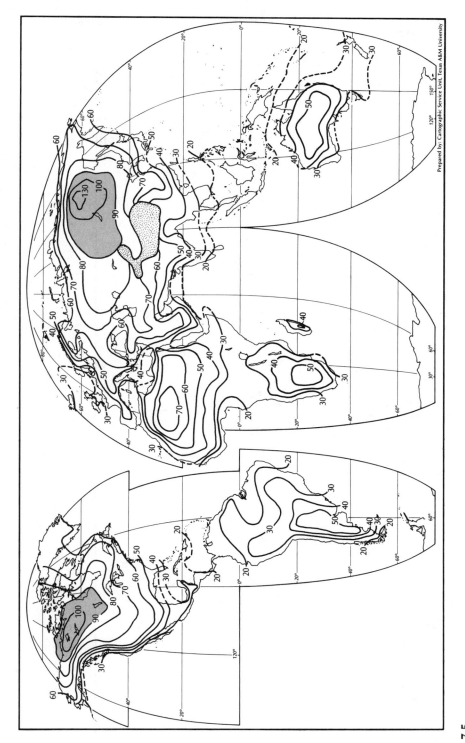

**FIGURE 7.5**
The mean annual temperature variation (°F).

155

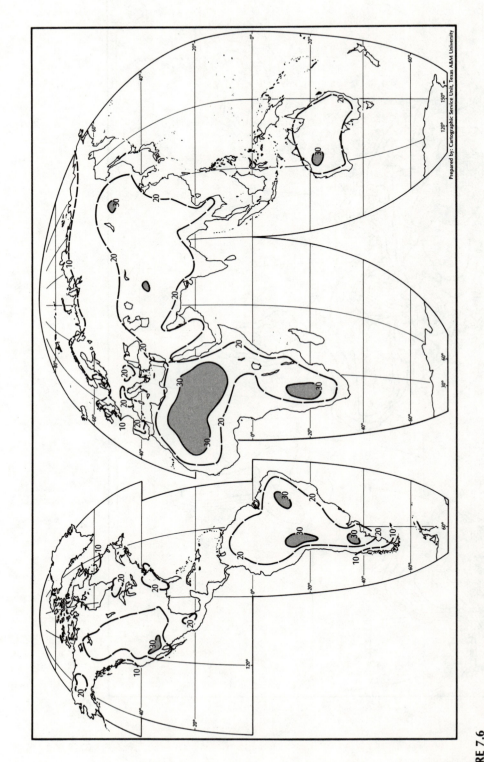

**FIGURE 7.6**
The mean annual diurnal range of temperature (°F).

## 7.2.2 TEMPORAL PATTERNS

The variation with the month, or seasons, of mean, mean maximum, and mean minimum temperatures and of temperature changes throughout the day also are of concern in the study of climate. Temporal variations were explained in Section 3.4 and illustrated by Figures 3.4 to 3.8 (seasonal) and 3.9 (diurnal).

To summarize, monthly mean temperatures coincide more or less with insolation. Exceptions occur in the rainy tropics, where insolation varies little from month to month and where changes of cloud cover, rain, and precipitation are more significant than those of radiation. Proximity to large water bodies reduces the annual range of temperature in mid-latitudes, especially on west coasts; continental areas, on the other hand, show a larger range. The drier the climate, the more pronounced are swings of both annual and diurnal range. In general marine climates have maximum and minimum monthly temperatures that lag highs and lows of insolation by about two months, while continental climates show a lag of approximately one month. In some areas—the Gulf coast of the United States, for example—summers show maritime characteristics, with a lag of one and one-half months, while in winter the lag is more typically continental, or about one month.

# 7.3
## PRECIPITATION

The two most important aspects of precipitation for climate in general, and for climatic classification in particular, are mean annual amount and the variation of mean monthly amounts within the year. And, just as with temperature, there are other aspects of interest. For precipitation, the counterparts of mean number of days above or below specified threshold temperatures and the length of the growing season—to name just two aspects (see Section 3.7)—are its intensity, its variability, the distinction among its forms (rain, snow, hail, etc.), and the number of days with precipitation in a year or a month. These aspects are considered later in this chapter.

### 7.3.1 SPATIAL PATTERNS

The controls of global precipitation patterns are not nearly as definitive as are the controlling factors of temperature. This is because temperature is a response to site-specific influences (e.g., insolation is latitudinally differentiated, and continentality depends on distance from a large water body), whereas precipitation-producing mechanisms are not nearly as specific with respect to location. In addition, whether or not precipitation occurs depends on a combination of events: first, air at or near saturation, and second, some mechanism that causes air to rise, cool, and condense the water vapor in it. Thus, the day-to-day prediction of the occurrence, form, and amount of precipitation is much more difficult than the prediction of temperature.

### Temperature

Temperature is one of the controls of the geographic variation of mean annual precipitation. This may seem surprising, but remember that the capacity of air to hold moisture—and thus the amount of water precipitable from it—increases with temperature. High-temperature areas and times thus have the potential for precipitation. The highest amounts of precipitation over short periods of time (an hour or a day) almost always occur in warm or hot situations, for example, during "cloudbursts." Conversely, where and when low temperatures are the rule, the amount of water vapor and thus the amount of precipitation is very limited. Therefore, the polar regions, the interiors of high-latitude continents in winter, and areas of high elevation experience little precipitation, even though there may be atmospheric mechanisms present to lift and cool the air.

Meteorologists express this potential for precipitation with the quantity **precipitable water.** Precipitable water is the equivalent depth, in centimeters or inches, of precipitation that would occur if all of the water vapor above a particular location could be condensed and would fall as rain or snow. Maps of mean precipitable water for January and July for Canada (Figure 7.7) demonstrate its seasonal range. Notice the general correspondence with temperature (see Figures 7.1 and 7.2); that is, the colder the location and time, the less is the precipitable water. It is particularly interesting that the Great Lakes do not act as a local moisture source to increase precipitable water in July, but instead reduce it slightly because of the somewhat cooler lake surfaces.

Another way in which temperature influences precipitation is through its role in promoting convection. Convection of thermal origin, which results from near-surface lapse rates being steepened (Section 5.8.1) by relatively intense surface heating, is a highly significant precipitation-producing mechanism in the tropics more or less throughout the year and in the mid-latitudes in spring and summer. The subtropics

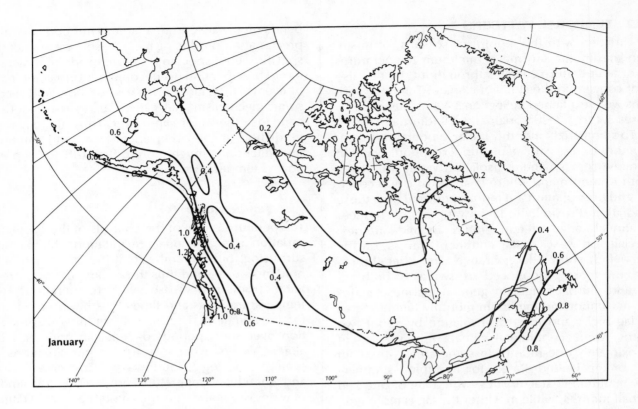
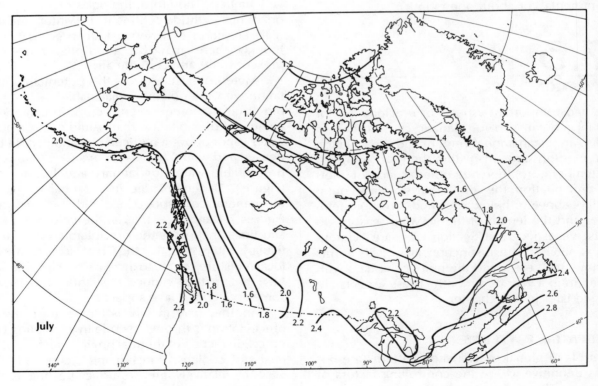

**FIGURE 7.7**
Mean precipitable water (cm) from 1957 to 1964 for Canada. (After J. E. Hay, 1970. From R. A. Bryson and F. K. Hare, *World Survey of Climatology: Climates of North America*. Vol. 11. Amsterdam: Elsevier Scientific Publishing, 1974.)

are intermediate in this respect. Of course, dry convection can occur with large near-surface lapse rates. Air may be forced to rise and cool, but without sufficient moisture clouds will not form. This again emphasizes that both factors—adequate moisture and a lifting mechanism—need to be present.

Thermal convection leading to precipitation also may occur at upper levels when shifting winds there promote a flow of cold air over warm, which results in steepened lapse rates. This is largely a mid-latitude mechanism, because only here is there likely to be sufficient horizontal temperature contrast. Synoptically (as in weather map analysis) this often occurs during the passage of short waves in the mid-troposphere, sometimes with no indication of a lifting mechanism apparent from surface analysis.

### Convergence/Divergence Areas

The bringing together, or convergence, of air invariably causes that air to rise (see Section 5.8.1). This is apparent on a global scale in the meeting of the trade winds at the ITCZ (see Section 4.6.2), in the inward and upward spiraling of air in low pressure centers, and in the convergence zones in troughs associated with these lows (see Figure 5.17). If sufficient moisture is available, convergence areas provide the mechanism required to cool air to saturation and initiate precipitation. The ITCZ is reasonably site-specific in that it is located at or within a few degrees of the equator in most longitudes (see Figures 4.12 and 4.13), although it is not a continuous trough—particularly over land areas—and migrates northward and southward with the seasons. The result is a circumequatorial belt of moderate to large annual precipitation in most longitudes. Thermal convection also may be involved, especially over land areas, and topographic variations may induce orographic precipitation.

On the other hand, cyclones are weather features found virtually everywhere except in the very lowest latitudes. They are most important as contributors to precipitation, however, in the mid-latitudes. Here, where they are an integral part of transient disturbances (see Section 4.6.3), they are most frequent. The latitudinal span in which such transient disturbances occur was identified in Chapter 4 as the polar front zone, the meeting place of warm, generally humid air on the equatorward side and cool, drier air poleward. Although the polar front zone is a mid-latitude feature, it is by no means fixed in location and tends, even more so than the ITCZ, to be discontinuous around the globe. Nevertheless, much of the precipitation received in the mid-latitudes results from the convergence areas associated with transient disturbances.

Much less frequent than mid-latitude disturbances, but still significant because of the large rainfall amounts which they produce, are storms of tropical and subtropical origin such as tropical depressions, easterly waves, and hurricanes and typhoons. They can affect vast areas of the oceans, where they originate, as well as littoral areas of the subtropics (see Figure 4.26). Indeed, it is not unusual for some marginal areas of the tropics to receive their average annual rainfall in one such storm.

In associating areas of relatively low pressure with convergence, we need to distinguish between the transient lows, which we have just discussed, and those of a semipermanent nature which are geographically fixed. In Figures 4.12 and 4.13 distinct climatological lows appear in July in the southwest United States and from northwest India to the Sahara. These are sometimes referred to as **thermal lows** because they appear to originate from thermal contrasts between the warm land and relatively cool water at this time of year (see Figure 4.10). Ostensibly, thermal lows should provide the mechanism for bringing rain to these areas, but there is an important difference between these cyclones and those that are transient. Transient lows have much greater vertical extent, and it is not at all unusual to find closed lows extending from the surface to the 300-mb, or even 200-mb, surfaces. Thermal lows, however, are almost without exception shallow features overlain by high pressure which is apparent as low as 850 mb (roughly 1500 m or 5000 ft). Over the southwest United States and the Sahara this high pressure is the eastern extension of the subtropical highs, which have great vertical as well as horizontal extent. This high pressure weakens eastward of northern Africa so that in northwest India the surface low extends through a deeper layer of the atmosphere (Figure 7.8). Here, during the rainy period of the monsoon, rainfall is abundant.

The other semipermanent cyclones are the subpolar lows (see Figures 4.12 and 4.13): the closed centers in the North Atlantic and North Pacific, and the circumglobal trough at about 60° to 70° in the Southern Hemisphere. These also are areas of generally large rainfall, given relatively low temperatures; but, unlike the geographically fixed thermal lows, subpolar lows are mostly a statistical composite resulting from the passage of transient cyclones through, and their origin within, these areas.

Whereas convergence is a mechanism for promoting upward air motion, divergence acts in the opposite way. Globally, the principal anticyclones are the subtropical highs. It is on their eastern sides that subsidence, which accompanies divergence, is most pronounced (see Figure 4.18). Some of the driest areas on the earth, in terms of mean annual precipita-

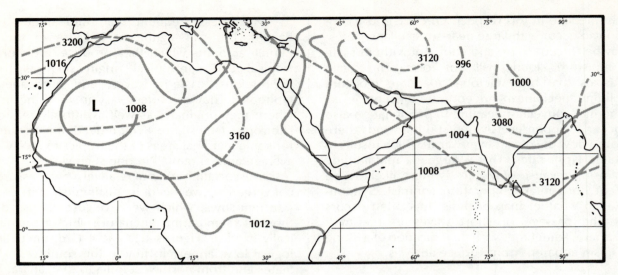

**FIGURE 7.8**
In July a broad area of low pressure at the surface extends from northwest Africa to northwest India. Relatively low pressure is maintained at 700 mb over India, but changes to high pressure (heights) over northwest Africa at this level. Solid lines are sea level isobars (mb), and dashed lines are heights of the 700-mb surface (m). (After H. van de Boogaard, *The Mean Circulation of the Tropical and Subtropical Atmosphere—July*. NCAR Technical Note. Boulder, Colo.: National Center for Atmospheric Research, 1977.)

tion, are located here. They include the southwest United States and northwest Mexico, coastal southern Peru and northern Chile, northwest and southwest Africa, and much of Australia.

Semipermanent anticyclones that inhibit the formation of precipitation are those over the Antarctic throughout the year and over northeastern Asia in winter. Transient highs, a part of weather disturbances of middle latitudes, also are unfavorable to precipitation, but their duration over particular areas is short. There are occasions, however, when long dry spells occur as the result of highs stagnating or moving very slowly. This is a fairly frequent event in fall in the eastern, and especially southeastern, United States. Transient highs are considered in connection with the potential for air pollution in Section 12.4.

### Mechanical Uplift

Mechanical uplift's role in promoting rising air was discussed in Section 5.8.1. It includes air forced upward and over topographic barriers and frontal discontinuities. Frontal discontinuities occur in convergence zones in extratropical depressions, and thus are usually located in troughs, which may be isolated from transient disturbances or extend from cyclones (see Figure 4.22, March 20). Uplift over fronts, therefore, contributes to precipitation only in the middle latitudes, and to a lesser extent in the high latitudes. Most such precipitation occurs in winter, when the horizontal temperature contrasts which promote *frontogenesis* (the formation of fronts) are most marked.

Uplift over topographic barriers (the orographic effect) is very site-specific. Large annual precipitation is the rule where mountains, or even hills, are located in areas to which prevailing winds advect moist air. In addition to simple uplift leading to cooling, condensation, and precipitation, the following processes may be involved in orography:

1. Initial uplift that triggers instability (i.e., an entire layer of air is made unstable, as explained in Section 5.8.1) so that the resultant precipitation is forced by both orography and thermal convection.

2. A slowing of the movement of cyclones and troughs due to friction, increasing local precipitation.

3. Topographic variations which funnel airflow, causing convergence and uplift.

Areas where the orographic effect is dominant include Cherrapunji, India, in the foothills of the Himalayas, whose mean annual precipitation is 11,430 mm (450 in.) and whose record year registered 26,467 mm (1042 in.); Mt. Waialeale on the island of Kauai in Hawaii, whose mean annual precipitation is 11,700 mm (460 in.); and a location in the Cameroon Mountains in Africa, whose mean annual precipitation is 10,300 mm (405 in.). Highest mean annual snowfall occurs where mountains lie athwart the prevailing westerlies in the middle-high latitudes, such as at Paradise Ranger Station on the slopes of Mt. Rainier in the state of Washington (14,800 mm, or 582 in., mean annual precipitation). There may of course be more rainy and snowy places than those just cited, but these locations are where long-term measurements have been made.

Orography can act to reduce as well as to enhance precipitation. This is referred to as the *rain shadow effect*, which was described in Section 5.8.1.

### Ocean Surface Temperatures

Contributions to the water vapor over continents by local sources—evaporation from water surfaces and transpiration—are largely inconsequential. Virtually all of this water vapor was obtained from evaporation over large bodies of water such as oceans, seas, and gulfs (see Figure 5.5). Evaporation from an ocean surface depends principally on its temperature and increases exponentially with it. For example, the increase in water vapor holding capacity of air warmed from 27° to 32°C (80° to 90°F) is 1.33 times the increase from 21° to 27°C (70° to 80°F). The actual amount of water evaporated depends of course on factors other than temperature, but globally there is a good correspondence between the two (see Figures 4.32, 4.44, and 5.6). We made this same point earlier in connection with air temperature over the continents.

Although ocean surface temperatures vary horizontally much less than the temperatures of air over land, the variation that does occur helps to explain precipitation. Because surface water and the air in contact with it have very nearly the same temperature, and because typical relative humidities are around 80 percent, the amount of water vapor available to be advected onto continents parallels the spatial and seasonal variation of water temperature. In locations where surface water is anomalously cool or warm (see Figure 4.51), there is a corresponding reduction or increase in both advected water vapor and in the precipitation that occurs. These associations are particularly noticeable in the middle and subtropical latitudes. For example, water evaporated from the relatively warm waters of the Kuro Siwo–North Pacific Drift and the Gulf Stream–North Atlantic Drift and advected by prevailing westerly winds onto northwestern North America and northwestern Europe, respectively, leads to very moist regimes. Annual precipitation may not be high—although it is high where the orographic mechanism applies, as in extreme western North America—but these areas have a high number of rainy days and overcast skies.

Other areas where ocean surface temperatures affect precipitation on the littoral are those where such temperatures are cooler than expected for the latitude (anomalously cool). In the subtropics these areas are found off the western coasts of continents in the span from about 25° to 40° (see Figure 4.51), where they coincide with the eastern sides of the subtropical highs. These cool surface waters are yet another reason for very low precipitation in these areas. Figure 7.9 summarizes these influences and contrasts the eastern and western sides of the subtropical highs (western and eastern sides of continents) with respect to subsidence and ocean surface temperatures. Also shown is the upwelling characteristic of and contributing to anomalously cool surfaces, and the height of the bottom of the **trade wind inversion.** The height of the bottom of the inversion marks the discontinuity between near-surface air with lapse rates forced by surface heating (thus tending to be unstable) and the

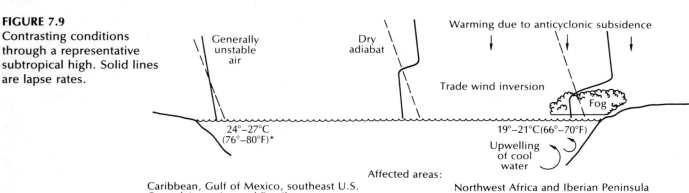

**FIGURE 7.9**
Contrasting conditions through a representative subtropical high. Solid lines are lapse rates.

Affected areas:
Caribbean, Gulf of Mexico, southeast U.S.
Coastal Argentina and Brazil
Western North Pacific and eastern Asia
 (Japan and coastal China)

Northwest Africa and Iberian Peninsula
Southwest Africa
Coastal California, Baja California
Southwest Australia

*Temperatures apply only to southern North Atlantic, but the east-west differences are typical.

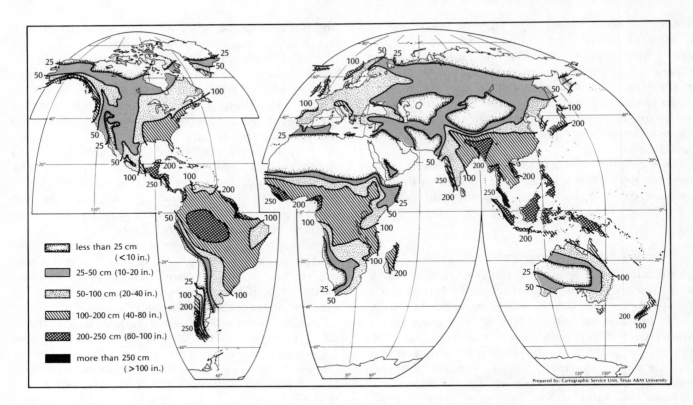

**FIGURE 7.10**
Average annual precipitation over the globe. (After A. Miller and J. C. Thompson, *Elements of Meteorology*, 3d ed. Columbus, Ohio: Charles E. Merrill, 1979.)

cloud-free air above, which is warmed and dried by subsidence. Perhaps surprisingly, fog is a frequent occurrence in these areas. When prevailing winds advect warm, moist air from the west over cooler water, that air is cooled from below, often to the point of saturation, and fog results.

Global mean annual precipitation is shown in Figure 7.10. The controls described in this section, weighted according to their varying influences from place to place, are generally sufficient to explain the broad-scale variations from dry to wet. However, there are locations with departures from expected amounts. In many instances these departures are explained by regional or local controls or by the particular configuration of land areas and their position with respect to the prevailing atmospheric circulation. A more detailed account of spatial precipitation variations, as well as of the other climatic elements, will be found in Chapters 9 and 10, which are devoted to smaller scale (meso- and micro-) influences.

## 7.3.2  TEMPORAL PATTERNS

There are two aspects to the temporal variation of precipitation. The first is seasonal variation, the changes in mean monthly precipitation among the months; and the second is diurnal variation, the changes in precipitation that occur within a day. These aspects are quite similar to temporal variations of temperature.

### Seasonal Variations

The primary control of changes in mean monthly precipitation within the year is the latitudinal shifting, with the seasons, of the principal zones of convergence: the ITCZ and the polar front zone. Remember that the latter is more diffuse and less site-specific than the former. Also, neither is continuous around all longitudes at any time, and the polar front zone occupies a wider latitude span over the continents than over oceans. A representative value for the displacement of the ITCZ is 20° of latitude, although this figure is much less over the Pacific Ocean and greater over the Indian Ocean (see Figures 4.12 and 4.13). The polar front zone shows a latitudinal shift of perhaps 10° to 15° on the average. For both convergence zones the maximum northward displacement occurs in July and the maximum southward displacement in January.

Now consider the timing of rainfall due to the presence of the ITCZ. Rain occurs in the very lowest

## SEC. 7.3 | PRECIPITATION

latitudes (about ±10°) more or less throughout the year. In higher latitudes in the Northern Hemisphere rainfall will be plentiful through this mechanism in summer, when the ITCZ is in the vicinity, but there will be a relatively dry period when the trough is displaced into the Southern Hemisphere six months later (summer in the Southern Hemisphere). Thus, in areas flanking the span in which rain occurs throughout the year, there are zones in which summers are wet and winters are dry. With distance poleward, to the latitude of the subtropical highs, summer rainfall decreases. The same reasoning can be applied to precipitation resulting from latitudinal shifting of the polar front zone. Adding the primary divergence zones centered at about 30° N and S and those over the poles completes the characterization of precipitation regimes, which are shown in Figure 7.11. This schematic diagram implies that the dry zones at about 30° (or a little less) are continuous around the globe. We know that this is not so because it is only eastward of the centers of the subtropical highs that aridity prevails. Thus, the schematic applies only to the western sides of continents. On the eastern sides summer rainfall is provided principally by localized convection leading to showers and thunderstorms in generally unstable air (see Figure 7.9), although larger-scale subtropical disturbances such as easterly waves and tropical storms are significant both in summer and fall.

Another exception to the schematic in Figure 7.11, and a secondary control of seasonal regimes of precipitation, involves seasonal reversals of circulation. Marked seasonal reversals occur only in the Northern Hemisphere, where continents are large enough to appreciably disrupt the zonal wind belts. In Asia, and to a lesser extent in North America, a general convergence toward these continents brings clouds and water vapor to them in summer. The lifting mechanisms are instability, expected in the heated

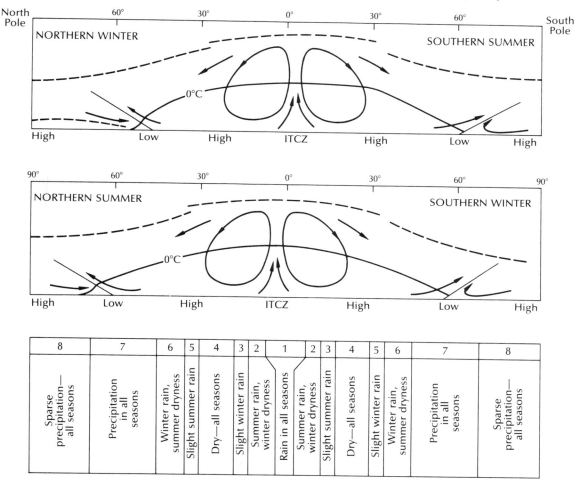

**FIGURE 7.11**
Schematic meridional cross section showing the main zones of ascending and descending motion and delineating the major seasonal regimes of precipitation. (From *Introduction to Meteorology*, 3d ed., by S. Petterssen. Copyright © 1969 by McGraw-Hill Book Company. Used with the permission of McGraw-Hill Book Company.)

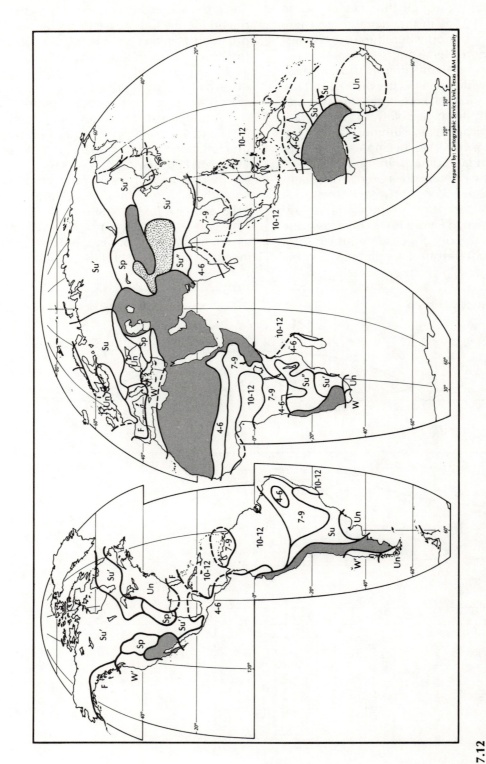

**FIGURE 7.12**
Regimes of annual precipitation. (See text for explanation.)

164

interiors, and transient disturbances, at least north of about 40°. Of course, orography also plays a localized role. In winter, when divergence prevails, the circulation is reversed (from continent to ocean), and the interiors are generally dry. The effects of the seasonal circulation reversal are augmented by very cold and warm interior temperatures in winter and summer, respectively, with consequences for the water vapor-holding capacity of air.

The combined effect of the latitudinal shifting of the principal zones of convergence and seasonal reversals of circulation, as well as of controls that apply only on smaller scales, is shown in Figure 7.12. This rather detailed map needs some explanation. First, it excludes areas of desert (diagonal lines) because precipitation regime means very little there. The criterion for exclusion is that precipitation, in inches, must be less than one fifth of the mean annual temperature, in °F. It may seem surprising that deserts are not defined simply as those areas receiving less than $x$ inches (or millimeters) annually. But remember that desert implies lack of water, which is a consequence of both precipitation and evaporation. In general the higher the temperature, the greater is the evaporation (see Section 5.6); thus, a combination of precipitation and evaporation is necessary to define deserts. This important relationship will be encountered again in Chapter 8 when we discuss climatic classifications.

Second, in the tropics the numbers used in Figure 7.12 represent the number of wet months, where *wet* is defined as more than 50 mm (2 in.) of precipitation per month. In the central Amazon, central Africa, and throughout the Indonesian Islands, for example, from 10 to 12 months are wet.

Third, outside the tropics areas of uniform precipitation are first delineated. *Uniform* (abbreviated UN in Figure 7.12) means that the ratio of the driest three consecutive months to the wettest three consecutive months is at least 0.5. For the remaining extratropical areas, the letters indicate the season in which precipitation is concentrated. For instance, in the Northern Hemisphere winter (W) is December to February, spring (Sp) is March to May, summer (Su) is June to August, and fall (F) is September to November. Season indicators without a prime (') mean that from about 30 to 40 percent of the annual precipitation occurs in that season. A single prime represents a concentration of from 41 to 60 percent, and a double prime ('') signifies a concentration greater than 60 percent. For practical purposes this means that no prime, a single prime, and a double prime indicate areas with concentrations of about one third, one half, and two thirds, respectively, in the indicated season.

In Figure 7.12 it is clear that with increasing distance from the equator the number of wet months generally decreases—from 10 to 12 in the aforementioned areas, to 7 to 9 months, to 4 to 6 months. In the latter two categories the center of these periods occurs shortly after the time of high sun, as is more or less expected from the model in Figure 7.11.

Farther poleward there is again general agreement with the model, which is modified by east-west differences across subtropical highs and circulation reversals over the continents. Here one finds summer concentration of precipitation in the interior of continents; uniform precipitation on the eastern sides of continents; and a progression, poleward of the deserts, from winter concentration to uniform or fall concentration. Notice that fall and spring concentrations cannot be explained in terms of the model with its modifications. These exceptions are considered in Section 8.3.

To summarize, there are three main precipitation regimes outside the tropics: uniform, winter maximum, and summer maximum. Smaller areas have a spring maximum, a fall maximum, or a double maximum (not indicated in Figure 7.12). In the tropics rain either coincides with the high-sun period or is more or less evenly distributed throughout the year, although there are some areas with a double maximum.

### Diurnal Variations

Diurnal variations of precipitation, when they occur, cannot easily be explained. We will describe two regimes that apply to global-scale variation, but there are many local (meso- and microscale) influences that tend to produce maxima during different times of the day. The first of these regimes is forced by thermal convection. Remember that temperature lags slightly behind insolation. Because this lag is even greater at heights of a few thousand meters above the surface, where rain forms, precipitation characteristically is at a maximum in the late afternoon or early evening. This regime is most evident in continental climates in summer, when and where maritime influences are minimal and diurnal temperature variation is most pronounced, and in land areas of the moist tropics throughout the year, where strong solar heating readily initiates convection in air that is conditionally unstable (see Section 5.8.1).

The diurnal variation of precipitation over the oceans is the second regime, which is opposite to the first regime. Over the oceans near-surface lapse rates change very little from day to night, but lapse rates at upper levels may be steepened. Radiative cooling of air at the top of the relatively shallow moist layer over the oceans is often sufficient to promote instability, and the precipitation that results tends to occur during the night or early morning. Some coastal areas also show this diurnal pattern.

Unlike convection, convergence and mechanical uplift are not diurnally forced. Therefore, in areas where these mechanisms predominate, there is no tendency toward systematic diurnal variations.

### 7.3.3 OTHER ASPECTS OF HYDROMETEORS

In the preceding sections we emphasized spatial patterns of mean annual, mean monthly, and, in the case of diurnal changes, mean hourly precipitation. These parameters help to distinguish the place-to-place variation of precipitation and climate. However, other characteristics of atmospheric hydrometeors also are important in climatology. Because fog is to be included in the following discussion, we use the more comprehensive term *hydrometeors* (defined in Section 5.8).

#### Rain Days

In many applications of weather and climate the amount of precipitation is not as important as the fact of its occurrence. That is, a construction crew may be prevented from working, a motion picture cannot be filmed, or a picnic is postponed simply because precipitation occurs, regardless of how much falls. In these cases the parameter of interest is a **rain day**, which also includes forms of precipitation other than rain. A rain day is defined by a certain threshold value, such as 0.01 in. or more, or 0.10 in. or more, in a day. Countries use different threshold values, which makes comparisons difficult; for example, when the metric system is used the value may be 1 mm, 0.1 mm, or 0.2 mm.

In general the mean number of rain days in a year or a month is proportional to mean precipitation for these periods, but there are some systematic departures from this rule. In areas or seasons in which thermal convection is the dominant precipitation-producing mechanism, the ratio of precipitation to rain days is comparatively large. When transient disturbances predominate, this ratio is smaller. Typically, in the former there may be long rainless periods and then sizeable amounts, while in the latter there are likely to be only small accumulations on days with rain.

Table 7.1 shows four major humid climatic types: tropical, subtropical, and marine and continental midlatitude. In the first two climates throughout the year, and in the continental climate during the warm six months, precipitation is exclusively or predominantly through convection, and ratios are typically from 10 to 13 when precipitation is given in millimeters. In marine climates, and in continental climates in the cool six months (October through March), where transient disturbances bring precipitation, ratios are lower. One of the highest ratios observed—73.4 for the six wettest (and consecutive) months—occurs at Cherrapunji, India.

The driest areas on the earth average only a very few rain days per year, of course. The rainiest, according to this criterion, are the highlands of Costa Rica (350+ days) and Bahia Felix in extreme southern Chile (348 days).

#### Intensity

Precipitation intensity refers to the amount that falls over relatively short periods of time, ranging from one minute to one or two days. Since excessive intensities invariably cause flooding, this aspect is of great importance to hydrology and dam design and also to agriculture. The highest values recorded during periods of 12 hours to one year have all been in tropical areas, where hot, humid air may be acted upon by all three mechanisms—thermal convection, convergence, and mechanical uplift (see Section 5.8.1)—although it is not necessary that all three be present at such times. The records for periods of less than 12 hours are in the temperate (extratropical) regions and are usually associated with fronts or squall lines in which extremely rapid moisture convergence, aided by thermal convection, occurs. It may be that these extremes are exceeded in the tropics, but the relatively sparse network of recording gages in those areas means that such occurrences may have been missed. The one-minute record is subject to appreciable error, but amounts of 2 to 3 cm (1 in.) are reasonably well substantiated. For comparison, an hour in which 10 to 12 cm (4 to 5 in.) of rain falls is generally considered a "cloudburst."

The maximum rainfall totals over various time intervals are shown in Table 7.2. The extremely close relationship between amount and time allows a simple calculation of the world maximum rainfall to be made for any period. The equation is approximately

$$R(\text{in.}) = 15\sqrt{D} \quad \text{or} \quad R(\text{mm}) = 375\sqrt{D}$$

where $D$ is the time period in hours. There is a rather amazing correspondence between the highest snow and rainfall totals for the same time period (Table 7.8).

#### Snow

Snow forms and reaches the ground when temperatures both within the cloud and between the cloud and ground are below freezing. In climatological

## TABLE 7.1

Mean monthly and mean annual values of precipitation and rain days (≥ 0.2 mm), and their ratios, for representative stations in four major climatic types. The first line for each location is mean precipitation (mm); the second line is number of rain days. The ratio is the mean value of precipitation amount over number of rain days.

| | J | F | M | A | M | J | J | A | S | O | N | D | Yr. | |
|---|---|---|---|---|---|---|---|---|---|---|---|---|---|---|
| **TROPICAL** | | | | | | | | | | | | | | |
| Georgetown, Guyana | 251<br>20 | 122<br>16 | 113<br>16 | 178<br>16 | 296<br>23 | 346<br>24 | 281<br>23 | 185<br>17 | 88<br>9 | 98<br>8 | 147<br>13 | 313<br>21 | 2418<br>206 | Ratio: 11.4 |
| Manila, Philippines | 23<br>6 | 11<br>3 | 17<br>4 | 32<br>4 | 128<br>12 | 252<br>17 | 414<br>24 | 437<br>23 | 353<br>22 | 195<br>19 | 138<br>14 | 68<br>11 | 2038<br>159 | Ratio: 12.9 |
| **SUBTROPICAL** | | | | | | | | | | | | | | |
| Tokyo, Japan | 48<br>6 | 73<br>7 | 101<br>10 | 135<br>11 | 131<br>12 | 182<br>12 | 146<br>11 | 147<br>10 | 217<br>13 | 220<br>12 | 101<br>8 | 61<br>5 | 1562<br>117 | Ratio: 13.4 |
| Birmingham, Alabama | 128<br>12 | 134<br>11 | 152<br>11 | 114<br>9 | 87<br>9 | 102<br>10 | 131<br>12 | 123<br>10 | 85<br>7 | 75<br>6 | 90<br>9 | 128<br>11 | 1349<br>117 | Ratio: 11.5 |
| **CONTINENTAL** | | | | | | | | | | | | | | |
| Peking, China | 4<br>2 | 5<br>3 | 8<br>3 | 17<br>4 | 35<br>6 | 78<br>8 | 243<br>13 | 141<br>11 | 58<br>7 | 16<br>3 | 10<br>2 | 3<br>2 | 618<br>64 | Cool 6 mo. ratio: 3.0<br>Warm 6 mo. ratio: 11.7 |
| Minneapolis, Minnesota | 18<br>8 | 20<br>7 | 39<br>11 | 47<br>9 | 81<br>11 | 102<br>13 | 83<br>10 | 81<br>10 | 62<br>9 | 40<br>8 | 36<br>8 | 22<br>8 | 631<br>112 | Cool 6 mo. ratio: 3.5<br>Warm 6 mo. ratio: 7.4 |
| **MARINE** | | | | | | | | | | | | | | |
| Central England | 65<br>15 | 48<br>14 | 44<br>15 | 49<br>13 | 56<br>14 | 48<br>12 | 68<br>13 | 67<br>15 | 58<br>13 | 70<br>17 | 67<br>16 | 60<br>17 | 640<br>174 | Ratio: 3.7 |
| Seattle, Washington | 132<br>19 | 99<br>15 | 84<br>16 | 50<br>13 | 40<br>11 | 36<br>9 | 16<br>5 | 19<br>6 | 42<br>8 | 83<br>14 | 127<br>17 | 138<br>19 | 866<br>152 | Ratio: 5.7 |

**TABLE 7.2** Greatest observed point rainfalls.

| Duration | Depth (in.) | Location | Date |
|---|---|---|---|
| 1 min | 1.23 | Unionville, Md. | July 4, 1956 |
| 8 min | 4.96 | Füssen, Bavaria | May 25, 1920 |
| 20 min | 8.10 | Curtea-de-Arges, Rumania | July 7, 1889 |
| 42 min | 12.00 | Holt, Mo. | June 22, 1947 |
| 4 h, 30 min | 30.8+ | Smethport, Pa. | July 18, 1942 |
| 12 h | 52.76 | Belouve, La Réunion | Feb. 28–29, 1964 |
| 24 h | 73.62 | Cilaos, La Réunion | Mar. 15–16, 1952 |
| 5 days | 151.73 | Cilaos, La Réunion | Mar. 13–18, 1952 |
| 31 days | 366.14 | Cherrapunji, India | July 1861 |
| 4 mo | 737.70 | Cherrapunji, India | April–July 1861 |
| 1 yr | 1041.78 | Cherrapunji, India | Aug. 1860–Jul 1861 |

records snow is usually combined with rain to give a single total, which is the case with the values of mean monthly and mean annual precipitation. A conversion of 10 in. or cm of snow to 1 in. or cm of rain is usually used, since on the average this much snow melts to 1 in. or cm of water. However, snow can be very dry, in which case a ratio of as much as 30 to 1 applies, or very wet, when only 3 in. or cm are the equivalent of 1 in. or cm of water. At some stations actual melted water equivalents are used.

It is difficult to generalize the geographical distribution of mean annual snowfall, but mostly it is significant in areas poleward of the −1°C (30°F) isotherm of a particular month (see Figure 7.1 as an example for January). Perhaps surprisingly, the highest single occurrences and largest mean annual amounts are found on the equatorward margins of snow climates. This occurs because of the relationship between temperature and moisture-holding capacity (see Section 5.3). Highest snowfalls usually occur in temperatures just below freezing. At temperatures well below this value (i.e., farther poleward), the capacity of air to hold water vapor is greatly limited. You may have heard the statement, "It's too cold to snow!" While this is not true in an absolute sense, the point is that at very low temperatures there is not very much water vapor available to be converted to snow.

Related factors govern both the frequency of snow occurrences and snow amounts because they have a bearing on either temperature or moisture supply. Elevation is the first such factor. Temperatures decrease upward, which means that while rain is falling at low elevations snow may fall on nearby mountains. This is particularly apparent in the western United States, the southern Andes, and the Alps, but also occurs in the British Isles, Scandinavia, and Japan, where elevation differences are not as pronounced. Another factor is orography. In many places the effects of both elevation and topography—the latter manifested where humid air is forced to rise over mountains or hills—combine to produce large snowfalls. Finally, large water bodies other than oceans, gulfs, and seas provide a local moisture source. The lee (downwind) sides of the Great Lakes are very snowy areas in winter. This is also the case at Lake Baikal in the USSR.

Figure 7.13 shows mean annual snowfall in the eastern United States. The western part of the country is not shown because the effects of elevation make the pattern extremely complex. Another snow characteristic of interest is the mean duration of snow cover, or the length of time during the cold season that snow is on the ground. Still another is the probability that there will be snow on the ground by a specified day. Figure 7.14 shows the probability of a "white Christmas."

It should be reemphasized that precipitation of any kind, especially snow, is a highly variable element of weather. Maps such as Figures 7.13 and 7.14 are statistical composites (averages) based on highly variable individual years. It is not at all unusual for an area with an average snowfall of 1016 mm (40 in.) to have only a few inches in some years and as much as 2540 mm (100 in.) or even more in other years. We will discuss climatic variability later in this chapter.

A final climatological aspect of snow is the "snowline," or the lower limit on mountains of permanent (year-round) ice and snow. This limit is determined not only by temperature—which, from a global perspective, varies with latitude—but also by the exposure of the slope to prevailing winds (normally the snowline will be higher on the leeward than on the windward side of a mountain); the amount of snow; and absorption of solar radiation, which is largely con-

## SEC. 7.3 | PRECIPITATION

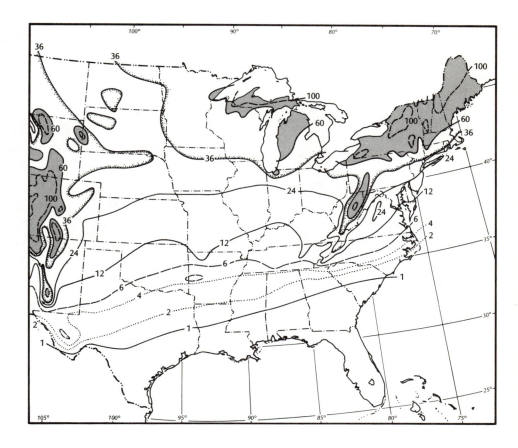

**FIGURE 7.13**
Mean annual snowfall (in.).
(From *Climatic Atlas of the United States*, Environmental Data Service, Environmental Science Services Administration, U.S. Department of Commerce, 1968.)

trolled by the presence or absence of clouds. Table 7.3 gives some average elevations for the snowline in both hemispheres.

**TABLE 7.3**
Mean altitude of the snowline at various latitudes.

| Latitude | Elevation (ft), Northern Hemisphere | Elevation (ft), Southern Hemisphere |
|---|---|---|
| 0°–10° | 15,500 | 17,400 |
| 10°–20° | 15,500 | 18,400 |
| 20°–30° | 17,400 | 16,800 |
| 30°–40° | 14,100 | 9,900 |
| 40°–50° | 9,900 | 4,900 |
| 50°–60° | 6,600 | 2,600 |
| 60°–70° | 3,300 | 0 |
| 70°–80° | 1,650 | 0 |

### Fog

In Section 5.8 we learned that clouds form either as the result of cooling of air to the point of saturation or from the addition of water vapor. Fog is essentially a stratus cloud with its base at the ground. It forms in much the same way, as the result of two general processes: cooling of near-surface air or the evaporation of water into it. Near-surface cooling results either when moist air is advected over a cooler surface (advection fog); is forced upslope (upslope fog); or when, during radiation conditions (no appreciable winds, energy exchanges principally in the vertical), the surface layer is cooled when the surface itself loses heat by upward radiation (radiation fog). Fogs that result from the addition of moisture include frontal fog and steam fog.

**Advection fog** is especially prevalent where warm, moist air moves from a warm to a cold ocean surface and is cooled from beneath to the point of saturation. This frequently happens off the Grand Banks of Newfoundland. There, with southerly flow, fog forms when air previously over the Gulf Stream is chilled as it moves over the relatively cold Labrador Current (see Figure 4.51). Advection fog also is frequent in the coastal areas of mid- and high latitude continents in winter, when relatively warm, moist air moves over these cooler continents. In summer it is very common along arid west coasts where cold water upwelling (see Section 4.8.2) occurs and onshore winds transport the fog inland. If air is forced gradually to ascend a slope of sufficient extent, adiabatic cooling can

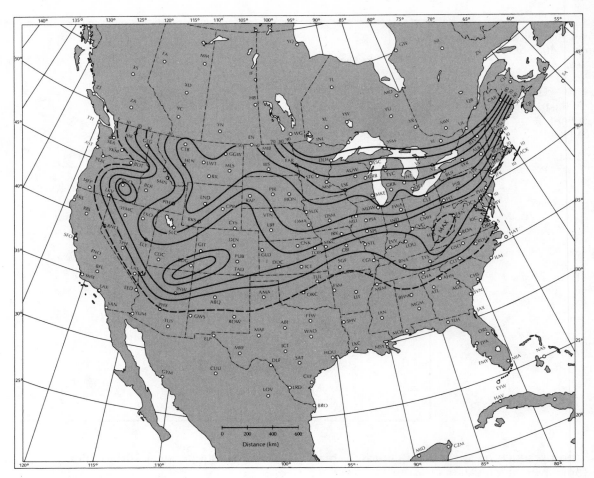

**FIGURE 7.14**
Probability of a "white Christmas" (2.5 cm or more of snow on ground). (From "Statistical Probabilities for a White Christmas," U.S. Department of Commerce news release, Washington, D.C., December 15, 1971.)

result in **upslope fog**. If there is too much mixing or if the air movement is rapid, the condensation will probably lead to clouds—and possibly to precipitation—rather than to fog. The high plains east of the Rockies and the eastern Ukraine (USSR) experience such upslope fogs.

**Radiation fog** is not as site-specific as other kinds, but it is restricted to land areas. A slight amount of vertical mixing is required for radiation fogs of appreciable depth. Conversely, if there is complete calm only dew or frost occurs, while appreciable mixing transports moisture through a much greater depth and reduces the likelihood of fog. During winter in very cold regions radiation fog can be formed of ice crystals and is therefore called **ice fog**.

Fog formed from the addition of water vapor by evaporation occurs along frontal boundaries (**frontal fog**) when rain formed in warm, moist air is evaporated into colder air beneath it and then condenses. **Steam fog** occurs when warm surface water is evaporated into colder air and then condenses. One familiar form of steam fog appears above pavements, roofs, and sidewalks after a rain shower on a hot day; another form occurs as people and animals exhale in cold weather. Steam fogs of great extent and depth are found in very cold climates when frigid air from land areas moves over warm water. This phenomenon also is known as *sea smoke*.

Table 7.4 summarizes the various kinds of fog, and Figure 7.15 shows its global distribution.

### Hail and Thunderstorms

Hail (see Section 5.8.4) is associated with the more general term *severe weather*, which also includes thunderstorms, high and gusty winds, lightning, and, on rare occasions, tornadoes. Thus, hail is a feature of severe weather, but it occurs only when an appreciable part of the cumulonimbus cloud (see Figure 5.23) extends above the freezing level. This means that hail

## SEC. 7.3 | PRECIPITATION

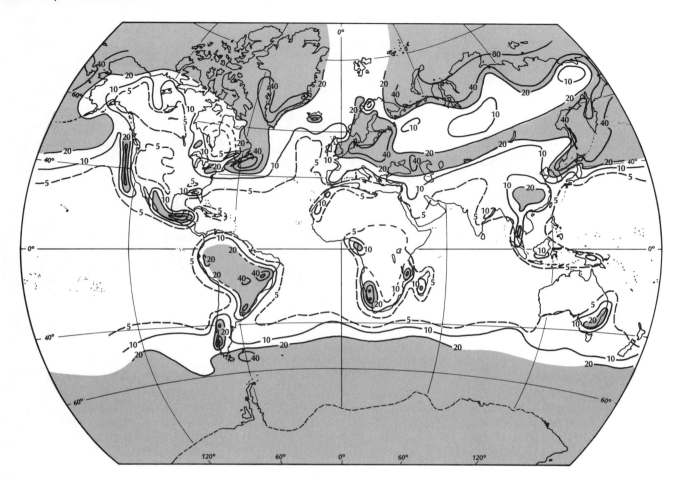

**FIGURE 7.15**
World distribution of mean annual number of days with fog. (From *Handbook of Meteorology* by F. A. Berry, E. Bollay, and N. R. Beers. Copyright © 1945 by McGraw-Hill Book Company. Used with the permission of McGraw-Hill Book Company.)

is mostly likely in the interiors of continents in mid-latitudes, where intense surface heating leading to instability combines with a relatively cold middle troposphere. Tropical and subtropical areas may have many days with thunderstorms (Figure 7.16), but temperature at upper levels are too warm, in general, for hail to occur. Tampa, Florida, for example, has many thunderstorms but very little hail, while in southeastern Wyoming and northeastern Nevada the ratio of hail days (Figure 7.17) to thunderstorm days is almost 1 to 5.

As would be expected, then, the main hail belt is between 30° and 50° N in North America and Eurasia. However, there are subtropical areas with appreciable hail occurrence, such as the plateau region of Mexico, northern India, and southern Africa (Figure 7.17). It appears likely that topographic effects are important in these areas. It may be that a superior reporting system is responsible for the fact that the most remarkable hail falls have been observed in the central plains

**TABLE 7.4**
Fog classification.

**Fogs formed when air is cooled to saturation**

*Advection fog:* Horizontal transport of air to a surface colder than the moist air.

*Upslope fog:* Upward transport of air leading to adiabatic cooling.

*Radiation fog:* The loss of heat by radiation from a land surface with concomitant cooling of both the surface and the near-surface air. (If temperature is well below freezing, *ice fog* occurs.)

**Fogs formed by the addition of moisture, with evaporation and then condensation occurring**

*Frontal fog:* Rain falls through a frontal discontinuity into initially unsaturated air, evaporates into it, and then condenses so that the lower layer becomes saturated.

*Steam fog:* Very cold air moves over warm water, promoting rapid evaporation and then condensation into that air.

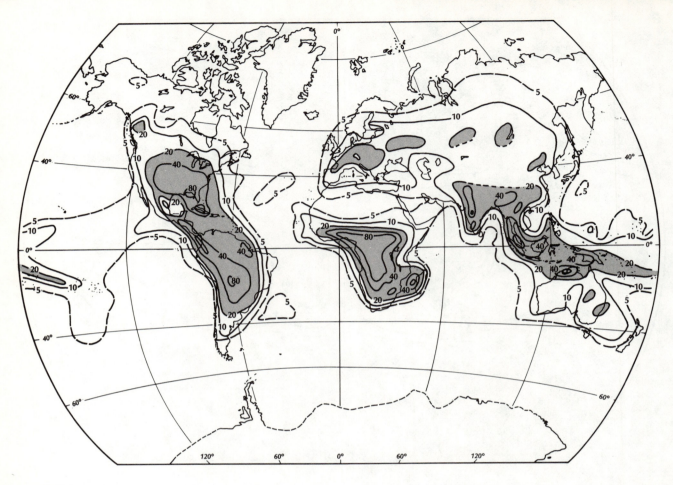

**FIGURE 7.16**
World distribution of mean annual number of days with thunderstorms. (From *Handbook of Meteorology* by F. A. Berry, E. Bollay, and N. R. Beers. Copyright © 1945 by McGraw-Hill Book Company. Used with the permission of McGraw-Hill Book Company.)

of the United States. Three times in this region (Iowa, Kansas, and Nebraska) hailstones exceeding 43 cm (17 in.) in circumference and weighing over 700 g (1.5 lb) have been recorded.

Late spring and early summer are the seasons of maximum hail and thunderstorm occurrence, although a secondary maximum in fall is often observed. Diurnally, hail and thunderstorms are most frequent from 1400 to 2200 local time.

# 7.4
## AIR MASSES

In Chapter 5 we learned that fronts constitute transition zones across which there are relatively abrupt changes of temperature, moisture, and density. This implies that the areas between fronts are homogeneous in these properties; in other words, there must be bodies of air in which temperature and moisture change only slowly over great horizontal distances. These are called **air masses.**

Because we want to know the kind of weather that will follow the passage of a warm or cold front, meteorologists have developed a classification system for air masses. This classification accomplishes two things: (1) it indicates both the temperature and moisture conditions in the air mass relative to the average weather at that location for that time of the year, and (2) it indicates the area of origin of the air mass. In connection with the latter, we recognize that air may exist for some period of time—a few days to perhaps even weeks—over areas of the earth's surface before being moved about by circulation systems such as lows and highs, and, in the upper air, major troughs and ridges. During the time such air resides over these areas, it acquires the characteristics of the sur-

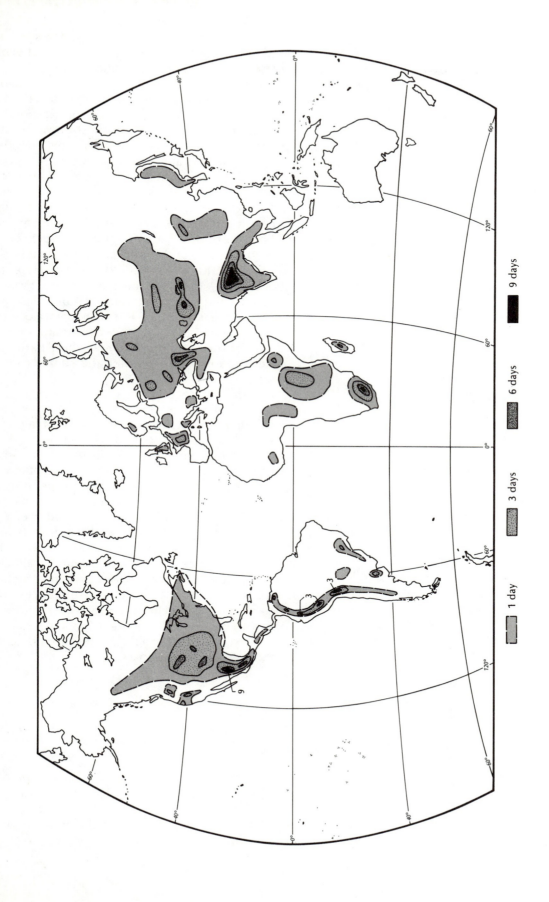

**FIGURE 7.17**
World distribution of mean annual number of days with hail. (Prepared by the U.S. Army Engineer Topographic Laboratories, 1972.)

face. Thus, air over the oceans and over the continents tends to be moist and dry, respectively. Also, continental air will tend to be warmer in summer and colder in winter than marine air in roughly the same latitude.

The letters assigned to air masses therefore indicate both temperature and moisture and its place of origin. A widely accepted system uses these letters:

**Temperature**
T (tropical): warm to hot
P (polar): cool to cold
A (arctic, or antarctic in the Southern Hemisphere): cold to very cold

**Moisture**
c (continental): dry
m (maritime): moist

The temperature letters refer to conditions relative to that time of year and the location. Thus, a cP air mass in Minnesota in January may have mean daily temperatures of $-23°$ to $-29°C$ ($-10°$ to $-20°F$) and $16°$ to $21°C$ ($60°$ to $70°F$) in July. In northern Texas these temperatures would be around $-7°$ to $-1°C$ ($20°$ to $30°F$) and $24°$ to $29°C$ ($75°$ to $85°F$) in these two months. To a lesser extent these seasonal and location modifiers apply also to humidity; an mT air mass is more humid in July than in January.

By combining the temperature and moisture indicators, six different air masses are possible, although other classifications are more complex and have more letters. In general, a cP or cA air mass is dry and cool to very cold; cT air is dry and warm to hot; mT humid and warm to hot; and so on.

An air mass classification is not complete without some indication of what happens to the surface layer of air while it is in transit. Poleward-moving air masses are cooled from beneath (although there are some exceptions, as when mT air moves poleward over the warm continents in summer and tends to be warmed), and equatorward-moving air masses tend to be warmed. These processes have important implications for lower layer stability. Remember from Section 5.8.1 that heating from below tends to reduce stability (or increase instability), while the reverse happens when surface air is cooled by contact with the ground. The letter k is added to the classification when the air is cooler than the underlying surface, and thus is being heated, whereas the letter w indicates the reverse. Therefore —k air masses tend to be unstable, and —w air masses stable. A cPk air mass is one that is dry and cool to cold and is being warmed from beneath, probably as the result of equatorward movement. Of course, air masses may also be moistened or dried out during such transit, but this is not explicitly indicated in the system explained here.

From a global perspective the two principal air masses are tropical and polar. As described in Chapter 5, these two air masses meet along the polar front zone, which shifts northward and southward with the seasons. In Figure 4.42 we may presume the air masses to be mT, and to a lesser extent cT, equatorward of the polar front zone; mP and cP between the polar and arctic front zones; and mA and cA poleward of the arctic front zone. At the polar front zone the two particular air masses that are dominant are cP and mT, but there may be battlegrounds, such as in the sub-Saharan area and in western Texas, where cT and mT air meet along so-called "dry lines." Here, as the letters imply, moisture differences are more pronounced than those of temperature. Figure 7.18 shows the areas of origin and principal trajectories of the air masses affecting the conterminous United States. Figures 7.19 and 7.20 show where particular air masses predominate in winter and summer in North America.

To end this section, we offer one very important qualifier. Nature seldom, if ever, produces the discrete, mutually exclusive categories of air masses which we have just described. There is instead a continuum of the properties that distinguish air masses, and the change from one to another is not always as abrupt as the foregoing classification system suggests. Also, there are many indeterminate situations, where the qualifiers "mixed," "modified," and even "indifferent" are added to the two or three primary letters. In the heyday of air mass analysis, the 1930s and 1940s, much effort was expended in detailing the climatological characteristics of air masses and constructing air mass calendars (tabulations of the frequency of occurrence of different air masses) for specific locations. Today we recognize that there is broad overlap in the air mass types and that the indicators are not as definitive as was earlier supposed.

# 7.5
## CLIMATIC ITERATIONS

Climate has been defined as the long-term average state of the atmosphere, usually for particular places. Variability also is a fundamental characteristic of climate, and periodicities such as those of temperature (diurnal and annual) are never repeated exactly from

**FIGURE 7.18**
Air mass source regions and trajectories for North America. (After Haynes, U.S. Department of Commerce. In A. N. Strahler, *Introduction to Physical Geography*. Copyright © 1965 by John Wiley & Sons, Inc. Reprinted by permission of John Wiley & Sons, Inc.)

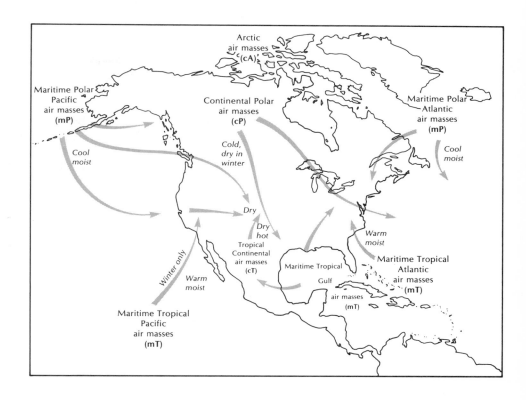

**FIGURE 7.19**
Winter air masses in North America. [After D. H. Brunnschweiler, "The Geographic Distribution of Air Masses in North America," *Viertljahresschrift der Naturforschenden Gesellschaft* 97 (1952): 42–49.]

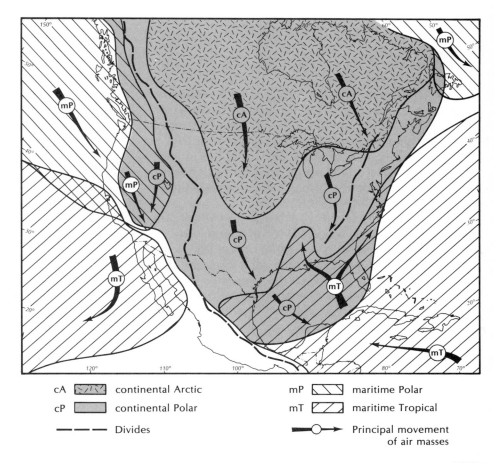

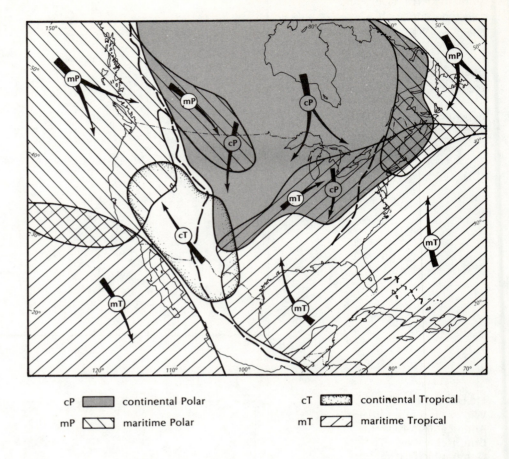

**FIGURE 7.20**
Summer air masses in North America. [After D. H. Brunnschweiler, "The Geographic Distribution of Air Masses in North America," *Viertljahresschrift der Naturforschenden Gesellschaft* 97 (1952): 42–49.]

day to day or year to year. Precipitation amounts are even more variable in this respect. Very long-term fluctuations of the climatic elements, chiefly temperature and precipitation, are treated in Chapter 11. In this section we consider short-term variations of these elements, or those that occur during the period for which climate is defined for most purposes—on the order of tens of years.

Relatively short-term variations are called **iterations**, or repetitions. Mean values of temperature or accumulated amounts of precipitation are first determined for a day, a week, a month, or a year, and then the variability among many of these periods is calculated. For example, we may want to know the variability of January mean temperatures, that is, how much the mean temperature of each of 30 or 40 Januarys varied over this period of years. Similarly, we can calculate annual precipitation variability by finding out how much precipitation accumulated over a year varies among the 20 or 30 years for which there are records.

Strictly, we should not call these iterations "climatic" because they occur within the period used to define climate. Since the state of the atmosphere for shorter periods is called *weather*, then these are weather iterations. But usage and tradition have established the term **climatic iteration,** so we will use it with these reservations.

If you need to review the more fundamental statistical concepts, Chapter 6 should be studied before proceeding with this section. In particular, review the terms *standard deviation, normal distribution,* and the other kinds of frequency distributions and their transformations.

### 7.5.1 TEMPERATURE

A fundamental observation for all climatic elements is that variability increases as the averaging period decreases. For example, there will be great variation among the mean temperatures of 30 January fifteenths, moderate variability among the means for the same number of Januarys, and relatively little variation in the mean temperature of each of 30 years. The same observation is true of space averaging. The more stations included in an average, the less will be the variation over time iterations. Thus, the variability of mean January temperatures over 30 years will be less for the average of 10 or 15 stations within a state or county than for any one of these stations.

## SEC. 7.5 | CLIMATIC ITERATIONS

Just as the mean temperature for a week, a month, or a year helps to distinguish temperature geographically, so also does the variability over many iterations of these periods. And, as with mean temperature, there are both spatial and temporal aspects of variability (Table 7.5). The measure of variability used—standard deviation of January and July temperature—increases with latitude and continentality (see Section 3.6 and Figures 3.11 and 3.12). Thus, Paris and Athens are very marine (the latter especially in winter), while Del Rio, at about 30° N, is moderately continental (compare with Figures 3.16 and 3.17). The annual variability is close to 1F° for all stations except Fort Yukon, which is at a very high latitude.

For every station January is more variable than July (all are Northern Hemisphere stations, so January is winter). This is reasonable because during winter horizontal temperature gradients are most pronounced and circulation systems most frequent and best developed. Finally, notice that the monthly standard deviations are almost always smaller than the annual (the exception is the Nigerian station), which agrees with the earlier statement that the variability decreases as the averaging period increases. Standard deviations for shorter periods such as a week or a day are not shown, but they would be higher than the monthly values.

In many applications we need to know by how much temperature can be expected to vary among many iterations. For example, in how many of the next five or ten growing seasons will temperatures be above the minimum required to grow a crop? If we use the frost-free period (loosely defined as the average length of time between the last occurrence in spring and the first in fall of a temperature of 0°C (32°F) or below), then Figure 3.21 gives some indication. The variability can be quantified by finding the standard deviation of the growing season length. Over the 30 years shown, the standard deviation is 15.5 days and the mean length is 181.8 days. Assuming that these growing season lengths are distributed normally, which is generally the case for mid-latitude areas, this means that about 68 percent of the years will have growing seasons between +1 and −1 standard deviation, or between 166 and 197 days. Ninety-five percent of the years will have growing seasons of between 151 and 213 days. This can be vital information in agricultural planning.

Another aspect of temperature variability is shown in Figure 7.21. A warming trend is very apparent. Can the temperature climate be changing? A small part of this increase reflects a slight warming in the Northern Hemisphere from the mid-nineteenth to mid-twentieth centuries, but most of it is due to the growth of the metropolitan area. Such increases have been recorded in many large cities all over the globe. At New York, the combined increase is about 2C° (3.5F°) in 90 years. Notice also that the average for the period from 1931 to 1960 is about 0.5C° greater than the average for 1921 to 1950. These 30-year periods are used by the U.S. National Weather Service to construct "normals," which are updated every 10 years. Unfortunately, this term implies that the temperatures so specified are normal and that everything else

**TABLE 7.5**
Variability of January, July, and annual temperatures for stations in various climates, expressed as standard deviation. The length of record is 30 years in all cases. The figure in parentheses is the mean.

| Station | Climate | Latitude, longitude | January (°F) | July (°F) | Annual (°F) |
|---|---|---|---|---|---|
| Sokota, Nigeria | Humid tropical | 13°10′ N, 5°16′ E | 2.7(75.9) | 1.2(82.3) | 1.3(83.2) |
| Del Rio, Tex. | Subtropical semidesert | 29°22′ N, 100°55′ W | 3.6(51.1) | 1.8(86.1) | 1.0(69.9) |
| Macon, Ga. | Humid subtropical | 32°42′ N, 83°39′ W | 4.9(48.6) | 1.5(80.6) | 0.9(64.7) |
| Athens, Greece | Dry summer subtropical | 37°58′ N, 23°43′ E | 2.9(49.0) | 1.6(82.4) | 1.0(64.9) |
| Valentine, Neb. | Mid-latitude humid continental | 42°52′ N, 100°33′ W | 5.7(20.6) | 2.8(74.3) | 1.3(47.1) |
| Paris, France | Mid-latitude marine west coast | 48°49′ N, 2°30′ E | 4.5(37.8) | 2.6(66.6) | 0.9(52.0) |
| Fort Yukon, Alaska | Arctic | 66°35′ N, 145°18′ W | 9.8(−18.6) | 2.5(61.4) | 2.2(20.3) |

**FIGURE 7.21**
Mean annual temperatures, by decades, for New York City, 1871–1960. Maxima (×) and minima (●) are the highest and lowest annual means in each decade.

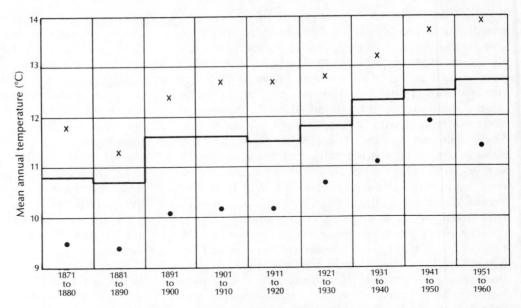

is abnormal. A preferable term—one that we use throughout this text—is *average*, or *mean*.

To illustrate the final aspect of the variability of climatic elements, we will consider the following ten values:

47    51    58    60    62    42    48    53    55    60

The mean of these values is 53.6, the standard deviation 6.25. Notice that as a measure of variability standard deviation takes no account of the order of the values. This is called **sequential variability**. **Consecutive variability** does take account of order, and usually is expressed as the mean of the absolute differences (differences without regard to sign) from one value to the next. In this example, consecutive variability is 5.9 (53/9). In some instances it is a more revealing and useful measure of variability than standard deviation. The values in our example could be precipitation accumulations, in millimeters, over 10 consecutive days or weeks. Or, they could be daily maximum temperatures over 10 consecutive days (i.e., the mean interdiurnal variability for this period). Figures 3.23 and 3.24 show this measure for the forty-eight states in January and July.

## 7.5.2 PRECIPITATION

The variability of precipitation for specified periods over many iterations of these periods is described in much the same way as temperature. However, there is a fundamental difference between these two elements. Temperature is a continuous variable, that is, there is always a value that can be specified, regardless of time or place. Precipitation, on the other hand, is discrete, at least for short time periods. There is almost always a finite amount accumulated over a period as long as a year, but at many places there will be no precipitation in a month, and the chance that no precipitation will be observed increases as the accumulation period decreases.

For single-station analysis this difference does not lead to difficulties, because the likelihood of receiving a specified range of precipitation in a given period can readily be calculated. For example, wheat farmers may want to know if they can count on just about the same amount of rainfall in each growing season, or if these amounts will vary widely among many such seasons. In this sense variability can be thought of as the opposite of dependability; that is, the more variable precipitation is, the less it is dependable. The standard deviation as an indication of variability is very helpful in this application. If we assume that rainfall amounts in the growing season are normally distributed, we can use the normal curve approximation to determine the probability of receiving between $X$ and $Y$ inches, or less or more, than $X$. Where and when such amounts are known not to be distributed normally, as will increasingly be the case with shorter time periods (e.g., months, weeks), we can apply transformations to make such distributions at least approximately normal. Alternatively, such theoretical distributions as the incomplete gamma can be applied, or, probability estimates can be made directly from the historical data, without assuming anything about the form of the frequency distribution.

We also may want to know how various areas or climates compare in the variability of precipitation over many iterations of specified periods. Such spatial

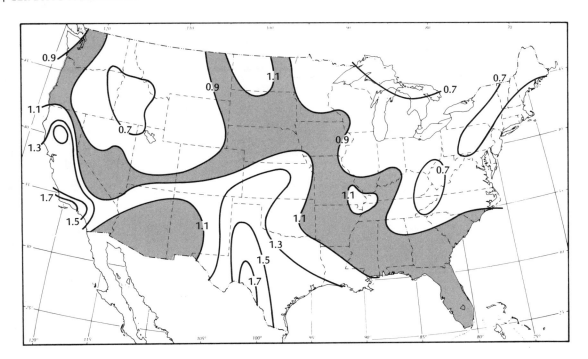

**FIGURE 7.22**
Anomalies of precipitation variability in the United States. (See text for explanation.)

analyses are difficult because of the discreteness of precipitation amounts and their very wide range, at least as compared with temperature. The standard deviation could be used, but since this measure of variability increases approximately with the mean (monthly or annual), spatial representations would show only that variability is essentially invariant with mean precipitation. Traditional maps of continental or global precipitation variability have used a measure called **relative variability,** which is the mean deviation (see Section 6.3.1) divided by the mean. This measure is useful when applied to limited areas, but, like standard deviation, is still dependent on the mean. Maps of large areas show little more than that the variability is high where precipitation is low, and vice versa (the opposite of using only the standard deviation).

We can attempt to account for this dependence of variability on the mean by finding the relationship between the two by linear regression, and then determining whether the variability for a particular location is greater or less than that expected from the regression. Figure 7.22 shows anomalies (departures from the expected relationship) of mean annual precipitation for the forty-eight states. The values of the isopleths are interpreted as follows. The variability in West Virginia is only 0.7 (70 percent) of the variability expected for all places in the country with comparable mean annual precipitation. Similarly, in southern California and south central Texas the variability is as much as 1.7 times the expected.

From Figure 7.22 it is apparent that variability increases as the precipitation becomes more concentrated during the year, as is the case along the middle and southern west coast, and is least in areas of uniform (year-round) precipitation, as in the eastern third of the country. This is reasonable, because in an area of uniform precipitation, a relatively dry spell can be balanced by a wet period (and vice versa) at some other time of the year. On the other hand, in a region with a single short wet season, a deficiency or surplus cannot be compensated at another time of the year.

It also is clear from Figure 7.22 that variability increases with the proportion of precipitation that occurs as the result of thermal convection (e.g., showers and thundershowers). Conversely, where precipitation is due mainly to transient disturbances, the variability is less. Since the proportion of precipitation due to convection increases with temperature, we can say in general that precipitation is more variable in hot than in cold climates.

Table 7.6 gives an indication of the extremes of precipitation encountered in various climates. Notice the greater variability at Los Angeles than at New York, for the reasons just given. London, which has a cool climate with uniform precipitation, has smaller extremes than the tropical stations in Africa.

**TABLE 7.6**
Averages and extremes of precipitation (mm). Notice that for maximum and minimum the annual totals are not the sum of the twelve monthly values.

| Station | | J | F | M | A | M | J | J | A | S | O | N | D | Annual |
|---|---|---|---|---|---|---|---|---|---|---|---|---|---|---|
| Douala, Cameroons (30 years) | Max. | 183 | 185 | 426 | 349 | 599 | 862 | 1154 | 1240 | 980 | 602 | 298 | 184 | 5328 |
| | Ave. | 57 | 82 | 216 | 243 | 337 | 486 | 725 | 776 | 638 | 388 | 150 | 52 | 4150 |
| | Min. | 1 | 5 | 58 | 130 | 141 | 226 | 277 | 248 | 315 | 259 | 36 | 4 | 3238 |
| Lagos, Nigeria (60 years) | Max. | 155 | 180 | 286 | 325 | 549 | 768 | 786 | 580 | 424 | 450 | 183 | 150 | 2934 |
| | Ave. | 40 | 57 | 100 | 115 | 215 | 336 | 150 | 59 | 214 | 222 | 77 | 41 | 1625 |
| | Min. | 0 | 0 | 5 | 34 | 90 | 138 | 2 | 2 | 10 | 75 | 4 | 0 | 1039 |
| Dar es Salaam, Tanzania (70 years) | Max. | 260 | 201 | 346 | 525 | 600 | 161 | 221 | 108 | 71 | 235 | 331 | 285 | 1531 |
| | Ave. | 71 | 64 | 120 | 280 | 303 | 35 | 33 | 25 | 29 | 49 | 79 | 91 | 1179 |
| | Min. | 1 | 1 | 12 | 44 | 1 | 1 | 1 | 1 | 1 | 2 | 5 | 1 | 438 |
| New York, N.Y. (1869–1960) | Max. | 190 | 166 | 231 | 240 | 227 | 234 | 269 | 276 | 304 | 320 | 244 | 190 | 1425 |
| | Ave. | 84 | 72 | 102 | 87 | 93 | 84 | 94 | 113 | 98 | 80 | 86 | 83 | 1076 |
| | Min. | 20 | 12 | 24 | 28 | 8 | 1 | 12 | 12 | 5 | 7 | 15 | 6 | 819 |
| St. Louis, Mo. (1869–1960) | Max. | 216 | 225 | 242 | 275 | 286 | 434 | 270 | 520 | 267 | 222 | 220 | 276 | 1748 |
| | Ave. | 50 | 52 | 78 | 94 | 95 | 109 | 84 | 77 | 70 | 73 | 65 | 50 | 897 |
| | Min. | 3 | 8 | 3 | 12 | 13 | 2 | 6 | 2 | T | 5 | 6 | 1 | 523 |
| Makindu, Kenya (50 years) | Max. | 214 | 175 | 650 | 822 | 203 | 41 | 8 | 7 | 29 | 327 | 518 | 603 | 1964 |
| | Ave. | 40 | 28 | 79 | 111 | 31 | 2 | 1 | 1 | 1 | 29 | 68 | 120 | 611 |
| | Min. | 0 | 0 | 0 | 0 | 0 | 0 | 0 | 0 | 0 | 0 | 0 | 0 | 67 |
| London, England (1841–1960) | Max. | 120 | 127 | 118 | 109 | 111 | 154 | 171 | 159 | 141 | 194 | 174 | 153 | 903 |
| | Ave. | 53 | 40 | 37 | 38 | 46 | 46 | 56 | 59 | 50 | 57 | 64 | 48 | 594 |
| | Min. | 10 | 1 | 1 | 2 | 6 | 3 | 4 | 4 | 3 | 4 | 7 | 9 | 319 |
| Los Angeles, Ca. (1877–1960) | Max. | 338 | 340 | 314 | 191 | 91 | 35 | 6 | 15 | 144 | 177 | 166 | 401 | 1024 |
| | Ave. | 78 | 85 | 57 | 30 | 4 | 2 | T | 1 | 6 | 10 | 27 | 73 | 373 |
| | Min. | T | 0 | 0 | T | 0 | 0 | 0 | 0 | 0 | 0 | 0 | 0 | 106 |
| Sallum, Egypt (35 years) | Max. | 67 | 37 | 59 | 10 | 23 | 1 | T | 0 | 9 | 73 | 227 | 70 | 326 |
| | Ave. | 12 | 12 | 12 | 1 | 3 | 0 | 0 | 0 | 1 | 5 | 28 | 21 | 95 |
| | Min. | T | T | 0 | 0 | 0 | 0 | 0 | 0 | 0 | 0 | T | T | 4 |

NOTE: T = trace.

## 7.5.3 CLIMATIC TYPES

Climatic classifications are based, almost invariably, on the average values of selected elements, usually temperature and precipitation. However, in any one year these values may not be near the average, and a station or area may not be classified in its average climate. For example, suppose we specify that an area is subtropical by using the criterion that the mean temperature of the coldest month must be above 6°C (43°F). In this case, as Figure 7.23 shows, all of the eastern United States was subtropical as far north as Delaware, Maryland, West Virginia, and Kentucky in 1950. However, in 1940 only Florida was subtropical. The isotherms for 1948 and 1949 show how the subtropical boundary progressed steadily northward from 1948 to 1950. The swing between 1940 and 1950 extremes is around 1100 km (700 mi).

When both temperature and precipitation are used to distinguish one climate from another, the displacement of boundaries can be even more extreme. In most classification systems dry areas are separated from humid areas by a combination of both elements. That is, a given location may be humid one year and arid the next either because the second year had less rain, or was warmer, or both. How much the humid-arid boundary migrates from one extreme to the other is shown in Figure 7.24. The determination of this boundary, which essentially coincides with the line along which precipitation equals evaporation, was based on a formula very similar to the one we will use in Chapter 8 (compare Figures 7.24 and 8.10).

# 7.6

## CLIMATIC EXTREMES

There is a natural inclination to be interested in unique phenomena. The popularity of the *Guinness Book of Records* attests to this. The purpose of this section on climatic extremes is not to gratify curiosity (although this might be commendation enough), but

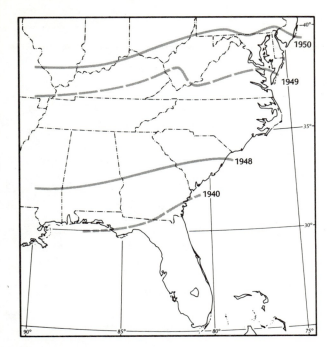

**FIGURE 7.23**
Variation of the subtropical boundary (the 43°F isotherm for the coldest month) among selected years in the eastern United States.

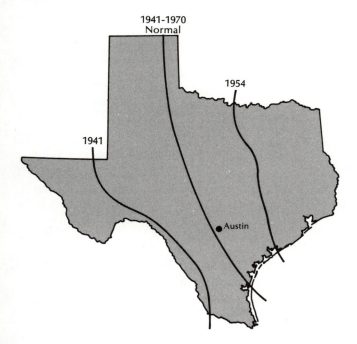

**FIGURE 7.24**
During a 30-year period (1941–1970), the humid-arid boundary was farthest west in Texas in 1941 and farthest east in 1954. These were very wet and very dry years, respectively. The mean boundary for this 30-year period also is shown.

to make us aware of the range of the climatic elements, as measured at the earth's surface. This is not to say that all values must lie within the limits given here, but certainly any new extremes should be carefully examined.

### 7.6.1 TEMPERATURE

As we would expect, many of the high temperature records have been reported at stations in the desert areas of the subtropics. Two of the stations at which records have been set (Death Valley, California, and Dallol, Ethiopia) are below sea level. As far as heat stress on humans is concerned, Dallol may be the most unpleasant place on the earth. Its proximity to the Red Sea gives it a very high absolute humidity in addition to the extreme heat, and even in its cooler months this area is not at all inviting. The highest mean monthly temperature experienced by a large city is the 37°C (99.5°F) recorded at Jacobabad, Pakistan, in June when the mean minimum is 29°C (85°F) and the mean maximum 46°C (114°F).

Oymyakon and Verkhoyansk, Siberia, are excellent examples of an extreme continental climate. Both have average July temperatures around 14°C (57°F) and average January temperatures of −50°C (−58°F). The latter is the lowest mean monthly temperature reported from any city.

Quito, Ecuador, has an extremely low annual range because of its location almost exactly on the equator and because of its high elevation. Heard and Macquarie Islands in the southern Indian Ocean and the Marshall Islands in the south Pacific are perfect illustrations of a marine climate. The very low daily range at the South Pole in December is due to the almost constant radiation experienced throughout the day.

### 7.6.2 PRECIPITATION

Some of the world records for intensity of precipitation have already been discussed in Section 7.3.3. The rainiest areas on the earth—at least in terms of highest annual total and highest annual means—are Cherrapunji, India, in the foothills of the Himalayas, and Kauai, the Hawaiian islands. The driest areas in this respect are northern Sudan and northern Chile. At Calama, Chile, in the Atacama Desert, rain has yet to be recorded in approximately 30 years of observation! In the United States the driest area is in extreme southeastern California and southwestern Arizona, where the average annual precipitation is 2 to 4 in. (50 to 100 mm).

**TABLE 7.7**
Extremes of temperature (°F).

| Measurement of Temperature | High Temperature | | Low Temperature | |
|---|---|---|---|---|
| | °F | Station | °F | Station |
| Absolute | 136.4 | Azizia, Libya (Sept. 13, 1922) | −126.9 | Vostok, Antarctica (Aug. 24, 1960) |
| Mean monthly maximum | 116 | Bou-Bernous, Algeria (July) Death Valley, Ca. (July) | −88 | Vostok, Antarctica (Aug.) |
| Mean monthly | 102 | Bou-Bernous, Algeria (July) Death Valley, Ca. (July) | −97 | Plateau Station, Antarctica (Aug.) |
| Mean monthly minimum | 90 | Dallol, Ethiopia (June) | −103 | Vostok, Antarctica (Aug.) |
| Mean annual | 94.5 | Dallol, Ethiopia | −72 | Pole of Cold (78° S, 96° E) |
| Extreme range | 192 (−94 to +98) | Verkhoyansk, Siberia | 24 (65 to 89) | Fernando de Noronha, Brazil (53 yrs) |
| Annual range | 121 (−60 to +61) | Oymyakon, Siberia | 0.8 | Quito, Ecuador, and Marshall Islands |
| Mean annual diurnal range | 39 (37 to 76) | Bishop, Ca. | 6 (31 to 37) (37 to 43) | Heard Island, Indian Ocean Macquarie Island, Indian Ocean |
| Mean monthly diurnal range | 50 (38 to 88) | Quincy, Ca. | 3.5 | South Pole (Dec.) |
| Extreme diurnal range | 100 (44 to −56) | Browning, Mont. (Jan. 23–24, 1916) | — | — |
| Daily minimum | 102 | Death Valley, Ca. | — | — |
| Daily maximum | — | — | −117 | Vostok, Antarctica |
| Extreme minimum | 70 | Dallol, Ethiopia Pt. Moresby, New Guinea | — | — |
| Average of coldest month | 88 | Dallol, Ethiopia (Jan.) | — | — |
| Average of warmest month | — | — | −27.6 | Vostok, Antarctica (Dec.) |

**Other Extremes**

Spearfish, S.D.: Temperature rose from −4° to 45° in 2 min (07.32, Jan. 22, 1943).

Kansas City: Temperature dropped from 76° at 11:00 A.M. to 14° at 6:00 P.M., and to 10° at 11:00 P.M. (Nov. 11, 1911).

Rapid City, S.D.: Temperatures on Jan. 22, 1943, were as follows:

| Time | 05.30 | 09.40 | 10.30 | 10.45 | 11.30 | 16.00 | 19.30 |
|---|---|---|---|---|---|---|---|
| Temp. | −5 | 54 | 11 | 55 | 10 | 56 | 5 |

On this day, at one time, there was a temperature of 52° at Lead while at Deadwood (600 ft lower) 3 mi away, it was −16°.

**Highest Average Monthly Temperature for Large Cities**

| | | | |
|---|---|---|---|
| N. Hemisphere | 99.5 | Jacobabad, Pakistan (114–85) | |
| | 92 | New Delhi, India (105–79) | |
| S. Hemisphere | 83 | Asunción, Paraguay (95–71) | |
| | 84 | Manaus, Brazil (92–76) | |

43 consecutive days with maximum over 120°: Death Valley, Ca. (July 6–Aug. 17, 1917).

162 consecutive days with maximum over 100°: Marble Bar, W. Australia (Oct. 30, 1923–April 7, 1924).

## SEC. 7.6 | CLIMATIC EXTREMES

Notice in Table 7.8 that all snow records are from the United States. It is possible that these records have been exceeded, perhaps in the Andes or the Himalayas, but at very snowy places in these mountains no observations have been made.

### 7.6.3 OTHER CLIMATIC EXTREMES

There are some other climatic extremes that merit attention. The mean global radiation—from sun and sky—received at the South Pole during 24 hours in late December is a record, averaging 40 cal cm$^{-2}$ h$^{-1}$, or 960 cal day$^{-1}$. Remember that although the sun is at an elevation of only 22° to 23° throughout the day, it never sets. Sunshine at Wadi Halfa, Sudan, is virtually 100 percent of the possible, and clouds are rare there.

The pressure extremes in Siberia are subject to errors of a few millibars because they are corrected to mean sea level using estimated temperatures (see Section 4.3.4). However, it appears that mean sea level pressure around 1100 mb could exist in the Turfan Depression of central Asia.

Record surface wind speeds have been reported by the observatory on the top of Mt. Washington in New Hampshire. Some phenomenal wind speeds have been recorded by the specially constructed anemometer there.

**TABLE 7.8**
Extremes of precipitation (in.).

| Measurement of Precipitation | Precipitation (in.) | Station |
|---|---|---|
| Highest annual total | 1042 | Cherrapunji, India (Aug. 1, 1860–July 31, 1861) |
| Highest annual means | 460 | Mt. Waialeale, Kauai, Hawaii (32 yr) |
| | 450 | Cherrapunji, India (74 yr) |
| | 405 | Debundscha, Cameroons (32 yr) |
| Lowest annual mean | 0.02 | Wadi Halfa, Sudan (39 yr) |
| | 0.03 | Arica, Chile (59 yr) |
| | | (Calama, Atacama Desert, Chile—rain never recorded) |
| | | Iquique, Chile—No rain for 14 consecutive yr |
| Highest monthly total | 366 | Cherrapunji, India (July, 1861) |
| Highest monthly mean | 106 | Cherrapunji, India (July) |
| Highest 5-day total | 150 | Cherrapunji, India (Aug., 1841) |
| | 115 | Jamaica (Nov., 1909) |
| Highest 1-day total | 73.62 | Cilaos, La Réunion (Mar. 16, 1952) |
| | 49.00 | Paishih, Taiwan (Sept. 10–11, 1963) |
| Highest 12-hr total | 52.76 | Belouve, La Réunion (Feb. 28–29, 1964) |
| Highest 4.5-hr total | 31.00 | Smethport, Pa. (July 18, 1942) |
| Highest 42-min total | 12.00 | Holt, Mo. (June 22, 1947) |
| Highest 20-min total | 8.10 | Curtea-de-Arges, Rumania (July 7, 1889) |
| Highest 1-min total | 1.23 | Unionville, Md. (July 4, 1956) |
| Highest average no. of rain days in a year | 180 | Bahia Felix, Chile (325 days) |
| | 128 | Cedral, Costa Rica (312 days) |
| | 191 | Ponape, Pacific Island (311 days) |
| Highest no. of rain days in a year | | Cedral, Costa Rica (1968)—355 days |
| | | Cedral, Costa Rica (1967)—350 days |
| | | Bahia Felix, Chile (1916)—348 days |
| Highest annual total (snow) | 1017 | Paradise Ranger Station, Mt. Rainier, Wa. (1970–71) |
| Highest annual means (snow) | 582 | Mt. Rainier, Wa. |
| | 575 | Crater Lake, Ore. |

**TABLE 7.8 (continued)**

| Measurement of Precipitation | Precipitation (in.) | Station |
|---|---|---|
| Highest monthly total (snow) | 390 | Tamarack, Ca. |
| Highest 12-day total (snow) | 304 | Norden Summit, Ca. (Feb. 1–12, 1938) |
| Highest 6-day total (snow, single snowstorm) | 174 | Thompson Pass, U.S.A. (Dec. 26–31, 1955) |
| Highest daily total (snow) | 76 | Silver Lake, Colo. (April 14–15, 1921) |
| Greatest depth on ground | 454 | Tamarack, Ca. (March 9, 1911) |

Hail: Coffeyville, Kan.—17.5 in. circum.; 1.67 lb (Sept. 3, 1970)
Potter, Neb.—17.0 in. circum.; 1.5 lb (July 6, 1928)
Cabo Raper, Chile, has 87 in. of rain per year, but has never had over 1.9 in. in one day (10 yr).
Concepcion, Chile, has 45 in. of rain per year, but has never had over 0.8 in. in one day.
Quseir, Egypt, has less than 0.1 in. of rain per year, but received 1.3 in. in one day (Nov., 19 yr).
Lobito, Angola, has 13.0 in. of rain per year, but has received 21.1 in. in one day (Mar., 19 yr).
Thrall, Texas, has 36 in. of rain per year, but has received 38.2 in. in one day (Sept., 1921, estimated 23.4 in./6 hr).

**TABLE 7.9**
Extremes of humidity.

| Measurement of Humidity | Humidity (°F or percent) | Station |
|---|---|---|
| Highest mean monthly dewpoint | 84°F | Assab, Ethiopia (June) |
|  | 83°F | Bahrain Island, Persian Gulf (August) |
| Extreme maximum dewpoint | 96.8°F | Khanpur, Pakistan (unverified) |
| Extreme minimum dewpoint | −150°F | South Pole |
| Mean annual relative humidity | 90% | Bear Island (74°31′ N, 19°01′ E; 20 yr) Decepcion, Antarctica (62°59′ S, 60°43′ W) |
| Mean monthly relative humidity | 97% | Ostrov Rudolfa, Franz Joseph Land (81°48′ N, 57°57′ E; 1 yr; July) |
|  | 94% | Mys Chelyuskin (77°43′ N, 104°17′ E; 8 yr; August) |

**TABLE 7.10**
Extremes of radiation.

| Measurement of Radiation | Radiation (ly or ly day$^{-1}$) | Station |
|---|---|---|
| Highest mean annual | 667 ly day$^{-1}$ | La Quiaca, Argentina (3459 m) (85% of extraterrestrial radiation at 22° S) |
| Highest mean monthly | 955 ly day$^{-1}$ | South Pole (2800 m; Dec.; 8 yr) |
| Highest in 1 h | 113 ly | Malange, Angola (1150 m) |
|  | 112 ly | Windhoek, Namibia (1700 m) |

**TABLE 7.11**
Extremes of sunshine.

| Measurement of Sunshine | Sunshine (h) | Station |
|---|---|---|
| Highest mean annual | 4300+ | Wadi Halfa, Sudan |
| Highest continuous | 60 | Antarctica (Dec. 9–12, 1911) |
| Highest day | 24 | Antarctica (Dec. 9–12, 1911) |
| Highest monthly | 14 h day$^{-1}$ | Syowa Base (69°00′ S, 39°35′ E) |
| Lowest mean annual | 500 | Laurie Island, (60°44′ S, 44°44′ W; 44 yr) |
|  | 550 | Argentine Island, (61°15′ S, 64°15′ W) |
| Lowest monthly | 5% possible | Argentine Island (June) |

**TABLE 7.12**
Extremes of pressure.

| Measurement of Pressure | Pressure (mb) | Station |
|---|---|---|
| Highest measured | 1081.8 (31.9 in.) | Sedom, Israel (1275 ft b.m.s.l.; Feb. 21, 1961) |
| Highest sea level estimated | 1079 (31.84 in.) | Barnaul, Siberia (Jan. 23, 1900) |
| Highest mean annual | 1021.7 | Minusinsk, USSR |
| Highest mean monthly | 1034.3 | Minusinsk, USSR; Troitsk Mine, USSR (Jan.) |
|  | 1046 | Minusinsk, USSR (Jan., 1954) |
| Lowest measured | 870 (25.62 in.) | Typhoon "Tip" (16°44′ N, 137°46′ E; Oct. 12, 1979) |

NOTE: Perhaps Turfan Depression, central Asia, has reached 1100 mb.

**TABLE 7.13**
Extremes of surface wind speed.

| Measurement of Wind Speed | Speed (mi h$^{-1}$) | Station |
|---|---|---|
| Highest absolute | 231 | Mt. Washington, N.H. (6288 ft; Apr. 24, 1934) |
| Highest in 1 h | 173 | Mt. Washington, N.H. |
| Highest in 1 day | 129 | Mt. Washington, N.H. |
| Highest annual average (USA) | 35.4 | Mt. Washington, N.H. |
| Highest annual average (world) | 43 | Cape Denison, Antarctica |
| Highest monthly average | 55 | Cape Denison, Antarctica (July, 1913) |

NOTES: In tornadoes speeds of 500 mi h$^{-1}$ have been calculated. At the coast of Commonwealth Bay, Antarctica, a number of gusts in excess of 200 mi h$^{-1}$ have been recorded.

**TABLE 7.14**
Extremes of fog.

| Measurement of Fog | Duration (h or days) | Station |
|---|---|---|
| Highest amount in 1 yr | 7613 h | Willapa, Wa. |
| Highest annual average | 2552 h | Cape Disappointment, Wa. |
|  | 1874 h | S. Orkneys, Antarctica |
|  | 1580 h | Moose Peak, Maine |
|  | 120 days | Grand Banks, Newfoundland |

**SUGGESTED READING**

CRITCHFIELD, H. J. *General Climatology*. Englewood Cliffs, N.J.: Prentice-Hall, 1974.

TREWARTHA, G. T. *The Earth's Problem Climates*. Madison: University of Wisconsin Press, 1961.

TREWARTHA, G. T., and L. H. HORN. *An Introduction to Climate*. New York: McGraw-Hill Book Company, 1980.

# 8 CLIMATES OF MACROSCALE AREAS AND THEIR CLASSIFICATION

**INTRODUCTION**
**HISTORY OF CLIMATE CLASSIFICATION**
**CLIMATE OF THE STANDARD CONTINENT**
Temperature Zones
Precipitation Zones
Topographic and Shape Effects
**CLASSIFICATION REQUIREMENTS**
**MARINE AND CONTINENTAL CLIMATES**
**KÖPPEN'S CLASSIFICATION**
**THORNTHWAITE'S CLASSIFICATION**
**OTHER IMPORTANT CLASSIFICATIONS**

## 8.1 INTRODUCTION

It has been estimated that more than 100,000 climatological stations around the world submit their data, on a more or less regular basis, to some local central agency for storage, analysis and, often, publication. Such a vast network generates an incredible amount of data every year, so much so that unless the information is collated and summarized its usefulness is limited. As an example of the amount accumulated, the National Climatic Center of the U.S. National Weather Service in Asheville, North Carolina, has over 75,000 reels of magnetic tape and adds about 10,000 reels annually.

The first step in most climatic analyses is preparing a summary of the data relating to the station(s). The summary enables the investigator to obtain an idea of the values involved. Typical summaries, as prepared in the United States for first-order stations and substations (cooperative), are shown in Tables 8.1 and 8.2. The format and contents can vary widely from country to country and even among agencies within a country.

Very often such a summary answers many of the questions concerning the climate of that station, but if a worldwide (or very large area) survey is to be undertaken, a bewildering quantity of summaries must be analyzed. To identify areas with similar or analogous climates and at the same time differentiate one climate from another, climatic classification systems have been developed and become ingrained in the science of climatology. These classification systems have the added advantage of reducing the mass of climatic data to a manageable form.

## 8.2 HISTORY OF CLIMATE CLASSIFICATION

The idea of classifying the climates of the known world has intrigued people for over two thousand years. In about 500 B.C. Parmenides suggested the earliest known classification, in which the torrid zone extended between the Tropics of Cancer and Capricorn, temperate zones were between the Tropics and the polar circles, and frigid zones extended from these to the poles. This approach, based on latitude alone, is known as the **solar climate classification.** Around 140 B.C. Hipparchus used a system in which the zones corresponded to the length of day at the

## TABLE 8.1
Annual summary of local climatological data, Amarillo, Texas.
### Meteorological Data for the Current Year

Station: AMARILLO, TEXAS    AMARILLO AIR TERMINAL    Standard time used: CENTRAL    Latitude: 35° 14' N    Longitude: 101° 42' W    Elevation (ground): 3604 ft    Year: 1975

| Month | Temperature °F — Averages | | | Temperature °F — Extremes | | | | Degree days Base 65°F | | Precipitation (water equiv.) — Total | Greatest in 24 hrs. | Date | Snow, Ice pellets — Total | Greatest in 24 hrs. | Date | Relative humidity, pct. (Local time) Hour 00 | 06 | 12 | 18 | Wind — Resultant Speed m.p.h. | Direction | Average speed m.p.h. | Fastest mile — Speed m.p.h. | Direction | Date | Percent of possible sunshine | Average sky cover, tenths, sunrise to sunset | Number of days — Sunrise to sunset Clear | Partly cloudy | Cloudy | Precipitation .01 inch or more | Snow, Ice pellets 1.0 inch or more | Thunderstorms | Heavy fog, visibility ¼ mile or less | Temperature °F Maximum 90° and above | 32° and below | Minimum 32° and below | 0° and below | Average station pressure mb Elev. 3604 ft m.s.l. |
|---|---|---|---|---|---|---|---|---|---|---|---|---|---|---|---|---|---|---|---|---|---|---|---|---|---|---|---|---|---|---|---|---|---|---|---|---|---|---|---|
| | Daily maximum | Daily minimum | Monthly | Highest | Year | Lowest | Date | Heating | Cooling | | | | | | | | | | | | | | | | | | | | | | | | | | | | | | |
| Jan. | 50.4 | 23.8 | 37.1 | 78 | 26 | 3 | 12 | 853 | 0 | 0.28 | 0.15 | 1-2 | 3.8 | 2.3 | 1-2 | 70 | 74 | 53 | 50 | 4.1 | 27 | 13.9 | 41 | 36 | 19 | 74 | 3.9 | 17 | 6 | 8 | 6 | 1 | 2 | 0 | 5 | 28 | 0 | 890.3 |
| Feb. | 46.7 | 23.8 | 35.3 | 73 | 13 | 7 | 9 | 826 | 0 | 1.33 | 0.49 | 21-22 | 11.7 | 5.5 | 28 | 80 | 64 | 59 | 28 | 2.6 | 25 | 13.9 | 38 | 17 | 26 | 58 | 6.4 | 15 | 6 | 7 | 10 | 2 | 3 | 0 | 1 | 23 | 0 | 889.3 |
| Mar. | 59.5 | 30.4 | 45.0 | 82 | 20 | 17 | 29 | 612 | 0 | 0.51 | 0.27 | 28 | 1.2 | 0.8 | 28 | 64 | 70 | 45 | 28 | 4.2 | 22 | 16.9 | 39 | 18 | 26 | 69 | 5.6 | 10 | 10 | 11 | 6 | 0 | 4 | 3 | 1 | 16 | 0 | 886.2 |
| Apr. | 69.6 | 40.1 | 54.9 | 87 | 2 | 19 | 7 | 323 | 24 | 1.02 | 0.57 | 10 | T | T | 12-13 | 60 | 73 | 41 | 45 | 8.4 | 20 | 16.8 | 43 | 18 | 26 | 73 | 4.3 | 11 | 10 | 9 | 7 | 0 | 7 | 2 | 1 | 6 | 0 | 887.2 |
| May | 78.9 | 50.1 | 64.5 | 88 | 25 | 37 | 2 | 67 | 61 | 2.47 | 1.11 | 22 | T | T | | 59 | 76 | 39 | 41 | 5.9 | 20 | 15.5 | 40 | 22 | 5 | 77 | 3.8 | 17 | 8 | 6 | 9 | 0 | 9 | 2 | 0 | 0 | 0 | 887.6 |
| June | 86.3 | 60.3 | 73.3 | 96 | 16 | 47 | 11 | 16 | 273 | 4.15 | 1.06 | 9-10 | 0.0 | 0.0 | | 70 | 81 | 49 | 42 | 7.8 | 16 | 14.8 | 40 | 15 | 7 | 75 | 4.0 | 17 | 13 | 0 | 5 | 0 | 8 | 2 | 9 | 0 | 0 | 889.3 |
| Jul. | 86.1 | 64.8 | 75.5 | 93 | 30 | 61 | 25 | 0 | 330 | 5.19 | 1.89 | 22-23 | 0.0 | 0.0 | | 79 | 86 | 54 | 39 | 6.4 | 18 | 11.5 | 35 | 26 | 30 | 65 | 6.0 | 10 | 12 | 9 | 14 | 0 | 14 | 2 | 9 | 0 | 0 | 892.7 |
| Aug. | 89.4 | 64.0 | 76.7 | 98 | 24 | 57 | 11 | 0 | 367 | 3.97 | 2.48 | 14-15 | 0.0 | 0.0 | | 65 | 78 | 40 | 39 | 9.2 | 14 | 13.1 | 30 | 01 | 26 | 86 | 2.5 | 21 | 8 | 2 | 5 | 0 | 5 | 1 | 19 | 0 | 0 | 893.0 |
| Sept. | 78.7 | 52.9 | 65.8 | 92 | 17 | 39 | 29 | 104 | 134 | 0.76 | 0.56 | 20-21 | 0.0 | 0.0 | | 60 | 73 | 40 | 36 | 7.2 | 28 | 13.1 | 30 | 01 | 27 | 69 | 4.0 | 14 | 8 | 8 | 5 | 0 | 4 | 1 | 5 | 0 | 0 | 894.3 |
| Oct. | 77.4 | 44.7 | 61.1 | 91 | 12 | 29 | 25 | 158 | 46 | 0.33 | 0.33 | 31 | 0.0 | 0.0 | | 44 | 58 | 27 | 31 | 3.3 | 28 | 13.1 | 30 | 03 | 27 | 98 | 1.3 | 24 | 6 | 1 | 3 | 0 | 1 | 1 | 0 | 0 | 0 | 892.0 |
| Nov. | 60.0 | 31.5 | 45.8 | 85 | 10 | 12 | 26 | 569 | 0 | 0.92 | 0.70 | 19 | 0.4 | 0.4 | 25 | 56 | 64 | 37 | 40 | 4.3 | 29 | 13.9 | 44 | 26 | 29 | 84 | 3.8 | 16 | 8 | 6 | 4 | 0 | 2 | 0 | 1 | 16 | 0 | 891.0 |
| Dec. | 55.0 | 25.7 | 40.4 | 72 | 10 | 9 | 18 | 757 | 0 | 0.15 | 0.13 | 28-29 | 1.8 | 1.6 | 28-29 | 58 | 62 | 38 | 38 | 3.6 | 24 | 12.0 | 35 | 27 | Nov. 29 | 83 | 3.9 | 17 | 7 | 8 | 3 | 1 | 0 | 3 | 0 | 25 | 0 | 892.3 |
| Year | 69.8 | 42.7 | 56.3 | 98 | Aug. 24 | 3 | Jan. 12 | 4285 | 1235 | 21.08 | 2.48 | Aug. 14-15 | 18.9 | 5.5 | Feb. 21-22 | 64 | 73 | 44 | 41 | 4.4 | 21 | 14.1 | 44 | 26 | | 76 | 4.1 | 181 | 96 | 88 | 76 | 4 | 51 | 29 | 44 | 114 | 0 | 890.4 |

### Normals, Means, and Extremes (1941–1970)

| Month (a) | Temperature °F — Normal | | | Temperature °F — Extremes | | | | Normal Degree days Base 65°F | | Precipitation in inches — Water equivalent | | | | | Snow, Ice pellets | | | Relative humidity, pct. (Local time) Hour 00 | 06 | 12 | 18 | Wind — Mean speed m.p.h. | Prevailing direction | Fastest mile Speed m.p.h. | Direction | Year | Pct. of possible sunshine | Mean sky cover, tenths, sunrise to sunset | Mean number of days — Sunrise to sunset Clear | Partly cloudy | Cloudy | Precipitation .01 inch or more | Snow, Ice pellets 1.0 inch or more | Thunderstorms | Heavy fog, visibility ¼ mile or less | Temperature °F Max. 90° and above | 32° and below | Min. 32° and below | 0° and below | Average station pressure mb Elev. 3604 ft m.s.l. |
|---|---|---|---|---|---|---|---|---|---|---|---|---|---|---|---|---|---|---|---|---|---|---|---|---|---|---|---|---|---|---|---|---|---|---|---|---|---|---|---|
| | Daily maximum | Daily minimum | Monthly | Record highest | Year | Record lowest | Year | Heating | Cooling | Normal | Maximum monthly | Year | Minimum monthly | Year | Maximum in 24 hrs. | Year | Maximum monthly | Year | Maximum in 24 hrs. | Year | | | | | | | | | | | | | | | | | | | | |
| | 15 | | | 15 | | | | | | 35 | 35 | | 35 | | 35 | | 35 | | 35 | | 14 | 14 | 14 | 14 | 14 | 15 | 34 | 34 | 34 | 34 | 34 | 34 | 34 | 34 | 34 | 34 | 34 | 34 | 34 | 3 |
| J | 49.4 | 22.5 | 36.0 | 79 | 1970 | −9 | 1963 | 899 | 0 | 0.54 | 2.33 | 1968 | T | 1967 | 1.74 | 1967 | 6.7 | 1960 | 6.7 | 1960 | 63 | 70 | 49 | 44 | 13.1 | SW | 62 | NE | 1953 | 69 | 5.1 | 13 | 7 | 11 | 4 | 1 | * | 3 | 0 | 4 | 27 | 2 | 890.4 |
| F | 53.0 | 26.4 | 39.7 | 88 | 1963 | 0 | 1971 | 708 | 0 | 0.56 | 1.83 | 1948 | T | 1943 | 1.28 | 1943 | 13.5 | 1971 | 13.5 | 1971 | 63 | 70 | 45 | 40 | 14.2 | SW | 70 | NW | 1956 | 68 | 5.1 | 11 | 7 | 10 | 4 | 1 | 1 | 2 | 0 | 2 | 22 | * | 890.6 |
| M | 60.0 | 31.2 | 45.6 | 94 | 1971 | 7 | 1967 | 601 | 0 | 0.77 | 3.99 | 1973 | T | 1950 | 2.27 | 1973 | 9.8 | 1961 | 9.8 | 1961 | 56 | 67 | 37 | 33 | 15.6 | SW | 72 | W | 1950 | 71 | 5.2 | 12 | 8 | 11 | 5 | 1 | 3 | 2 | 1 | 1 | 15 | 0 | 886.0 |
| A | 70.9 | 42.1 | 56.5 | 98 | 1965 | 18 | 1973 | 275 | 20 | 1.23 | 3.74 | 1942 | 0.19 | 1951 | 1.57 | 1942 | 6.4 | 1947 | 5.1 | 1947 | 53 | 65 | 35 | 33 | 15.5 | SW | 74 | SW | 1942 | 73 | 4.9 | 12 | 8 | 10 | 6 | * | 6 | 2 | 3 | 1 | 3 | * | 887.7 |
| M | 79.2 | 51.9 | 65.6 | 99 | 1974 | 30 | 1970 | 81 | 99 | 2.83 | 9.81 | 1951 | 0.01 | 1960 | 6.75 | 1966 | 5.1 | 1953 | 5.1 | 1953 | 57 | 70 | 37 | 35 | 14.8 | SW | 84 | SW | 1949 | 73 | 5.0 | 11 | 8 | 12 | 8 | * | 9 | 2 | 6 | 0 | * | 0 | 887.9 |
| J | 88.0 | 61.2 | 74.6 | 104 | 1970 | 43 | 1970 | 10 | 298 | 3.45 | 10.73 | 1965 | 0.01 | 1965 | 6.15 | 1960 | 0.0 | | | | 61 | 73 | 43 | 38 | 14.4 | S | 75 | SW | 1974 | 77 | 4.3 | 13 | 12 | 5 | 8 | 0 | 9 | 1 | 12 | 0 | 0 | 0 | 889.5 |
| J | 91.4 | 65.9 | 78.7 | 104 | 1972 | 54 | 1943 | 0 | 425 | 2.95 | 7.59 | 1960 | 0.12 | 1943 | 4.09 | 1946 | 0.0 | | | | 65 | 76 | 43 | 42 | 12.4 | S | 66 | SW | 1948 | 78 | 4.6 | 13 | 12 | 6 | 8 | 0 | 10 | 1 | 21 | 0 | 0 | 0 | 892.7 |
| A | 90.4 | 64.7 | 77.6 | 104 | 1964 | 52 | 1972 | 0 | 391 | 2.93 | 7.55 | 1974 | 0.39 | 1945 | 3.42 | 1947 | 0.0 | | | | 65 | 76 | 40 | 46 | 12.0 | S | 65 | E | 1942 | 78 | 4.2 | 15 | 10 | 6 | 7 | 0 | 9 | 1 | 16 | 0 | 0 | 0 | 892.7 |
| S | 82.9 | 56.7 | 69.8 | 100 | 1970 | 36 | 1971 | 20 | 164 | 1.93 | 5.02 | 1950 | 0.24 | 1941 | 3.45 | 1947 | T | 1945 | T | 1945 | 71 | 80 | 46 | 42 | 13.0 | S | 68 | NE | 1949 | 75 | 3.8 | 16 | 9 | 7 | 6 | 0 | 4 | 1 | 6 | 0 | 1 | 0 | 893.1 |
| O | 72.9 | 46.1 | 59.5 | 94 | 1973 | 25 | 1970 | 206 | 36 | 1.83 | 7.64 | 1941 | 0.00 | 1952 | 3.45 | 1948 | 3.9 | 1970 | 2.9 | 1970 | 62 | 71 | 41 | 40 | 13.0 | S | 59 | NW | 1949 | 73 | 3.8 | 17 | 8 | 6 | 5 | * | 2 | 2 | 2 | 0 | 3 | 0 | 892.9 |
| N | 60.0 | 32.5 | 46.1 | 85 | 1975 | 12 | 1961 | 561 | 0 | 0.53 | 2.26 | 1961 | T | 1970 | 1.29 | 1960 | 13.6 | 1971 | 12.2 | 1971 | 65 | 72 | 47 | 47 | 13.2 | SW | 68 | SW | 1970 | 73 | 4.3 | 15 | 8 | 7 | 4 | 1 | * | 3 | 0 | 3 | 13 | 0 | 891.0 |
| D | 51.5 | 25.5 | 38.5 | 76 | 1961 | −3 | 1959 | 822 | 0 | 0.73 | 4.52 | 1959 | T | | 3.11 | 1943 | 8.5 | 1943 | 7.4 | 1943 | 64 | 69 | 48 | 49 | 13.7 | SW | 62 | NE | 1953 | 67 | 4.4 | 13 | 8 | 10 | 4 | 1 | * | 3 | 0 | 3 | 26 | * | 890.6 |
| Yr | 70.8 | 43.9 | 57.4 | 104 | July 1970 | −9 | Jan. 1963 | 4183 | 1433 | 20.28 | 10.73 | June 1965 | 0.00 | Oct. 1952 | 6.75 | May 1951 | 13.6 | Feb. 1971 | 13.5 | Feb. 1971 | 62 | 72 | 44 | 40 | 13.7 | SW | 84 | SW | May 1949 | 73 | 4.6 | 161 | 104 | 100 | 68 | 4 | 48 | 25 | 64 | 12 | 106 | 2 | 890.4 |

Means and extremes above are from existing and comparable exposures. Annual extremes have been exceeded at other sites in the locality as follows: Highest temperature 108 in June 1953; lowest temperature −16 in February 1899; maximum monthly snowfall 28.7 in February 1903; maximum snowfall in 24 hours 20.6 in March 1934.

(a) Length of record, years, through the current year unless otherwise noted, based on January data.
0 Through 1974
* Less than one half.
T Trace.

NORMALS - Based on record for the 1941-1970 period.
DATE OF AN EXTREME - The most recent in cases of multiple occurrence.
PREVAILING WIND DIRECTION - Record through 1963.
WIND DIRECTION - Numerals indicate tens of degrees clockwise from true north. 00 indicates calm.
FASTEST MILE WIND - Speed is fastest observed 1-minute value when the direction is in tens of degrees.

SOURCE: NOAA, Environmental Data Service, National Climatic Center, Asheville, N.C.

## TABLE 8.2
Climatological summary for Boulder, Colorado.

### Means and Extremes for Period 1951–1973

| Month | Temperature (°F) | | | | | | | | | Mean number of days | | | | Precipitation totals (inches) | | | | | | Snow, sleet | | | | | | Mean number of days | | |
| | Means | | | Extremes | | | | | | Max | | Min. | | | | | | | | | | | | | | .10 or more | .50 or more | 1.00 or more |
| | Daily maximum | Daily minimum | Monthly | Record highest | Year | Day | Record lowest | Year | Day | 90° and above | 32° and below | 32° and below | 0° and below | Mean | Greatest monthly | Year | Greatest daily | Year | Day | Mean | Maximum monthly | Year | Greatest depth | Year | Day | | | |
|---|---|---|---|---|---|---|---|---|---|---|---|---|---|---|---|---|---|---|---|---|---|---|---|---|---|---|---|---|
| Jan. | 46.0 | 21.4 | 33.7 | 72 | 53 | 9 | −22 | 63 | 12 | 0 | 5 | 26 | 2 | .63 | 1.52 | 62 | .98 | 62 | 08 | 10.1 | 25.1 | 62 | 15.0 | 62 | 08 | 2 | 0 | 0 |
| Feb. | 48.4 | 24.1 | 36.3 | 79 | 54 | 8 | −15 | 51 | 1 | 0 | 3 | 23 | 0 | .81 | 1.53 | 56 | 1.03 | 57 | 28 | 12.0 | 25.6 | 65 | 12.0 | 71 | 21 | 2 | 0 | 0 |
| Mar. | 52.4 | 27.0 | 39.7 | 79 | 71 | 26 | −3* | 60 | 3 | 0 | 3 | 22 | 0 | 1.54 | 3.86 | 70 | 1.48 | 52 | 21 | 19.5 | 56.7 | 70 | 27.0 | 70 | 31 | 4 | 1 | 0 |
| Apr. | 62.3 | 35.8 | 49.1 | 83* | 60 | 21 | 5 | 59 | 10 | 0 | 0 | 10 | 0 | 2.11 | 6.85 | 57 | 3.31 | 57 | 02 | 12.0 | 44.0 | 57 | 19.0 | 59 | 09 | 5 | 1 | 0 |
| May | 72.1 | 45.8 | 59.0 | 93 | 54 | 20 | 22 | 54 | 2 | 0 | 0 | 1 | 0 | 3.21 | 9.27 | 57 | 3.37 | 69 | 07 | 1.3 | 6.6 | 73 | 4.0 | 73 | 01 | 6 | 2 | 1 |
| June | 82.0 | 54.4 | 68.3 | 104 | 54 | 23 | 30 | 51 | 2 | 7 | 0 | 0 | 0 | 2.11 | 5.34 | 69 | 2.65 | 63 | 16 | .1 | 2.2 | 51 | | | | 5 | 1 | 0 |
| July | 87.9 | 60.3 | 74.1 | 104 | 54 | 11 | 42 | 72 | 5 | 14 | 0 | 0 | 0 | 1.82 | 5.20 | 65 | 1.49 | 69 | 17 | .0 | | | | | | 5 | 1 | 0 |
| Aug. | 86.2 | 58.8 | 72.5 | 101 | 69 | 8 | 43 | 72 | 25 | 10 | 0 | 0 | 0 | 1.58 | 7.49 | 51 | 3.06 | 51 | 03 | .0 | | | | | | 4 | 1 | 0 |
| Sept. | 73.1 | 49.8 | 63.9 | 97* | 59 | 6 | 22 | 71 | 19 | 2 | 0 | 1 | 0 | 1.66 | 4.89 | 61 | 1.64 | 71 | 17 | 1.7 | 21.0 | 71 | 18.0 | 71 | 17 | 4 | 1 | 0 |
| Oct. | 67.9 | 40.4 | 54.1 | 90 | 53 | 1 | 10 | 69 | 13 | 0 | 0 | 6 | 0 | 1.20 | 5.39 | 69 | 1.61 | 51 | 05 | 5.9 | 49.3 | 69 | 13.0 | 69 | 12 | 3 | 1 | 0 |
| Nov. | 54.0 | 29.0 | 41.6 | 79 | 52 | 4 | −2 | 51 | 2 | 0 | 2 | 19 | 0 | 1.04 | 2.15 | 72 | .85 | 70 | 13 | 11.7 | 26.8 | 72 | 15.0 | 72 | 01 | 3 | 1 | 0 |
| Dec. | 47.4 | 23.5 | 35.5 | 75 | 65 | 4 | −16 | 72 | 6 | 0 | 3 | 25 | 1 | .68 | 1.54 | 51 | .55 | 72 | 29 | 10.6 | 31.4 | 67 | 9.0 | 72 | 29 | 3 | 0 | 0 |
| Year | 65.4 | 39.2 | 52.3 | 104+ | | Jul 11 | −22 | 63 | Jan 12 | 33 | 16 | 133 | 3 | 18.39 | 9.27 | May 57 | 3.37 | May 69 | 7 | 84.9 | 56.7 | Mar 70 | 27.0 | Mar 70 | 31 | 45 | 10 | 1 |

*Also on earlier dates.

### Freeze Probabilities

#### Probability of Later Date in Spring (mo/da) than Indicated

| Temp. | .10 | .20 | .30 | .40 | .50 | .60 | .70 | .80 | .90 |
|---|---|---|---|---|---|---|---|---|---|
| 32 | 5/20 | 5/14 | 5/10 | 5/ 6 | 5/ 3 | 4/29 | 4/26 | 4/22 | 4/16 |
| 28 | 5/ 6 | 5/ 1 | 4/26 | 4/23 | 4/19 | 4/16 | 4/12 | 4/ 8 | 4/ 2 |
| 24 | 4/26 | 4/20 | 4/15 | 4/11 | 4/ 7 | 4/ 4 | 3/31 | 3/26 | 3/19 |
| 20 | 4/15 | 4/ 9 | 4/ 5 | 4/ 2 | 3/30 | 3/27 | 3/24 | 3/20 | 3/15 |
| 16 | 4/10 | 4/ 5 | 4/ 1 | 3/28 | 3/25 | 3/22 | 3/19 | 3/15 | 3/10 |

#### Probability of Earlier Date in Fall (mo/da) than Indicated

| Temp. | .10 | .20 | .30 | .40 | .50 | .60 | .70 | .80 | .90 |
|---|---|---|---|---|---|---|---|---|---|
| 32 | 9/23 | 9/29 | 10/ 3 | 10/ 7 | 10/10 | 10/13 | 10/17 | 10/21 | 10/27 |
| 28 | 10/ 3 | 10/ 9 | 10/13 | 10/17 | 10/20 | 10/24 | 10/27 | 11/ 1 | 11/ 7 |
| 24 | 10/12 | 10/20 | 10/25 | 10/29 | 11/ 2 | 11/ 6 | 11/11 | 11/16 | 11/23 |
| 20 | 10/20 | 10/27 | 11/ 1 | 11/ 5 | 11/ 9 | 11/12 | 11/17 | 11/21 | 11/28 |
| 16 | 10/28 | 11/ 4 | 11/ 9 | 11/13 | 11/16 | 11/20 | 11/24 | 11/29 | 12/ 5 |

#### Probability of Longer than Indicated Freeze Free Period (days)

| Temp. | .10 | .20 | .30 | .40 | .50 | .60 | .70 | .80 | .90 |
|---|---|---|---|---|---|---|---|---|---|
| 32 | 185 | 176 | 170 | 164 | 159 | 154 | 149 | 143 | 134 |
| 28 | 200 | 194 | 190 | 186 | 183 | 180 | 176 | 172 | 166 |
| 24 | 238 | 227 | 220 | 214 | 208 | 202 | 196 | 189 | 179 |
| 20 | 250 | 240 | 234 | 228 | 223 | 217 | 212 | 205 | 196 |
| 16 | 260 | 251 | 245 | 240 | 235 | 230 | 225 | 219 | 211 |

### Precipitation with Probability Equal or Less than

| Lvl. | Jan. | Feb. | Mar. | Apr. | May | June | July | Aug. | Sept. | Oct. | Nov. | Dec. |
|---|---|---|---|---|---|---|---|---|---|---|---|---|
| 0.05 | 0.09 | 0.15 | 0.40 | 0.28 | 0.80 | 0.47 | 0.63 | 0.10 | 0.08 | 0.15 | 0.33 | 0.12 |
| 0.10 | 0.14 | 0.22 | 0.54 | 0.45 | 1.09 | 0.66 | 0.80 | 0.20 | 0.17 | 0.25 | 0.43 | 0.18 |
| 0.20 | 0.23 | 0.34 | 0.76 | 0.76 | 1.56 | 0.97 | 1.05 | 0.40 | 0.36 | 0.42 | 0.57 | 0.28 |
| 0.30 | 0.32 | 0.45 | 0.96 | 1.05 | 1.97 | 1.25 | 1.26 | 0.62 | 0.58 | 0.58 | 0.69 | 0.37 |
| 0.40 | 0.41 | 0.56 | 1.15 | 1.36 | 2.37 | 1.52 | 1.45 | 0.86 | 0.84 | 0.76 | 0.81 | 0.47 |
| 0.50 | 0.51 | 0.68 | 1.35 | 1.69 | 2.80 | 1.82 | 1.66 | 1.14 | 1.14 | 0.95 | 0.94 | 0.57 |
| 0.60 | 0.62 | 0.81 | 1.58 | 2.09 | 3.29 | 2.15 | 1.89 | 1.48 | 1.51 | 1.18 | 1.07 | 0.69 |
| 0.70 | 0.76 | 0.97 | 1.84 | 2.57 | 3.83 | 2.53 | 2.13 | 1.91 | 1.99 | 1.46 | 1.22 | 0.83 |
| 0.80 | 0.94 | 1.19 | 2.18 | 3.21 | 4.55 | 3.04 | 2.45 | 2.52 | 2.67 | 1.83 | 1.42 | 1.01 |
| 0.90 | 1.26 | 1.55 | 2.76 | 4.30 | 5.78 | 3.90 | 3.00 | 3.54 | 3.84 | 2.47 | 1.75 | 1.32 |
| 0.95 | 1.54 | 1.86 | 3.22 | 5.28 | 6.77 | 4.61 | 3.41 | 4.55 | 5.01 | 3.04 | 2.01 | 1.59 |

SOURCE: NOAA, Environmental Data Service, National Climatic Center, Asheville, N.C.

summer solstices. He called these zones *klimata* because they were based on the inclination, or slope, of the solar beam—this is the origin of the word climate. Other classifications, their originators now unknown, also used latitude to distinguish climatic zones. One known to sailors was the "seven climes" division, defined by latitudes through Meroe (17°), Syene (24°), Alexandria (31°), Rhodes (36°), Borysthenes (45°), and the Riphaean Mountains (48°). Another such classification, attributed to Ptolemy (90–168 A.D.), used latitude bands of 0 to 15°, 15 to 22°, 22 to 29°, 29 to 35°, 35 to 40°, 40 to 48° and 48 to 62°.

History does not record any further development in the science of climate classification until in the twelfth century the Arab scientist Idrisi proposed seven climatic zones (once again the "magic" seven), each having ten divisions. It was not until suitable instruments were developed and reliable instruments had been used to report data for many years in different locations that further steps could be taken. In the early eighteenth century Jedediah Morse refers to twenty-four climatic zones between the equator and the pole. Modern climatology is generally recognized as beginning at the end of the eighteenth century, especially when the explorer and scientist Alexander von Humboldt introduced the isotherm and, in 1817, mapped the mean annual isotherms for much of the Northern Hemisphere. He described the departures of these lines from the "classical" latitude concept and advanced explanations for their patterns based upon ocean currents and land-sea configuration.

In the nineteenth century many botanists, who were becoming climatologists, were concerned with mapping the various climatic elements. In 1842 J. R. Hind developed a sixteen-zone classification that depended upon four classes of mean annual temperature—two of precipitation and two of seasonal variation—which was considered original because it combined two elements. This was at about the time when Wilhelm Mahlmann and Heinrich Wilhelm Dove were developing isothermal maps. James Henry Coffin began collecting wind data in the 1850s, continuing to do so until his death in 1873. Independently Matthew Fontaine Maury was embarking on a similar task, but including sea current and temperature data. In 1890, Oscar Drude combined the patterns of thermal and vegetation zones, giving a bias to climatology that still exists. At the turn of the century the first real mathematical system of classification, a system relating plant distribution to climatic divisions, was advanced by Vladimir Köppen. In 1900 Ernest Ravenstein developed a temperature–relative humidity classification—an early base for human comfort studies. Köppen's classification (see Section 8.6) evolved over a period of about twenty years, starting in 1918, and since then has been the most common classification used.

In 1931 C. W. Thornthwaite devised his first classification based, like Köppen's, on temperature and rainfall but using calculated indices as the variables. Other systems used climatic elements that were measured directly (see Section 8.7). In 1948 he introduced another classification based upon an estimate of potential evapotranspiration and the resulting water deficit or surplus.

The subject of climatic classification is still under study and every year sees at least one new system. Some are concerned with mapping an index related to human comfort or stress, others with an index of use in agroclimatology. Unfortunately, as H. E. Landsberg, an eminent climatologist, has said, "Much of this effort has been rather futile. The multiplicity of atmospheric parameters, their continuity in space, and their fluctuations in time do not permit a meaningful universal taxonomy." Perhaps this is a little harsh, but unless a classification is devised with a specific problem in mind (for example, to examine the interrelationship between climate and human comfort, or between climate and building design), it is unlikely to be more than an academic pursuit.

# 8.3

## CLIMATE OF THE STANDARD CONTINENT

We have described in earlier chapters how the climate of an area is determined by certain basic controls. Using this knowledge, it is possible to devise the likely climatic zonation that would occur over a large area. For simplicity at this stage, we will use the concept of the **standard continent,** a landmass stretching from pole to pole on which topographic variations are negligible. We will consider topographic features once the zonation is established.

The standard continent technique helps us to develop a model of simple zonations which we may then compare with the climatic zones occurring on the continental masses of the "real world." In this chapter, the variations in the climatic elements are discussed only in general terms; this method is not a substitute for the more rigorous study of climatic modeling that is discussed in Chapter 11.

For convenience and realism the standard continent is depicted as having a shape in which the width at each latitude is proportional to the actual landmass at that latitude. Thus, Figures 8.1 through 8.5 will have somewhat of a spinning-top shape.

## SEC. 8.3 | CLIMATE OF THE STANDARD CONTINENT

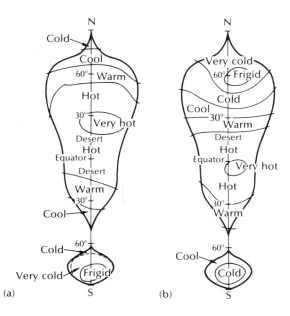

**FIGURE 8.1**
(a) Temperature patterns in summer (Northern Hemisphere) for the standard continent. (b) Temperature patterns in winter (Northern Hemisphere) for the standard continent. (From J. F. Griffiths, *Applied Climatology*, London: Oxford University Press, 1976. Reproduced by permission.)

### 8.3.1 TEMPERATURE ZONES

The first element to be considered is temperature. The seasonal pattern of temperature over the earth is influenced mainly by solar radiation (latitude), although altitude, land-sea configuration, ocean currents, and air masses also play a role. By a subjective combination of these effects, it is possible to devise the thermal patterns shown in Figure 8.1.

The summer and winter patterns (for the Northern Hemisphere) are shown in Figure 8.1(a) and (b), respectively. Notice that on the western side of the continent in the high-latitude region the effect of the warm ocean currents is smaller in summer than in winter. This occurs because the insolation, which shows a basically zonal distribution, is great at the time of high sun and tends to mask the effect of the ocean currents. In the winter hemisphere, when the most pronounced land-sea temperature differences occur in high latitudes, relatively warm currents have a marked moderating effect on the west coast of the model continent. This is indicated by a poleward deflection of the isotherm crossing the land-sea boundary there. In the summer hemisphere, on the other hand, when the most marked land-sea temperature differences occur in the subtropical latitudes, cool ocean currents result in cooler temperatures in coastal areas. This is indicated by an equatorward deflection of the isotherms in these latitudes (see Section 4.8.2).

### 8.3.2 PRECIPITATION ZONES

There are five major determinants of seasonal patterns and amounts of precipitation:

1. Radiation via latitude through its general effect on temperature. Because low latitudes are warm and high latitudes are cool, there is greater potential for large amounts of rain in the tropical regions.
2. Proximity (measured along prevailing air mass trajectories) to large bodies of water (continentality).
3. Station location with respect to convergent or divergent air flows, and the seasonal shifts that occur in these.
4. Orographic influences.
5. The amplifying or inhibiting effects of warm or cold offshore currents, respectively.

From subjective and qualitative analysis of the above controls (except for orographic influences), the patterns shown in Figure 8.2(a) can be deduced. Combining the summer and winter zonation suggests the annual pattern in Figure 8.2(b), but there are two regions that require special consideration. We have mentioned how, in winter, cyclones moving eastward are usually blocked or deflected by the dominant high pressure over the continent. As spring approaches there is both a weakening and a poleward shift of this high, a combination that allows some of the moisture-laden air masses coming from the west to penetrate further inland. In addition, the warming trend permits air to hold more moisture, increasing the potential for precipitation above that existing in winter. Therefore, an area with a spring or summer maximum of precipitation generally is found in continental interiors (see Section 7.3.1).

The second area is in the high latitudes along the western seaboard. As fall progresses, the land cools and convective activity occurs over the relatively warm ocean. These convective cells then move inland and, together with precipitation associated with the equatorward shift in position of the polar front at this time, tend to give a fall maximum of rainfall.

If we consider the rainfall regimes of each of the four seasons we can develop a composite picture of the annual regime on the standard continent [Figure 8.2(b)]. In the tropical region the number of wet

**FIGURE 8.2**
(a) Rainfall patterns in winter (Northern Hemisphere) for the standard continent. (b) Annual rainfall patterns for the standard continent. (From J. F. Griffiths, *Applied Climatology*, London: Oxford University Press, 1976. Reproduced by permission.)

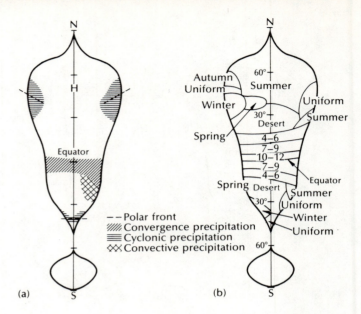

months is indicated because, in this zone of little temperature variation, the terms summer, autumn, winter, and spring do not have the meaning ascribed to them in temperate (extratropical) climates.

### 8.3.3 TOPOGRAPHIC AND SHAPE EFFECTS

In considering the standard continent we have omitted topography, but some allowance can be made through a few generalizations. For example, the orientation of long, narrow mountain ranges will have a major effect on climate. If the range runs approximately north-south, then meridional movement of air masses will not be impeded. In other words, cold, continental air from the interior can flow equatorward in winter, while the hot, tropical air of summer can sweep poleward. In South America, for instance, the Andes do not inhibit meridional flow, so the *friagem* or *sur* (cold air masses) from Argentina occasionally reach into Brazil. In North America the Rockies, mainly running north-south but with great width, have more complicated effects. Their orientation does not prevent *northers* (outbreaks of cold air) from reaching as far south as the Gulf states or even Central and South America, but they do shelter California from similar phenomena. The north-south range will, however, restrict or inhibit zonal (east-west or west-east) flow and result in more clearly defined climatic boundaries, such as the division between marine and continental climates, and the creation of rain shadow (arid) areas.

If the mountain ranges run east-west, they act as efficient barriers to the meridional flow of tropical and polar air masses. They therefore restrict both outward flow from the winter anticyclone, keeping the interior extremely cold, and the poleward movement of tropical air. This configuration can be seen in Eurasia, where the Alps, the Anatolian-Iranian ranges, and the Tibetan massif all exhibit a basically east-west pattern. In this example, zonal flow is hardly affected and oceanic influences can be detected far inland.

The problem of the different shapes of the landmasses can also be solved, in part, by imagining the zonation of the standard continent to be printed on rubber that can easily be deformed along the latitudes. The various zones will then change in relative size. For instance, if there is little latitudinal extent of the landmass around the tropical region (as in North and Central America), then a small desert area will result. Again, if the landmass in higher latitudes is very large, then the winter temperatures will be extremely cold, as in Siberia.

The influences of two features are not shown easily on the standard continent: inland lakes of great area and large water bodies that are almost landlocked. Such inland water bodies are the North American Great Lakes, the Mediterranean and the Black seas, and the Gulf of Mexico, to name a few. Each has its own influence on the surrounding climate and will be described separately in Chapter 9.

## 8.4 CLASSIFICATION REQUIREMENTS

Climatic classifications, of which more than 100 exist, generally have been devised by nonmeteorologists who have had a specific problem in mind when developing their zonation of climates. To be effective, a classification should exhibit four important characteristics:

1. It should be able to collate the mass of climatic data into a manageable and meaningful form.
2. It should be easy to apply.
3. It should be directed toward limited, well-defined objectives based on the atmospheric parameters.
4. It should be based on meteorological principles.

The fourth requirement is particularly relevant in a text of this nature, which stresses meteorological principles.

As has been explained earlier, the most basic controls of climate and weather are solar radiation, the interface (surface or volume) that receives the radiation, topography, and the land-sea configuration. A climatic classification devised by consideration of these parameters alone would be a **genetic classification** because it would be based on the basic controls that are the origin, or genesis, of the climate. A genetic classification is explanatory in nature and distinct from the descriptive/applied type of classification discussed in Section 8.2 and described later in Sections 8.6, 8.7, and 8.8. Unfortunately, we are presently a long way from developing such a true genetic classification. Classifications based on air masses are often erroneously referred to as genetic because there is a tendency to perceive weather (and climate) as generated by air masses. Actually, air masses are themselves only a manifestation of the basic controls. Such classifications, first developed in the 1930s, have seen recurrences of popularity from time to time.

If, however, we use the air mass (see Section 7.4) approach to the classification of climate, it is convenient to divide the land areas of the globe into three groups: (1) low-latitude areas in which tropical air masses dominate; (2) middle-latitude areas in which both tropical and polar air masses play a role; and (3) high-latitude areas in which polar or arctic air masses dominate.

Consideration of the first group shows that a number of subgroups exist. These subgroups are

1. Areas in which mT (moist) air masses prevail throughout the year.
2. Areas in which mT and cT (dry) air masses both play a role.
3. Areas in which cT or $cT_s$ (continental tropical subsiding) air masses prevail throughout the year.
4. Areas in which $mT_s$ air masses are often experienced.

Naturally, in the second subgroup the resulting climate depends upon the seasonal patterns of mT and cT dominance. Similar subgroups also can be devised for the other two land area groups.

Although this method of identifying areas with analogous climates is meteorologically desirable, it has one disadvantage—it is difficult to quantify the modification of air masses. Thus we must introduce climatic elements for which quantitative measurements exist as the determinants of inter-climatic boundaries. Temperature and precipitation are often used to identify the climatic zone divisions. We must remember that these boundaries are really transition zones of variable width and not the clear-cut lines they appear to be in map presentations. Also, a worldwide classification generally disregards small differences in order to indicate the broad similarities.

# 8.5

## MARINE AND CONTINENTAL CLIMATES

Earlier (Section 3.4) we described how the nature of the interface plays a major role in determining the type of climate experienced by a region. This association is most pronounced when examining the very different characters of the climates designated as marine and continental (Table 8.3). The fundamental differences between these climates are a result of variations in the thermal properties of water and land substances (see Section 2.3.4.).

**TABLE 8.3**
Marine versus continental climates.

| | |
|---|---|
| Annual temperature range | M < C |
| Diurnal temperature range | M < C |
| (Mean April temp.) − (Mean October temp.) [N. Hemisphere] | M < C |
| Sunshine | M < C |
| Summer rainfall (as % of annual) | M < C |
| Radiation | M < C |
| Time lag of maximum and minimum temp. | M > C |
| Cloud (autumn and winter) | M > C |
| Rainfall | M > C |
| Relative humidity | M > C |

Incident radiation can penetrate water deeply, and the warmed layers are subject to a great degree of mixing. Over land these characteristics are modified considerably. Because of these differences, the upper-layer water temperatures respond to the radiation flux more slowly and to a lesser degree than do the top layers of the soil. In the Northern Hemisphere, therefore, the mean April temperatures are greater in continental than in marine climates, while the reverse is true in October. For the Southern Hemisphere the months must be exchanged.

Over the oceans evaporation is a continuous process, but over land this is not always so. Thus marine air masses have greater relative humidity than do continental air masses, and marine areas experience greater annual rainfall, more clouds, and less sunshine and radiation than does the contiguous littoral zone. An interesting trait of the continental regions is that, due to the colder winter and warmer summer temperatures (relative to the adjacent marine region), a greater percentage of the annual precipitation occurs in the summertime than it does in marine climates.

## 8.6 KÖPPEN'S CLASSIFICATION

No introductory climatology text can ignore Köppen's classification, for it is still, after sixty years, the most widely used system. Many scientists have disagreed with certain threshold values chosen by Köppen and have introduced their own—a line-shifting technique. Others have tried new approaches (see Thornthwaite's classification, Section 8.7). Since Köppen's classification has stood the test of time, it is certainly worthy of consideration. By using metric units as Köppen did, we will see the simplicity of some of his thresholds. However, because Köppen often made only qualitative statements, such as "if the precipitation is concentrated in summer then _____," it remained for others to suggest the definition of a "summer concentration." Remember that Köppen was looking for major climatic types corresponding to the chief vegetation types of the continents. He was not looking at meteorological principles or controls. The major vegetation types of his system used the broad relationship outlined in Table 8.4.

Without attempting to review all of Köppen's reasoning, which sometimes is not very clear, let us simply define the thresholds of the major moist types:

A: Mean of coldest month is at least 18°C (64°F).

C: Mean of coldest month is between 18° and −3°C (64° and 27°F), and at least one month has a mean above 10°C (50°F).

D: Mean of coldest month is less than −3°C (27°F), and at least one month has a mean above 10°C (50°F).

E: Mean of all months is below 10°C (50°F).

It is inherent in the system that zones A, C, and D are moist or humid—all are forest climates. Zone B is the dry zone. Köppen realized that the effectiveness of precipitation is related to the evaporative loss, which generally increases with temperature. Therefore, a climate with summer concentration of precipitation is more likely to be arid than a climate with winter concentration if both have the same mean annual precipitation ($\bar{R}$). Köppen suggested that if $\bar{R}$ (in cm) is less than $\bar{T}$ (mean annual temperature in °C) and there is winter concentration of precipitation, the climate is arid (labelled BW). For a summer concentration, the criterion for BW classification was $\bar{R} < \bar{T} + 14$. For an even distribution of precipitation the average was used, $\bar{R} < \bar{T} + 7$.

For semiarid (BS) climates Köppen allowed twice as much precipitation as for BW climates. A graphical presentation of the criteria is given in Figure 8.3(a). When these criteria are expressed in English units, the equations appear unwieldy. For example, for BW climates with winter concentration, $\bar{R}$ must be less than $0.22 (\bar{T} - 32)$. For summer concentration $\bar{R}$ must be less than $0.22 (\bar{T} - 7)$.

Seasonal concentration is determined by comparing the precipitation in the six cooler months (October to March in the Northern Hemisphere) with the amount in the six warmer months. If the cooler six months receive more than 70 percent of $\bar{R}$, then there is said to be a winter concentration. If the warmer six months receive more than 70 percent of $\bar{R}$, there is a summer concentration. If neither condition holds, there is an even distribution. Notice that this concept becomes confused in the tropical areas, where it is not correct to identify October to March with winter or summer, or even a cool or warm season.

**TABLE 8.4**
Köppen's major climatic types.

| Vegetation Types | Climatic Types | Major Type Letter |
|---|---|---|
| Megathermal (much heat) | Tropical rainy climates | A |
| Xerophilous (dry) | Dry climates | B |
| Mesothermal (medium heat) | Warm temperature, rainy climates | C |
| Microthermal (little heat) | Cold snow-forest climates | D |
| Ekistothermal | Snow climates | E |

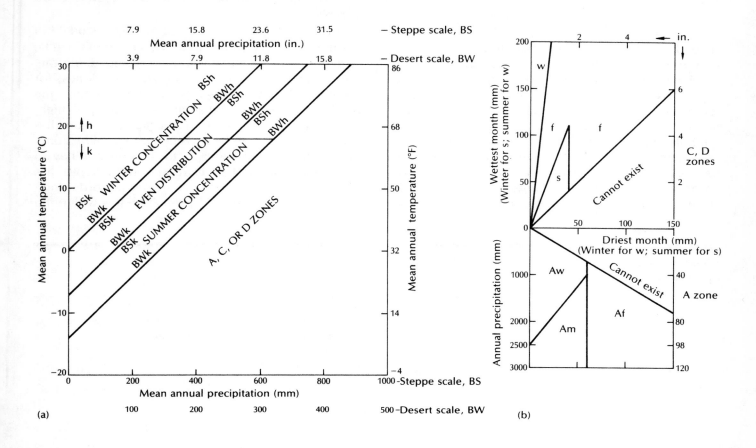
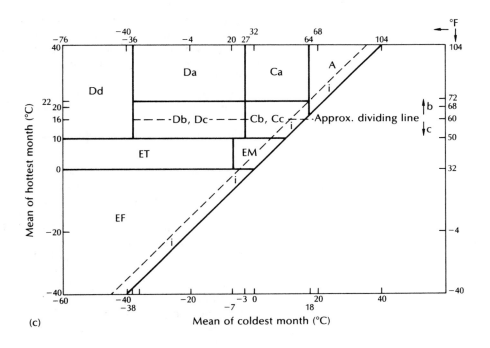

**FIGURE 8.3**
(a) Graph for determination of BS and BW climates. (b) Graph for determination of second-order subdivisions (precipitation). (c) Graph for determination of third-order subdivisions (temperature).

The major humid types (A,C,D,E) are followed by second-order and third-order subdivisions. The second-order subdivisions relate to the seasonality of the precipitation, the third-order to thermal features. The criteria are listed in Table 8.5, and Figures 8.3(b) and (c) can assist in determining graphically the classification of a station. Köppen devised a number of other symbols, but in practice these are rarely used.

There are three additions to Köppen's classification, accepted here as improvements, that require little or no extra symbolism. James A. Shear has suggested an EM (polar maritime) zone, in which the mean of the coldest month is above −7°C (20°F), but the mean of the warmest month is below 10°C (50°F), as in Köppen's E zone. Second, it is useful to indicate the major type to which an arid station belongs; for instance, BWh(A) means a hot desert located in the tropical (A) zone. A third addition identifies the highland areas—those regions in which altitude alone has caused the station to be classified into a colder zone. Köppen made suggestions for the H zone, but they do not assist in differentiating areas. Nairobi, Kenya, has a Cb (strictly a C–b, but since no second order indicator exists it is written Cb) climate but is within the

**TABLE 8.5**
Köppen's climatic classification.

| 1st Order | Subdivisions 2nd Order | 3rd Order | Characteristics |
|---|---|---|---|
| A | | | Mean temperature of coldest month > 18°C |
| | f | | Every month has at least 6 cm precipitation |
| | m | | $\bar{R} > 250 - 25$ (amount in driest month) |
| | w | | $\bar{R} < 250 - 25$ (amount in driest month) |
| | | i | Mean annual temperature range < 5C° |
| B* | | | |
| | | h | $\bar{T} > 18°C$ |
| | | k | $\bar{T} < 18°C$; warmest month > 18°C |
| | | k' | $\bar{T} < 18°C$; warmest month < 18°C |
| C | | | One or more months with mean temperature < 18°C; none < −3°C; at least one > 10°C |
| | s | | Summer dry; precipitation of driest month in warm season < ⅓ wettest winter month and < 4 cm |
| | w | | Winter dry; precipitation of driest month in winter season < 1/10 wettest summer month |
| | f | | Humid, when s and w do not apply |
| | | a | Hot summer; warmest month > 22°C |
| | | b | Warm summer; warmest month < 22°C; 4 to 12 months > 10°C |
| | | c | Cool summer; warmest month < 22°C; 1 to 3 months > 10°C |
| D | | | One or more months with mean temperature < 18°C; one or more < −3°C; at least one > 10°C |
| | s | | Summer dry; precipitation of driest month in warm season < ⅓ wettest winter month and < 4 cm |
| | w | | Winter dry; precipitation of driest month in winter season < 1/10 wettest summer month |
| | f | | Humid, when s and w do not apply |
| | | a | Hot summer; warmest month > 22°C |
| | | b | Warm summer; warmest month < 22°C; 4 to 12 months > 10°C |
| | | c | Cool summer; warmest month < 22°C; 1 to 3 months > 10°C |
| | | d | Severe winter; coldest month < −38°C |
| E | | | No mean monthly temperature > 10°C |
| | | F | All monthly means < 0°C |
| | | T | Not all monthly means < 0°C |

$\overline{R}$ = mean annual precipitation (cm); $\overline{T}$ = mean annual temperature (°C)

*(a) Winter precipitation = 70% $\overline{R}$ received in 6 cooler months (October to March in N. Hemisphere)
(b) Summer precipitation = 70% $\overline{R}$ received in six warmer months
(c) When neither (a) nor (b) applies

| If (a) holds and if | Resultant Zone |
|---|---|
| $\overline{R} > 2\overline{T}$ | A, C, D |
| $\overline{R} < 2\overline{T}$ but $> \overline{T}$ | BS (steppe) |
| $\overline{R} < \overline{T}$ | BW (desert) |

If (b) holds and if

| | |
|---|---|
| $\overline{R} > 2(\overline{T} + 14)$ | A, C, D |
| $\overline{R} < 2(\overline{T} + 14)$ but $> (\overline{T} + 14)$ | BS |
| $\overline{R} < (\overline{T} + 14)$ | BW |

If (c) holds and if

| | |
|---|---|
| $\overline{R} > 2(\overline{T} + 7)$ | A, C, D |
| $\overline{R} < 2(\overline{T} + 7)$ but $> (\overline{T} + 7)$ | BS |
| $\overline{R} < (\overline{T} + 7)$ | BW |

tropics; using a thermal "correction" of 0.55C° per 100 m (3F° per 1000 ft), its climate can be written as Cb(HA) or Cb(A), indicating a tropical highland climate with Cb characteristics. Figure 8.4 gives the zonal divisions after some simplification and generalization. The percentages of the world's land areas included in each zonation are

A: 19.9 (Af—8.5; Am—1.9; Aw—9.5)
BS: 14.3
BW: 12.0
C: 15.5 (Cf—6.2; Cs—1.7; Cw—7.6)
D: 21.3 (Df—16.5; Dw—4.8)
ET: 6.9
EF: 10.1

## 8.7
### THORNTHWAITE'S CLASSIFICATION

A critic of Köppen's classification, Thornthwaite disliked the arbitrariness of the limits, the complexity of the system, and the use of temperature to define the major types. In 1931 Thornthwaite introduced a system based on climatic efficiency, retaining the botanical concept but using plant communities instead. He introduced two indices: temperature efficiency (*TE*) and precipitation effectiveness (*PE*).

Temperature efficiency is given by

$$\frac{1}{4}\sum_{1}^{12}(\overline{T}_i - 32)$$

where $\overline{T}_i$ is the mean temperature (°F) in month *i*. As can be seen, this measure considers only values of $\overline{T}_i$ above freezing as contributing to the development of the plant; for a value of $\overline{T}_i$ below 32°F, $(\overline{T}_i - 32)$ counts as zero.

Precipitation effectiveness is given by

$$115 \sum_{i=1}^{12} \left[ \frac{\overline{r}_i}{(\overline{T}_i - 10)} \right]^{10/9}$$

where $\overline{r}_i$ is the mean precipitation (in.) in month *i*. The *PE* is an estimate of the precipitation's usefulness for plant growth, regarded by Thornthwaite as a function of the ratio of precipitation to evaporation. The expression is rather cumbersome because an empirical relationship is used for the evaporation value. Notice that $\overline{T}_i$ must exceed 10°F, and for low values (below about 20° to 25°F) the *PE* ratio for that month can give a misleadingly high number.

*TE* ranges from zero at the poleward limit of the tundra (the plains of low-growing vegetation poleward of the tree line) to 128 at the edge of the tropical rainforest (Table 8.6). When *TE* is below 32 no *PE* value is calculated. The *PE* index is really a precipitation/evaporation ratio, expressed basically in terms of the element (temperature) that Thornthwaite suggested Köppen had overused (Table 8.6). The *PE* index is qualified further by use of a letter indicating its seasonal distribution: *r* for year-round precipitation, *s* for summer drought, *w* for winter drought, and *d* for

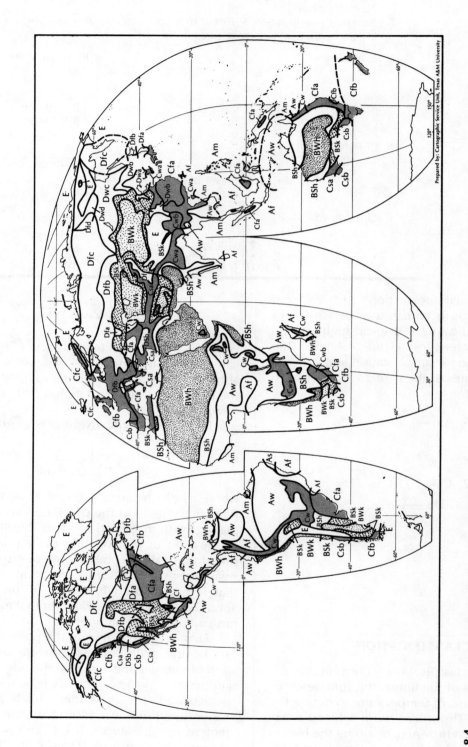

**FIGURE 8.4**
Köppen's classification of climatic zones of the world.

## SEC. 8.8 | OTHER IMPORTANT CLASSIFICATIONS

**TABLE 8.6**
Grouping of parameters in the Thornthwaite classification.

| TE Value | Symbol | Description | PE Value | Symbol | Description |
|---|---|---|---|---|---|
| > 127 | A′ | Tropical | > 127 | A | Perhumid |
| 64–127 | B′ | Mesothermal | 64–127 | B | Humid |
| 32–63 | C′ | Microthermal | 32–63 | C | Subhumid |
| 16–31 | D′ | Taiga | 16–31 | D | Semiarid |
| 1–15 | E′ | Tundra | < 16 | E | Arid |
| 0 | F′ | Frost | | | |

NOTE: The percentages for each area are A—2.3; B—16.1; C—21.1; D—15.2; E—15.3; D′—12.8; E′—5.9; F′—11.3.

year-round drought. In Table 8.6 the nomenclature associated with this system is given, together with the percentage of land areas that fall into specific categories. Occasionally, Thornthwaite used a fourth symbol to indicate the percentage concentration of TE during the three summer months; for example, a indicates 25 to 34 percent and e represents 100 percent.

A simplified world map using this system is shown in Figure 8.5. In 1948 Thornthwaite devised another classification using potential evapotranspiration calculated from an empirical equation. A moisture index is developed, from which a seasonal pattern of months of water surplus or deficit is produced. A problem with the method is that the parameters involved in indices used are not universal, so the system should be applied with caution.

In the years since their publication both classifications have received praise and criticism. Köppen was quite clear that he wished to "arrange in review a mass of facts overlarge and difficult to master," adding that "the selection of criteria involved must naturally be individual, but it suffices if it is instructive. If the classification has helped the reader to a mastery of the facts it has done its duty and thenceforth loses its significance for him."*

Thornthwaite was outspoken about Köppen's classification, commenting that "it is not simple; it is unnecessarily difficult and complicated because it is so unsystematic, using such a miscellany of definitions."† This is quite true, as a glance at Table 8.5 proves, especially when the letters w and f can have two definitions.

The weakness of the Thornthwaite classification is centered on the evaporation term, which is expressed in terms of temperature and precipitation but completely ignores wind speed and humidity. However, Thornthwaite had taken a far-reaching and notable step forward in introducing a contrived, but important in specific application, index. It was a great advance toward the ideal of a classification based on the heat balance/water balance relationship, for which temperature and precipitation are often used as substitutes.

Attempts to compare the two classifications are rather pointless because they were derived with different purposes in mind. Although with the complexity of the patterns of the climatic elements no two classifications will agree completely, each will make some aspect a little clearer and lead to a better understanding of the climatic principles involved.

# 8.8
## OTHER IMPORTANT CLASSIFICATIONS

Although there are numerous climatic classifications, it is not our intent to comment upon or list them all. However, we should be aware of some that are of specific interest in certain applications.

A recent and important development in climatic classification is attributed to L. R. Holdridge. Initially he devised the concept of the life zone to apply to tropical countries, where it has been used in estimating forest productivity and in determining land-use policy. A **life zone** is defined as a division of the earth's climate which supports a distinct set of plant associations. A **plant association** is a dominant community of plants which, in its native natural state, has a physiognomy distinct from that of all other plant associations. There are four types of plant associations:

1. Climatic association—plant community growing on zonal soil in a zonal climate. (There is only one life zone in each climatic association.)
2. Edaphic association—plant community growing on azonal (not sufficient time, horizon not well differentiated) or intrazonal (bedrock and relief dominate; marshes) soil.

---
*Arthur A. Wilcock, "Köppen After 50 years," *Annals of the Association of American Geographers* 58(1968): 21.
†Ibid., p. 22.

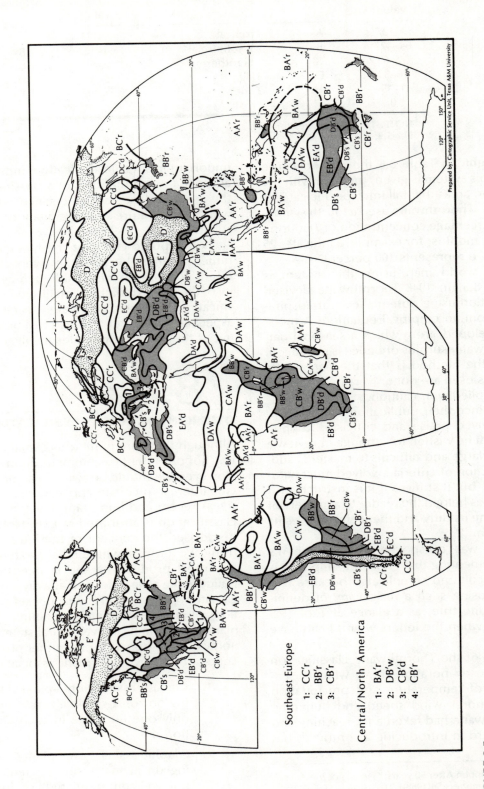

**FIGURE 8.5**
Thornthwaite's classification of climatic zones of the world.

## SEC. 8.8 | OTHER IMPORTANT CLASSIFICATIONS

3. Atmospheric association—plant community growing in an azonal climate (e.g. Mediterranean climate, monsoon climate, cloud forest, mountain ridges).
4. Hydric association—plant community growing where soil is covered with water for all, or almost all, of the year.

**Zonal soils** have easily identifiable horizons and a close correlation with climate. A **zonal climate** has a normal average distribution of precipitation, a distribution dependent upon the **annual mean of the biotemperature** ($T_{bio}$) and the **potential evapotranspiration ratio** (P.E.R.) The mean annual $T_{bio}$ is actually the annual sum of mean hourly temperatures in °C, divided by (24 × 365), where values < 0°C and > 30°C both count as zero. The equation for the P.E.R. is

$$T_{bio} \times 58.93/\text{annual mean precipitation}$$

where the annual mean precipitation is measured in millimeters. The P.E.R. is actually the amount of water that potentially could be used by normal mature vegetation at a site with zonal soil in a zonal climate. It will approximate the potential evapotranspiration only in regions with zonal soil and zonal climate. A normal average distribution of precipitation is given by

| P.E.R. | ≥ 16 | 8 | 2 | 1 | > 0.5 | ≤ 0.5 |
|---|---|---|---|---|---|---|
| Effectively dry months | 12 | 10 | 6 | 4 | 2 | 0 |

This means that a station with a P.E.R. of ≥ 16 must have twelve effectively dry months or it does not exhibit a normal average distribution of precipitation. An effectively dry month occurs when soil moisture reserve is down to 5 percent of the annual precipitation.

In Table 8.7 this system is shown. For example, an area with $T_{bio}$ = 9°C and an annual precipitation of 1500 mm could be (1) a cool temperate wet forest; (2) a warm temperate montane wet forest; (3) a subtropical montane wet forest; or (4) a tropical montane wet forest. The decision as to which is correct is made by increasing $T_{bio}$ at a rate of 6°C per 1000 m elevation to obtain the basal or sea-level temperature and zone. If, as an example, the area had been at 2000 m, then $T_{bio}$ would be 9°C + (2 × 6)°C = 21°C—in the subtropical belt.

The frost, or critical temperature, line is the one great anomaly. As Holdridge remarks, "On opposite sides of the critical temperature lines the taxonomic lists of plant species are markedly distinct. Cultivated plants are likewise generally different on either side of the line. The change in land use in the humid tropics where coffee is cultivated up to the top of the premontane altitudinal belt and then gives way sharply to dairy pastures or to grain or potato cultivation is a clearcut example of the significance of the line."*

During the past few years the need for climatic classification based on or applied to human requirements has grown. W. J. Maunder developed such a classification for conditions in New Zealand. He iden-

---
*L. R. Holdridge, *Life Zone Ecology* (San Jose, Costa Rica: Tropical Science Center, 1967), p. 23.

**TABLE 8.7**
The life zone classification.

| Latitude (Altitude Equivalent) | Mean Annual $T_{bio}$ (°C) | Mean Annual Precipitation (mm) | | | | | | | |
|---|---|---|---|---|---|---|---|---|---|
| | | > 8000 | 4–8000 | 2–4000 | 1–2000 | 500–1000 | 250–500 | 125–250 | < 125 |
| | | ( F | O | R | E S | T ) | | | |
| Tropical | 24–30 | Rain | Wet | Moist | Dry | Very dry forest | Thorn woodland | Desert scrub | Desert |
| Subtropical (Premontane) | 18–24 | | Rain | Wet | Moist | Dry forest | Thorn woodland | Desert scrub | Desert |
| Warm temperate (Lower montane) | 12–18 | | Rain | Wet | Moist | Dry forest | Thorn steppe | Desert scrub | Desert |
| Cool temperate (Montane) | 6–12 | | | Rain | Wet | Moist forest | Steppe | Desert scrub | Desert |
| Boreal (Subalpine) | 3–6 | | | | Rain | Wet forest | Moist forest | Dry scrub | Desert |
| Subpolar (Alpine) | 1.5–3 | | | | | Rain tundra | Wet tundra | Moist tundra | Dry tundra |
| Polar (Nival) | < 1.5 | | | | | | | | |

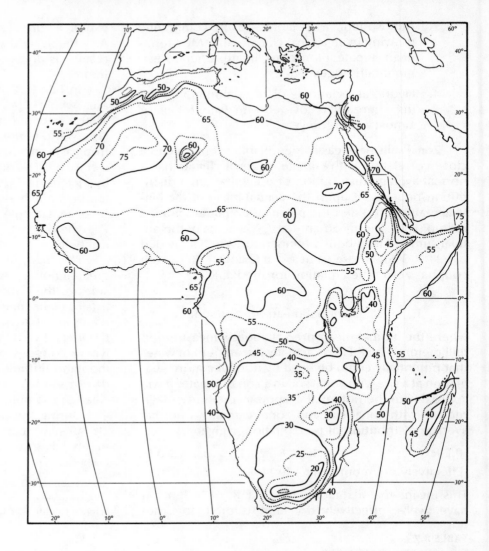

**FIGURE 8.6**
Still-air temperatures (SAT) in °F for July, nighttime. [From W. H. Terjung, "The Geographical Application of Some Selected Physio-Climatic Indices to Africa," *International Journal of Biometeorology* 11 (1967): 5–19.]

tified certain aspects of a number of climatic elements that would influence human awareness and judgment of the "pleasantness" of the ambient atmosphere. Each aspect was given a rating from 1 through 5—with 1 as the best—and a composite index, $X$, was derived. The composite index is defined as

$$X = (3R_1 + 3R_2 + 2R_3) + (4S_1 + 3S_2) \\ + (2T_1 + T_2 + T_3 + T_4 + T_5) \\ + (5H_1) + (2W_1 + 2W_2)$$

where  $X$ = Human climatic index
$R_1$ = Mean annual rainfall
$R_2$ = Mean annual duration of rain
$R_3$ = Percentage of rainfall 9 P.M.–9 A.M.
$S_1$ = Mean annual duration of bright sunshine
$S_2$ = Mean winter duration of bright sunshine
$T_1$ = Mean annual degree-days (>60°F)
$T_2$ = Mean number of days with screen frost per year
$T_3$ = Mean daily maximum temperature of coldest month
$T_4$ = Mean annual maximum temperature
$T_5$ = Mean number of days with ground frost per year
$H_1$ = "Humidity index"
$W_1$ = Mean number of days with wind gusts 40 mi h$^{-1}$ and over
$W_2$ = Mean number of days with wind gusts 60 mi h$^{-1}$ and over

The most "favorable" climate will have an index of 30, and the least "favorable" an index of 150. For New Zealand the values ranged from 60 at Nelson to 101 at Invercargill. As examples of the rating system, consider $R_2$, $T_3$, and $W_1$. A value of 355 to 446 hr scores 1 for $R_2$, the threshold values being 562, 707 and 891. Over 891 hr scores 5. For $T_3$ a temperature of above 13°C (56°F) scores 1, the relevant class limits being 11°, 9° and 7°C (52°, 48°, and 44°F). In the case of $W_1$, below 15 merits 1, while the other thresholds are 45, 90, and 150.

## SEC. 8.8 | OTHER IMPORTANT CLASSIFICATIONS

**FIGURE 8.7**
Computer grouping of Australian climatic stations into 14 regions. [From the Institute of Australian Geographers, "Climatic Classification of Australia by Computer," *Australian Geographical Studies Journal* 9 (1971): 10–11.]

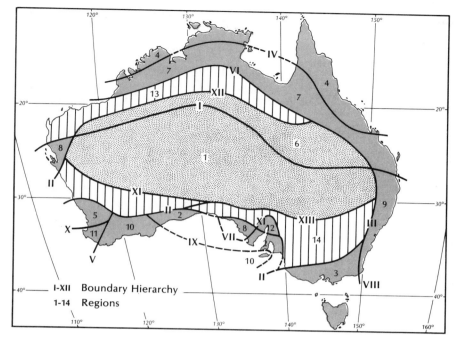

A different approach was introduced by W. H. Terjung, who applied four different indices of human physiological comfort to the whole of Africa, using data from 600 stations. Because many approximations were needed to derive the maps, the original paper should be consulted for details. However, one index, the **still-air temperature** (SAT), will be taken as an example. In this index the cooling effect of the wind is calculated by a nomogram, similar to that for the wind chill factor in Section 14.2.3, and then subtracted from the actual dry-bulb temperature. Figure 8.6 shows the map of SAT for nighttime conditions in July. An SAT of 24°C (75°F) is rather uncomfortably warm for nighttime, while an SAT below −4°C (45°F) is quite chilly. In January SAT values in the northern Sahara are as low as 25°F.

We have said that a climatic classification ideally should refer only to climatic data and avoid any subjective judgment in delimiting boundaries. The method of factor analysis (correlating $X$ climatic elements among themselves and identifying composites of some of the $X$ elements that explain a large percent of the variance—a technique very similar to principal components analysis, referred to in Section 6.6.2) has been used by Geoffrey McBoyle to identify climatic zones in Australia. Using 66 stations and 20 variables, he calculated the three most important indices to be

1. "Cool and moist"—a function of mean annual temperature, average daily mean temperature in January, average daily maximum temperature in January, average daily minimum temperature in January, mean annual relative humidity (afternoon), average relative humidity (afternoon) in July, average precipitation in July, and average number of rain days in July.

2. "Tropical summer wet"—a function of average daily mean temperature in July, average daily minimum temperature in July, average relative humidity (afternoon) in January, mean annual precipitation, average precipitation in January, and temperature range (January–July).

3. "Mediterranean"—a function of precipitation ratio (July/January) and rain day ratio (July/January).

These three indices accounted for nearly 87 percent of the variance (48 percent, 32 percent, and 6 percent, respectively). From this approach McBoyle devised boundaries in Australia corresponding to 10, 14, and 28 regions; those for 14 regions are given in Figure 8.7. In the 10-region pattern boundaries XIII, XII, XI, and X are removed. The boundaries actually indicate where the most rapid change of climate occurs. McBoyle's method, while statistically sound, does make the meteorological interpretation of the indices extremely difficult. It is likely that with the increased use of computers many similar types of statistical techniques will be used to devise climatic classifications.

**SUGGESTED READING**

GENTILLI, J. *A Geography of Climate.* Crawley: University of Western Australia, 1952.

HARE, F. K. "Climatic Classification." In *Climate in Review,* ed. G. McBoyle. Boston: Houghton Mifflin, 1973.

HOLDRIDGE, L. R. *Life Zone Ecology.* San Jose, Costa Rica: Tropical Science Center, 1967.

LANDSBERG, H. E. "Roots of Modern Climatology." *Journal of the Washington Academy of Sciences* 54 (1964): 130–41.

MAUNDER, W. J. "A Human Classification of Climate." *Weather* 17 (1962): 3–12.

SHEAR, J. A. "The Polar Marine Climate." *Annals of the Association of American Geographers* (1964): 310–17.

———. "A Set-Theoretic View of the Köppen Dry Climates." *Annals of the Association of American Geographers* 56 (1966): 508–15.

TERJUNG, W. H. "Physiologic Climates of the Conterminous United States: A Bioclimatic Classification Based on Man." *Annals of the Association of American Geographers* 56 (1966): 141–79.

THORNTHWAITE, C. W. "The Climates of the Earth." *Geographical Review* 23 (1933): 433–40.

# 9
# REGIONAL CLIMATES

**INTRODUCTION**
**HOT, HUMID CLIMATES (A CLIMATES)**
Hot, Wet Climate (Af)
Hot, Short Dry Season (Am)
Hot, Wet and Dry Seasons (Aw)
**DRY CLIMATES (B CLIMATES)**
Hot (Low Latitude) Steppe (BSh)
Cold (Mid-Latitude) Steppe (BSk)
Hot (Low Latitude) Desert (BWh)
Cold (Mid-Latitude) Desert (BWk)
**MID-LATITUDE CLIMATES (C CLIMATES)**
Mid-Latitude, Uniform Precipitation, Hot Summer (Cfa)
Mid-Latitude, Uniform Precipitation, Warm Summer (Cfb)
Mid-Latitude, Uniform Precipitation, Cool Summer (Cfc)
Mid-Latitude, Dry Winter, Hot Summer (Cwa)
Mid-Latitude, Dry Winter, Warm Summer (Cwb)
Mid-Latitude, Dry and Hot Summer (Csa)
Mid-Latitude, Dry and Warm Summer (Csb)
**THE HIGH LATITUDE CLIMATES (D CLIMATES)**
High Latitude, Uniform Precipitation, Hot Summer (Dfa)
High Latitude, Uniform Precipitation, Warm Summer (Dfb)
High Latitude, Uniform Precipitation, Cool Summer (Dfc)
High Latitude, Uniform Precipitation, Extremely Cold Winter (Dfd)
High Latitude, Dry Summer (Ds)
High Latitude, Dry Winter, Hot Summer (Dwa)
High Latitude, Dry Winter, Warm Summer (Dwb)
High Latitude, Dry Winter, Cool Summer (Dwc)
High Latitude, Dry and Extremely Cold Winter (Dwd)
**COLD CLIMATES (E CLIMATES)**
Polar Marine Climate (EM)
Tundra Climate (ET)
Frost Climate (EF)
**HIGHLAND CLIMATES (H CLIMATES)**

## 9.1
### INTRODUCTION

The number of climatic zones we identify naturally depends on the classification we use. In areas with little relief, one zone will blend gradually into the contiguous zones. We have decided to use the Köppen classification in describing regional climates because of its worldwide and prolonged use. In the Köppen system some authors identify 11 "distinct" zones, while others make a case for the existence of 25 to 30 zones. In this chapter we describe the climates of zones that occur over sufficient land area to make them important, giving climatic data for representative stations in each. Many variations in temperature and precipitation within each region are possible, as the examples illustrate, but we try to give the broad picture for each region. Because a number of the zones have wide latitudinal extent and/or have a transition from marine to continental exposure, you should read the relevant section and those of neighboring climatic areas to understand the type of variation that does exist in the region. In all cases it is important to know the station's geographical position to appreciate the proximity of different climatic zones.

A working knowledge of the world's climates is essential for any meteorologist or geographer, who must understand the results of the atmospheric processes; the effects of climate on vegetation, animal life, and the people's life styles and customs; and the stresses induced by most climates. Even to the nonscientist a study of regional climates is challenging; it is like a travelog—a glimpse of distant places. For convenience of presentation, we will start with the tropical zone and work poleward, leaving the highland areas until last.

## 9.2
### HOT, HUMID CLIMATES (A CLIMATES)

In the original Köppen classification this zone was called the "tropical forest climates." Its main criterion was that the mean temperature of the coldest month exceeded 18°C (64.4°F). All the tropics are often thought to be humid because they are equated with the A climates. However, much of the tropics falls in the semiarid (BS) or arid (BW) climates.

The A zone extends in a continuous belt around the equator, mostly between 15° N and 15° S (Figure

8.5). The only real discontinuity is in eastern Africa, where a large B region intervenes. In addition, there are some large areas within the geographical tropics in which an A climate does not exist due to the effect of elevation. Such areas are mentioned in Section 9.7 under the highland climates.

Generally, the only symbols added to stations in this zone are f, m, and w, which refer to the relationship between mean monthly and mean annual precipitation. Sometimes the symbol i, indicating a mean annual temperature range of less than 5C° (9F°), is added.

An Af climate (f comes from the German *feucht*, meaning moist) is defined as one in which the driest month has a mean precipitation of at least 60 mm (2.4 in.). If this is not the case, then it may be an Am climate, where the mean precipitation of the driest month is more than 100 − 0.04 (mean annual precipitation, mm), or 3.94 − 0.04 (mean annual precipitation, in.); or an Aw climate, where the mean precipitation of the driest month is less than the result of the equation given for Am climates. (See Figure 8.4 for a graphical presentation.)

### 9.2.1 HOT, WET CLIMATE (Af)

The phrase "hot and humid" perhaps best summarizes the Af climate. The adequate, sometimes abundant or superabundant, rainfall plus the maritime air masses ensures high absolute and relative humidity. Fortunately, the trade winds bring some relief to most Af regions so that, if one can benefit from these, the conditions are tolerable. In towns and cities, where the wind flow is changed, the densely packed homes create very stressful interior conditions. Air conditioning is pleasant but not a necessity; good ceiling fans are very desirable. Clothing should be light and not tight. The radiation load is not too great, as the humid and cloudy atmosphere absorbs much of the direct insolation, but nevertheless a shady, well-ventilated location is the most pleasant. Vegetation is luxuriant, growing rapidly in this ideal plant environment, and a major problem is controlling its prolific growth. As for agriculture, it is often difficult to get any machinery into the sodden fields, and impassable roads and flooding rivers add to the transport problems.

This climate occurs in four main areas: much of Central America and western Colombia, the northern sector of the Amazon basin, the equatorial belt of west central Africa, and the Indonesian Islands. In addition, there are small Af areas along the southeastern coast of Brazil and eastern Madagascar where the onshore trade winds bring year-round moisture. Most of the Hawaiian Island group falls into this zone, but

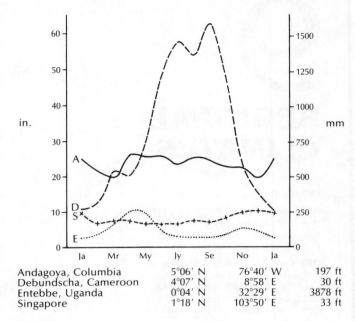

| | | | |
|---|---|---|---|
| Andagoya, Columbia | 5°06′ N | 76°40′ W | 197 ft |
| Debundscha, Cameroon | 4°07′ N | 8°58′ E | 30 ft |
| Entebbe, Uganda | 0°04′ N | 32°29′ E | 3878 ft |
| Singapore | 1°18′ N | 103°50′ E | 33 ft |

**FIGURE 9.1**
Precipitation in Af climates.

on the leeward (southwest) sides of the islands Am and even Aw climates are usually found.

In the Af climate mean annual rainfall amounts can vary greatly, with some stations having less than 1500 mm (60 in.) while others exceed 10,000 mm (400 in.). Also, the precipitation may be reasonably uniformly distributed or with a pronounced peak or peaks (Figure 9.1). The classification does not indicate the time of maximum precipitation, although most stations record their highest monthly fall(s) shortly after the period(s) of high sun. For example, at Entebbe, Uganda, the highest monthly averages are in April/May and November/December while the sun is in the zenith in late March and late September. The rainfall, usually coming from convective causes, is mostly associated with the movements of the ITCZ, the extreme positions of which are seen in Figure 9.2. (More detail is shown in Figures 4.14 and 4.15.) In West Africa fast-moving squall lines can bring intense falls and high winds. Many parts of this climate report a great number of thunderstorm days; Kuala Lumpur, Malaysia, records over 180. Rainfall often has its maximum occurrence in the late afternoon or early evening.

Temperatures are less variable than the precipitation, with many stations showing only small annual ranges (less than 3C° or 5F°) and little variations in monthly extremes, as Figure 9.3 for Singapore shows. It is clear from this figure how uniform, both by month and hour, is the temperature at such a location. However, some stations near the edge of the

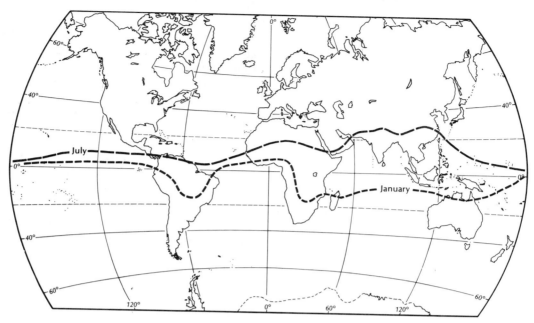

**FIGURE 9.2**
Average extreme positions of the ITCZ.

geographical tropics can exhibit a relatively large annual range; for example, Santos, Brazil, has an annual range of 7C° (13F°). Daily maximum temperatures are rarely much above 32°C (90°F), but with the small daily temperature range in coastal locations there is little alleviation of the enervating conditions, save for the welcome air movement due to the sea breeze and/or the trade winds.

In the humid air of the Af climates insolation is depleted appreciably, although in periods of lower rainfall, or when a slightly drier air mass moves in, the sunshine can be quite intense. Cloudiness is generally high, and the net radiation at night is low. Weather conditions change relatively little, and slowly, from day to day, thus the term *weather* is almost synonymous with *climate*. An exception can occur in West Africa when the *harmattan*—the dry, dusty wind from the Sahara situated to the north—reaches even to the coast. But this is generally of short duration.

The Af areas, with their high temperatures and abundant rainfall, produce a luxuriant forest known as the tropical (or equatorial) rainforest, or *selva*. The two- or three-tiered vegetation, together with lianas and epiphytes, usually reduces light at ground level to a low value so that little vegetation or underbrush grows.

### 9.2.2 HOT, SHORT DRY SEASON (Am)

In its practical impact on life the Am climate differs little from the Af. The short dry season is seldom sufficient to bring much significant change. Because the Am climate is really intermediate between the Af and the Aw, their respective influences and characteristics usually occur as their boundaries are approached from within an Am region.

In this region there is a short dry season, basically determined by one or more months having a mean precipitation of less than 60 mm (2.4 in.) (Figure 9.4). The mean annual precipitation can, however, be as high as or higher than stations in the Af climate. Köppen gave the name "tropical monsoon" to this zone, which is misleading because many Am areas do not

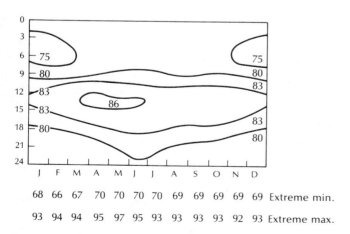

**FIGURE 9.3**
Thermoisopleths for Singapore (°F).

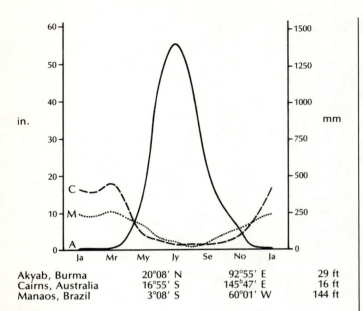

**FIGURE 9.4**
Precipitation in Am climates.

exhibit the seasonal reversal of wind, while other hot areas that do have monsoonal characteristics are not included in this zone. The main Am locations are in northern Brazil and the Guianas, Burma, and much of the Philippines (see Figure 9.5). Other areas are found along the West African coast, southwestern India, and the northeastern coast of Queensland, Australia. The southern tip of Florida also is included in this climate. Often Am climate is located where high land backs the tropical coastal plain so that orographic ascent of the moist air mass occurs. However, convective rainfall is generally the rule.

Like the Af climate, annual precipitation amounts can vary greatly. The seasonal distribution can show

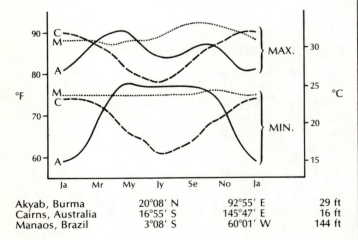

**FIGURE 9.5**
Temperature in Am climates.

one peak—the most usual—or two maxima. Mean monthly temperatures generally exhibit only a small annual range but can have a single or a double maximum (Figure 9.5). If a double maximum occurs, the mean annual temperature range is normally very small. Notice that stations near the limit of the geographical tropics can have a mean annual temperature range of around 8C° (14F°). Daily maximum temperatures rarely exceed about 33°C (92°F), and there are often cooling sea breezes.

In the short dry season the reduction in cloudiness and humidity increases the insolation so that the hottest season usually precedes the onset of the rainy season. This is the period of crop sowing and also when some of the deciduous trees, such as teak, shed many of their huge leaves. As in the Af areas, the best clothing is light and loose.

### 9.2.3 HOT, WET AND DRY SEASONS (Aw)

The Aw climate has two distinct seasons: a wet and a dry of varying length depending on location. The wet season has the characteristics of the Af climate, but the dry season brings very different conditions. The dry period occurs at the time of low sun (the tropical winter), so some of these months are relatively cool.

During the dry season many of the trees shed their leaves, the grasses die, and the smaller rivers cease to flow. This is the area of the flat-topped acacia trees and the savanna lands (also called *campos* or *llanos*). The relatively clear skies, except when extensive grassland burning is practiced, allow the radiation to pass readily so that the daily range of temperature is increased and some cool night temperatures occur. However, light clothing is still worn. As the rainy season approaches monthly rainfall amounts and temperatures increase, with maximum temperatures being reached just prior to the onset of the rainy season proper.

Köppen called this the "winter dry zone," but the more general and more accurate term is the "wet and dry zone" since *winter* can be misleading when applied to the tropics. The dry season lasts normally for three to six months and is characterized in all but a few locations by some months with a mean precipitation below 25 mm (1 in.). The Aw climate comprises over half of the A zone and is found in central Brazil, Colombia and Venezuela, north central and south Africa, the Indian subcontinent, southeast Asia, northern Australia, western Central America, and the Caribbean.

Mean annual precipitation totals are less than in the Af or Am areas, with amounts from 750 to 1750 mm

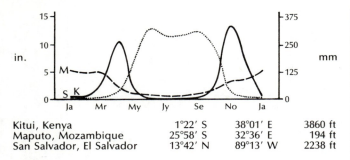

| Kitui, Kenya | 1°22′ S | 38°01′ E | 3860 ft |
| Maputo, Mozambique | 25°58′ S | 32°36′ E | 194 ft |
| San Salvador, El Salvador | 13°42′ N | 89°13′ W | 2238 ft |

**FIGURE 9.6**
Precipitation in Aw climates.

(30 to 70 in.) being common. In most cases there is one season of maximum precipitation, but some areas have two peaks of a moderate or an extreme nature (see the Kitui data in Figure 9.6). The Aw areas are normally located poleward of the Af region, and their precipitation pattern varies from the rainy months of an Af climate to the dryness of a BS (semiarid) or even a BW (arid) climate. At the time of the high sun the ITCZ influences these zones, and they experience the wet season of the year—a period that gets shorter as one moves poleward. At this time severe thunderstorms occur, with Darwin, Australia, and parts of Nigeria, western Madagascar, and Thailand reporting from ten to twenty per month. Annual rainfall totals show marked variation from year to year. In Brazil there are years when strong zonal flow prevents the moist northwesterly flow from penetrating south of the Amazon River.

Because they extend farther poleward than the Af or Am areas, the Aw areas usually have larger mean annual temperature ranges, reflecting the variation from dry to wet season. Annual means vary from about 20° to 28°C (68° to 82°F), and the mean annual temperature range goes from 2 to 11C° (4 to 20F°). As with the Am climate, the hottest month often immediately precedes the onset of the rainy season, with many stations recording 32° to 38°C (90° to 100°F) (Figure 9.7). In the dry season night temperatures may drop below 16°C (60°F), as the low dewpoint and calm, clear conditions allow appreciable loss of long-wave radiation.

There is a unique region along the coast of eastern Brazil around 10° S in which the maximum rainfall occurs during the low-sun (cool) season. This is known as an As zone, a "summer-dry" area, but for practical purposes it can be included in the adjacent Aw area. It is in this region that heavy fog is reported on over 60 days of the year, especially on the coastal slopes. However, radiation fogs are common (90 days annually) in much of the Amazon basin.

## 9.3

### DRY CLIMATES (B CLIMATES)

The dry climates cover more than a quarter of the world's land area. They can be found at all latitudes from 50° N to 50° S, but the main concentration is between 15° to 30° N and S. The BW (arid or desert) zones are generally flanked by a BS (semiarid or steppe) zone. The major BW zones in the Northern Hemisphere are the Sahara-Nubian-Arabian-Thar desert, the Gobi desert, and the Turkestan desert. In the Southern Hemisphere the major zones are located in western Australia, Namibia (South-West Africa), and the northern Chile–coastal Peru area. Except for tiny areas in Spain, Europe has no B climates.

The B climates have six fairly general climatic characteristics: (1) little precipitation; (2) great variation in precipitation amounts and distribution; (3) low relative (but not necessarily absolute) humidity; (4) a potential evapotranspiration far exceeding the precipitation, often in every month; (5) much insolation (few clouds); and (6) high average wind speeds.

The major subdivisions of the B zone are the S (steppe) and W (W is from the German *wüst*, meaning desert) zones; the letter B never stands alone. In addition there is the qualifying symbol h or k, indicating a mean annual temperature of above or below 18°C (64.4°F), respectively. It has been suggested that each station in the B zones should also be given the symbol A, C, or D (the definitions are given in Section 8.6) to denote its temperature pattern. This is a useful refinement that will be used here, together with a symbol, in the C or D zones, for the precipitation pattern. The symbols for precipitation pattern are s for summer

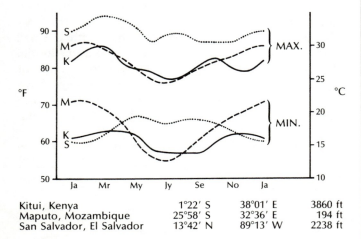

| Kitui, Kenya | 1°22′ S | 38°01′ E | 3860 ft |
| Maputo, Mozambique | 25°58′ S | 32°36′ E | 194 ft |
| San Salvador, El Salvador | 13°42′ N | 89°13′ W | 2238 ft |

**FIGURE 9.7**
Temperature in Aw climates.

(high-sun) dry; w for winter (low-sun) dry; or f for neither.

### 9.3.1 HOT (LOW LATITUDE) STEPPE (BSh)

The BSh climate, generally surrounding the low-latitude deserts, is characterized by two seasons—one wet and one dry—with pronounced year-to-year rainfall variability. Although normally located between 10° and 35° latitude, a BSh area can extend across the equator, as in eastern Africa. The reasons for this are given in Section 9.3.3 on BWh climates.

Precipitation has to be less than about 860 mm (35 in.) because of Köppen's criterion, but actually most stations have around 380 to 630 mm (15 to 25 in.). The BSh areas to the equatorward side of the low latitude deserts have their rainy season at the time of high sun (or summer in the Köppen classification), when convection is maximum and the ITCZ is in the vicinity (see Section 8.3.2). Locations poleward of the low-latitude deserts (see data for Agadir) have winter rains brought by mid-latitude cyclones and fronts (Figure 9.8). The variability of annual rainfall totals in the BS areas can be seen from Table 9.1, which shows how unreliable the rains actually are. The stations have between 50 and 75 years of record. The totals for individual months show even more percentage variation.

The annual temperature range is often quite high in this climate (8 to 14C°, or 15 to 25F°), and stations near to the poleward limit have a distinct winter or cool season. Those stations nearer to the equator can have small annual ranges of less than 6C° (10F°) (Figure 9.9).

In those regions with summer rain the clouds tend to reduce the solar radiation load, but the evaporation rate remains high and the precipitation efficiency is low. The vegetation includes xerophytic shrubs, stunted woodland (except along river beds) of broad-leafed and deciduous plants, and some grass.

### 9.3.2 COLD (MID-LATITUDE) STEPPE (BSk)

Since BSk climates, with a mean annual temperature of less than 18°C (64°F), are seldom found at less than

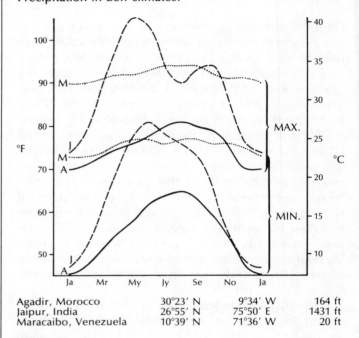

Agadir, Morocco 30°23' N 9°34' W 164 ft
Jaipur, India 26°55' N 75°50' E 1431 ft
Maracaibo, Venezuela 10°39' N 71°36' W 20 ft

**FIGURE 9.8**
Precipitation in BSh climates.

Agadir, Morocco 30°23' N 9°34' W 164 ft
Jaipur, India 26°55' N 75°50' E 1431 ft
Maracaibo, Venezuela 10°39' N 71°36' W 20 ft

**FIGURE 9.9**
Temperature in BSh climates.

30° latitude, they rarely occur in the Southern Hemisphere. By far the largest BSk areas surround the BWk climates of Asia and North America, with a small region in the hills of northern Morocco and Algeria. These areas, generally further removed from moist air masses than are the low-latitude steppe climates, have only some 200 to 400 mm (8 to 16 in.) of annual precipitation (Figure 9.10). Those areas with a precipita-

**TABLE 9.1**
Extremes of annual precipitation of steppe areas.

| Station | Maximum | Minimum | Climatic Zone |
|---|---|---|---|
| Jaipur, India | 1015 mm (36.39 in.) | 269 mm (10.60 in.) | BSh [Cwa (A)] |
| Monterrey, Mexico | 1126 mm (44.32 in.) | 228 mm (8.98 in.) | BSh (Cfa) |
| Salisbury, Rhodesia | 1401 mm (55.17 in.) | 552 mm (21.74 in.) | BSh [Cwb (A)] |
| St. Louis, Senegal | 1239 mm (48.80 in.) | 144 mm (5.67 in.) | BSh (Aw) |
| Cordoba, Argentina | 930 mm (36.61 in.) | 438 mm (17.24 in.) | BSk (Cwa) |
| San Diego, Ca. | 633 mm (24.93 in.) | 151 mm (5.93 in.) | BSk (Csb) |

## SEC. 9.3 | DRY CLIMATES (B CLIMATES)

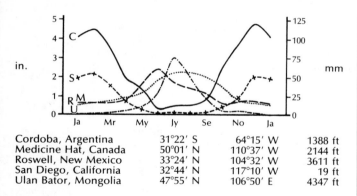

| | | | |
|---|---|---|---|
| Cordoba, Argentina | 31°22′ S | 64°15′ W | 1388 ft |
| Medicine Hat, Canada | 50°01′ N | 110°37′ W | 2144 ft |
| Roswell, New Mexico | 33°24′ N | 104°32′ W | 3611 ft |
| San Diego, California | 32°44′ N | 117°10′ W | 19 ft |
| Ulan Bator, Mongolia | 47°55′ N | 106°50′ E | 4347 ft |

**FIGURE 9.10**
Precipitation in BSk climates.

tion concentration in winter are normally affected by some cyclones and frontal activity. In the areas with summer rain much precipitation stems from convective activity in the moist air masses that can reach these regions. As with the BSh climate, yearly variation in rainfall can be appreciable (see Table 9.1). Winter precipitation is often in the form of snow in the areas of higher latitude.

Because of the large latitudinal extent of the BSk area (20° to 55°), mean monthly and mean annual temperatures can show wide variations (Figure 9.11). For example, at Ulan Bator, Mongolia, the mean of the coldest month is −26°C (−14°F), while at Mogador,

Morocco, it is 14°C (57°F). Means of the warmest month and extreme temperatures show similar variation. In the North American and Asian locations outbreaks of cold air, some severe, and blizzards are common in winter. Regions that are affected by maritime air masses show less extreme conditions.

### 9.3.3 HOT (LOW LATITUDE) DESERT (BWh)

The tropical deserts are located usually in the 15° to 30° latitude belt in the area of air subsiding from the poleward edge of the Hadley cells. They are mostly located on the western sides of the continents, where dry subsiding continental air masses are dominant much of the year.

Mean annual precipitation amounts can range from almost zero, such as at Aswan, Egypt, and in the northern Atacama desert of South America, to as much as 350 mm (14 in.). Generally the rainfall is convective and local, so its reliability is extremely low. However, some BWh climates do exhibit a seasonal pattern, generally with a maximum at the time of high sun. One exception is Kandahar, Afghanistan, which has some precipitation from winter depressions.

The intense radiation during the high-sun season causes surface convection, often with well-developed dust devils. The subsiding air aloft generally prevents

**FIGURE 9.11**
Mean monthly maximum and minimum temperatures of the hottest and coldest months in BSk zones.

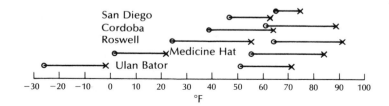

○ Mean minimum
× Mean maximum

TEMPERATURE (°F)

| | Coldest Month | | Warmest Month | |
|---|---|---|---|---|
| | Max. | Min. | Max. | Min. |
| Cordoba | 89 | 17 | 114 | 42 |
| Medicine Hat | 62 | −51 | 108 | 36 |
| Roswell | 88 | −19 | 106 | 53 |
| San Diego | 85 | 25 | 94 | 54 |
| Ulan Bator | 21 | −47 | 92 | 34 |

| | | | |
|---|---|---|---|
| Cordoba, Argentina | 31°22′ S | 64°15′ W | 1388 ft (423 m) |
| Medicine Hat, Canada | 50°01′ N | 110°37′ W | 2144 ft (653 m) |
| Roswell, New Mexico | 33°24′ N | 104°32′ W | 3611 ft (1101 m) |
| San Diego, California | 32°44′ N | 117°10′ W | 19 ft (6 m) |
| Ulan Bator, Mongolia | 47°55′ N | 100°50′ E | 4347 ft (1325 m) |

**FIGURE 9.12**
Mean monthly maximum and minimum temperatures of the hottest and coldest months in BWh zones.

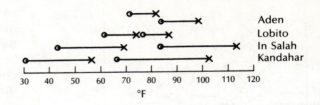

|  | TEMPERATURE (°F) | | | |
|---|---|---|---|---|
|  | Coldest Month | | Warmest Month | |
|  | Max. | Min. | Max. | Min. |
| Aden | 86 | 61 | 106 | 79 |
| In Salah | 88 | 26 | 122 | 73 |
| Kandahar | 70 | 14 | 108 | 53 |
| Lobito | 85 | 53 | 94 | 66 |
| Aden, S. Yemen | 12°50′ N | 45°01′ E | 22 ft (7 m) | |
| In Salah, Algeria | 27°12′ N | 2°28′ E | 919 ft (280 m) | |
| Kandahar, Afghanistan | 31°36′ N | 65°40′ E | 3462 ft (1055 m) | |
| Lobito, Angola | 12°22′ S | 13°32′ E | 4 ft (1 m) | |

the development of thunderstorms, but, if these do occur, the dry air (relative humidity is often below 10 percent) causes much of the precipitation to evaporate before reaching the ground. Electrical discharge displays, especially in mountainous areas, can be quite spectacular.

The world's highest temperatures have been recorded in BWh areas: Azizia, Libya, with 58°C (136.4°F), and Death Valley with 57°C (134°F). (See Section 7.3.) Remember that these temperatures are measured in the shade. Surfaces exposed to the intense insolation from the often cloudless skies will reach much higher values; 82°C (180°F) on the sand surface is not uncommon. Due to the low moisture content of the air, daily temperature ranges are high, often reaching 22C° (40F°) or more during the low-sun period. Temperatures are expected to exceed 38°C (100°F) sometime during the year, although the eastern coast of Somalia is an exception. Here temperatures above 38°C (100°F) are rare, and the mean annual temperature range is very small (4C°, or 7F°)—much different from most BWh regions in which a range of 11 to 22C° (20 to 40F°) is expected (Figure 9.12).

Somalia is an unusual BWh area because it is located on the east coast of a continent. The reasons for its extreme aridity are basically that both the moist northeasterly trade winds and the southwesterly flow into the monsoonal low over the Indian subcontinent (June to September) run parallel to the coast. This means that the region is rarely influenced by moisture-laden air masses.

### 9.3.4 COLD (MID-LATITUDE) DESERT (BWk)

The BWk climates are much less extensive than the BWh. Only in the Turkestan-Gobi-Mongolia area and Patagonia are there large regions of this climate type. Smaller areas can be found in North America (mainly in Nevada), in northern Chile, and southern Africa. These regions are not affected by the trade winds; their aridity is due to high pressures with subsiding air, which is in some cases reinforced by cool water upwelling offshore, such as the Peruvian and Benguela currents.

Because mean annual temperatures are lower than in the low latitude deserts, BWk stations must have relatively little precipitation. Annual falls of around 125 mm (5 in.) are average. What little rain there is can be concentrated in the high-sun or low-sun seasons or be fairly evenly distributed. Those areas marginal to the Mediterranean type of climate (Csb) have winter rains from cyclonic disturbances, while areas close to the tropical zone (Lobito, Angola) or in continental interiors (Kucha, China) have summer (or high sun) convective showers.

Mean annual temperatures vary widely in the BWk climate, and stations can have a thermal classification of A (rare), C, or D. The mean annual temper-

# SEC. 9.4 | MID-LATITUDE CLIMATES (C CLIMATES)

## FIGURE 9.13
Mean monthly maximum and minimum temperatures of the hottest and coldest months in BWk zones.

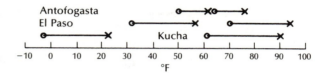

○ Mean minimum
✗ Mean maximum

|  | TEMPERATURE (°F) | | | |
|---|---|---|---|---|
|  | Coldest Month | | Warmest Month | |
|  | Max. | Min. | Max. | Min. |
| Antofogasta | 79 | 41 | 85 | 57 |
| El Paso | 77 | −6 | 105 | 56 |
| Kucha | 36 | −13 | 99 | 52 |
| | | | | |
| Antofogasta, Chile | 23°42′ S | 70°24′ W | 308 ft (94 m) | |
| El Paso, Texas | 31°48′ N | 106°24′ W | 3920 ft (1195 m) | |
| Kucha, China | 41°40′ N | 83°06′ E | 3182 ft (970 m) | |

ature range is generally large, 18C° (30F°) or more, and can exceed 39C° (70F°). Some stations have only a mild winter (low-sun period), but others (especially in the D classification) experience severe cold stress (Figure 9.13). As is typical of the arid zones, diurnal temperature ranges at interior stations are high, especially during the warmer months.

# 9.4
## MID-LATITUDE CLIMATES (C CLIMATES)

The C climates were originally given the name "mesothermal," denoting medium or middle heat, by Köppen. Where found at lower elevations, these climates are generally located between 30° and 60° latitude. Where warm ocean currents play an important role, the C zone is found further poleward; for example, Reykjavik, Iceland, at 64° N and Bodø, Norway, at 67° N are both in a C zone.

Using the two subdivisions—the first for precipitation regime, the second for summer temperatures (see Table 8.5)—there are nine possible C climates, but the combinations Cwc and Csc do not exist. The C climates occur on all six continents (not on Antarctica) and are characteristically located (1) on the western sides of continents, between 35° and 60° to 70°, their inland extent being influenced by interior highlands (narrow belts in North and South America, far inland in Europe); (2) on the eastern sides of continents in lower latitudes, around 20° to 40°, or (3) in some interior regions where altitude or cold air flow in winter lowers the temperature below the A zone threshold.

There is a special type of climate called Mediterranean which has hot, dry summers and warm, moist or wet winters. Although this climate appears on all six continents, it was first identified in the Mediterranean Sea area, which is by far the largest region with this type of climate. In some publications the Cs climates have been equated with the Mediterranean climates. This is not strictly true but, except for some areas in the northwestern United States and in the Middle East, the parallelism is not greatly in error.

The terms we will use for the seven different C climates are

1. Cfa: mid-latitude, uniform precipitation, hot summer.
2. Cfb: mid-latitude, uniform precipitation, warm summer.
3. Cfc: mid-latitude, uniform precipitation, cool summer.
4. Cwa: mid-latitude, dry winter, hot summer.
5. Cwb: mid-latitude, dry winter, warm summer.
6. Csa: mid-latitude, dry and hot summer.
7. Csb: mid-latitude, dry and warm summer.

## 9.4.1 MID-LATITUDE, UNIFORM PRECIPITATION, HOT SUMMER (Cfa)

In the Cfa climate summers are hot and humid and winters are relatively mild with sufficient moisture. As an example, consider the seasonal pattern of the climate at Houston, Texas (Figure 9.14). In winter there are a number of short periods during which northerly air dominates and cool, clear conditions result. Snow is almost unknown, but in the same zone farther to the north some heavy falls are experienced but do not persist. Soon, however, the southerly flow off the Gulf of Mexico returns, bringing warm days, more clouds, and higher humidity. In spring frontal activity can lead to thunderstorms, which can be severe as the two air masses (cP and mT) meet. Farther to the north, where frontal passages are common, this is the tornado season. Summer brings an almost constant flow of southerly, humid air which, allied to afternoon temperatures often in excess of 35°C (95°F), at times makes conditions more enervating than within the Af zone and makes air conditioning very desirable. In August and September the coast can be affected by hurricanes, dumping large amounts of rain and bringing damaging high winds, but in many years none reach this region. Fall is often delightful, with cooler temperatures, slightly less rainfall than the other seasons, and much sunshine.

Generally the Cfa climate is found on the eastern sides of continents between the tropics and about 40° latitude. In some locations (the United States and China) the zone extends a long way inland. The U.S. region is the most extensive, but large areas are found also in eastern China, South Korea and Japan, Uruguay, southern Brazil and eastern Argentina, and eastern Australia. Small areas occur around Durban in South Africa, the shores of the Black Sea, the Po Valley of Italy, and parts of Spain.

Annual precipitation amounts can vary greatly in this climate. The lowest amount is around 380 mm (15 in.), although in the United States most stations exceed 630 mm (25 in.). The highest amounts occur in Japan, with many stations reporting 1750 mm (70 in.) or more (Figure 9.15).

Unfortunately, the criteria for identifying a station as f, or one of uniform rainfall, allow much variation. Houston, Texas, has little seasonal variation, but Hengchow, China, has over five times as much rain in June as in December. Some stations in western Kansas have six to eight times as much precipitation in

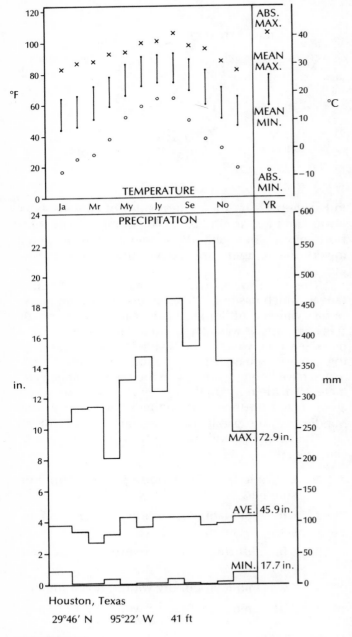

**FIGURE 9.14**
Temperature and precipitation characteristics for Houston, Texas (Cfa climate).

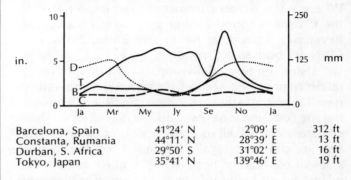

**FIGURE 9.15**
Precipitation in Cfa climates.

## SEC. 9.4 | MID-LATITUDE CLIMATES (C CLIMATES)

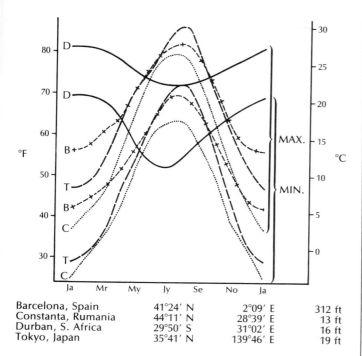

| Barcelona, Spain | 41°24' N | 2°09' E | 312 ft |
| Constanta, Rumania | 44°11' N | 28°39' E | 13 ft |
| Durban, S. Africa | 29°50' S | 31°02' E | 16 ft |
| Tokyo, Japan | 35°41' N | 139°46' E | 19 ft |

**FIGURE 9.16**
Temperature in Cfa climates.

June as in December. Near the steppe borders of this zone rainfall amounts vary greatly from year to year and severe droughts are expected. Generally summer is the wettest period, but in the Northern Hemisphere the maximum may occur in any month from March through October. The precipitation at this time is mostly the result of convective activity, but frontal storms can occur, especially in spring and early summer. In the United States the Cfa zone is infamous for the frequency of tornadoes. In the southern sector the greatest frequency is in early spring, but activity in the northern section reaches its height in late spring or early summer. Hurricanes can bring unusually large amounts of rain to the eastern part of the U.S. Cfa zone during June to September. Their equivalent in the Pacific, the typhoon, brings 20 to 30 percent of Japan's rainfall during May to October.

Hot summers, characteristic of the Cfa climate, can be very enervating if humidity is high because mT air dominates. The interiors, although less humid, have higher temperatures. In winter there is a pronounced north-south gradient, and on the edges of the D zone some stations have reported temperatures well below −18°C (0°F) (Figure 9.16). The Cfa zones of the Southern Hemisphere are not subject to such intensely cold outbreaks of polar or arctic air and are much milder. In the United States these outbreaks, often referred to as *northers*, can drop temperatures 28C° (50F°) or more in a few hours. Sometimes the cold front will be felt as far south as Central America, even to Panama. After the passage of a cold front the skies become clear, the air drier, and minimum temperatures drop.

The Cfa climates receive about 60 percent of the possible hours of sunshine on the average—a value that does not vary much from season to season.

### 9.4.2 MID-LATITUDE, UNIFORM PRECIPITATION, WARM SUMMER (Cfb)

There is no truly typical station to represent all the Cfb zones, but a brief description of the climate of London, England, will give a good idea of the seasonal pattern (Figure 9.17). Winter is not particularly cold or snowy, but the dampness, the cloudy skies, and wind create both physiological and psychological discomfort. An outbreak of colder air from the east, bringing crisp weather and sunshine, is welcome. Spring is a delightful time with comfortable temperatures, not too much rain, and emerging vegetation. In summer temperatures above 27°C (80°F) are not common, but there are many daylight hours. Rain is frequent and threatens outdoor activities, but it is mostly of a convective nature and clears rapidly. Sometimes, how-

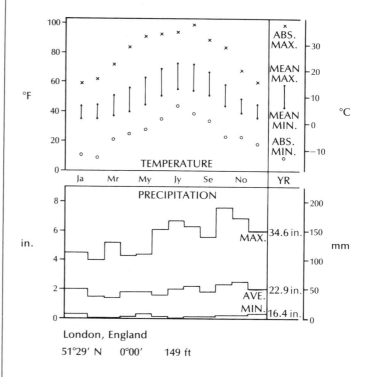

London, England
51°29' N   0°00'   149 ft

**FIGURE 9.17**
Temperature and precipitation characteristics for London, England (Cfb climate).

ever, frontal passages occur frequently over a period of weeks. In autumn some short, warm periods can occur but, on the average, this is the wettest season of the year. The climate cannot be described as harsh, but it is certainly quite variable.

The Cfb zone, with cooler summers than experienced in the Cfa climates, is generally located poleward of the latter zone. The lower summer temperatures are often the effect of altitude or maritime air masses. By far the largest Cfb zone is located in Europe, where it stretches from northern Greece to the central Norwegian coast, and from eastern Bulgaria to central Spain. In the United States the zone appears in Massachusetts, Appalachia, and a belt from the Alaskan panhandle to northwestern Washington. In the Southern Hemisphere it is found in southern Chile, around Port Elizabeth in South Africa, and in a large area that includes New Zealand, Tasmania, and southeastern Australia.

There is no real dry period in this zone, but the season of maximum rainfall varies from late fall and winter near the coasts to summer farther inland. Annual amounts are about 625 to 1250 mm (25 to 50 in.), but where topographic effects supplement cyclonic activity these can rise to 3800 to 5100 mm (150 to 200 in.) (Figure 9.18). Cyclonic storms, often with gale force winds, affect these zones throughout the average year, and spells of cloudy, rainy weather are common. In winter the precipitation is often in the form of sleet and snow, especially at high elevations. When the mountains are near the coast annual snow totals of 250 to 500 cm (100 to 200 in.) are reported.

The maritime influence tends to reduce both annual and diurnal temperature ranges, but Cfb areas near the D or A zones reflect the proximity of these climates. For instance, Boston, Massachusetts, has an annual range of 24C° (44F°) (Figure 9.19), but in South Africa Port Elizabeth's range is only 7C° (12F°). Some locations can experience hot air from the tropics or the continental interior during short spells in sum-

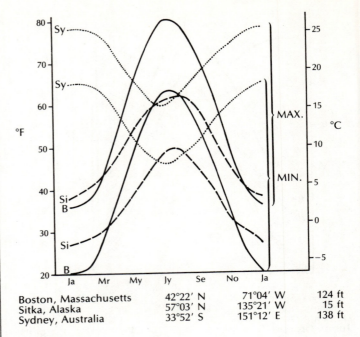

**FIGURE 9.19**
Temperature in Cfb climates.

mer, and temperatures may top 38°C (100°F). Although average conditions give mild winters, cold air outbreaks (cP or cA) can drop temperatures to −18°C (0°F).

### 9.4.3 MID-LATITUDE, UNIFORM PRECIPITATION, COOL SUMMER (Cfc)

Reykjavik, the capital of Iceland, is a good example of Cfc climate, a climate in which the marine influence is at a maximum (Figure 9.20). Winters are not particularly cold, with mean temperatures hovering around freezing. The unpleasant aspects are the dampness, heavy rain (or sleet, occasionally snow), fogs, overcast skies, the winds, and the short hours of daylight. Spring comes late and summer is short. A "warm" summer's day would reach 18°C (64°F), but spells of rain and cold air (4°C, or 40°F) can occur even in that season. Fall is wet and squally, then winter sets in once more. In sheltered spots with good soil coniferous forests are found, but otherwise tundra dominates. The people support themselves by fishing and raising hardy crops such as barley and cabbages. Glasshouse crops are produced, and around Reykjavik good use is made of geothermal energy (hot springs) to heat homes and greenhouses so that even tropical plants, including bananas, are being raised.

The Cfc zones are really poleward extensions of the Cfb climates and occur in coastal Norway, southern Alaska, and the tip of South America. In addition,

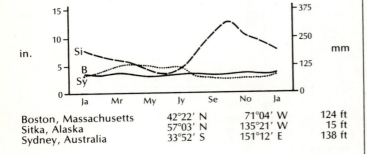

**FIGURE 9.18**
Precipitation in Cfb climates.

### SEC. 9.4 | MID-LATITUDE CLIMATES (C CLIMATES)

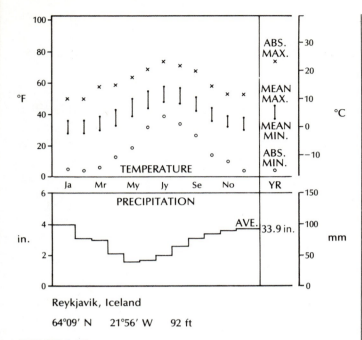

**FIGURE 9.20**
Temperature and precipitation characteristics for Reykjavik, Iceland (Cfc climate).

parts of southern Ireland and the Faröe and Shetland Islands (north of Scotland) are in this zone.

The precipitation is fairly uniformly distributed through the year, although, as at Reykjavik, some months may receive twice as much as others. Mean annual totals can vary greatly; for example, in Chile Punta Arenas receives 400 mm (16 in.), while Cabo Raper receives 2200 mm (87 in.). Generally amounts range from 750 to 1250 mm (30 to 50 in.). There is often a slight tendency toward a winter maximum, with some of the precipitation coming as sleet or snow. Winter is punctuated with stormy spells, heavy precipitation, and high winds, but summer may bring periods of clear, sunny days. These are most welcome as heavy clouds, fog, and rain are the norm here. On such sunny days temperatures may reach 21°C (70°F), but 18°C (65°F) is considered a warm day. Temperatures below −7°C (20°F) are not common, but the combination of high humidity, high winds, and little sunshine makes temperatures even a little above freezing feel most unpleasant.

### 9.4.4 MID-LATITUDE, DRY WINTER, HOT SUMMER (Cwa)

Most of the Cwa zone has a continental climate, and for this reason it may help to study the seasons at New Delhi, India, as a typical location (Figure 9.21). Winter is quite pleasant, with afternoon temperatures at or

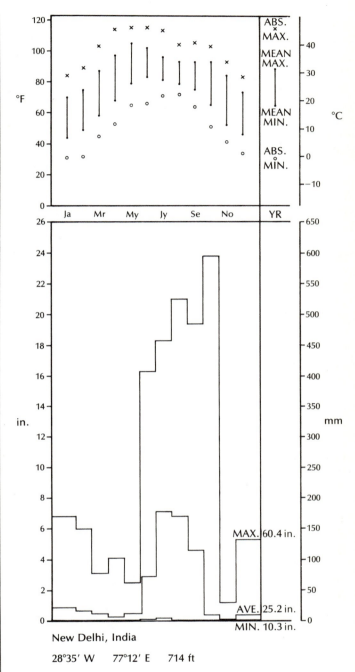

**FIGURE 9.21**
Temperature and precipitation characteristics at New Delhi, India (Cwa climate).

above 21°C (70°F) and plenty of sunshine. However, nights are cool and can occasionally reach near freezing. This places some stress on the lightly clad people, for rarely are their homes equipped for efficient heating. In addition, after the occasional cyclonic rains the air is quite damp and fog persists for many hours. By March some days reach 35°C (95°F) or more, and in

May the highest temperatures, 43°C (110°F) or above, are experienced. The nights can be uncomfortably hot, but fortunately the relative humidity is less than 40 percent. When the monsoon rains come, daytime temperatures drop, but night temperatures remain very high because of the cloud layer's blanketing effect. Rains fall only about one day in four in this region—contrary to usual image of the monsoon—and can be extremely intense. By mid-September, on the average, rainfall decreases rapidly, the air flow from the continent dominates, and cloud cover decreases. Afternoons remain unpleasantly hot until late in October, but night temperatures fall rapidly as winter approaches.

The Cwa climate occurs mostly on the poleward edge of the A zone, where either altitude or increased latitude contributes to the appearance of a cool season [the mean of coldest month is less than 18°C (64°F)]. The area most removed from an A zone is in the Shantung province of China at about 37° N.

Annual rainfall amounts are about 1000 mm (40 in.) but can range from 750 mm (25 in.) to over 2000 mm (80 in.). Those areas influenced by monsoon circulation show the greatest concentration of amount in a short period. In northeastern China, where annual totals of rainfall are relatively low, there is great year-to-year variability.

In the interior areas of low elevation summers are extremely hot, with afternoons exceeding 38°C (100°F) expected. This temperature, together with the high dewpoint at the time, creates very unpleasant conditions. Coastal or island locations, such as Hong Kong, have slightly lower temperatures, but the relative humidity is greater. The cool season is a pleasant relief from the enervating summer. In the Indian area and Hong Kong temperatures in January will fall to 4° or 7°C (40° or 45°F) at night. In China January is quite different, for cold, dry air (cP) can break out of Siberia and plunge temperatures way below freezing in association with strong winds. In Shantung province temperatures around −18°C (0°F) have been recorded.

## 9.4.5 MID-LATITUDE, DRY WINTER, WARM SUMMER (Cwb)

A Cwb climate occurs only rarely, normally at high elevations and in small areas. The important zones are found in some of the higher regions of Africa (around Addis Ababa, for example), in parts of India and China, and in areas from Mexico City southward along the Andean chain.

The variation in mean annual precipitation totals is a feature of this zone. For example, Mexico City receives about 750 mm (30 in.); Guatemala City over

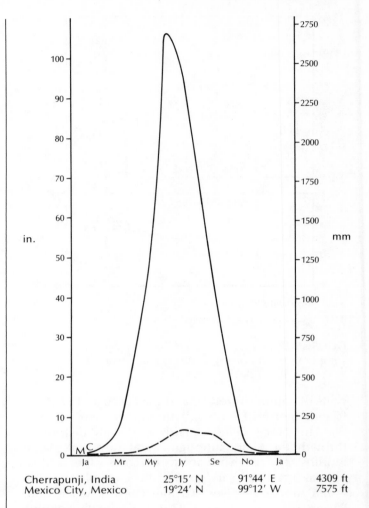

| | | | |
|---|---|---|---|
| Cherrapunji, India | 25°15' N | 91°44' E | 4309 ft |
| Mexico City, Mexico | 19°24' N | 99°12' W | 7575 ft |

**FIGURE 9.22**
Precipitation in Cwb climates.

1250 mm (50 in.); Darjeeling, northeast India, around 3000 mm (120 in.); and Cherrapunji, northeast India, 10,800 mm (425 in.) (Figure 9.22). At Cherrapunji the monthly mean for each of the months May through September exceeds Mexico City's annual average. Both the Indian stations mentioned are influenced by the monsoon rains, which are reinforced by orographic ascent. It is unusual for a Cwb station to have less than 70 percent of its precipitation in the warmer six months. Precipitation is often extremely intense.

In highland areas the seasonal distribution of precipitation is generally the same as that of lowland stations in the same climatic region—only the amount varies. If the highland site is near the zone of maximum precipitation, then amounts can be large; if it is in a rain shadow area, the amount will be small. This means that it is rather misleading to select a single station and consider its yearly pattern of climate as representative of the Cwb zones as a whole. Another facet is that the w symbol is indeterminate in equatorial regions. Nairobi, Kenya, for example, is a Cb climate with its dry periods at the times of low sun (winter). If winter is a period of low sun, then at the equator there are two "winters." We suspect that this is not what Köppen had in mind when he delineated Cwb climates.

The mean annual temperatures usually are about 16°C (60°F), and temperatures exceeding 24° to 27°C (75° to 80°F) are seldom experienced. However, during winter some of the stations at higher elevations or near the poleward limits of this zone can experience freezes. The mean annual range of temperature is, characteristically, around 8 to 11C° (15 to 20F°), but this may be less in the tropical highlands that qualify as Cb climates.

### 9.4.6 MID-LATITUDE, DRY AND HOT SUMMER (Csa)

There are three main areas where the Csa climate is found: along the shores of the Mediterranean Sea and into Iran, in the Central Valley of California, and in southern Australia. Depending upon the averages used, Salt Lake City, Utah, is sometimes designated Csa. Its January mean temperature is given variously as −3.3°C or −2.8°C (26° or 27°F), while its precipitation pattern is poised between an "s" and an "f" (see Section 9.5.1).

These regions are often quite popular as summer holiday resorts because most of the zone is close to either the sea or higher (and therefore cooler) land, or both. The beautiful city of Athens, Greece, is such a location (Figure 9.23). Winters are very pleasant, but even in the afternoon temperatures are only around 13° to 14°C (55° to 57°F), with an occasional temperature of 18°C (65°F). The daily range is not as great as in much of this zone because of the maritime location, but a few days dip below freezing. Rainfall amounts are small but give rise to some rather raw conditions, especially when cold winds blow from the northeast. In April rainfall frequency decreases to three days per month, and temperatures regularly begin to reach above 21°C (70°F). The temperatures continue to rise steadily, and by July maxima are above 32°C (90°F). These temperatures, with the intense insolation from an almost cloudless sky, create conditions suited more to beach life than to walking or working in the city or countryside. September also can be quite hot, but October and November (although averaging rain on one day in six) have very comfortable temperatures ranging mainly from 10° to 24°C (50° to 75°F).

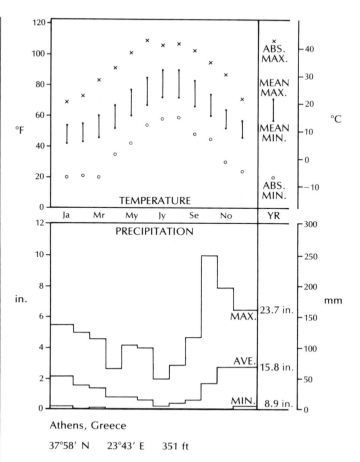

**FIGURE 9.23**
Temperature and precipitation characteristics for Athens, Greece (Csa climate).

Precipitation in the Csa zone is generally about 500 to 750 mm (20 to 30 in.) annually and is concentrated, sometimes almost wholly, in the winter months. For instance, 85 percent occurs during winter in Perth, Western Australia, and 70 percent in Istanbul, Turkey. Summertime conditions, when the Cs climates are under the influence of the subsiding air of the high-pressure cells, are not conducive to rainfall. In winter the equatorward movement of the pressure belts puts these areas under the control of the westerlies and their frontal activity. Periods of precipitation seldom last long, but occasionally a severe outbreak of cold continental air will bring snow.

Summers are hot, with the temperature expected to reach 35° to 38°C (95° to 100°F) on many days and even higher near the dry zone boundary. Freezes are not common in most areas, but in interior regions they are normal in mid-winter. In summer the diurnal temperature range is large [about 17C° (30F°)], the sky is often almost cloudless, and relative humidity is low.

## 9.4.7 MID-LATITUDE, DRY AND WARM SUMMER (Csb)

Like the Csa zone, the Csb zone is found in areas with Mediterranean climate, but not frequently in the lands actually bordering the Mediterranean Sea. The Csb zones are located on the western edges of the continents, around latitudes 35° to 40°, and are quite small in extent. Typical areas are along the coast from California to Washington, central Chile, northern Portugal, the Cape Province of South Africa, and south of Perth, Australia.

The climate of San Francisco, California, is reasonably typical of the Csb zones (Figure 9.24). In winter temperatures normally range from about 4° to 16°C (40° to 60°F), but occasionally they will reach 21°C (70°F) or fall to near freezing. Days are quite sunny, but about one day in three has a small amount of rain. Relative humidity is rather high and leads to a raw feeling in the atmosphere during many hours of the day. About twelve foggy days (half the annual total) occur in winter. Winds are predominantly northerly, and some light snow can settle for a short time. Spring is a delightful season, when the wind changes to easterly, rainfall amounts decrease, afternoon temperatures are slightly above 17°C (60°F), and evenings are mild. The area receives about 70 percent of possible hours of sunshine. Summers are warm, but only a few days experience temperatures over 29°C (85°F), generally when the air flow is from the east. Rainfall is negligible. September is the hottest month, with plenty of insolation and almost no rain. Cool afternoons begin in November, when rainfall frequency increases to about one day in four. The extreme range of temperature in an average year is around 4 to 32C° (40 to 90F°), but with most days in the 7° to 21°C (45° to 70°F) range, this is a very agreeable climate.

Precipitation amounts and seasonal distribution are very similar to those of the Csa zone, but generally the summer dry spell is not so intense. However, where a pronounced offshore, cool ocean current is instrumental in keeping the summer temperatures below the 22°C (72°F) threshold (such as around San Francisco), the summer rainfall is low. Winter precipitation is associated with the seasonal dominance of the prevailing westerlies and their frontal storms. Amounts in the wettest months normally range from 75 to 150 mm (3 to 6 in.). Snow is unusual, although it does occur at higher elevations.

Summer temperatures are very pleasant because even on the hotter days the coastal location brings a cooling breeze. Exceptions to this occur when hot air comes from the east and temperatures of as high as 38°C (100°F) are reported. Winters are mild and freezing temperatures are rare, except in special locations. For example, Marseilles, France, almost a Cfa climate, is affected by the *mistral,* the cold north wind down the Rhone Valley, and can reach −10°C (14°F). Some areas in Turkestan, near the D zone, also are likely to experience extreme cold.

Insolation is quite plentiful in summer, except along coastal belts near cold currents, where fogs are common. The fogginess of the Golden Gate area of San Francisco is famous, but Cape Disappointment in northwest Washington averages over 2500 hours of fog annually—the world record. Although there is more cloudiness in winter, spells of sunny days are quite frequent.

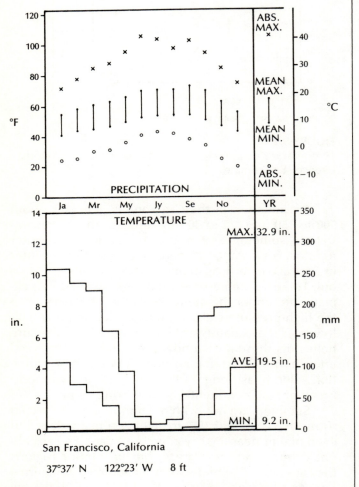

**FIGURE 9.24**
Temperature and precipitation characteristics for San Francisco, California (Csb climate).

## 9.5
### THE HIGH LATITUDE CLIMATES (D CLIMATES)

The major difference between a D and a C zone is the mean temperature of the coldest month—above −3°C (27°F) for a C, below it for a D. Generally this means that the D zones are contiguous to C zones, except where they abut on a semiarid BSk zone. In the Southern Hemisphere, especially in the high latitudes where D climates are generally located, there is very little land area. Because of the marine nature of this hemisphere, the area where the coldest month is below −3°C (27°F) will have a warmest month only a little higher—certainly below 10°C (50°F)—and will therefore be an E climate. Thus, there are really no D zones in the Southern Hemisphere.

Summers in nearby C and E zones can be very similar to summers in D zones, but normally the D climate has more severe winters and a short spring and fall. In most D stations summer is the wettest season, although a few areas influenced by maritime conditions have a very uniform seasonal distribution of precipitation. In the parts close to the steppe climates rainfall can be very irregular and sparse. Snow amounts vary greatly; 10,000 mm (400 in.) fall in some of the Sierra Nevada, whereas only a few inches fall in the continental interiors where the bitterly cold air can hold only little moisture. Where there is little snow to blanket the soil, the ground becomes frozen to a great depth. Permanently frozen ground (*permafrost*) can be 50 m (150 ft) or more deep in some areas. In summer the top few feet thaw in many places, producing a number of shallow extensive swamps (ideal mosquito-breeding areas) and a vast expanse of soggy ground that is extremely difficult to traverse.

The cold air of winter forms the intense high-pressure cells over Canada and the USSR. These tend to block or repel cyclonic activity in the zone so that winters often are dry. Most summer rainfall stems from frontal conditions.

### 9.5.1 HIGH LATITUDE, UNIFORM PRECIPITATION, HOT SUMMER (Dfa)

The Dfa zone is found almost exclusively in the United States, stretching from central Kansas and northern South Dakota eastward in a narrowing band to the western edge of Lake Erie and also including a small area around western Massachusetts. Other small areas are found in the eastern Ukraine, eastern Rumania, and some highlands in central Asia and the Sierra Nevada–Cascade range in the United States.

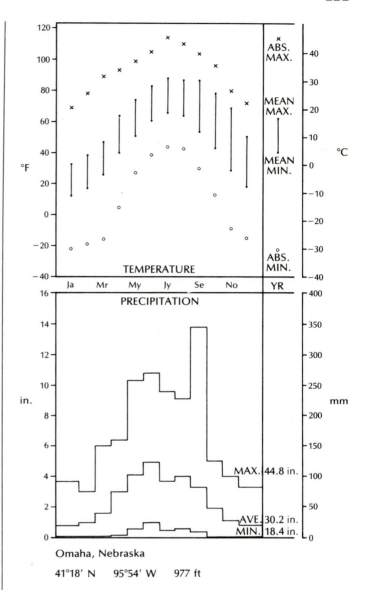

**FIGURE 9.25**
Temperature and precipitation characteristics for Omaha, Nebraska (Dfa climate).

Omaha, Nebraska, can be used as a typical city for this region (Figure 9.25). In winter the maximum temperature is below freezing for about 35 days, but a dozen days report minima of less than −18°C (0°F). There is a total of over 3500 heating degree days (see Section 15.6)—about 60 percent of the annual average. Precipitation is light (75 mm, or 3 in.), but about 50 cm (20 in.) of snow is expected during November to March, with freezes and snow continuing into April. The northerly air flow can bring blizzards, and travelers' warnings are often necessitated. By May, the warming is rapid and days with temperatures above 24°C (75°F) are recorded. Precipitation increases, and

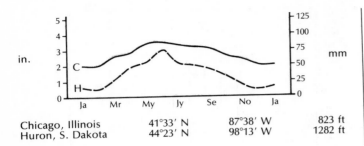

Chicago, Illinois    41°33' N    87°38' W    823 ft
Huron, S. Dakota    44°23' N    98°13' W    1282 ft

**FIGURE 9.26**
Precipitation in Dfa climates.

thunderstorms average about eight to ten each month from May to August. Summers are warm, with hot days [some as warm as 32°C (90°F)] and cool nights. Rain falls about one day in three from both thundershowers and from the more gentle rains associated with cyclonic storms. There are many hours of sunshine, and the radiation is quite intense. Fall is a transition season with very pleasant temperatures, decreasing precipitation, and fewer thunderstorms. Temperatures can decrease rapidly in October. November very often brings the first snow, but during one year 18 cm (7 in.) fell in October.

In the Dfa zone the annual precipitation is around 625 to 875 mm (25 to 35 in.), although Salt Lake City—which in some climatic summaries just makes the classification—has only 375 mm (15 in.). This latter value is typical of the stations in the USSR. The time of maximum precipitation is late spring and early summer, with a definite winter minimum that is not sufficient for a w classification (Figure 9.26).

In all of the central United States and Canada the time of maximum precipitation tends to come later as one moves northward and westward. Warm-season precipitation is usually convectional, but some frontal activity brings days with steady rain. Like the Cfa zones to the south and the Dsb zone to the north, Dfa zones report severe storms and tornadoes every year, for here the tropical and polar air masses meet. The cold air masses in winter bring cold, clear weather, with occasional high winds—a relief from the gloomy days with snow or rain. Snow is reported on about ten to fifteen days over much of the area.

Temperatures in summer are high, and when mT air invades the zone, the climate has subtropical characteristics. However, cP invasions bring some relief and occur much more frequently than they do farther south. Some temperatures above 38°C (100°F) are expected each year, in contrast with the −22° to −29°C (−10° to −20°F) of the winter months (Figure 9.27). In the USSR the extreme lows are around −34°C (−30°F). The mean of the coldest month in all Dfa zones is normally around −12° to −7°C (10° to 20°F).

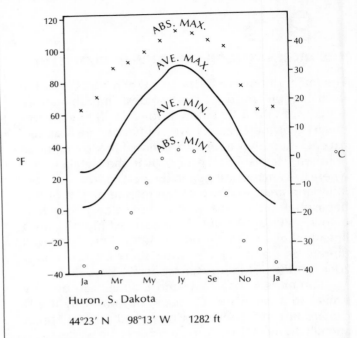

Huron, S. Dakota
44°23' N    98°13' W    1282 ft

**FIGURE 9.27**
Temperature characteristics for Huron, South Dakota (Dfa climate).

Diurnal ranges of temperature vary from about 14 to 18C° (25 to 32F°).

### 9.5.2 HIGH LATITUDE, UNIFORM PRECIPITATION, WARM SUMMER (Dfb)

There are two large bands of Dfb climates: one encompassing much of southern Canada and the northern United States, the other extending from Poland to central Siberia. In addition, the Dfb zone is found in northern Japan and South Island, New Zealand, at high elevations, and also in the Pyrenees, the Caucasus, eastern Anatolia, and Colorado, among others.

The twin cities of Minneapolis–St. Paul, Minnesota, are a good example of this type of climate (Figure 9.28). Most winter days (66 on the average) remain below freezing, with one third of the days reporting minima less than −18°C (0°F). The number of heating degree-days is in excess of 4500, or about 55 percent of the annual total. Snowfall is around 114 cm (45 in.), but the maximum occurs in March (25 cm, or 10 in.), a month which is really a continuation of winter. By late April freezes are minimal and temperatures can exceed 21°C (70°F). Summers can produce some very hot days; temperatures over 40°C (104°F) have been reported in every month from May to September. Thunderstorms occur about one day in four, and precipitation averages some 75 to 100 mm (3 to 4 in.) per month. Temperatures decrease rapidly during the fall,

## SEC. 9.5 | THE HIGH LATITUDE CLIMATES (D CLIMATES)

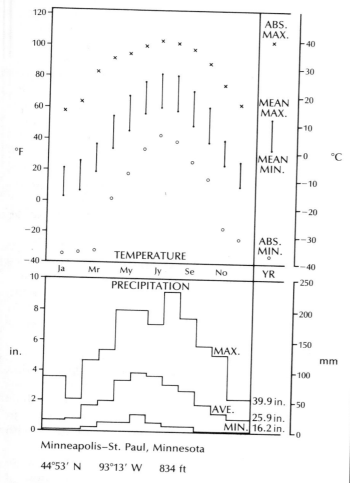

**FIGURE 9.28**
Temperature and precipitation characteristics for Minneapolis–St. Paul (Dfb climate).

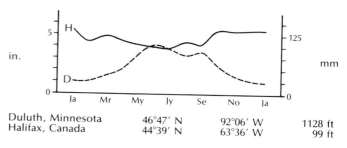

**FIGURE 9.29**
Precipitation in Dfb climates.

with freezes in late October or early November and snow later in the month as winds shift back to the northwest.

In the Dfb zone annual precipitation totals vary greatly, from around 375 to 625 mm (15 to 25 in.) near the dry borders to over 1250 mm (50 in.) at locations near marine influences (Figure 9.29). In addition, the seasonal distribution shows large variation, from the great uniformity of marine areas (with a slight fall or winter maximum from frontal activity) to the spring or summer maximum of the interiors. Precipitation results from air mass showers and cyclonic storms (transient disturbances) in summer and from the latter in winter. Skies are cloudier and drizzle is more frequent than in the Cfa zone. There are many days of snow, and continuous snow cover is the rule for November to March in much of the area. In the United States the Dfb zone includes the "snow belt" of western New York State, northwestern Pennsylvania, and northeastern Ohio, where unusually heavy falls result as the west to northwesterly flowing air masses pick up moisture from the Great Lakes, while these remain at least partially unfrozen.

Most stations have a warmest month with a mean around 16° to 18°C (60° to 65°F), but some reach 21°C (70°F) (Montreal, Canada, and Concord, New Hampshire). The warmest period (means above 10°C, or 50°F) does not last more than five months, in spite of an expected 32°C (90°F) extreme maximum at most locations (Figure 9.30). A few areas, located in the far interiors, have monthly means below −18°C (0°F). Maritime stations can have mean February (coldest month) temperatures above −7°C (20°F), but even

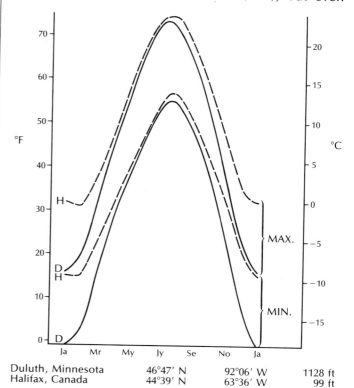

**FIGURE 9.30**
Temperature in Dfb climates.

these experience temperatures as low as −29°C (−20°F) when a fast-moving cA air mass breaks out. Inland in the United States and the USSR values below −51°C (−60°F) have been recorded.

### 9.5.3 HIGH LATITUDE, UNIFORM PRECIPITATION, COOL SUMMER (Dfc)

There are two large areas of Dfc climate: one extending from Alaska to Newfoundland, the other from Norway to eastern Siberia. In addition, an area around the Kamchatka Peninsula (western USSR) also qualifies. Occasionally the zone reaches to about 45° N, with a northern limit of 70° N. Its longitudinal extent is great—165° E to 55° E, 10° W to 130° W and 155° W to 180° W—so it is found at almost 70 percent of all longitudes.

Within the United States, Fairbanks, Alaska (Figure 9.31), gives a good example of this region. This is a truly continental climate, as indicated by the extreme temperature domain [37° to −54°C (99° to −66°F)]. Winters are very cold, with rarely a day above freezing, and over 90 percent of the minima are below −18°C (0°F). The number of heating degree days per year is over 14,000, but less than half of these are in winter. Even summer accumulates 700 units. There is a warming in spring, but winter really extends through mid-April; only in May do afternoons reach 16°C (60°F) and freezes become rare. Every month has recorded one freeze and at least a trace of snow. In summer the winds, averaging around 10 km/h (6 mi/h), are out of the southwest; a few thunderstorms occur; rainfall averages about 50 mm (2 in.) per month; and some warm days are reported. The expected yearly high is 30°C (86°F). Fall is quite short, for October sees −18°C (subzero) temperatures and snow return. Annual precipitation is about 25 cm (10 in.). A rather depressing aspect, to those unaccustomed to it, is the long night; from mid-November to late January there are less than six hours of daylight, and at Christmas it is as low as three and a half hours.

Over most of the Dfc zone annual precipitation is small, generally 375 to 500 mm (15 to 20 in.), due mostly to the low temperatures. However, along western coasts influenced by a warm ocean current values above 1000 mm (40 in.) are reported (Figure 9.32). In these areas the seasonal distribution is uniform, with a tendency for a slight maximum in the fall.

In the interior stations there is a summer concentration, with almost two thirds of the annual amount coming in April to September. The summer precipitation is generally frontal but of short duration, so there is much insolation. Winter precipitation is normally of dry snow, but the marine areas will have a preponderance of rain and sleet with some heavy snow.

Although the mean temperature of the hottest month is rarely above 16°C (60°F), the interior stations expect some days above 30°C (80°F) (Figure 9.33). Manitoba has reached 43°C (108°F) in July, and even

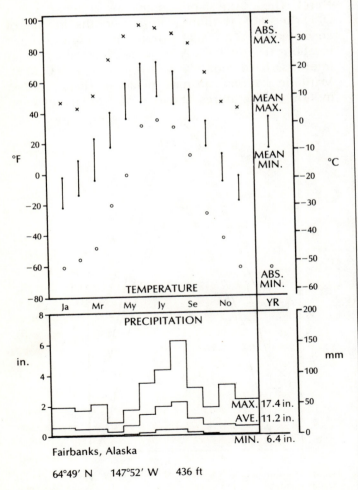

**FIGURE 9.31**
Temperature and precipitation characteristics for Fairbanks, Alaska (Dfc climate).

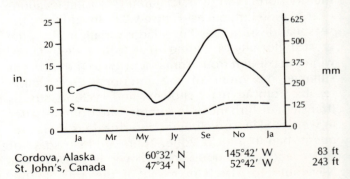

**FIGURE 9.32**
Precipitation in Dfc climates.

### SEC. 9.5 | THE HIGH LATITUDE CLIMATES (D CLIMATES)

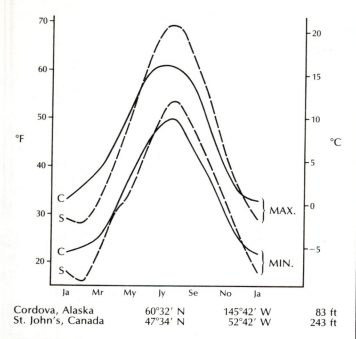

Cordova, Alaska   60°32′ N   145°42′ W   83 ft
St. John's, Canada   47°34′ N   52°42′ W   243 ft

**FIGURE 9.33**
Temperature in Dfc climates.

Fairbanks has recorded 37°C (99°F), as has Akmolinsk in the Kazakstan area of the south central USSR. Generally, six or seven months have means below freezing, but means of the coldest month vary greatly, from −3°C (26°F) to around −26°C (−15°F). This indicates the pronounced marine/continental differences; for example, in the marine belt an annual low near −18°C (0°F) is expected, while interior stations average around −40°C (−40°F). However, at such times air movement is light.

The Dfc zone is the area of the **taiga**, the northern coniferous forest. In the less forested regions, toward the tundra or the steppes, blizzards (North America) or *buran* or *purgas* (Siberia) can bring high winds and blowing snow sufficient to reduce visibility drastically and cause loss of human and animal life.

#### 9.5.4 HIGH LATITUDE, UNIFORM PRECIPITATION, EXTREMELY COLD WINTER (Dfd)

There are two zones of Dfd climate, both in northeastern Siberia and separated by the Dwd zone. This is a sparsely inhabited region whose inhabitants lead a rigorous existence. There is some deciduous woodland, especially in the western zone.

Precipitation is low, often less than 250 mm (10 in.) annually; even July and August, the wettest months, average only around 50 mm (2 in.). Zyryanka,

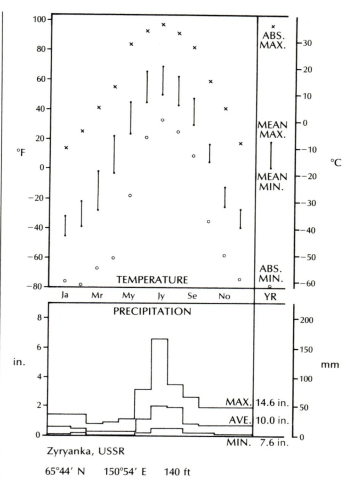

Zyryanka, USSR
65°44′ N   150°54′ E   140 ft

**FIGURE 9.34**
Temperature and precipitation characteristics for Zyryanka, USSR (Dfd climate).

USSR, is typical of this climate (Figure 9.34). Frontal activity brings most of the precipitation. Most months in the dry winter period average about 8 to 15 mm (0.3 to 0.6 in.), but this is sufficient to yield moist conditions because evaporation is very low under such extremely cold temperatures. The winter precipitation, mostly in the form of snow, is associated with cyclonic disturbances that can reach Dfd areas more readily than the Dwd zone, which experiences a drier winter.

Summer temperatures average above 10°C (50°F) for two or three months, but a few days generally reach 28° to 30°C (84° to 86°F). However, a freeze can occur during any month. Winters are almost as bone-chilling as in the Dwd zone. Temperatures around −57°C (−70°F) occur, with values below −46°C (−50°F) expected any time from November to April (Figure 9.34). There is relatively little cloudiness at any time of the year.

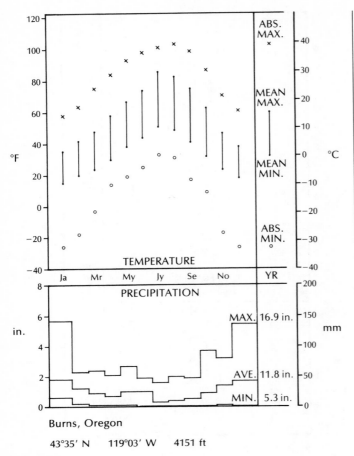

**FIGURE 9.35**
Temperature and precipitation characteristics for Burns, Oregon (Ds climate).

## 9.5.5 HIGH LATITUDE, DRY SUMMER (Ds)

The Ds zones are very unusual and are mostly sparsely populated regions. They tend to occur on the western sides of a continent and conterminous with a Cs zone but at higher elevations. A small Dsa region is found in the northern parts of Turkey, Iran, and Iraq at elevations of about 1900 to 2200 m (6000 to 7000 ft).

Precipitation is low, around 375 mm (15 in.) annually; summers are extremely dry, and the winter and spring months receive about 50 mm (2 in.) each. The mean July temperature is around 22°C (72°F), and daily maxima can exceed 32°C (90°F). January means are near −4°C (25°F), but when cold air masses flow out of central Asia the temperature can plummet to −18° to −29°C (0° to −20°F).

In Oregon, south of the Columbia River, there is a small Dsb area (Figure 9.35). The Cascade Mountains play an important role in determining the precipitation, which amounts to about 375 mm (15 in.) annually. The seasonal pattern is similar to the Csb zone

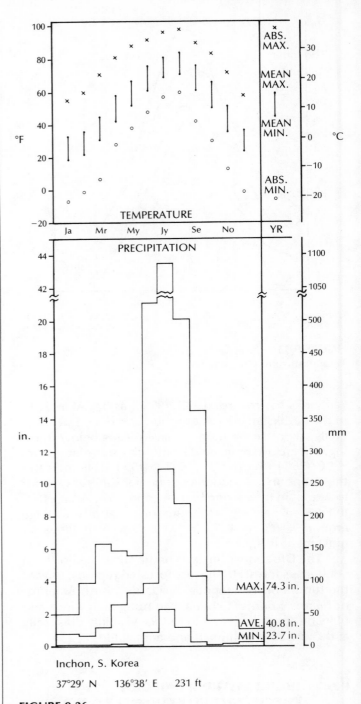

**FIGURE 9.36**
Temperature and precipitation characteristics for Inchon, South Korea (Dwa climate).

in that area. This zone, at an elevation of above 900 m (3000 ft), is strictly a highland region (see Section 9.7). There is a wide range of temperatures, with some days near 38°C (100°F) and occasional minima down to −23° to −29°C (−10° or −20°F). Another small Dsb area is found in eastern Turkey.

## 9.5.6 HIGH LATITUDE, DRY WINTER, HOT SUMMER (Dwa)

The Dwa climate is rare, appearing only in eastern Asia around latitude 40° N. It is sometimes called the Manchurian climate. At Inchon, South Korea (Figure 9.36), winter winds are mainly from the north and northwest so, in spite of its proximity to the sea, the air is quite dry. Temperatures hover near −4°C (25°F), although they occasionally will drop to about −18°C (0°F) or climb to 13°C (55°F). The average precipitation is small, but as much as 15 to 30 cm (6 to 12 in.) of snow on the ground have been reported. Snow cover generally persists for about 30 days. In spring the southeast winds begin, and by May temperatures can reach 27°C (80°F); precipitation is still light. In summer winds are dominantly from the east to the south; days are warm [around 28°C (82°F) in the afternoons] to hot [37°C (98°F)]; the humidity increases; and the rainy period begins. This is the typhoon season, when high winds can occur and cause severe damage. Fog is most prevalent (14 days in the three months) at this time, but sunshine averages about 8 hours per day. In fall the northerly flow returns, usually in October, and by early November freezes are expected.

In the Dwa zone annual precipitation is around 675 mm (25 in.), which is similar to the drier parts of the Dfa climates, although parts of Korea have received up to 1000 mm (40 in.). Winter is distinctly dry with little cloud, and only about 15 percent of the precipitation falls from October through March, when winds are blowing out of continental Asia. When a monsoonal reversal of winds occurs in the warm season, convective and orographic rainfalls result. Typhoons can affect the area and cause serious flooding.

Temperatures show a very wide variation, from summer maxima of 35° to 38°C (95° to 100°F) to winter minima of around −18° to −23°C (0° to −10°F). The extreme values are generally higher than in the Dfa zones, but the mean temperatures of the winter months are lower by some 6 to 10C° (10 to 18F°). However, in Korea the influence of the ocean moderates these extremes.

## 9.5.7 HIGH LATITUDE, DRY WINTER, WARM SUMMER (Dwb)

The Dwb climate is found only in eastern Asia, with the most important region located in a band to the north of the Dwa (Manchuria) zone. Although population density is not great, there are some large cities in the zone—Vladivostok, USSR, for example (Figure 9.37). In winter the cold, dry, continental winds dominate and bring arctic air, clear skies, and many snowy

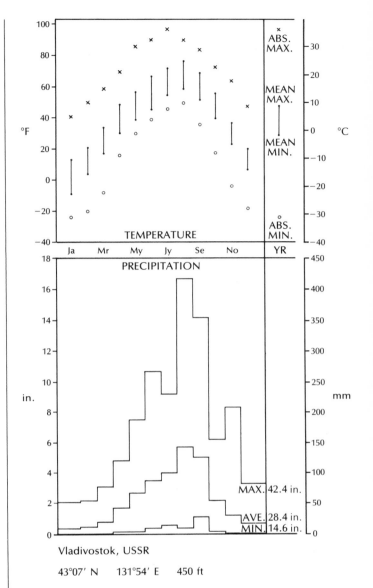

**FIGURE 9.37**
Temperature and precipitation characteristics for Vladivostok, USSR (Dwb climate).

periods. Temperatures are seldom above freezing at any time from December through February, while minima average around −18°C (0°F). There is little precipitation, but snow covers the ground for about 150 days and does not melt until late April. Wind speeds are high, and occasional blizzard conditions exist. The port is actually closed by ice for nearly four months. In spring temperatures rise rapidly, reaching around 13°C (55°F) in mid-May. Precipitation amounts remain low, but cloudiness and humidity are on the increase. Fogs are reported about 80 days per year, with 20 of these occurring in July when the monsoon wind reversal brings cool, moist maritime air to the region.

Precipitation is at a maximum in August and September, but the skies then clear and a short autumn occurs, which is followed by the high winds (averaging 27 km h$^{-1}$, or 17 mi h$^{-1}$) that presage winter.

The mean annual precipitation over the Dwb zone is about 375 to 625 mm (15 to 25 in.), at least 75 percent of it coming during April to September. August, the wettest month, receives some 90 to 115 mm (3.5 to 4.5 in.) due to convective activity which develops in the flow of moist air inland—the monsoonal reversal. In winter the subsiding air from the Siberian high pressure region reduces precipitation drastically; snowfall is light, while many days are very cold and clear.

Summer temperatures reach 27° to 29°C (80° to 85°F) on a number of afternoons during July and August, but after September temperatures drop rapidly and most stations have mean monthly temperatures below freezing from November to April or May. Even at Vladivostok, where marine influence is at a maximum, temperatures around −26°C (−15°F) are reached and −34°C (−30°F) is anticipated inland.

## 9.5.8 HIGH LATITUDE, DRY WINTER, COOL SUMMER (Dwc)

The Dwc climate is found only in Asia—in a small area to the northwest of Peking, a sector of northern Mongolia, and a larger area in the eastern USSR, stretching from west of Lake Baikal to the coast. Irkutsk, near Lake Baikal, has an extremely continental cimate (Figure 9.38). The winter is frosty, with temperatures ranging between −12° and −26°C (10° and −15°F). Snow amounts are not great, but snow cover lasts for over 200 days and does not clear until the end of April. Steam fog, formed when very cold air drifts across relatively warm water, occurs on over 100 days annually, with December and January each reporting about 200 fog hours. The air is generally calm and skies are often clear. Temperatures rise about 0.6C° (1F°) every two days during the spring, but freezes are usual until mid-May. Only about 95 days per year are freeze-free. Spring winds are from the west or northwest. Precipitation reaches a maximum in summer and can be quite intense; occasionally a 24-hour fall exceeds the winter precipitation total. Summer temperatures can reach 29°C (85°F) or more on a few days. Fall comes and goes rapidly, for freezes can occur in August.

For the Dwc zone annual precipitation totals are around 250 to 375 mm (10 to 15 in.)—less than in the Dwb zone—with about 80 percent or more concentrated in the warmer six months. Interior stations show July or August maxima of about 75 mm (3 in.), while coastal stations have August or September max-

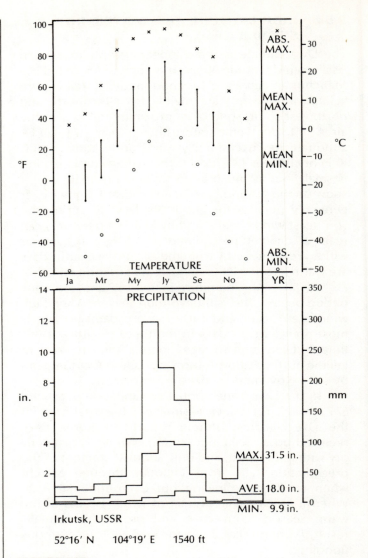

**FIGURE 9.38**
Temperature and precipitation characteristics for Irkutsk, USSR (Dwc climate).

ima. Much of the precipitation is convective, but cyclonic rain occurs near the coast. The number of rain days is small, and summer cloudiness is only about 50 percent. When the effect of the strong Siberian anticyclone is dominant (October to April), precipitation is light and in the form of snow. Clear skies and cold winds are normal conditions at this time.

The warmest month has a mean of only about 16°C (60°F), but temperatures of 32°C (90°F) or even 38°C (100°F) are reported from the interior stations. Near the coast 27°C (80°F) constitutes a hot day. Temperatures rise by about 17C° (30F°) from mid-March to mid-May and fall 19 to 22C° (35 to 40F°) between September and November. Most of the area can expect temperatures from −37° to −40°C (−35° to −40°F) sometime during the winter. Along the coast conditions can seem particularly severe, for many raw and foggy days are the rule. Freezes are likely in all months except July and August.

### 9.5.9 HIGH LATITUDE, DRY AND EXTREMELY COLD WINTER (Dwd)

The zone of Dwd climate exists only in northeastern Siberia. This is perhaps the most inhospitable area of permanent habitation on the earth's surface. In the taiga of eastern Siberia the continental climate reaches its extreme at Oymyakon (Figure 9.39), where the annual temperature range exceeds 64C° (115F°). The absolute temperature range is 104C° (187F°)! Maritime air masses are almost unknown. Winter is very severe, but skies are cloudless, snow amount is small, and winds are almost calm. The mean daily temperature is below −42°C (−44°F), but the thermometer can plunge below −62°C (−80°F). Incidentally, such low temperatures must be measured with special thermometers because mercury and some alcohols freeze! In this area, the coldest in the USSR, we find unusual phenomena, as we do in other extremely cold regions. The "whisper of the stars" (the settling of tiny ice particles produced by sublimation) and the sound of water vapor freezing in the breath (compared to the rustle of poured grain) are typically heard when temperatures fall below about −50°C (−58°F). The cold season lasts at least 7 to 8 months, and the snow cover exists for 230 days, melting only in mid-May. Summer brings a little precipitation and some very hot days, often around 32°C (90°F). In August, as in April, the daily range averages around 25C° (45F°). Only in July are there no freezes. Fall is very transient, with light snows occurring in September and the temperature dropping over 30C° (54F°) in two months. Fogs occur on about 50 percent of the days in the cold season, and result from the intense radiational cooling of the air.

For the whole zone the annual precipitation is low, from 125 to 250 mm (5 to 10 in.), and even the wettest months seldom have over 50 mm (2 in.) of rain, generally of cyclonic origin. There are few clouds even in the wet season. Winter brings little moisture, and snow cover is very light. The precipitation at all seasons is frontal in nature, since convectional activity is minimal.

The annual temperature range is extreme—in excess of 56C° (100F°)—and about 5 months have means below −18°C (0°F). January often has days of −57°C (−70°F) weather, while July and August can bring 28° to 30°C (84° to 86°F), a variation of 83 to 89C° (150 to 160F°). From mid-September to mid-November the mean temperature falls over 33C° (60F°), or 0.6C° (1F°) daily. Higher land in the area (above some 600 m, or 2000 ft) falls in the E zone.

## 9.6
### COLD CLIMATES (E CLIMATES)

The E climate, with no month having a mean temperature above 10°C (50°F), is found either in the high latitudes or at high elevations in any latitude. Because there is no true warm season, there is no agriculture, and population density is very low.

There is a surprising diversity of climates within the E zones since they may be found at any latitude. Sometimes there is appreciable rainfall of 1000 mm (40 in.) or more, and sometimes there is very little. In areas where altitude is the dominant factor, the climate is broadly similar in seasonal temperature and precipitation patterns to nearby stations at lower ele-

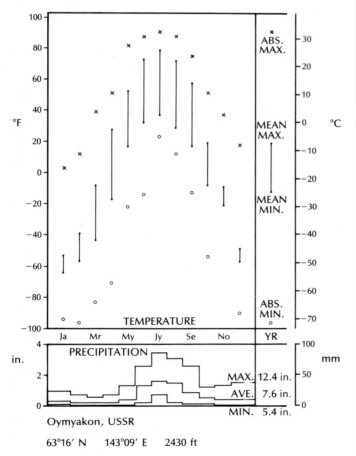

**FIGURE 9.39**
Temperature and precipitation characteristics for Oymyakon, Siberia (Dwd climate).

vations—naturally, temperatures are less and often precipitation is greatly reduced.

In this section we identify three distinct zones:

1. EM: the polar marine climate. The mean of the warmest month is above 0°C (32°F); the mean of the coldest month is equal to or above −7°C (20°F).
2. ET: the tundra climate. The mean of the warmest month is above 0°C (32°F); the mean of the coldest month is below −7°C (20°F).
3. EF: the frost climate. The mean of the warmest month is equal to or below 0°C (32°F).

### 9.6.1 POLAR MARINE CLIMATE (EM)

The polar marine climate is really a subdivision of Köppen's ET zone, but it is sufficiently different and important to rate a subclassification. It is clear that, with a mean coldest month of −7°C (20°F) and a mean warmest month of at most 10°C (50°F), all EM stations must have an annual temperature range of less than 17C° (30F°). There are three large EM zones: (1) in the Bering Sea and including most of the Aleutian Islands; (2) in the North Atlantic Ocean, stretching from near the southern tip of Greenland across most of Iceland and to the northern tip of Norway; and (3) a circumpolar band in the Southern Hemisphere, generally between 50 and 60° S, and thus over the oceans. Some small EM areas are found at high elevations, especially in the tropics, where the low annual temperature range satisfies a necessary criterion. For example, on Mt. Kilimanjaro in Tanzania and Kenya the EM zone occurs from around 3000 to 4500 m (10,000 to 15,000 ft). El Alto, Bolivia (see Table 9.2), is also an EM location, as is the station at 1300 m (4400 ft) on Ben Nevis in Scotland. Locations at high elevations but far removed from oceans make the term *marine* rather misleading.

Annual precipitation in the EM zones occurring at low elevations is generally about 500 to 1000 mm (20 to 40 in.), but local relief can influence this greatly, as shown by Evangelist's Isle (Figure 9.40). Generally the wettest season is winter, although all stations would qualify for an "f," as used in Köppen's C and D zones. Sleet is common and some of the precipitation occurs as snow in all EM climates, but the amount varies greatly from 10 to 65 percent of the total.

Temperature variations at a station are small because of the marine or equatorial influence; Evangelist's Isle, for example, has an annual range of only 4C° (8F°) (Figure 9.41). The differences among the stations are also small, although some sites have a coldest month with a mean of 4°C (40°F), and others have recorded −6°C (21°F). A few stations, mainly in the Southern Hemisphere, have no monthly means below freezing, and freezes at some sites are actually an unusual occurrence.

The EM zones have high winds and much cloud, averaging 80 percent coverage. However, in the equatorial EM areas the windy, cloudy periods are not as persistent as in the lowland locations.

### 9.6.2 TUNDRA CLIMATE (ET)

The tundra zone is found almost exclusively in the Northern Hemisphere. In the marine Southern Hemisphere it exists only in small regions between the EM and EF climates. The large ET areas are on the northern limits of Eurasia and North America, with other regions in Tibet and on high mountains.

Precipitation varies from less than 250 mm (10 in.) at many low-level stations, to 750 to 1000 mm (30 to 40 in.) at stations near the EM zones, and up to 1250 mm (50 in.) or more in the elevation-induced ET zones of the mid-latitudes. The seasonal distribution shows a maximum in summer (Figure 9.42), with a fall maximum in some areas.

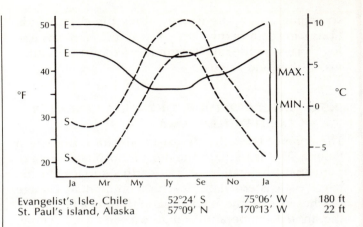

Evangelist's Isle, Chile   52°24' S   75°06' W   180 ft
St. Paul's Island, Alaska   57°09' N   170°13' W   22 ft

**FIGURE 9.41**
Temperature in EM climates.

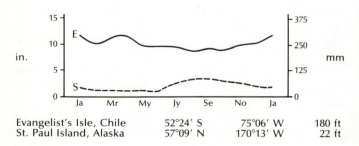

Evangelist's Isle, Chile   52°24' S   75°06' W   180 ft
St. Paul Island, Alaska   57°09' N   170°13' W   22 ft

**FIGURE 9.40**
Precipitation in EM climates.

**TABLE 9.2**
Climatic data for highland stations.

| Station (Lat., long., elev.) | Average (°F or in.) | J | F | M | A | M | J | J | A | S | O | N | D | Year | Climatic Zone |
|---|---|---|---|---|---|---|---|---|---|---|---|---|---|---|---|
| Asmara, Ethiopia (15°17' N, 38°55' E, 7628 ft) | Daily max. | 74 | 76 | 77 | 78 | 78 | 78 | 71 | 71 | 74 | 72 | 71 | 71 | 74 | BSk(A) |
| | Daily min. | 44 | 46 | 48 | 51 | 53 | 53 | 53 | 53 | 55 | 53 | 50 | 49 | 51 | |
| | Prec. amt. | 0.1 | 0.1 | 0.4 | 1.5 | 1.5 | 1.3 | 6.7 | 5.0 | 1.3 | 0.3 | 0.4 | 0.1 | 18.4 | |
| | No. of rain days | 0.4 | 0.8 | 3 | 5 | 5 | 5 | 17 | 14 | 5 | 2 | 2 | 1 | 60 | |
| Nairobi, Kenya (1°16' S, 36°48' E, 5971 ft) | Daily max. | 77 | 79 | 77 | 75 | 72 | 70 | 69 | 70 | 75 | 76 | 74 | 74 | 74 | Cb(A) |
| | Daily min. | 54 | 55 | 57 | 58 | 56 | 53 | 51 | 52 | 52 | 55 | 56 | 55 | 55 | |
| | Prec. amt. | 1.5 | 2.5 | 4.9 | 8.3 | 6.2 | 1.8 | 0.6 | 0.9 | 1.2 | 2.1 | 4.3 | 3.4 | 37.7 | |
| | No. of rain days | 5 | 6 | 11 | 16 | 17 | 9 | 6 | 7 | 6 | 8 | 15 | 11 | 117 | |
| Guatemala City, Guatemala (14°37' N, 90°31' W, 4855 ft) | Daily max. | 73 | 77 | 81 | 82 | 84 | 81 | 78 | 79 | 79 | 76 | 74 | 72 | 78 | Cwb(A) |
| | Daily min. | 53 | 54 | 57 | 58 | 60 | 61 | 60 | 60 | 60 | 60 | 57 | 55 | 58 | |
| | Prec. amt. | 0.3 | 0.1 | 0.5 | 1.2 | 6.0 | 10.8 | 8.0 | 7.8 | 9.1 | 6.8 | 0.9 | 0.3 | 51.8 | |
| | No. of rain days | 4 | 2 | 3 | 5 | 15 | 23 | 21 | 21 | 22 | 18 | 7 | 4 | 145 | |
| El Alto, Bolivia (16°30' S, 68°12' W, 13,468 ft) | Daily max. | 55 | 55 | 57 | 57 | 55 | 55 | 55 | 57 | 57 | 61 | 61 | 57 | 57 | ET(A) |
| | Daily min. | 37 | 37 | 37 | 36 | 32 | 28 | 28 | 30 | 32 | 36 | 37 | 37 | 34 | |
| | Prec. amt. | 4.6 | 4.3 | 2.6 | 1.0 | 0.6 | 0.1 | 0.2 | 0.5 | 1.7 | 1.1 | 1.7 | 3.8 | 23.4 | |
| | No. of rain days | — | — | — | — | — | — | — | — | — | — | — | — | — | |
| Lincoln, Neb. (40°49' N, 96°42' W, 1181 ft) | Daily max. | 34 | 37 | 49 | 63 | 73 | 83 | 89 | 87 | 79 | 67 | 50 | 38 | 62 | Dfa(C) |
| | Daily min. | 14 | 17 | 28 | 41 | 51 | 61 | 66 | 64 | 56 | 43 | 30 | 19 | 41 | |
| | Prec. amt. | 0.6 | 0.9 | 1.2 | 2.5 | 4.0 | 4.2 | 3.9 | 3.5 | 2.9 | 2.0 | 1.2 | 0.8 | 27.7 | |
| | No. of rain days | 5 | 6 | 7 | 10 | 11 | 11 | 9 | 9 | 9 | 7 | 5 | 6 | 95 | |
| Cheyenne, Wyo. (41°09' N, 104°49' W, 6139 ft) | Daily max. | 36 | 38 | 44 | 53 | 62 | 74 | 80 | 79 | 71 | 58 | 46 | 39 | 57 | Dfb(C) |
| | Daily min. | 15 | 16 | 22 | 29 | 38 | 47 | 53 | 52 | 43 | 32 | 23 | 18 | 32 | |
| | Prec. amt. | 0.4 | 0.6 | 1.0 | 1.9 | 2.4 | 1.6 | 2.1 | 1.6 | 1.2 | 1.0 | 0.5 | 0.5 | 14.8 | |
| | No. of rain days | 6 | 6 | 8 | 10 | 12 | 9 | 11 | 10 | 6 | 6 | 5 | 5 | 94 | |
| Ben Nevis, Scotland (56°47' N, 5°0' W, 4400 ft) | Daily max. | 28 | 27 | 27 | 31 | 37 | 44 | 45 | 44 | 42 | 34 | 32 | 29 | 35 | EM(C) |
| | Daily min. | 21 | 20 | 21 | 24 | 29 | 36 | 38 | 37 | 35 | 28 | 26 | 22 | 28 | |
| | Prec. amt. | 18.7 | 15.1 | 17.0 | 10.2 | 8.3 | 7.8 | 11.3 | 14.0 | 16.9 | 14.8 | 16.0 | 21.2 | 171.3 | |
| | No. of rain days | — | — | — | — | — | — | — | — | — | — | — | — | — | |
| Mt. Washington, N.H. (44°16' N, 71°18' W, 6262 ft) | Daily max. | 19 | 18 | 19 | 29 | 40 | 51 | 55 | 54 | 46 | 38 | 27 | 18 | 35 | EF(C) |
| | Daily min. | 1 | 3 | 5 | 17 | 28 | 39 | 44 | 43 | 35 | 26 | 15 | 5 | 22 | |
| | Prec. amt. | 4.3 | 4.2 | 5.9 | 5.9 | 6.2 | 9.1 | 10.3 | 8.6 | 8.4 | 7.5 | 6.6 | 5.2 | 82.2 | |
| | No. of rain days | 18 | 16 | 18 | 18 | 18 | 17 | 17 | 16 | 15 | 13 | 19 | 19 | 204 | |
| Eismitte, Greenland (70°53' N, 40°42' W, 9843 ft) | Daily max. | −33 | −42 | −29 | −14 | 6 | 13 | 19 | 11 | 4 | −23 | −33 | −28 | −12 | EF(D) |
| | Daily min. | −53 | −64 | −51 | −37 | −18 | −9 | 1 | −13 | −20 | −42 | −57 | −46 | −34 | |
| | Prec. amt. | 0.6 | 0.2 | 0.3 | 0.2 | 0.1 | 0.1 | 0.1 | 0.4 | 0.3 | 0.5 | 0.5 | 1.0 | 4.3 | |
| | No. of rain days | — | — | — | — | — | — | — | — | — | — | — | — | — | |

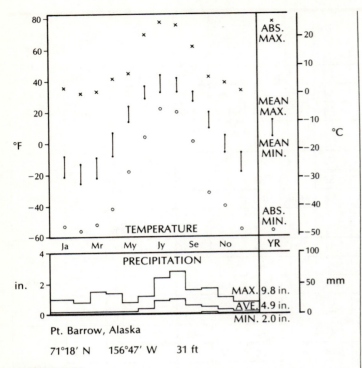

**FIGURE 9.42**
Temperature and precipitation characteristics for Point Barrow, Alaska (ET climate).

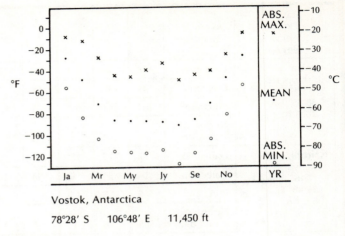

**FIGURE 9.43**
Temperature and precipitation characteristics for Vostok, Antarctica (EF climate).

The biggest temperature difference between EM and ET climates is in the mean of the coldest month. Also, the mean annual temperature range is greater in the ET than the EM stations; this is characteristic of the more continental nature of the tundra zone.

As with the EM climates, cloudiness is great, especially in summer. Winds reach gale force often, but the usual winter conditions are fairly calm days with clear, intensely cold weather.

### 9.6.3 FROST CLIMATE (EF)

The inhospitable frost zones are located mainly in the polar regions, especially in Antarctica and the Greenland interior, but such a region also exists in the highlands of west China and Tibet. The EF zone is found in highlands in many latitudes, even at the equator.

The low temperatures do not permit large amounts of precipitation; maxima are 125 to 250 mm (5 to 10 in.). Most precipitation is in the form of snow and ice pellets. Blizzards, often violent, are expected in winter and even in summer.

Temperatures, while always low, show very wide variation. In Antarctica and central Greenland annual means are usually below −18°C (0°F), with Vostok, Antarctica, holding the world's record, for annual mean temperature, of −55°C (−67°F). This station also has experienced the lowest temperature ever recorded, −88°C (−127°F) (Figure 9.43). The highest temperature ever recorded in Antarctica was 14°C (58°F) at Esperanza (63° 24′ S, 57° W). Perhaps the "hot spot" of the EF climate is the Jungfraujoch in Switzerland, where the coldest month has a mean of −14°C (6°F) and the annual mean is −8°C (18°F).

## 9.7

### HIGHLAND CLIMATES (H CLIMATES)

There are many areas around the world where the elevation plays a dominant role in determining the climate. For example, Nairobi, Kenya, has a Cb climate, but it is clearly in the humid tropical, or A, zone. Its elevation of 1820 m (5970 ft) reduces the temperature so that the thermal criterion for A [mean temperature of coolest month over 18°C (64°F)] is not met. If elevation causes a station to be placed in a different climatic zone from what it would be at sea level (at that longitude and latitude), then the zonal letter (A,C,D,E) corrected to sea level should be indicated. Elevation is "corrected" at the rate of 0.55C° per 100 m (3F° per 1000 ft). For Nairobi the mean temperature of the coolest month is 16°C (60°F), but the elevation "correction" of 10C° (18F°) gives a value of 26°C (78°F). Therefore, Nairobi's climatic zonation could be written as Cb(A) or Cb(HA). Data are given in Table 9.2 for nine stations to show the varying climates that can exist in the highlands of the world.

Although many combinations of symbols are theoretically possible in a highland region, only a few actually are found. In the highlands of the A zone, especially the equatorial area, the terms *summer* and *winter* are not specific, so the corresponding symbols generally are not used. The most common classifications are Cb (Nairobi); Cwb (Guatemala City); and ET (El Alto, where the elevation is extreme). A D zone is very unlikely because it requires a mean annual temperature range of at least 13C° (24F°), and this could only occur on the extreme poleward edge of the tropics and in a dry climate. There are some regions of BSk (Asmara, Ethiopia), but BWh regions are rare. In C zones the most common highland classifications are Dfa (Lincoln, Nebraska); Dfb (Cheyenne, Wyoming); ET; and EF (Mt. Washington, New Hampshire), while the D zone has some EF areas (Eismitte, Greenland).

## SUGGESTED READING

KOEPPE, C. E. and G. E. DeLONG. *Weather and Climate.* New York: McGraw-Hill Book Company, 1958.

RUMNEY, G. R. *Climatology and the World's Climates.* New York: Macmillan, 1968.

*World Survey of Climatology.* Vols. 4, 5, 6, 7, 8, 10, 11, 12, 13, and 14. Amsterdam: Elsevier Scientific Publishing, 1969 to 1980.

# 10
# SMALL-SCALE CLIMATES

**INTRODUCTION**
**MESOSCALE CONTROLS**
Land-Sea Configuration
Topography
Lake Effects
**MICROSCALE CONTROLS**
Bare Surface Characteristics
Topography
Vegetation
Constructions

## 10.1
### INTRODUCTION

In earlier chapters we saw how the main climatic controls—solar radiation, atmospheric circulation, land-sea configuration, and topography—determine weather and climate. All regions of the earth have a climate, even those high in the atmosphere or deep in the soil or water. We identify areas that experience similar climates by considering the extent of this similarity. In Chapter 8 we discussed worldwide macroscale climate and its classification, but other scales must also be studied. Often the terms *mesoclimates* and *microclimates* are used; these can be differentiated by their linear dimensions, and sometimes by height and time as well. Table 4.1 shows how scales of atmospheric motion are similarly distinguished. In one definition microscale applies to horizontal distances of less than 2 km (1.25 mi), mesoscale from 2 to 500 km (1.25 to 300 mi), and macroscale above 500 km (300 mi). Strictly, there are no such climates as micro-, meso-, or macroscale climates; instead, it is the controls of climate that act at these scales.

If the climate of an area is different from that immediately adjacent to it, that usually is because the two regions experience different energy patterns. For example, part of a lawn may be shaded from direct solar radiation for many hours, while a short distance away no such shading occurs. A region's climate is the result of controls acting on all three scales, but the relative importance of the roles they each play varies greatly with time and place. For example, in the lawn analogy, the role of the shading device will be greatly altered at night when there is no solar radiation. Another example would be a straight wall that shelters a region from northerly winds; clearly, the wall will give little protection from westerly or easterly winds.

## 10.2
### MESOSCALE CONTROLS

In the first chapters we learned climate is determined by certain controls and that this applies to climate on all scales. However, the climate of a given area is often influenced by mesoscale conditions to the extent that it is significantly different from the overall, macroscale climate of the area.

**FIGURE 10.1**
Schematic development of a sea breeze. Double arrows indicate horizontal pressure gradient forces; single arrows, resultant air motions. (From *Understanding Our Atmospheric Environment* by Morris Neiburger, James G. Edinger, and William D. Bonner, W. H. Freeman and Company. Copyright © 1973.)

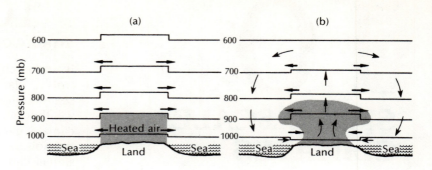

## 10.2.1 LAND-SEA CONFIGURATION

When equal amounts of solar radiation fall on neighboring water and land surfaces, there is a great difference in the resulting temperature patterns. The thermal capacities of water and the various land surface substances are quite different, and radiation can penetrate more deeply into water than into land (see Section 2.3.4). Let us consider the case of a flat island and its surrounding ocean, assuming that initially the surface pressures are the same over land and water. At sunrise the land temperature begins to respond rapidly to the insolation, and within a short period (generally an hour or so) the pressure surfaces will appear as shown in Figure 10.1(a). There are "steps" in the pressure patterns, denoting different spacing between isobaric surfaces when over land or sea, because the air density is inversely related to the temperature and it takes a larger depth of warm air to yield a certain pressure than is the case with cooler air. Above the island the pressure at any elevation (in the lower layers of the atmosphere) is higher than it is over the oceans. The outflow caused by the horizontal pressure gradient forces above the island would lead to an outward movement (from the land) in the air aloft. Notice that there are as yet no horizontal pressure gradient forces at sea level. The outward movement would cause air to accumulate at upper levels over the water with a corresponding decrease of sea level pressure over the land. Such pressure changes would, in turn, develop a pressure gradient force in the lower levels to accelerate the air toward the land and produce the sea breeze (water to land). The complete result is shown in Figure 10.1(b). Remember that this process is continuous, starting slowly and then accelerating until in the mid-afternoon the sea breeze is at its maximum. Along the shores of a very large landmass, the process is illustrated by only one side of the diagrams in Figure 10.1.

At night the reverse process occurs. The land surface cools more rapidly than the sea surface, because in the latter case the cooling water sinks and causes cooling to be spread through a deep layer of water. At this time the isobaric surfaces over the land become depressed relative to those over the sea. The acceleration of the air that now results takes the air from the sea to the land so that the sea level pressure rises over the land. A seaward pressure gradient is developed, and a land breeze results.

On overcast days with little insolation the sea breeze may not develop at all, while on cloudless mornings the sea breeze may be felt a few hours after sunrise. If the land surface is dry, medium or dark colored, and of low thermal conductivity, the surface will heat rapidly and the land-sea pressure difference necessary to induce air flow will soon develop. Forests and woodlands near the coast tend to reduce the onset of the sea breeze, as does a wide swath of marsh or swamp land. Topography is also important, for hills near the coast will reduce or prevent the shallow air flow. A very gentle slope beneath the water will also reduce the thermal and pressure differentials in much the same way as do marshy conditions. The general circulation plays an important role because its air flow may reinforce or reduce the sea breeze, depending upon the direction of the prevailing wind and the coastline. Generally the sea breeze is approximately at right angles to the coastline.

The sea breeze is usually only gentle, between 5 and 10 km h$^{-1}$ (3 and 6 mi h$^{-1}$), and of fairly shallow depth, 100 to 300 m (300 to 1000 ft). In light of all the variables just mentioned, however, the sea breeze shows considerable variation around the world. Since the sea breeze circulation has to be generated, thrive, and die within about 8 to 10 hours, it cannot have great horizontal or vertical dimensions. In this respect there is a significant difference between the sea breeze and the somewhat related monsoon circulation (see Section 4.6.2). In the latter case the complete cycle is measured in months and not hours.

Land breezes are generally of less intensity than sea breezes, and speeds of 2 to 5 km h$^{-1}$ (1 to 3 mi h$^{-1}$)

**FIGURE 10.2**
*Gemini XI* photograph of India and Sri Lanka (September, 1966) showing sea breeze circulation and convective clouds. (Courtesy of NASA.)

and depths of 100 m (300 ft) or less are typical. The lesser speed and depth of the land breeze is due to the smaller temperature variation between the land and sea surfaces at night as compared with the daytime difference. All of the preceding statements also apply to large lakes, but these need to be quite large—on the order of 250 km² (100 mi²).

The sea breeze often has an interesting outcome. As the breeze brings moist, warm air onshore, a horizontal discontinuity in both temperature and humidity can develop a sea-breeze front. This front, added to the convergence resulting from differential frictional effects on the wind (see Figure 5.22), may induce an upward vertical motion sufficient to produce convective clouds (Figure 10.2) and often showers. These showers normally occur a short distance inland from the coast, but their effect can extend for many kilometers if the prevailing general circulation flow reinforces the movement inland. The cloud cover and rain then tend to reduce the land temperature, resulting in a diminution or cessation of the sea breeze, an example of a negative feedback situation (see Section 11.13).

The air brought inland by the sea breeze causes both a drop in temperature and an increase in relative humidity. This change in the elements is generally greater in middle latitudes and where a cold current flows offshore. In the tropics the difference is less because the sea usually is also warm. The moist air mass and increasing cloudiness over the land both tend to reduce the incoming radiation—negative feedback again—so the strength of the sea breeze can vary. To a lesser degree the same is true of the land breeze, because the moister, upper return flow reduces the outgoing long-wave radiation from the land.

In regions where there are frequent and close changes of coastline direction, the almost unidirectional onshore flow will give rise to regions of high convective rainfall interspersed with relatively dry areas. This condition is seen in some of the islands of southeastern Asia and especially in New Guinea.

A cause of some mesoscale climatic differences is the increased frictional drag effect on the air as it flows from sea to land. This can cause greater turbulence, often enhanced by thermal differences between the two surfaces, that may trigger precipitation if the macroscale controls are favorable. When the wind is flowing generally parallel to the coastline, this coastal drag can cause low level (below 500 m, or 1500 ft) zones of convergence or divergence. The effect is often most noticeable in the trade wind belt. In the Northern Hemisphere, when the land lies to the left of the prevailing wind direction, a divergent flow results and subsidence is induced. If the land lies to the right of the wind direction, convergence and

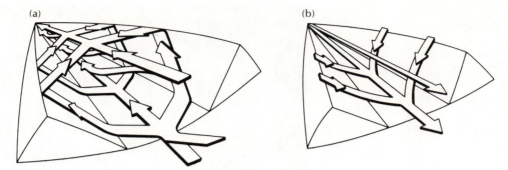

**FIGURE 10.3**
Schematic illustration of flow in a valley. In the early afternoon (a) the valley and upslope winds are occurring. About midnight (b) the flow is down the valley and the slope. (After F. Defant, in *Compendium of Meteorology*, ed. T. F. Malone, p. 665. Boston: American Meteorological Society, 1951.)

ascent occur. In the Southern Hemisphere these conditions are reversed. In middle and high latitudes the Coriolis force can be a factor in the distance of inland penetration of the sea breeze.

### 10.2.2 TOPOGRAPHY

In regions where pronounced elevation changes occur over short distances, we can of course expect variation in climates. One such mesoscale regime involves the mountain-valley wind flows. This process is rather similar to the sea breeze process, but now air near the high land is heated by insolation to a greater degree than that at the same height above sea level in the free air over the valley. The resultant pressure gradients then produce an upslope movement of air. The flow is generally three-dimensional, with movement up the valley and up the slope. Figure 10.3(a) shows the early afternoon situation, while Figure 10.3(b) depicts the nocturnal flow. At night air cooled by contact with the mountains begins to descend into the valley, causing a drop—sometimes appreciable—in temperature. In places where there is also cold air drainage and channeling, the mountain wind can be very strong.

In areas where mountains are near the sea, the sea-land and mountain-valley circulation can combine to give rise to a pronounced diurnal wind variation. Los Angeles, California, is an area where this process occurs. Because in winter the land is cooler than the sea for many hours, a large temperature difference develops, causing the land breeze to become stronger and more prolonged than the sea breeze. In summer the inland areas of mountain and desert become very hot, so a strong sea breeze forms and at night the land breeze is almost nonexistent.

An air flow that results from topographic variation and also from adiabatic expansion and contraction of an air mass is the well-studied case of the **foehn wind**. This is the name by which the hot, dry wind is known in the Alps. In North America the name **chinook** (from an Indian tribe of the northwest) is applied to a similar wind. The foehn has many names around the world—*zonda* in western Argentina and *koembang* in Java, Indonesia. The hot, dry and dusty Santa Ana of southern California is also related in many ways to these winds (see Section 12.2 and Figure 12.2).

The flow pattern for chinook winds is shown in Figure 10.4. As moist air ascends, it may reach a condensation level above which the latent heat released will reduce the rate of cooling of the air. This moist adiabatic process rate actually varies with temperature and pressure, but in the following example it is assumed to average 0.5C° per 100 m (2.7F° per 1000 ft). When the air mass begins to descend on the leeward side of the range, it will increase in temperature (adiabatic warming due to compression) by about 1.0C° per 100 m (5.4F° per 1000 ft) so that, if it returns to its original elevation, a net gain of around 0.5C° per 100 m (2.7F° per 1000 ft) will result. Therefore, saturated air starting at 10°C (50°F) and cooling at the moist

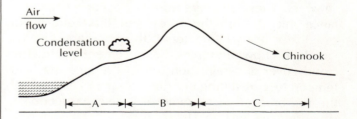

**FIGURE 10.4**
The chinook (foehn) phenomenon. In zone A, moist air cools by expansion (5.4F°/1000 ft). In zone B, cooling is retarded by the release of the latent heat of condensation (approximately 3F°/1000 ft). Heating by compression (5.4F°/1000 ft) occurs in zone C, resulting in dry, warm air.

rate can reach the same level on the lee side of a mountain 3000 m (10,000 ft) high as much drier air with a temperature of 25°C (77°F). These outbreaks of hot, dry air can be quite stressful. If they occur at a time when a thick layer of snow covers the ground, problems of rapid melting and runoff may become serious, although evaporation into the dry air is appreciable. The chinook wind can affect an area in the same way as a front, causing extremely rapid temperature rises. The record is a rise of 27C° (49F°) in 2 minutes at Spearfish, South Dakota. The foehn wind phenomenon is intimately related to synoptic scale flows, as it needs a circulation of sufficient strength and depth to force the air up and over the large range within a short time span.

The next type of air flow that is induced or affected by topographic features is that of the **fall wind**—a strong and cold downslope wind. As in the case of *katabatic winds* (winds moving down an incline), the air is descending, but often it is initially so cold and the descent insufficient to cause great compressional heating that it arrives, generally at a coast, as a strong, chilling wind. The best known cases of fall winds are the *bora*, a winter phenomenon of the northern Adriatic, and the *mistral* of the southern Rhône valley. However, the *papagayo* (parrot) of the Pacific coast of Guatemala and Nicaragua and the *tehuantepecer* of southern Mexico are famous—or infamous—in those regions. In Commonwealth Bay, Antarctica, some fall winds have had bone-chilling temperatures and winds in excess of 240 km h$^{-1}$ (150 mi h$^{-1}$). A characteristic feature of many of these winds is that local topography causes funneling or channeling of the flow with consequent increase in intensity of the air movement. This is then called a *jet-effect wind*.

### 10.2.3 LAKE EFFECTS

Large lakes, as mentioned in Section 10.2.1, can generate their own land-water diurnal circulation pattern. A special climatic effect of some large lakes is the snow belt on their leeward (downwind) sides. In fall and early winter cool air moving across Lakes Erie and Ontario takes up water and then precipitates much of it in the form of snow in a localized region (see Figure 7.13). When the lakes become completely or mostly frozen later in winter, the region no longer experiences anomalously high snowfalls. Because large inland water bodies act as a local moisture source, one might think that showers would be more frequent over and in the vicinity of such lakes. However, since the water surface temperatures during the day often are considerably lower than the surrounding land surface temperatures, convection is suppressed, and precipitation over lakes is actually less than over the surrounding areas, at least during the warm season. The suppression of clouds over water during the daytime is seen in Figure 10.2. In the winter lakes are a source of both heat and water, and the few available studies suggest that precipitation over and around the lake, certainly on the lee side, is greater than that over areas farther downwind.

In some regions, especially the tropics, convective cells can form over lakes during darkness and bring frequent nighttime storms. Such a situation is experienced near Entebbe, Uganda, on the northwestern shores of Lake Victoria. Here, during the northeast monsoon (October through May), the warm waters of the lake lead to instability that is possibly reinforced by the downslope air drainage from the highland on the eastern edge of the lake, around Kisimu and Kericho in Kenya, and frictional convergence on the downwind shore. In the southern half of the lake, where no comparable highland exists on the eastern shore, the nighttime thunderstorm is rare.

## 10.3
### MICROSCALE CONTROLS

Microscales come in many different sizes—from a square kilometer or a square mile, through the area of a backyard, down to conditions within a crack in a log or between blades of grass. As explained earlier, the word *microclimate* is rather misleading because nearly all so-called microclimatic studies have been concerned with short time periods (hours, days, or weeks)—hardly a climatic study.

Naturally the number of microscale climates are almost infinite and all cannot be described. We can better understand these variations if we consider the causes or controls under four headings:

1. Bare surface characteristics
2. Topography
3. Vegetation coverage
4. Constructions

Normally, a particular microscale climate is the effect of more than one of these controls.

### 10.3.1 BARE SURFACE CHARACTERISTICS

The nature of the underlying surface plays an important role in determining the patterns of climatic elements. Because of its role in mixing parcels of air, the first element studied will be wind flow.

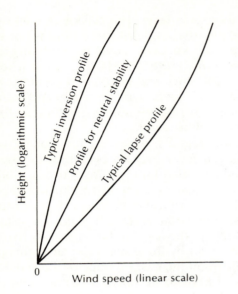

**FIGURE 10.5**
Shape of the wind profile as a function of atmospheric stability.

## Wind Patterns

We learned in Section 4.3.4 that the frictional drag exerted by the earth's surface promotes turbulence, and slows the wind and turns it toward low pressure, relative to the geostrophic (or gradient) wind. Characteristically, then, wind speed increases, in part because friction decreases, with height. Both aspects of the wind also are affected by the nature of the surface: the more irregular the surface, the more frictional drag is exerted.

Temperature stratification also influences the change of wind speed with height. When temperature decreases with height relatively rapidly—creating a superadiabatic lapse rate (see Section 5.8.1), or, simply, lapse conditions—convection currents promote vertical exchange and turbulent mixing. This leads to a smoothing of the wind profile; that is, the increase in horizontal wind speed with height is gradual. At the other extreme, inversion conditions, with stable stratification, are associated with relatively rapid increases of wind speed with height. These relationships are shown in Figure 10.5. Notice that the height scale is logarithmic.

Theoretical considerations enable us to derive an equation which relates the mean wind speed to the height above ground, a thermal stratification parameter, a frictional parameter, and a feature related to the height of the surface irregularities or vegetation cover. For many purposes, however, wind speed can be related only to the first two of these. This is the **power law,** which says that wind speed is proportional to $Z^\beta$. $Z$ is height above the surface, and $\beta$ ranges from 0.07 for inversion conditions, to 0.15 for neutral stability (when the lapse rate equals the dry adiabatic process rate), to 0.35 when the lapse rate is superadiabatic.

## Soil and Air Temperatures

The amount of heat absorbed by the soil surface depends upon certain physical characteristics. In a soil with a large albedo, much heat is reflected and relatively little is absorbed. In addition, if a surface is wet, some heat is used in the evaporative process and less remains for surface heating (see Section 2.3.2). It can be shown that, with some approximations and reasonable assumptions, the range of temperature at a certain depth decreases exponentially with the thermal diffusivity, $\alpha$, of the soil. $\alpha$ is the ratio of the thermal conductivity to the specific heat which is related to the heat capacity (see Section 2.3.4). Also, the time lag (the delay in the heat reaching a particular depth) is proportional to $\alpha^{-1/2}$. For most practical purposes, we find that the daily cycle is negligible at a depth of about 30 cm (1 ft) (time lag of extremes is then about 12 hours), while the annual cycle is almost insignificant at a depth of 600 cm (20 ft) (time lag is 6 months). Of course, actual values are dependent upon soil type and condition. Figures 10.6 and 10.7 depict some typical variations of temperature with depth.

Nighttime cooling of a surface depends upon $\alpha$, the length of the night, the cloud characteristics, and the amount of water vapor in the air. Wind flow is important but has not been included in theoretical expressions. Studies in Germany have shown that freezing conditions penetrated only one third as fast in humus as in sand; for example, at 20 cm (8 in.) the temperatures were −0.5°C (31°F) under humus and −7.0°C (19.5°F) in sand. Shallow mulches reduce dras-

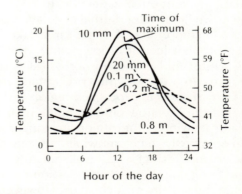

**FIGURE 10.6**
Schematic pattern of soil temperature (daily cycle) at various depths.

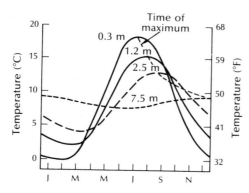

**FIGURE 10.7**
Schematic pattern of soil temperature (annual cycle) at various depths.

tically the thermal fluctuations (see Section 13.7.2). A case in Iowa in July showed that with dry surfaces on a sunny day the temperature at 2.5 cm (1 in.) in bare soil went from 20° to 42°C (68° to 108°F); under a leaf mulch the temperature was about 21° to 26°C (70° to 79°F).

Just above the surface the temperature gradient can be astounding. One summer in the Arabian desert the following conditions were recorded:

1. 77°C (171°F) at surface; 49°C (120°F) at 5 cm (2 in.) ≈ 55 × 10³ Γ
2. 71°C (160°F) at surface; 57°C (135°F) at 12.5 mm (0.5 in.) ≈ 110 × 10³ Γ

where Γ is the dry adiabatic lapse rate of 1C° per 100 m (5.4F° per 1000 ft). If the lapse rate is 3.4 Γ, air density is constant with height (**autoconvective lapse rate**); when it exceeds this value, density increases with height. It is when this value is exceeded that mirages, caused by the refraction upward of light rays, are seen. The nature of the surface—more specifically, its physical characteristics such as albedo and thermal diffusivity—plays an important role in determining the temperature lapse rate in the lower levels. For example, in June in England tar macadam showed a range of 32.6C° (58.7F°), from 42.6° to 10.0°C (108.7° to 50°F); sandy soil showed a range of 26C° (46.8F°), from 35.1° to 9.1°C (95.2° to 48.4°F); and grass showed a range of 16.0C° (28.8F°), from 29.3° to 13.3°C (84.7° to 55.9°F). During this time shelter air temperature varied from 21.8°C (71.2°F) to 7.5°C (45.7°F). The importance of insolation on the vertical gradient of temperature during daytime is depicted in Figure 10.8. The patterns shown in this schematic diagram are of conditions in high latitudes (about 50° N), where winter insolation is very much less than summer insolation. This fact is seen to affect the gradient of temperature near the ground, especially in the lowest 15 m (50 ft).

### Humidity

The humidity of the atmosphere's lower layer is determined by the exchange of moisture between the atmosphere and the surface. Under most situations the flux of water is upward and is especially large during the day if the surface contains adequate moisture. Because of this pattern, the vertical profile of the atmosphere's water vapor content is generally as shown in Figure 10.9.

The water vapor pressure is normally at a minimum around sunrise, but the humidity begins to increase rapidly as the evapotranspiration process develops. Soon after noon the convective activity will induce mixing with descending parcels of drier air, and the humidity will show a slight decrease. Later, as the surface cools, stability develops and moisture is concentrated in the lower layers, causing a secondary maximum. If radiative cooling at night is substantial, dewfall (condensation) may occur and an inversion of humidity develop.

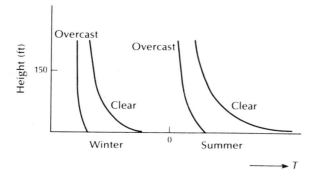

**FIGURE 10.8**
Schematic diagram of temperature gradients for a site at about 50° N.

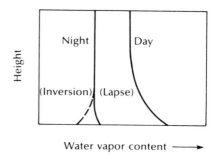

**FIGURE 10.9**
Idealized mean profiles of water vapor concentration near the ground surface.

## 10.3.2 TOPOGRAPHY

In this section we will examine the small-scale effects of topography on climate. We will discuss two facets of topography: the influences of slopes and the depression/rise complex. This aspect is often called the exposure climate.

### Slopes

The most important climatic role of slopes is the way in which they change the insolation pattern. For example, at the equinoxes at latitude 30° N the sun's rays at midday hit a horizontal surface at a 60° inclination, but a south-facing 30° slope will receive them perpendicularly. On a north-facing 30° slope the rays will impinge at a 30° angle. These slopes need only to be of small dimension (even the slopes in a plowed field will suffice) to receive their own distinct insolation pattern.

Figure 10.10 shows the direct beam insolation during clear days on flat surfaces at 30° N. The transmission of the atmosphere is assumed to be 0.7; that is, 70 percent of the perpendicular solar beam is transmitted to the surface (see Section 2.2.2 and Figure 2.6). In July the east- and west-facing slopes receive the most insolation and, perhaps surprisingly, the north-facing slopes have almost as much direct radiation as the corresponding south-facing slopes; this is due to their greater receipt of insolation during the four or five hours after sunrise and before sunset. In January the sun is never seen in the northern half of the sky and the southern slopes are the most favored. Such graphs can be constructed for any latitude. Because they indicate the direct solar radiation impinging on a flat surface, they are extremely helpful in gauging optimal orientations for walls or roofs. Similar graphs can be drawn for both direct and diffuse radiation if measurements of these two components on a horizontal surface are available.

Studies for Potsdam, Germany (52° N), showed that in January a horizontal surface received only 7 percent of July's insolation. For a 30° north-facing slope this is reduced to zero, and for a 30° south-facing slope it is increased to 19 percent. For vertical surfaces in July all orientations received less insolation than the 30° north-facing slope.

Absorption of radiation can have a considerable effect upon slope temperatures. In the Northern Hemisphere south-facing slopes will normally experience higher ground and air temperatures than those of other orientations. Some exceptions to this do occur. Notice from Figure 10.10 that at 30° latitude east- and west-facing slopes should have the highest temperatures in July. Also, keep in mind that the near-surface wind circulation can enhance or moderate these temperature differences. The thermal differences are best developed in calm or nearly calm air, and minimized when circulation is well developed.

### Depressions and Knolls

The amount of outgoing long-wave radiation with a clear sky is at a maximum to the zenith, falling to about 95 percent at a 45° angle of elevation, 50 percent at 10°, and zero at 0°. This has important implications for the radiation balance in depressions and on knolls, even those of very small dimensions. These topographic variations also affect the time of real sunrise and sunset in their immediate vicinity.

The redistribution of radiation results in special temperature patterns, as was shown by a study in late August in Germany. The ridges used, oriented north-south, were only 13 to 15 cm (5 to 6 in.) high, separated by 25 to 30 cm (10 to 12 in.), with slopes of about 45°. Near sunrise the top was the coldest spot. Four hours later a point on the eastern slope had reached the highest value, 30°C (86°F), and by mid-afternoon the maximum temperature was on the western slope. Maximum differences, at the same time, approached 11°C (20°F). A similar study on east-

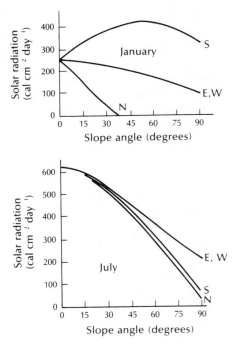

**FIGURE 10.10**
Variations in solar radiation intensity on sloping surfaces at 30° N with different orientations.

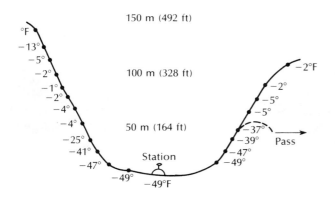

**FIGURE 10.11**
Temperatures on a chosen night in the Gstettneralm frost hollow. (From E. L. Hawke, "Frost Hollows," *Weather* 1 (1946): 41–45.)

west oriented ridges showed little temperature difference from level ground conditions.

In Section 10.2.2 we described fall winds, a particular case of katabatic, or downslope, winds. Foehns are warm katabatic winds, and fall winds are cold. If the descending cold air is trapped, such as in a depression, a frost hollow results. **Frost hollows,** or cold air pockets, occur in any depression, but they are most severe in those that are deep and fairly steep. At times of dry, clear, and calm conditions—such as occur during a stagnant anticyclone—the long-wave radiative loss can be appreciable at night, leading to a partial stagnation of air in the enclosed valley or depression. If cold air drainage (katabatic flows) also occurs, the effect is magnified. The frost hollow temperature pattern is influenced by the local relief, the depth and slope angle relationship, the heat flow from the soil, the reduction of turbulent heat exchange between the cool air at the ground and the warmer air above, and the aforementioned shortening of effective day length.

A famous frost hollow is located in the Austrian Alps at Gstettneralm. The temperatures on the calm, clear, and dry morning of March 31, 1931, are shown in Figure 10.11. Notice that although the temperature at the base of the hollow was −45°C (−49°F), temperatures on the slope were some 27C° (48F°) warmer. The very cold air drains away over an outlet or pass, the lowest point in the hills surrounding the hollow. The extreme cold of the valley bottom compared to the warmer slopes is an important feature to consider when planting cold-sensitive crops and trees.

For example, many fruit orchards are located on intermediate slopes to avoid the relative cold of the lower parts of valleys and depressions. Thus they are away from the area of maximum cold air drainage, yet not so high as to experience the lower temperatures that occur at high elevations. The cold air drainage phenomenon often will cause fog pockets to form as the air is cooled below its dewpoint, but wind movement is sufficient to prevent dew formation. Sometimes this downslope movement of cold air is obstructed by fences, buildings, or embankments.

A similar phenomenon is the **heat hollow.** In this case the depression must be of such dimensions and orientation as to allow much radiation to enter but reduce turbulent exchange with the lower layers of air. Since the hot, low density air will tend to rise, this phenomenon will not be as extreme, except at the soil-air interface, as the frost hollow. The intense heat can have devastating effects on low cover vegetation.

### 10.3.3 VEGETATION

Because of the wide diversity of plant types, we will consider three divisions scaled according to the height of the vegetation. Each division will be treated separately so that we can see its influence upon the local climate. Generally there is great intermingling among the three types. Tall vegetation, over 2 m (6 ft) in height, consists mainly of trees and the taller shrubs. Medium-height plants, from about 0.3 to 2 m (1 to 6 ft) tall, are shrubs and some flowers. The third type is ground or low cover plants, grasses, ivy, clover, and similar vegetation.

#### Tall Vegetation

The main impacts of tall vegetation are on the elements of radiation, air flow, temperature, and humidity. Trees will determine their own patterns of shade, thus altering thermal patterns greatly. These shade areas will change drastically when deciduous trees lose their leaves. In addition, the radiation received within the vegetation has a different spectral distribution from that of the insolation falling on the outer canopy. Radiation is not transmitted uniformly through a leaf. Normally solar radiation reaches a maximum in the yellow-green; however, most of the incident radiation (60 to 80 percent) penetrates a leaf in the near infrared. With a high sun only about 10 percent of the insolation in the visible band penetrates a stand of trees in full foliage. Without foliage the percentage increases to about 40. These percentages tend to increase with the degree of cloudiness. Hardwoods in full leaf exclude as much light as conifers.

Air flow within a dense tree stand is usually extremely light, with the crown acting as a very effi-

**FIGURE 10.12**
Wind flow in a forest clearing.

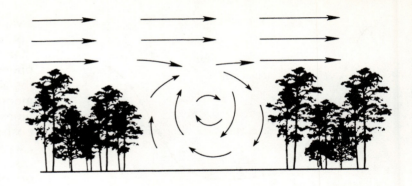

cient frictional barrier. A study in an oak stand in Germany showed that when the trees were in full leaf the average speed through the ground to mid-crown layer was about 15 percent of that in the free air above the top canopy. Before leafing occurred, this value was some 40 percent. In tropical rainforests the speed in the trunk space (volume between the ground and the lower leaves) is about 5 percent of that in the free air. Wind circulation can occur in forest clearings if they are large enough and may exhibit a flow at ground level 180° different from the wind direction above the crowns (Figure 10.12).

Air temperatures within the trunk space show smaller diurnal variations than those in the adjacent open country. Some studies suggest the reduction is only about 1 or 2C° (2 to 4F°) during winter but increases to around 4C° (7F°) in summer, when the foliage is dense. In the crown and treetop area, at the main radiation interface, the diurnal fluctuation is greatest, as it would be at the soil surface in open exposures. Diurnal temperature ranges measured in a fir plantation near Munich, Germany, were 20C° (36F°) at the crown and only 13C° (23F°) at the ground.

Humidity, as measured by the water vapor pressure, shows some fluctuation with height, with two maxima—in the crowns and at the surface. In late afternoon there is a tendency for a minimum in the crown as the stomates close. Relative humidity generally shows only little change with height during the night, although there is a mid-afternoon minimum in the crowns.

Trees cause a definite redistribution of rainfall. Rain patterns that would be reasonably uniform in the open environment become more discrete. Large amounts of rain run down the trunks and much is concentrated around the drip line, the line around the outer perimeter of the foliage. Narrow-leafed (needle) trees give a different pattern from broad-leafed varieties, with less trunk drainage and more rain being held back by the crowns.

### Medium-Height Vegetation

For medium-height vegetation, a most important differentiation is whether the leaves are basically horizontal or vertical. The vertical type includes the *graminacae*, grasses and cereal crops. It is easier for the high sun of summer to penetrate deeply into vertically orientated plants than into those where horizontal leaves dominate. This fact has implications in determining temperature patterns on a fairly calm day within vertical and horizontal vegetation. If the vegetation is tall and dense, then around noon the horizontal type will shade the ground effectively so that the coolest temperature will be found there and the maximum will occur at or near the top of the plants [Figure 10.13(a)]. With vertical growth the sun's rays can penetrate much deeper, so the ground is nearly as warm as the temperature in the lower half of the plant. At night in the horizontal stand the cold air from the leaves cascades downward, and the ground generally

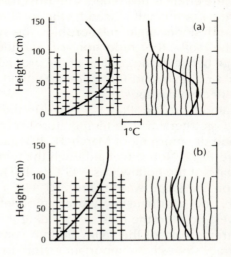

**FIGURE 10.13**
Schematic diagram of temperature patterns during (a) midday and (b) night for horizontal (left) and vertical (right) vegetation.

## SEC. 10.3 | MICROSCALE CONTROLS

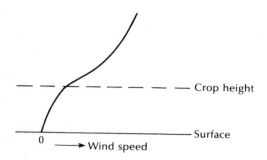

**FIGURE 10.14**
Schematic representation of wind speed distribution within and above a crop canopy.

becomes the coolest area [Figure 10.13(b)]. In the case of vertical growth the minimum is recorded at the effective radiating surface (interface)—within the plants—since the density of vegetation normally prevents the cold air from settling to ground level.

Air flow in medium-height vegetation follows the pattern shown in Figure 10.14. Within the vegetation air movement is slight, especially in the lower region, but then it increases rapidly above the vegetation until, at about two or three times the vegetation height, it approaches the speed of the undisturbed flow.

Within the vegetation there is a release of water vapor, the transpiration from the plants, especially during the daylight hours. Most of the few studies conducted show that around noon the water vapor pressure differential between the air above and the air within the canopy is at a maximum. During the hours of darkness the difference is very small. The daily pattern of relative humidity is approximately the same within and above the canopy, but values are greater in the vegetation.

### Low-Growing Vegetation

The radiation penetrating the leaves of ground-cover vegetation will have its spectral intensity altered radically (see Section 13.2), becoming very rich in the yellow-green. Some vegetation that forms a very dense cover, such as clover (with its horizontal leafing), effectively cuts out all light from reaching the soil.

Air flow within the vegetation depends greatly upon the foliage density, but, from a practical viewpoint, there is little effect upon the wind movement at a small height above the plants.

Temperature patterns through the vegetation can show some rapid alterations in small height changes.

Again the orientation of the leaves—horizontal or vertical—is most important, as is the leaf density. Generally conditions approximate those over bare ground, but the temperature range is not as great. Low vegetation thus acts very much like a mulch (see Section 13.7.2).

### 10.3.4 CONSTRUCTIONS

As with vegetation there are various scales of construction (artificial changes). We will consider only four in this text: fences, single buildings, building complexes, and cities. The latter two are discussed in Chapter 14.

#### Fences

Air flow around a fence of height $H$ is dependent upon the degree of openness. A solid fence, while

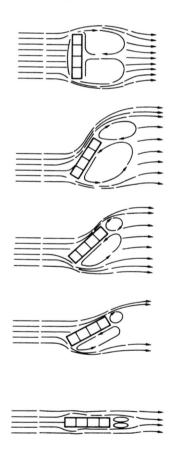

**FIGURE 10.15**
Air flow around a rectangular building. (From B. H. Evans, "Natural Air Flow Around Buildings," Texas Engineering Experiment Station Research Report No. 59. College Station: Texas A&M University, 1957.)

inducing almost calm conditions immediately in its lee, affects air flow only some 10 to 15$H$ downwind. However, the most efficient types have about 50 percent openness and significant effects can be distinguished at a distance of 25$H$ (see Figure 13.4). The presence of a fence mainly has an effect on the air flow, but it will also act as a shading device, altering both radiation and temperature patterns. Stone or brick walls can absorb much radiation, retaining heat that is later reradiated.

Studies of shelter belts, snow fences, or windbreaks are quite numerous and much information has been obtained concerning their effect on the climatic elements in their vicinity (see Section 13.7.3).

### Single Buildings

The major impacts of single buildings upon the microscale climate are through the changes they induce in the air flow and the radiation distribution (shading effects). Alterations in air flow, such as channeling or sheltering, modify the amount of air mixing, creating or dampening turbulence. In this way many of the values of the climatic elements are redistributed.

Figure 10.15 shows the types of disturbance to the air flow that occur with even a single, simple (rectangular) building. It is clear that the pattern alters greatly with changes in the wind direction, so the constantly varying wind direction creates a complex composite picture. With buildings of other shapes the patterns are even more complicated.

The shading caused by a house can be calculated quite simply by use of a solar chart or sun path diagram (showing the sun's altitude and azimuth by latitude, day, and time) and a shadow angle protractor. The different radiation patterns play important roles in determining the temperatures around the house. These temperatures are also changed from what they would be in the absence of a building due to the absorption, reflection, and reradiation caused by the materials.

Humidity patterns also are disturbed by the redistribution of precipitation around the building. The moisture is concentrated along the drip line of the eaves and to a greater degree at the downspouts when guttering is used.

**SUGGESTED READING**

GEIGER, R. *The Climate Near the Ground.* Cambridge, Mass.: Harvard University Press, 1965.

OKE, T. R. *Boundary Layer Climates.* London: Methuen & Co., Ltd., 1978.

ROSENBERG, N. J. *Microclimate: The Biological Environment.* New York: John Wiley & Sons, 1974.

# 11 CLIMATIC CHANGES

INTRODUCTION
ASSESSING PAST CLIMATES
THE PRECAMBRIAN ERA
THE PALEOZOIC ERA
THE MESOZOIC ERA
THE TERTIARY PERIOD
1.8 MILLION YEARS BEFORE PRESENT TO 100,000 YEARS BEFORE PRESENT
100,000 TO 10,000 YEARS BEFORE PRESENT
10,000 TO 1,000 YEARS BEFORE PRESENT
THE LAST 1000 YEARS
THE LAST 100 YEARS
THE SUSPECTED CAUSES OF CLIMATIC CHANGE
Solar Output
The Sun-Earth Path
Atmospheric Composition
Surface Changes
MODELING THE CLIMATE
FORECASTING FUTURE CLIMATES

## 11.1
### INTRODUCTION

Climate is not a static feature of the environment—it is almost as fickle as the weather. With the excellent modern system of communication we are soon apprised of conditions almost anywhere on the globe, and climatic anomalies have had their share of publicity. But we must not consider our present concerns with possible climatic change as something new. The French Secretary of State for the Interior wrote to all the provincial mayors, "For several years we have witnessed atmospheric cold spells, unexpected variations in the seasons, severe storms and extraordinary flooding." He "invited" the mayors to begin research into these anomalies. The year? 1821.

In 1896 the report of the chief of the U.S. Weather Bureau stated,

> The extraordinary period of drought which reached its culmination in the autumn of 1895 created a feeling of apprehension in many localities in regard to the stability of climatic conditions over a large extent of territory. A feeling of unrest was also created by the attempt to show that the changed conditions were a result of man's agency in the breaking up and the cultivation of the soil. In order to meet the call for information on the subject a brief study of the rainfall records collected and preserved in the files of the Weather Bureau was made. It was shown clearly from the investigation made that periods of alternating wet and dry weather were characteristics of the seasons forty and fifty years ago, and that there was no general law governing the recurrence of years of drought or abundant rainfall.

Perhaps the first phenomenon in recent years that lent credence to the idea of changing climate was the extreme drought in the southern fringe of the Sahara, the Sahel, in the early 1970s. Although this region had experienced more severe droughts in the past (as from 1913 to 1917), they had not been so widely publicized. Next came the bad wheat harvest in the USSR in 1972; the unusual freeze in Brazil in 1974 that destroyed much of the coffee crop; very limited monsoon rains over the Indian subcontinent; the very hot and dry summers of northwestern Europe in 1975 and 1976; the severe winters of 1976–77, 1977–78, and 1978–79 in large areas of the United States, especially east of the Rockies; and the torrential rains associated with the retreat of the 1978 monsoon in India and Southeast Asia.

**FIGURE 11.1**
Estimates of the changes in prevailing surface air temperature (°C) since 1870 for the whole earth; latitudes 0° to 80° N (averages); and latitudes 0° to 60° S (averages). (From H. H. Lamb and H. T. Mörth, "Arctic Ice, Atmospheric Circulation and World Climate," *The Geographical Journal* 144(1978): 17.)

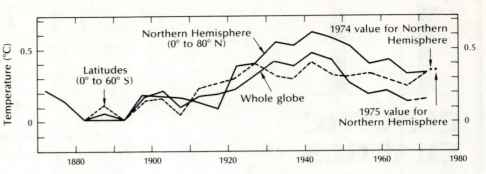

Also adding weight to the climatic change idea were the investigations during the 1960s which indicated there had been a cooling of the land areas in the mid- and high latitudes of the Northern Hemisphere by a few tenths of a degree beginning in the early 1940s (Figure 11.1). These well-publicized variations have caused scientists and many others to become extremely interested in the subject of climatic change and its implications.

These changes are actually an inherent feature of the atmosphere, and some in the scientific community have been aware of this characteristic for many decades. The idea of an inconstant climate is reinforced when climatologists refer to the average value of an element, such as temperature, over a 30-year period as a *climatic normal*. This implies that fluctuations from this value are abnormal when, in fact, they are characteristic (Figure 11.2).

In the same year that the French Secretary of the Interior wrote to the provincial mayors, Ignatz Venitz, an engineer, used geomorphological evidence to suggest that the Alpine glaciers were actually remnants of

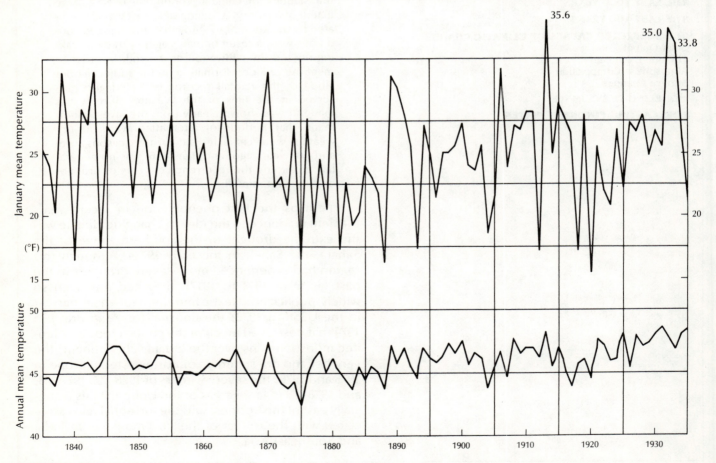

**FIGURE 11.2**
Mean temperature at Blue Hill Observatory, Boston, Massachusetts, from 1835 to 1935.

more extensive ice fields. This was a shattering thesis because previously scientists had assumed climate at a site was constant. Using hindsight, it is difficult to imagine how this concept could have prevailed. In the early years of the earth's formation—with a different type of atmosphere, violent changes of landforms, and the gradual development of soil and vegetation—climate was subject to varying controls and would have had to change. Now we are realizing that apparently everything is in a state of continuous flux, except what we consider to be the natural laws (and we may find surprises here also). Continents are not static, the atmosphere alters in composition, and the solar output is not constant—why should climate not change?

A problem that faces us as we study climatic variations is one of semantics. This is a difficulty because climate appears to be varying on all time scales. H. E. Landsberg has suggested four basic designations, related to duration, for climatic variations:

1. Climatic revolution: over $10^6$ years.
2. Climatic change: $10^4$ to $10^6$ years.
3. Climatic fluctuation: $10^1$ to $10^3$ years.
4. Climatic iteration: less than 10 years.

The revolutions could be caused, for example, by continental drift; the changes, by the earth's orbital variations; the fluctuations, by solar emission changes or deep-ocean circulation; and iterations, by terrestrial or air-sea interactions. However, in this text we shall use the general terms of *climatic variation* and *climatic change* without implying any time scale, unless this is specified. Let us look first at how we try to estimate what the climate has been.

## 11.2
### ASSESSING PAST CLIMATES

Many methods have been used to assess past climates. For climates of the distant past (more than about 500 million years ago) the most useful techniques are via a study of landforms, such as the evidence or absence of glaciation or the incidence of coastal or river terraces, and through deductions based on the location of the various landmasses. Earlier ice ages are usually identified from landform and sedimentary evidence, but these features are not always easy to interpret and analyze correctly. However, rocks older than about 2.8 billion years generally tend to be metamorphosed to such an extent that the identification of glacial origin becomes unlikely. Tectonic, topographic, and paleomagnetic observations are helpful in determining the position of the continents at various times. Figure 11.3 shows the estimated continental positions about 340 million years before the present time, which is written as 340 MYBP. Of course, these are only best estimates of positions; later information may cause modifications. The latitudinal drift of various sites is shown in Figure 11.4. It is

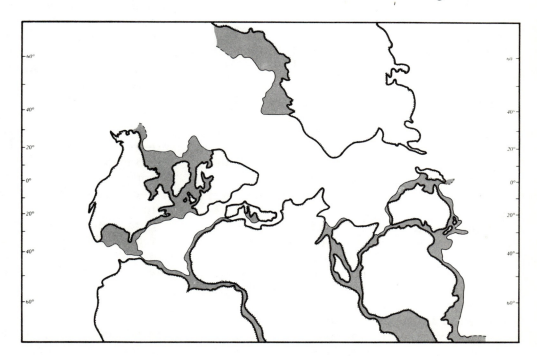

**FIGURE 11.3**
Location of continents about 340 ± 30 MYBP. The edges of the present continents are shown together with the 1000-m depth contour where necessary to show the linkage of landmasses. (From A. G. Smith, J. C. Briden, and G. E. Drewry, "Phanerozoic World Maps," In *Organisms and Continents Through Time*, ed. N. F. Hughes. London: The Palaeontological Association, 1973.)

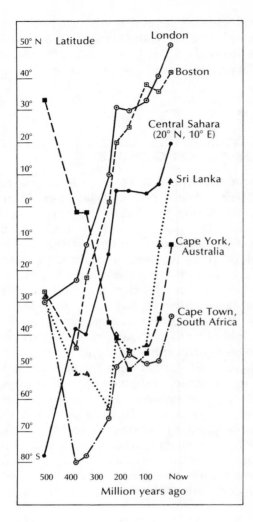

**FIGURE 11.4**
Latitudinal changes at six locations during the past 500 million years. (From R. E. Newell, "The Earth's Climatic History," *Technology Review*, December 1974, pp. 30–45.)

clear that all these sites, as with almost any other location, should have experienced drastic climatic changes. Some locations, especially the Central Sahara and the region of the present Cape Town, South Africa, experienced little latitudinal change from 150 to 220 MYBP, whereas others, such as the area around what we identify now as Boston, have been subject to pronounced and continuous changes.

Other forms of geomorphological evidence include weathering patterns of rocks, deposits of windblown (aeolian) sands, studies of sea and lake levels, salt deposits (occurring at times of rapid evaporation), and the investigation of varves. **Varves** are the deposits on the floor of glacial lakes. During a warm (summer) period a light-colored layer of silt settles from suspension when surplus water from ice- and snowmelt yields a thick layer of sediments, while in winter a darker layer of finer, often organic, material is laid down. The constitution and color of these deposits indicate temperature and rainfall patterns. This technique has been particularly useful in Alaska and Scandinavia.

Another method for estimating past climate uses the **uniformitarianism principle.** This concept states that if a particular biological or geological situation exists under known climatic conditions at the present time, then its existence in the past predicates the idea that climatic conditions were then as they are now. In other words, there is uniformity with time. Some of the most publicized methods of climatic assessment use this technique: pollen studies (palynology), dendroclimatology (tree-ring analysis), and oxygen-18 isotopic analysis. The tree-ring method is only effective over the past 10,000 years—the Holocene period—and then only in certain areas.

**Palynological evidence** uses pollen counts taken from core samples of sediments from lakes or bogs. The changes in pollen-type distribution with depth in the core can give a measure of the climatic changes. Under the best conditions this method can be used to assess climate as far back as 200,000 to 300,000 years before present.

**Dendroclimatology** relates tree-ring width to the weather conditions. For most species of trees the width of the annual growth rings varies with weather and numerous other conditions in a very complex and interrelated manner. Some idea of the complexity is shown in Figure 11.5, where the aspects of low precipitation and high temperature during the growing season are shown to lead eventually to the growth of only a narrow ring in trees in arid areas. In the lower right-hand portion of the figure, the note "From B" links the effects of low precipitation and high temperature occurring prior to the growing season. This latter relationship is as complicated as that shown in Figure 11.5. Actually, corrections to Figure 11.5 are necessary because the annual or seasonal radial increase decreases with age. Different species of trees give information on different periods or seasons; some emphasize the effects of temperature, others those of precipitation. The most useful information is deduced from trees under environmental stress, for these are more susceptible and responsive to climatic or weather changes. The bristlecone pine, a very slow-growing tree, is an excellent indicator and has been used to extend climatic records back some 8000 years.

Recently, **isotopic analysis** of the wood within tree rings (using the ratio of deuterium to hydrogen in the wood cellulose) has indicated that temperatures at the time of maximum recent glaciation (22,000 to 14,000

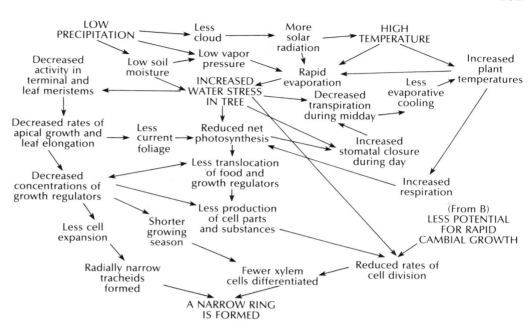

**FIGURE 11.5** Climate–tree ring relationships. [Reproduced with permission from H. C. Fritts, *Tree Rings and Climate*. Copyright © 1976 by Academic Press Inc. (London) Ltd.]

years before present) in some of the ice-free sections of the United States were milder than at present. If this fact is verified, it could be another indication that not all regions of a continent experience similar climatic changes concurrently.

Cores from the ocean bottoms have been used to examine possible temperature patterns, with the time scale obtained from radioactive dating and other methods. In cores of ice or sediments from the ocean bed, the ratio of the heavy oxygen isotope ($^{18}O$) to the lighter isotope ($^{16}O$) is measured. A standardized form of this ratio, $\delta^{18}O$, is related to temperature in an interesting manner. Because the vapor pressure of $H_2^{16}O$ is greater than that of $H_2^{18}O$, atmospheric water vapor exhibits a smaller ratio of heavy isotopes than the water from which it has evaporated. As the amount of ice on a continent increases, the oceans become isotopically heavier and there is an enrichment of $^{18}O$ in the calcite content of ocean foraminifera. The cooler the air, the lighter is the rain, so a measured $\delta^{18}O$ can be converted to a corresponding temperature. A very large $\delta^{18}O$ indicates a cold climate. It is estimated that the complete range of $\delta^{18}O$ from the glacial to interglacial period is about 0.22 percent. Among the findings from such studies is the fact that glacial buildup developed slowly but disappeared much more rapidly. In addition, the $\delta^{18}O$ of the glacial ice is clearly a function of the temperature at the time the snow occurred and began to form the ice sheet. We shall return to this topic and some important related findings in Section 11.7. A summary of many paleoclimatic data sources is shown in Table 11.1.

Other methods using written records (legal documents, diaries, ships' logs, harvest data, and phenological information) have also increased our knowledge of more recent climatic conditions, especially in northwestern Europe. Instrumental records only apply over the past few hundred years, and for the first hundred or so of these a degree of circumspection is essential.

To help you appreciate the time scale with which we are dealing, the age of the earth, 4.6 billion years, will be represented by one year so that one second is equivalent to about 146 years.

## 11.3  THE PRECAMBRIAN ERA

The Precambrian Era comprises about seven-eighths of the earth's life span, and in our concentrated time it covers the period from January 1 to November 16. Very little is known of the climate of this entire period, during which cooling processes began, cataclysmic upheavals occurred, and the earth's atmosphere began to form and evolve in the late Precambrian into something approaching what we have now. However, there were some major differences in the atmosphere, such as little ozone to filter out the intense ultraviolet radiation and perhaps an absence of carbon dioxide.

In the next few sections we will use the names given to various geological eras, periods, and epochs.

**TABLE 11.1**
Characteristics of paleoclimatic data sources.

| Proxy Data Source | Variable Measured | Continuity of Evidence | Potential Geographical Coverage | Period Open to Study (YBP) | Minimum Sampling Interval (yr) | Usual Dating Accuracy (yr) | Climatic Inference |
|---|---|---|---|---|---|---|---|
| Layered ice cores | Oxygen isotope concentration, thickness (short cores) | Continuous | Antarctica, Greenland | 10,000 | 1–10 | ±1–100 | Temperature, accumulation |
| | Oxygen isotope concentration (long cores) | Continuous | Antarctica, Greenland | 100,000+ | Variable | Variable | Temperature |
| Tree rings | Ring-width anomaly, density, isotopic composition | Continuous | Mid-latitude and high-latitude continents | 1,000 (common) 8,000 (rare) | 1 | ±1 | Temperature, runoff, precipitation, soil moisture |
| Fossil pollen | Pollen-type concentration (varved core) | Continuous | Mid-latitude continents | 12,000 | 1–10 | ±10 | Temperature, precipitation, soil moisture |
| | Pollen-type concentration (normal core) | Continuous | 50° S to 70° N | 12,000 (common) 200,000 (rare) | 200 | ±5% | Temperature, precipitation, soil moisture |
| Mountain glaciers | Terminal positions | Episodic | 45° S to 70° N | 40,000 | — | ±5% | Extent of mountain glaciers |
| Ice sheets | Terminal positions | Episodic | Mid-latitude to high latitudes | 25,000 (common) 1,000,000 (rare) | — | Variable | Area of ice sheets |
| Ancient soils | Soil type | Episodic | Lower and mid-latitudes | 1,000,000 | 200 | ±5% | Temperature, precipitation, drainage |
| Closed-basin lakes | Lake level | Episodic | Mid-latitudes | 50,000 | 1–100 (variable) | ±5% ±1 | Evaporation, runoff, precipitation, temperature |
| Lake sediments | Varve thickness | Continuous | Mid-latitudes | 5,000 | 1 | ±5% | Temperature, precipitation |
| Ocean sediments (common deep-sea cores, 2–5 cm/1000 yr) | Ash and sand accumulation rates | Continuous | Global ocean (outside red clay areas) | 200,000 | 500+ | | Wind direction |
| | Fossil plankton composition | Continuous | Global ocean (outside red clay areas) | 200,000 | 500+ | ±5% | Sea-surface temperature, surface salinity, sea-ice extent |
| | Isotopic composition of planktonic fossils; benthic fossils; mineralogic composition | Continuous | Global ocean (above $CaCO_3$ compensation level) | 200,000 | 500+ | ±5% | Surface temperature, global ice volume; bottom temperature and bottom water flux; bottom water chemistry |
| (rare cores, >10 cm/1000 yr) | As above | Continuous | Along continental margins | 10,000+ | 20 | ±5% | As above |
| (cores, <2 cm/1000 yr) | As above | Continuous | Global ocean | 1,000,000+ | 1000+ | ±5% | As above |
| Marine shorelines | Coastal features, reef growth | Episodic | Stable coasts, oceanic islands | 400,000 | — | ±5% | Sea level, ice volume |

SOURCE: Reproduced from *Understanding Climatic Change* (1975) with the permission of the National Academy of Sciences, Washington, D.C.

Table 11.2 lists these names and includes an idea of the time intervals and some remarks on important phenomena. Because landmasses are likely to have floated around, becoming supercontinents and then splitting apart, evidence of climatic conditions during much of this period is mainly deduced from glaciations and landforms. The earliest of the glacial or ice ages, for which there is the strongest evidence, are the Sturtian (750 MYBP) and the Varangian (675 MYBP). The glaciation can be identified in Australia, Central Africa, Eastern Asia, Greenland, and Scandinavia, with some indications of more than a single ice age. Other ice ages are recognized at 950 MYBP (the Gnejsö, in Greenland and Scandinavia) and at 2300 MYBP (the Huronian, evidenced by Canadian deposits).

# 11.4

**THE PALEOZOIC ERA**

In our concentrated time scale we are now in the period between November 16 and December 12. It has been suggested by some scientists that during the Paleozoic Era and the Precambrian Era there has been

**TABLE 11.2**
Geologic time scales.

| Era | Period | Epoch | MYBP | Remarks |
|---|---|---|---|---|
| PRECAMBRIAN | | | | Little definite climatic information. Late Precambrian glacial around 6 or 7 × 10⁸ yr before present. |
| | | | 570 | |
| PALEOZOIC | Cambrian | | | |
| | | | 500 | |
| | Ordovician | | | Often called older Paleozoic. |
| | | | 440 | |
| | Silurian | | | |
| | | | 390 | |
| | Devonian | | | Glaciation widespread. Rise of land plants. |
| | | | 340 | North America and Eurasia drifting towards Gondwana. First reptiles. |
| | Carboniferous | | | |
| | | | 270 | Permo-Carboniferous. Ice age. |
| | Permian | | | |
| | | | 230 | |
| MESOZOIC | Triassic | | | First mammals appear. Pangaea, a single supercontinent, exists. |
| | | | 190 | |
| | Jurassic | | | |
| | | | 140 | Northern landmass (Laurasia) splitting from Gondwana. |
| | Cretaceous | | | Warm climate. No ice caps. |
| | | | 65 | North America and Europe still joined; also Australia and Antarctica. |
| CENOZOIC | Tertiary | Paleocene | | |
| | | | 54 | |
| | | Eocene | | Modern mammals appear. Development of circum-Antarctic current. |
| | | | 36 | |
| | | Oligocene | | |
| | | | 22 | |
| | | Miocene | | Glaciation on Alaskan Mtns. |
| | | | 5.5 | Western Antarctic ice sheet. Australopithecus appears. Eastern Antarctic ice sheet is very large. |
| | | Pliocene | | |
| | | | 1.8 | |
| | Quaternary | Pleistocene | | Primitive humans appear. |

a cycle of about 250 million years in climatic patterns. This cycle could be related to the **cosmic year**, the period of revolution of the solar system around the center of gravity of the Milky Way, which is about 225 million years duration. During its orbit through the galaxy the solar system traverses regions of interstellar plasma or particles of varying overall density which could cause fluctuations in the intensity of solar radiation received at the top of the earth's atmosphere and trigger certain climatic controls. The concept is somewhat similar to that proposed by certain meteorologists who suggest that annually recurrent mete-

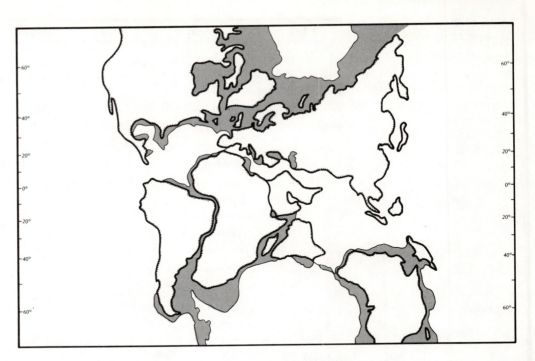

**FIGURE 11.6**
Location of continents about 100 ± 10 MYBP. The edges of the present continents are shown together with the 1000-m depth contour where necessary to show the linkage of the landmasses. (From A. G. Smith, J. C. Briden, and G. E. Drewry, "Phanerozoic World Maps." In *Organisms and Continents Through Time*, ed. N. F. Hughes. London: The Palaeontological Association, 1973.)

oric showers can cause periods of increased rainfall.

Although there was widespread glaciation in the Devonian Period, the most intense glaciation since the last Precambrian ice age occurred in the Permo-Carboniferous time, about 260 MYBP (corresponding to about December 10). This intensely cold spell must have affected the lives of the small-brained reptiles of all sizes that had then evolved. Recent evidence has begun an intriguing controversy about whether some of these animals were actually warm-blooded mammals. From around 250 MYBP a rapid (from the viewpoint of the geological time scale), steady warming occupied the next 50 million years.

## 11.5
### THE MESOZOIC ERA

On our special time scale the Mesozoic Era occurs between December 12 and 26. As is to be expected, when we get closer to the present, more information or clues become available, especially knowledge of the land-sea configuration (see Figures 11.3 and 11.6). In the Triassic Period the Pangaean supercontinent existed, and temperatures were high enough and land positions sufficiently equatorward so that no ice caps were present. At the end of the Jurassic Period, as the landmass of Laurasia began splitting from Gondwana, there was an apparent slight cooling. However, the whole era is characterized, from present evidence, by high temperatures in lands that are now in northern latitudes. There is relatively good agreement among researchers about a cool spell at about 70 MYBP (around December 25) in the Cretaceous Period, and some believe another occurred at 90 MYBP (December 24).

## 11.6
### THE TERTIARY PERIOD

On our compressed time scale, the Tertiary Period (the earliest period of the Cenozoic Epoch) extends from December 26 to about 8:30 P.M. on December 31. During much of this time the continents had a basically meridional (north to south) configuration, and it is thought that no circumpolar ocean current existed in either hemisphere. However, as Pangaea began to break apart, the Indian Ocean widened and deepened so that by about 30 MYBP the Antarctic circumpolar current was established.

Geologic evidence indicates that the long cooling trend, known as the Cenozoic climate decline, began about 55 MYBP, although there have been some relatively short-lived warming spells. Evidence also suggests that around 35 MYBP substantial cooling of the Antarctic waters occurred. In the Oligocene Epoch (22 to 36 MYBP) the global climate was cooling, and ice reached the edge of the Antarctic continent in the region of the Ross Sea.

In the early Miocene Epoch (15 to 20 MYBP) there is evidence that the lower and middle latitudes were quite warm, but that the higher latitudes of the South-

## 11.7
### 1.8 MILLION YEARS BEFORE PRESENT TO 100,000 YEARS BEFORE PRESENT

The time from 8:30 P.M. to 11:49 P.M. on December 31 on our concentrated time scale covers the period from 1.8 MYBP to 100,000 years before present (YBP). Some of the methods for estimating climates over this period are the isotopic composition ($^{18}O/^{16}O$ ratio) of Pacific Ocean plankton from deep-sea cores, the chemical composition of calcium carbonate in an equatorial Pacific core (Figure 11.7), the composition of plankton from the Caribbean foraminifera, and sequences of soil types in Czechoslovakia.

From around 1.0 to 0.7 MYBP these techniques indicate some pronounced, but not always synchronous, fluctuations. However, from 700,000 to 450,000 YBP the records tend to converge, and from about 450,000 to 100,000 YBP they indicate very similar variations in all areas. The evidence suggests that during this period ice ages have occurred at approximately 100,000-year intervals, with smaller (more irregular, but perhaps equally important) interglacial intervals superimposed on this pattern.

A pioneer in the study of this particular period was Yugoslavian meteorologist Milutin Milankovitch. In the 1920s he used knowledge of variations of earth-sun geometry to calculate the solar radiation incident at the top of the atmosphere for certain latitudes in both the Northern and Southern Hemispheres. Recently more refined calculations have extended the data from 4 MYBP to 120,000 YBP. A sample pattern is given in Figure 11.8. Such data have definite bearing on the global climate, although their complete significance has still to be determined.

Recently a record of ocean cores covering the past 450,000 years has been analyzed, and it has been found that most of the fluctuations (85 percent of the variance) can be related to cycles of three lengths: 100,000, 42,000, and 22,000 years. The dominant 100,000-year component (accounting for 50 percent of the variance) is in phase with, and of a period very close to, the cycle of changes in the **eccentricity** of the earth's orbit, or the amount the path deviates from a circle [Figure 11.9(a)]. The 42,000-year period (25 percent of the variance) is the same length as the cycle of **obliquity,** or tilt, of the earth's axis [Figure 11.9(b)], which varies from 21.8° to 24.4°. The 22,000-year cycle (10 percent of the variance) has about the same periodicity as the **precession of the equinoxes,** which is related to the wobble of the earth's axis and the change in the day on which the earth is closest to the sun—the **perihelion** [Figure 11.9(c)]. The perihelion is now January 5.

From this strong evidence, although based on data from only two cores, it is concluded by the investigators that changes in earth-sun geometry are the fundamental causes of the succession of Quaternary ice ages (Figure 11.10). This idea was first proposed by J. F. Adhemar in 1842, refined in 1860 by James Croll, and then saw a resurgence of interest in the early

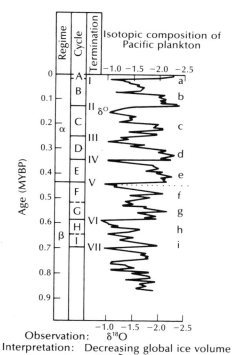

**FIGURE 11.7**
Climatic record of the last 1,000,000 years. This oxygen-isotope curve of a Pacific deep-sea core is interpreted as reflecting global ice volume. The relatively rapid and high-amplitude fluctuations indicate sudden deglaciations and are designated as the terminations I to VII. These occur at approximately 10,000-year intervals. [Reproduced from *Understanding Climatic Change, A Program in Action* (1975), p. 159, with the permission of the National Academy of Sciences, Washington, D.C.]

**FIGURE 11.8**
Calculated radiation distribution from 20,000 years before to 119,000 years after 1950 A.D. Values (in ly/day) are of differences from the 1950 A.D. standard values. (From A. D. Vernekar, "Long-Period Global Variations of Incoming Solar Radiation," *Meteorological Monographs* 12 (1971): 1–21.)

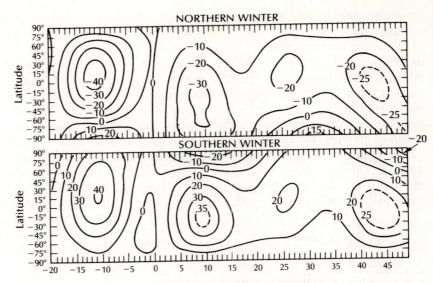

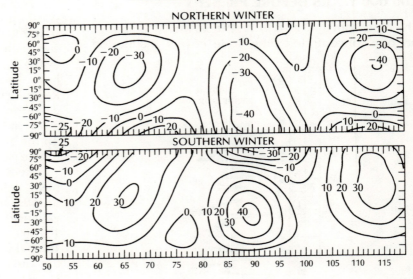

twentieth century. Extrapolation from these findings (and others in Sections 11.8, 11.9, and 11.10) suggests that the long-term trend over the next few thousand years is in the direction of colder winters and, in higher latitudes, increasing glaciation of the Northern Hemisphere. However, such a deduction does not take into account any anthropogenic effects that may halt or even reverse this trend. The most recent studies, using cores covering 2 million years, identify a 413,000-year cycle that is also related to a periodicity in the eccentricity of the earth's orbit. It must be realized that the variations in the earth's orbit are caused by perturbations in which all the other planets play a role. These variations are not of a simple cyclic nature.

## 11.8

### 100,000 TO 10,000 YEARS BEFORE PRESENT

We are now up to 11:59 P.M. on December 31 on our shortened time scale. It appears that around 100,000 YBP there was a very rapid warming, culminating in a global temperature as high as any in the past million years. Then, about 75,000 YBP and from 22,000 to 14,000 YBP there were intense glacial maxima, which were sometimes abruptly terminated. After 14,000 YBP, there was a warming trend that culminated around 6000 YBP [Figure 11.11(a)], but in northern Europe there was an advance of glaciers at 11,000 YBP.

**FIGURE 11.9**
Important components of earth-sun geometry.

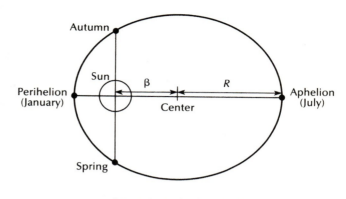

(a)

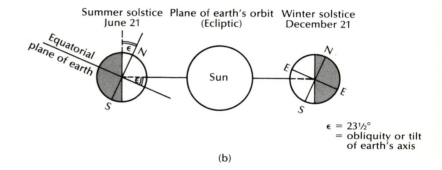

(b)

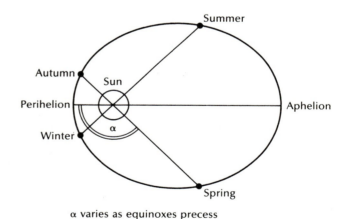

(c)

As more data become available for study from deep-sea cores and lake or bog deposits, it appears that the patterns of ice accumulation and decline are not contemporaneous for the various ice sheets; in fact, they vary around the Northern Hemisphere by some 8000 years. Because of this complexity and the fact that the Milankovitch curves show low extraterrestrial radiation during summertime in the Northern Hemisphere, the period beginning about 18,000 YBP has been selected for special study by scientists involved in the CLIMAP project. CLIMAP (Climate: Long-range Investigation Mapping and Prediction) is an investigation by a multi-institutional consortium of scientists who study long-term climatic changes and patterns. The basic findings so far suggest that at 18,000 YBP there were huge ice sheets over some land areas which reached a thickness of 3 km (2 mi). Forests gave way to grasslands or deserts, while ocean temperatures poleward of 40° N and S and in the central Pacific were some 4 to 8C° (7 to 14F°) cooler than at the present time. July air temperatures showed a decrease of 5 to 11C° (9 to 20F°) compared with current conditions along most latitudes; but some areas (e.g., northern India) were 25C° (45F°) cooler, while a few regions were actually warmer than now. It has been estimated that around $44 \times 10^6$ km² ($18 \times 10^6$ mi²) of land—nearly three times the present amount—were ice covered at the glacial maximum. This was sufficient to lower the ocean about 140 m (450 ft) compared with today's level, for about 9 percent of all the earth's water was then locked in glaciers or ice packs.

## 11.9
### 10,000 TO 1000 YEARS BEFORE PRESENT

The last 10,000 years are referred to as the Holocene Period; climatically, they correspond generally to a warmer spell. Evidence for this has been deduced from palynological studies, oxygen-isotope studies, sea-level records, and tree-ring research (see Section 11.2). By 7000 years ago most of the ice conditions had reached their present state, but growing evidence indicates that there were in fact several quite widespread cooling and warming periods within the early Holocene [Figure 11.11(b)]. For example, there was a

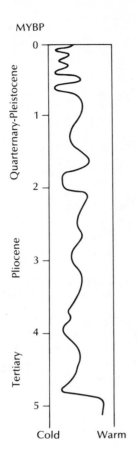

**FIGURE 11.10**
Generalized temperature trends for the past 5 million years based mainly on deep-sea sediment cores. (From J. Gribbin, ed., *Climatic Change*. New York: Cambridge University Press, 1978.)

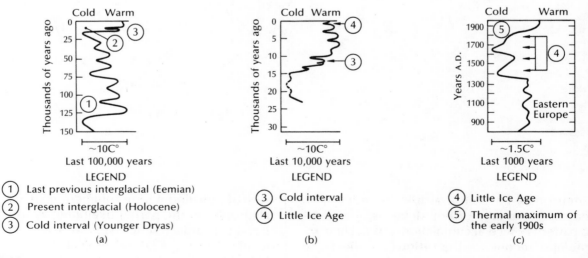

**FIGURE 11.11**
Mid-latitude temperature changes during the last (a) 100,000 years; (b) 10,000 years; and (c) 1000 years. [Reproduced from *Understanding Climatic Change* (1975), p. 130, with the permission of the National Academy of Sciences, Washington, D.C.]

glacial advance (named the Younger Dryas event) in Europe around 10,000 YBP that established itself in a century or less and lasted for approximately 700 years—a most abrupt climatic phenomenon.

The period from 7000 to 5000 years ago was marked by temperatures warmer than those existing today, but since then all areas of the world have been characterized by declining temperatures. Many regions experienced especially cold intervals around 5300, 2800, and 350 years ago. This period of warm temperatures, when the Northern Hemisphere was 0.5 or 1C° (1 or 2F°) higher than at present and when extraterrestrial radiation was high during the Northern Hemisphere's summer, is often called the **climatic optimum**. During this time some of the earliest civilizations of which we have record flourished. Indirect records indicate cooler and wetter periods from about 4000 to 3000 years before present, especially in Europe and the Near East, with a warm, dry spell in the western Mediterranean from 2300 to 1600 YBP (300 B.C. to 400 A.D.). However, cold, wet periods then increased in frequency from 1300 to 300 B.C. and from 400 to about 800 or 900 A.D. In 800 to 801 A.D. the Black Sea was frozen, and in 829 A.D. ice formed on the Nile—a phenomenon that occurred again about 200 years later. Some indirect records dating as far back as 3000 B.C. have survived from China. Starting then and lasting for 2000 years the climate appears to have been 2C° (3 to 4F°) warmer than at present. But, around 500 B.C. and again at 150 B.C. and 500 A.D., there were cold periods. On our comparative time scale we are now at about 7 seconds before midnight on December 31.

# 11.10
**THE LAST 1000 YEARS**

During historical times people have used old documents, diaries, crop records (especially those of wheat and grapes), and specific measurements, such as the level of the Nile floods, to assess climatic change. It is only during the past 100 years that reasonably reliable observations representing a large area of the world land surface have been made, although some specific sites have records (generally of precipitation) extending back 200 years or more. One must exercise care because there are some apparently useful observations that can be misleading. For instance, there were many ice fairs held on the river Thames in London during the sixteenth to mid-nineteenth centuries, suggesting perhaps a more benign climate in the last 100 years. However, because the channel of the river has been changed by banking and power plants now feed hot water into the flow, strict comparisons are not possible. Another example is the Yellow River (Hwang Ho) in China, where during the first six centuries A.D. only two bad floods were recorded on the lower reaches. Over 900 floods have occurred since then. But it is not only rainfall that causes floods; bad agricultural practice over the years has led to the silting of the river bed.

By using manuscript and observational records from Europe and tree-ring data from the western United States, it has been shown that the climates of western Europe and North America are reasonably similar with regard to major fluctuations. The early medieval period of 800 to 1200 A.D. has been called the **little climatic optimum** because of the relative warmth of most of the Northern Hemisphere [Figure 11.11(c)]. Conditions were warm, with few severe storms, and during this time Greenland and Iceland were settled and developed. The interval from about 1430 to 1850 A.D., known as the **Little Ice Age,** saw a global temperature decrease of about 1.5C° (3F°). However, evidence suggests that this Little Ice Age began one or two centuries earlier in eastern Asia and reached its coldest level in 1650 to 1700 over most of the Northern Hemisphere. In the Little Ice Age many glaciers increased almost to the limit they reached in the last ice age, about 20,000 YBP. The increased glaciation during the Little Ice Age was due to more meridional circulation than at present, which caused short, wet summers and long, severe, snowy winters. In Europe there were some particularly cold spells between 1300 and 1750 A.D., with famines and severe storms frequently reported. As is found so often in the study of climatic changes, there can be an apparent discrepancy between two areas not far removed. For example, while Europe as a whole had warm conditions from 1100 to 1400 A.D., Iceland had its warm period from about 750 to 1200 A.D. after which ice off the coast began to increase significantly (Figure 11.12). In the late thirteenth century the U.S. Southwest and parts of Central America suffered severe droughts.

An appropriate comment by Nigel Calder points out the complexity of climate change:

> The bald phrase 'Little Ice Age' conceals an enormous variability in both space and time. Abundant historical evidence about the Little Ice Age makes obvious the dangers of vague generalizations about earlier periods—or with future periods, for that matter. Winters in China and Japan were actually milder during Europe's Little Ice Age because weaker winds from the west allowed them to benefit more from the warm Pacific air.*

---

*Nigel Calder, *The Weather Machine* (London: British Broadcasting Corporation, 1974), p. 17.

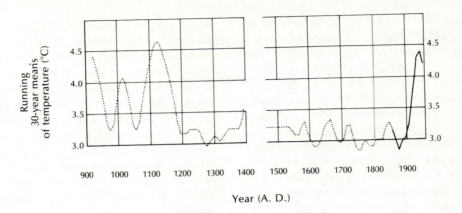

**FIGURE 11.12**
Mean annual temperature in Iceland over the past millennium. The solid line indicates actual values, and the dotted line shows estimated values for the variation of mean temperature. The ten-year mean temperature from 1971 to 1980 is 3.6°C, and the 30-year mean for 1951 to 1980 is 3.9°C, showing a rather definite trend toward the climate before the "hot" period which occurred from 1925 to 1965. [From P. Bergthórsson, "An Estimate of Drift Ice and Temperature in Iceland in 1000 Years," *Jokull*. Vol. 19 (1969): 94-101.]

## 11.11

### THE LAST 100 YEARS

During the last 100 years, which corresponds to the last second of our time scale, the most accurate measurements of the atmosphere have been made. Because many of the long-period stations have been absorbed into or become affected by cities, however, these measurements reflect urban-induced changes that can only be estimated very approximately. Attempts have been made to assess average conditions for the land areas of the Northern Hemisphere by using a number (usually about 80 to 120) of stations, but this method has its own inherent inaccuracies. Nevertheless, records indicate a gradual warming trend from about 1880 to 1940, a decrease from 1940 to 1960 or 1965, and no discernible trend from 1965 to the present (see Figure 11.1). Data for the Southern Hemisphere, less numerous and less representative than those for the Northern Hemisphere, do not show the same fluctuations. Statements such as, "We are living in one of the warmest periods of the past million years," while apparently true, must be qualified by other comments like, "Perhaps the last million years are themselves aberrant."

A problem that further complicates global, or very large-scale, analysis is not all areas report drought, extreme precipitation, high or low temperatures at the same time—something we know if we follow world weather reports. The Great Plains may be experiencing very dry conditions, while the western United States or eastern Europe may be suffering torrential rains. Similarly, there are often pronounced and important differences between seasons, when the deviations from their respective averages are considered. This feature of disharmony or noncontemporaneity can lead to spurious results when dissimilar areas are averaged together. The real problem is, How do we identify or define a homogeneous area? Oklahoma, Kansas, and Nebraska may act as a unit in one year, but in the next it is Arkansas and Texas with which Oklahoma appears to be linked.

## 11.12

### THE SUSPECTED CAUSES OF CLIMATIC CHANGE

We have seen in earlier chapters that the climatic system is extremely complicated. Its linkage to many features is seldom direct or simple, but if we consider its separate components, we can begin to understand the causes of climatic change. A schematic representation of the components of the climatic system is shown in Figure 11.13. The diagram depicts the coupling between land features (soil, snow, ice, urban areas) and the atmosphere, between water surface and the atmosphere, and between the ocean and basin features and the atmosphere. The linkages are shown more specifically in Figure 11.14(a) by the steps

# SEC. 11.12 | THE SUSPECTED CAUSES OF CLIMATIC CHANGE

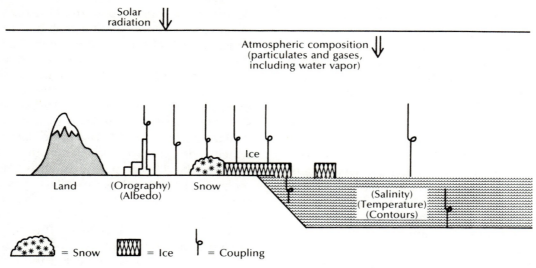

**FIGURE 11.13**
Schematic illustration of the components of the climatic system.

**FIGURE 11.14**
Interrelationships of climatic causes: (a) radiation relationships and (b) effects on the atmosphere.

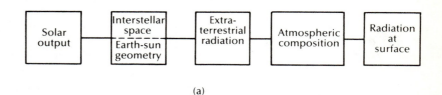

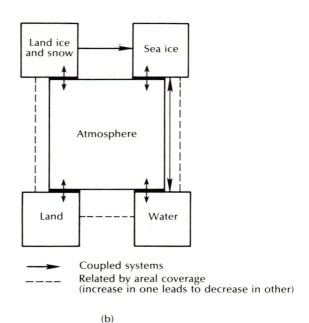

or controls between solar output and the final receipt of radiation at the surface. Figure 11.14(b) illustrates how certain global surface characteristics can affect the atmosphere and thus its composition.

## 11.12.1 SOLAR OUTPUT

In general climatology the tacit general assumption is often made that the solar output is invariable (see Section 2.2.1), although we know that this is not true. For example, apparently there are random changes in the ultraviolet intensity, and sunspots alter, by very small amounts, the solar output in some regions of the solar spectrum. It is also possible that the solar output may alter systematically and/or irregularly over very long periods of time. Measurements of the solar intensity have not been made over enough of a time span or with sufficient accuracy to determine these long-term characteristics.

**Sunspots** are internal eruptions in the sun that are identified only when they reach the surface. Because they are relatively cool, they appear there as dark spots. Observations of their size and number have been made, with varying accuracy, for some 300 years. From these observations the sunspot number is calculated for each day, and variations in the mean monthly and annual value are studied. As is well known, there is an apparent periodicity of about 11 years (the Schwabe cycle) in the pattern of the sunspot number. However, we cannot assume that this 11-year period is constant or that the sunspot number always reaches the same maximum, although the minimum is generally almost zero (Figure 11.15). It can be seen from the pattern in Figure 11.15 that the period between extremes can range from 8 to 17 years—hardly a constant. Emphasis is sometimes placed on the double sunspot cycle of 22 to 23 years (the Hale cycle), during which there are two reversals of the solar magnetic field. In the 11-year period there is one reversal of magnetic polarity. Paradoxically it has been shown that the solar constant, or solar parameter, increases as the annual sunspot number increases from zero to about 60 and then decreases. The maximum difference is about 2 percent.

Recent work has supported the contention that from 1645 to 1715 (called the Maunder minimum after

**FIGURE 11.15**
Annual mean sunspot number from 1610 to 1975. (Copyright © 1961 and 1976 by M. Waldmeier, Swiss Federal Observatory, Zurich, Switzerland. Reproduced by permission.)

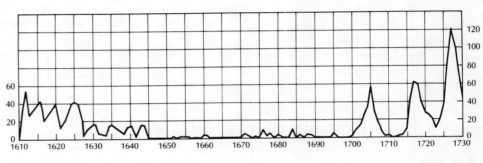
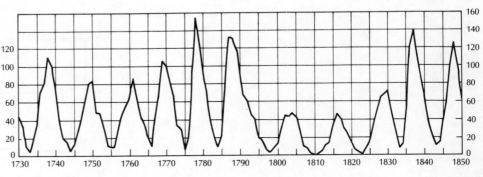
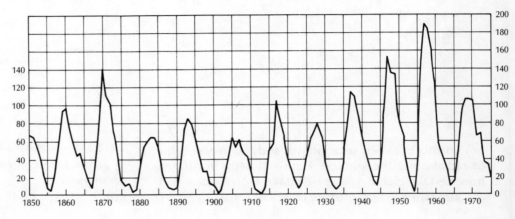

## SEC. 11.12 | THE SUSPECTED CAUSES OF CLIMATIC CHANGE

its nineteenth-century identifier) there were consistently few or no sunspots. Another quiet period for sunspots, the Spörer minimum, spanned the years 1460 to 1550. Also, nitrogen levels (of ammonia and nitrate) in Antarctic ice have been linked to sunspot activity, and preliminary studies suggest this technique may yield data on further and earlier sunspot irregularities. Increasing amounts of information lend proof to the fact that the sun is an inconstant star and may have periodic or aperiodic quirks of which we are presently unaware. In addition, sunspots exhibit a special latitudinal distribution which may have implications for climatic change. That is, a sunspot at solar latitude 30° may have different effects upon the electromagnetic field than one of the same size (intensity) at solar latitude 10°.

### 11.12.2 THE SUN-EARTH PATH

In its journey from the sun to the outer limits of the earth's atmosphere solar radiation undergoes variations in its intensity. The first cause of this variation is the presence of interstellar gases or particles whose density may vary during the cosmic year (see Section 11.3) and the sidereal year (time of one complete revolution around sun from perihelion to perihelion: 365 days, 6 hours, 9 minutes, 9.5 seconds). Second, the parameters of the earth-sun geometry each play a role in determining the radiation receipts. Three of these parameters were described in Section 11.7—the eccentricity of the earth's orbit, the change in the tilt of the earth's axis, and the precession of the equinoxes. To these must be added the shorter period change of the earth-sun distance during the year. This presently varies from 0.983 to 1.017 times the mean distance, but these values alter with the eccentricity of the orbit.

### 11.12.3 ATMOSPHERIC COMPOSITION

Once the solar beam begins to pass through the atmosphere, numerous particles and gases become important. Among the most important of these appear to be the concentrations and distributions of dust, water vapor, and carbon dioxide. Naturally, there are some others—witness the recent chlorofluorocarbons (from spray cans) investigations. It is likely that other problem areas will be identified as knowledge of the physical and chemical composition of the atmosphere develops.

#### Dust

The total global atmospheric content of small particles at any one time is estimated at $4 \times 10^7$ tons, of which about 25 percent is derived from human activities. The term *small* refers to sizes less than 0.5 mm (0.02 in.) in diameter. These particles remain in the atmosphere for long periods, while larger particles are soon deposited out. The accepted, but approximate, mean time that small particles remain in the atmosphere is about one or two weeks. The annual weight of small particles fed into the atmosphere from human activities is about $4.5 \times 10^8$ tons, but a very large percentage of this soon settles out rapidly. These activities include especially chemical reactions (sulfates and nitrates), combustion (fly ash and smoke), and agricultural practices (grass fires, slash-burn, disturbance of natural ground cover).

The major sources of natural small particles are from the sea, windblown dust, chemical processes, and, occasionally of great magnitude, volcanoes. The volcanic input can double the average particle loading (e.g., Krakatoa in 1883, and Gunung Agung in 1963) and result in a definite loss of atmospheric transparency (by about 1.5 to 2 percent), which has been linked with changes in temperature patterns. The connection between temperature over large areas and the atmospheric particulate concentration cannot be determined with much reliability from present data, but a cooling effect is likely as the load increases. Part of the difficulty in estimating the effect is the complexity of the feedback processes. Increased numbers of particles will deplete solar radiation, but this in turn will result in decreased evaporation, which could lead to less cloud and thus allow more solar radiation to penetrate the atmosphere, perhaps offsetting the original direct cooling process. It is suggested that the sand clouds that occur on the west side of Africa reduce the radiation by 100 ly day$^{-1}$, saving 40 cm (15 in.) of evaporation annually from a water body such as the Lac de Guiers in Senegal. Equally important are the radiative properties of the particulate matter in the balance of atmospheric counterradiation. For large particles the absorbed heat can be instrumental in affecting the thermal conditions of the atmosphere, causing a net warming effect.

#### Water Vapor

Although water vapor absorbs primarily in the wavelengths of terrestrial radiation, there is also absorption within the solar beam (see Section 2.2.2). This reduction of solar radiation can be quite appreciable under certain circumstances; for example, the depletion due to 25 mm (1 in.) of precipitable water is about 26 percent of the solar intensity over the whole spectrum when the sun is in its zenith and increases to 37 percent with a solar altitude of 30°. With precipitable water of 6 mm (0.25 in.), the values become 19 percent and 27 percent, respectively. Therefore, when the water vapor content of the atmosphere changes, as it is doing all the time, the solar radiation received at the surface alters in response.

### Carbon Dioxide

Carbon dioxide occurs naturally in the earth-atmosphere system, but the increased burning of fossil fuels since the beginning of industrialization now accounts for one tenth to one eighth of the present level of 330 parts per million (ppm) by volume. Measurements indicate an annual increase in carbon dioxide of around 1 ppm due to the burning of fossil fuels.

When the carbon dioxide level rises, the atmosphere absorbs more of the long-wave (terrestrial) radiation, thereby augmenting atmospheric counter-radiation ($I\downarrow$ in Figure 2.15). This increases the atmospheric (greenhouse) effect and should result in a rise in the overall temperature of the lower atmosphere. It is now thought that, due to the carbon dioxide increase, a warming of 0.6C° (1F°) will occur in the global surface temperature by 2000 A.D.

At high altitude [about 10 km (6 mi)] the effect of the changing carbon dioxide concentration becomes negligible, but since the lower layers are becoming warmer, the lapse rate will increase, leading to greater instability. Some mathematical studies suggest that changing the vertical lapse rate of temperature will cause a relocation of the subtropical anticyclones. Figures used in a substantiation of the argument indicate that a 1 percent increase in the atmospheric greenhouse effect would cause, via the latitudinal shift in the semipermanent subtropical high pressure cells, a reduction in West African rainfall associated with the ITCZ of about 100 mm (4 in.).

However, as with dust, there are complex mechanisms involved. Burning fossil fuels also release more water vapor into the air, which may increase cloudiness and thus reflect more solar radiation back to space. The carbon dioxide content of the atmosphere has been increasing over millions of years, with concomitant thermal changes. It is only in the last few decades that anthropogenic alterations have magnified its effect.

## 11.12.4 SURFACE CHANGES

In Chapter 2 we learned how the nature of the surface (interface) receiving the radiation plays a fundamental role in determining the resultant climate. If this surface is changed, in character or distribution, the climate will alter. In this section we will consider some surface changes on different scales.

### Land-Sea Configurations and Topography

Over periods of less than about 5 million years large-scale changes in continental slopes and positions are negligible. On the geological time scale the movement of landmasses (the continental drift theory of Alfred Wegener first propounded in the 1920s, or the more modern tectonic plate theory) leads to different climates on a continent as that continent moves relative to the equator and the poles (see Section 11.5). The major mountain-forming processes, generally associated with plate movement, also play a role in determining air mass trajectories with related alterations to the climate.

### Ocean Conditions

Another likely cause of a changing climate is the oceans. Air and water are closely coupled, particularly with respect to latent and sensible energy flows. Because of the large disparity between the thermal capacities of water and air (about 3000 to 1), a relatively small change in water surface temperature can lead to much larger changes in air temperature. This means also that a set of conditions within the atmosphere is much more short-lived than a set within water. However, there are eddy patterns within the oceans that lead to a degree of mixing in the upper layers so that thermal conditions some meters below the water surface can eventually influence the climate in the atmosphere above. From MODE (Mid-Ocean Dynamic Experiment) we gained an appreciation of the complexity of these movements; superimposed on the assumed long-period patterns of the ocean (its climatic picture) were short-term, small complicated patterns (comparable with the synoptic or weather picture). It appears that many thermal cells (energy sources or sinks) exist at any time, generally with their movements unnoticed and/or unmeasured. Each of these cells could affect appreciably an overlaying air mass. We have a rapidly growing knowledge of the surface movements of ocean currents and thermal pools, mainly obtained from satellite imagery, but the subsurface characteristics are largely unknown and can appear at the surface quite unexpectedly.

The typical vertical temperature distribution in the deep oceans is shown in Figure 11.16(a) and Figure 4.43. In the surface layer mixing brings about an almost isothermal condition, but below this the temperature decreases rapidly with depth—a feature called a **thermocline.** The thermocline, a stable region, resists vertical movements strongly so that generally the layer tends to insulate the deep water thermally from the atmosphere. In some locations, such as the high latitudes, and because of seasonal variations in the depth and strength of the thermocline [Figure

# SEC. 11.12 | THE SUSPECTED CAUSES OF CLIMATIC CHANGE

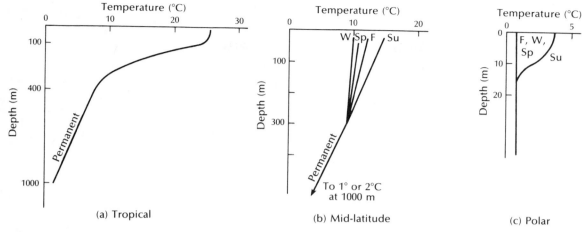

**FIGURE 11.16**
Typical seasonal thermal patterns. Transient thermoclines are not shown.

11.16(b) and (c)], the deep water thermal conditions can sometimes influence the atmosphere directly. The seasonal thermocline is often absent in winter and is most pronounced in mid- to late summer. More than one thermocline may exist at once, in the same way that more than one thermal inversion may exist in the atmosphere. Near the surface the transient thermocline can occur, especially in the warmer seasons, but its temperature differential is small (1 or 2C°) and it rarely persists for more than a week. The seasonal thermocline is deeper and is maximal in late summer, when it is closest to the surface. In tropical latitudes the permanent thermocline, which extends to a depth of about 1000 m (3000 ft), can have a temperature decrease of 20 to 25C° (36 to 45F°).

Movements within the oceans tend to be slow compared with those in the air. Evidence suggests that circulation to the surface from the abyssal depths may take almost 1000 years. Therefore, it is not beyond the realms of scientific feasibility that the impact of an undersea volcanic disturbance feeding great heat into the water may not be felt at the surface, and in the air, for many hundreds of years.

## Land Surface Conditions

In Chapter 2 we described the role of the surface energy balance in determining the climate. Certain characteristics of the surface—its albedo, heat capacity and conductivity, frictional characteristics, and the availability of water—lead to specific patterns of temperature, humidity, air flow, and other climatic elements. It is clear, then, that a change in the nature of the surface of a large area of the land can have climatic repercussions. Most of us have experienced such a change when we walk from a grassy area onto a black-topped parking lot or roadway.

A major contributor to changes in surface conditions is land use, an anthropogenic cause. Very often overgrazing is the culprit, especially in semiarid regions where precipitation is only marginally sufficient to sustain grass coverage. There are many examples of this overgrazing plague: the Sahel, large parts of Eastern Africa, much of the Mediterranean area, and the northwestern part of the Indian subcontinent (the Thar desert). People are responsible, through the agents of cattle and the voracious goat. Another example is the United States, where overgrazing in the prairie lands first changed the area's ecology, and then in the 1930s poor agricultural practice helped lead to the Dust Bowl conditions. One of the more outspoken climatologists, Reid Bryson, believes that the extra dust taken up from the bare surface can have a significant effect on convective cells forming over the unprotected earth. Solar radiation is obscured by the dust, and the stability of the air is increased.

Deforestation is another dangerous practice, for often the resulting climate is too severe for the regeneration of small saplings that need to be protected in their early stages of growth. When trees are removed the surface albedo is increased, moisture is not easily retained, and wind speeds are increased now that the frictional drag of the trees is gone. Because of these effects, the rapid development (deforestation) of the Amazon Basin is being watched with some trepidation by climatologists.

The changes due to urbanization are numerous and will be described in detail in Section 14.3.

### The Cryosphere

The **cryosphere** consists of three components: land ice, sea ice, and snow. The great difference between these and ocean or land surfaces is in their albedos. For the oceans typical albedos are 0.04 to about 0.10, and for land surfaces 0.05 to around 0.35 (see Figure 2.11). The albedo for freshly fallen snow is near 0.80, and for ice the albedo is from 0.60 to 0.80. If the earth's mean surface albedo were to increase with time, such as would occur if ice and snow cover increased, a cooling of the earth-atmosphere system could be expected. It has been calculated that if sea ice were replaced by water, the global albedo would be reduced by about 1 percent, almost the same as if continental snow were absent in winter (Northern Hemisphere).

The land ice, mainly in the caps of Antarctica and Greenland, contains enough water to raise the mean sea level by about 65 m (215 ft) and is thought to vary on a time scale of thousands of years. The global sea ice cover, about 7 percent in northern summer and 4 percent in northern winter, has a distribution that is sensitive to climatic change on a much smaller time scale—less than a few years. The seasonal snow cover varies appreciably both within the year and between years. In August and September the snow and ice cover in the Northern Hemisphere is about $10 \times 10^6$ km$^2$ ($4 \times 10^6$ mi$^2$), while in January and February it can reach $60 \times 10^6$ km$^2$ ($23 \times 10^6$ mi$^2$). In November 1967 the coverage was only $28 \times 10^6$ km$^2$ ($11 \times 10^6$ mi$^2$)—some 55 percent of the November 1972 amount. Some idea of the seasonal change in the Southern Hemisphere is seen in Figure 11.17. One school of thought suggests that ice ages are triggered almost entirely by terrestrial-based factors, themselves the result of continental glaciation over relatively small regions.

The feedback mechanisms between ice cover and temperature are not simple. Models have shown that the latent heat component must be put in the feedback loop, in the form of appropriate precipitation and its change of state, if results are to be meaningful. The whole climatic system—atmosphere, hydrosphere, cryosphere, lithosphere, and the biomass—are coupled by biological, chemical, and physical processes and present a complex system that is apparently undergoing changes on all time scales.

A problem that has been of great concern to climatologists is, What particular features are conducive to glaciation? The primary requirement for the initiation of glaciation is a great amount of land in the higher latitudes. Ocean-covered polar regions will not lead to large, growing areas of ice. Another fundamental requirement is a heavy winter snowfall fol-

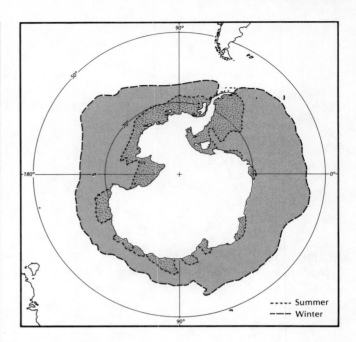

**FIGURE 11.17**
Ice limits in Antarctica in summer and winter.

lowed by a summer in which not all the snow has a chance to melt. For this to happen, there must be cool summers with relatively mild or warm winters, since very cold air masses hold little moisture and give only small amounts of snow. This situation appears most likely to occur when there is much land in the middle and high latitudes, as occurs at present at least in the Northern Hemisphere. Other land-sea configurations that would speed glaciation occur when ocean currents are diverted from warming areas and when impacting plates lead to orogeny (mountain building) and volcanism so that a lot of particulate matter is fed into the air, helping to reduce radiation incident at the surface. With respect to earth-sun geometry, three conditions—small obliquity of the ecliptic (reduced axial tilt of the earth), perihelion occurring in summer in the Northern Hemisphere, and a small orbital eccentricity—will assist in causing glaciation, given the present position of the continents.

Consider a possible pulsation of the climatic pattern when many of the preceding conditions have been met. In a period of major glaciation the cooler temperatures mean that less infrared radiation is lost to space, and solar radiation begins to dominate. Also, the cold air over the continental interiors blocks movement of pressure cells and leads to intense pressure ridges and troughs being oriented mainly north-south. This meridional flow assists heat transport between the tropics and the poles. Winters are still cold and snow amounts are low, so the warmer summers lead to melting which exceeds the fall during the previous cold season. Exposed earth has a much lower albedo (about 0.3 to 0.5) than ice and snow (0.9), so more heat is absorbed and the warming continues at an increasing rate until an interglacial period is underway.

A summary of the suggested causes of climatic changes described in this section is presented in Table 11.3.

## 11.13 MODELING THE CLIMATE

We have seen earlier in this chapter that the climate is not constant and that our knowledge of what causes variations to occur is extremely limited and very sketchy. In other words, there is neither a comprehensive theory to explain the fluctuations nor physical models that can simulate completely the whole climatic system. What we need, then, are quantitative mathematical models that will help us to understand the system and that will offer a technique for obtaining some predictive capacity. The interest in the development of climate models began in the 1960s, and models have grown in complexity as computer capabilities and capacities have increased.

Our knowledge of the main determinants (controls or factors) of climate—solar radiation, the composition of the atmosphere, and the surface characteristics of the globe—is not complete. We have good knowledge of solar radiation patterns, although the variations in the spectral distribution of its output are not well known. Information on the variability of the atmosphere's composition is very limited, particularly with respect to the quantity and distribution of dust, water vapor, and carbon dioxide, to name a few components. The basic surface characteristics of the globe and land and water configuration are known, but input from LANDSAT (a land-scanning satellite) is necessary to keep track of the changes as snow or ice melts or appears, as deciduous forests shed or grow their leaves, and as crops develop or are harvested.

Climate modeling also is made complicated by the coupled or interacting processes that occur. There are many such processes, but we will mention only

**TABLE 11.3**
Suggested causes of climatic change.

| Suggested Cause | Remarks or Periodicity (yr) |
|---|---|
| **SOLAR OUTPUT** | |
| Ultraviolet radiation | Random (?) |
| Sunspots | ca. 11 or 22 |
| **SUN-EARTH PATH** | |
| Passage through interstellar "dust" cloud | ? |
| Cosmic year (interstellar gases) | $225 \times 10^6$ |
| Eccentricity of earth's orbit | $10^5$ |
| Obliquity of earth's axis | $42 \times 10^3$ |
| Precession of equinoxes | $21 \times 10^3$ |
| Sun-earth radius | 1 |
| **ATMOSPHERIC COMPOSITION** | |
| Dust: Volcanic | Random |
|     Anthropogenic | Gradually increasing |
| Water vapor: Natural | Not identifiable |
|     Anthropogenic (burning) | Short lived |
| Carbon dioxide: Natural | Gradually increasing |
|     Anthropogenic (burning) | Rapidly increasing |
| **SURFACE CHANGES** | |
| Land-sea configuration | Important only for $> 10^6$ or $10^7$ |
| Topography | Important only for $> 10^6$ or $10^7$ |
| Ocean conditions | From $10^{-1}$ to $10^3$ |
| Land conditions (anthropogenic) | About 1 to 10 |
| Cryosphere: Land ice | $10^3$ and longer |
|     Sea ice | Seasonal and longer |
|     Snow | Seasonal and longer |

two. If a substance increases in temperature, the thermal energy radiated will also increase and will tend to return the temperature to its equilibrium value. This is a temperature-radiation coupling with a stabilizing feature called **negative feedback.** In the case of a cooling of the globe an increase in ice cover leads to a greater albedo and less absorption of insolation, causing a further lowering of temperature—a **positive feedback.**

Climate models are based on the fundamental hydrodynamic and thermodynamic equations in the form of a set of coupled (dependent) nonlinear three-dimensional partial differential equations. These equations are then solved by iterative techniques, but they also make use of such features as a given initial state of the earth-atmosphere system, particular boundary conditions, and a pattern of solar radiation input. In all cases the equations are solved at a certain pattern and number of grid points; the greater the number of grid points, the more detailed will be the result. For a global pattern on a 5° grid, there will be 2592 points. This method of grid points assures continuity, but phenomena of less than grid-scale size are not included in the model. The idea is that by this means we shall be able to make some deductions concerning the average character of the weather although being unable to determine the weather at a chosen spot on a chosen day.

The so-called **mechanistic models** are directed toward understanding the dependence on other parameters of the mechanism used. The earliest models were of this type. They showed a large temperature sensitivity to changes in the solar parameter, especially in the polar regions. The models were run until a steady-state condition of the system resulted. In these early models the albedo-temperature feedback played an important role; that is, a decrease in solar radiation leads to a reduction in surface temperature, which results in an increase in albedo (through increased snow and ice cover), less radiation absorption, and a further decrease in temperature. Later refinements showed that, using the present value of the solar parameter, these models have only two mathematically stable temperature solutions—the present climate and an ice-covered earth. In another model the Southern Hemisphere was shown to be less sensitive than the Northern Hemisphere to climatological perturbations because of the former's larger fraction of ocean coverage. In a model developed by an Australian meteorologist the equations are balanced to satisfy a condition of minimum entropy production. With this single criterion, values of mean annual zonally averaged temperature and precipitation are obtained that resemble present climatic patterns.

The **simulation models** usually include as many interacting physical variables as practically possible and come in a great variety of types. It is difficult to trace cause-and-effect relationship from these complex models. In some of these approaches a **general circulation model** (GCM) is used; a GCM integrates the basic equations of motion in various degrees of complexity. One of these GCMs was tested by imposing changes of 2C° (4F°) on the ocean surface temperatures; case (a) involved the whole global ocean, and case (b) covered two zonal strips centered at 5° S and 15° N. In case (a) there was a response in cloudiness at a 3-km (2-mi) height which was opposite to the change in temperature; that is, an increase in cloud corresponded to a decrease in temperature, and vice versa. In case (b) the temperature change at 15° N led to an increase in cloudiness over the strip and some decrease over adjacent zones; at 5° S the only significant change was a decrease in cloud over adjacent zones. Experiments such as these led to a better understanding of the cloudiness-temperature feedback relationship and improvement in later GCMs.

A very difficult problem facing the developer of any climatic modeling technique is that of result verification. Even the most complex global models incorporate many simplifications, such as constant albedo, and these assumptions may completely invalidate the findings or at least alter them significantly from the truth. Models that try to assess climatic changes over very long time spans (centuries, millennia, or more) may be particularly sensitive to assumptions that could be approximately true at some period and much in error at others. A basic problem is that the climate models are linked to the presently observed climate by empirical constants—values that may change as controls and resulting conditions alter.

In summary, we echo the words of W. Lawrence Gates, a major researcher in climate modeling:

> It is generally believed that both the dynamical and statistical approaches will play prominent roles in future climatic modeling and research. The models of minimum parameterization and maximum resolution (small grid point separation) such as the general circulation models are at the same time the models of maximum computational demands, while the reverse is true of the essentially statistical or averaged models. And while a dynamical model may permit more insight into the mechanics of climatic change, the averaged or parameterized models more directly address the essentially statistical nature of climate. A judicious combination of dynamics and statistics is therefore indicated, with the use of a hierarchy of models to explore as wide a variety of climatic variation as possible.

The need for greater understanding of the dynamics of climate assumes particular importance when we envisage the possibly serious environmental and agricultural impacts of future climatic variations, both natural and anthropogenic. At the present time we know far too little about such changes, and we are thus ill prepared either to project or to cope with their consequences. Among the many specific questions to be answered are: What long-term effects would the removal of the arctic sea ice have on the world's climate? And what will happen as man continues to change the character of the land's surface, to pollute the environment, and to emit waste from his expanding energy consumption?*

## 11.14
### FORECASTING FUTURE CLIMATES

In this section we will be concerned with climates ten or more years into the future. What can we say about these from our present knowledge? The answer is, unfortunately, very little. The two most physically sound cause-and-effect relationships would indicate that, from a consideration of earth-sun geometry, we are entering a cooling period, with an ice age of indeterminate severity likely within 10,000 years. In contrast, the steady anthropogenic increase of carbon dioxide should lead to a warming trend during the next 60 to 100 years unless fossil fuel burning ceases or decreases drastically.

The real problem of climate forecasting is that the greater the extent to which believable records are analyzed, the more obvious it becomes that there are variations on all time scales. The variation of climate from year to year is generally as much as might occur through drastic climate change. For example, in central Texas the fall of 1977 was about 6C° (11F°) warmer than the fall of 1976, a change much greater than those in 20- or 30-year means which are only of tenths of a degree. Expressed statistically, the noise level or high-frequency fluctuations (1-year cycle) contribute very significantly to the total variance. Sometimes there are suggestions of cycles or periodicities in the data, but they are seldom significant enough to produce a practically important improvement in forecasting. If such cycles have a period of around 11 or 22 years, then a relationship with sunspots is often suggested.

---

*W. Lawrence Gates, "An Essay on Climate Dynamics," *Bulletin of the American Meteorological Society* 57(1976): 546.

Many forms of graphical or statistical analysis have appeared in past and recent literature with extrapolations concerning the future. Without an understanding of the cause-and-effect relationships, however, such "forecasts" are even less reliable than a similar analysis of stock prices undertaken with regard only to the numbers themselves and not accounting for the current economic situation, company policies, demand, available money, and so on. A form of statistical analysis can reduce a complicated curve into cyclic components. In Figure 11.18 a mean surface temperature curve during the past 10,000 years is shown. The complicated curve has been broken down into five components with periods of 100, 200, 2500, 20,000, and 100,000 years. The shorter-period fluctuations account largely for the rate of change, while the longer-period ones have the largest amplitudes and can account for the major swings between interglacial periods and the ice ages. Much more research needs to be undertaken to increase our knowledge of the workings of the atmosphere, but it is unlikely that any developments capable of improving significantly our limited climatic forecasting ability will occur in less than 10 or 20 years.

Occasionally large-scale schemes for climatic modification have been proposed, even though all the implications via known and suspected feedbacks have not been studied or even thought out. Just two of these will be sufficient illustration of such schemes.

In 1960 there was a proposal to increase the precipitation in the semiarid regions of Africa south of the Sahara. The scheme was to dam the Congo River near Kinshasa, thereby causing a large inland basin to fill with water, which in turn would make the Ubangi River flow north and fill Lake Chad. These two bodies of water would cover 3 million km$^2$ (1,160,000 mi$^2$). It was thought that the extra humidity thus injected into the atmospheric system would be recycled within the zone. Although the logistical and political problems involved are clear, the climatic results of such an undertaking are not so obvious, although they are unlikely to achieve the desired outcome.

The second proposal was the damming of the Ob and Yenisei rivers in Siberia to produce a lake of 250,000 km$^2$ (97,000 mi$^2$). The lake could be used for irrigation of the steppe region and prevent the drying up of the Aral and Caspian seas. Calculations show that if this plan were instituted, about 15 to 20 percent of the river flow (fresh water) into the Arctic Ocean would be terminated, causing an increase in the salinity and a reduction of the ice cover. This latter effect likely would change the flow of jet streams and cyclonic movements over the whole Northern Hemisphere so that feedback mechanisms could have pro-

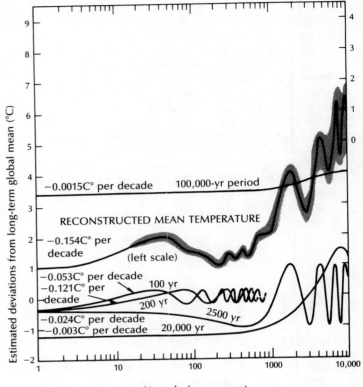

**FIGURE 11.18**
A depiction of the mean surface temperature changes during the past 10,000 years analyzed into five cyclic components. (From J. Gribbin, ed., *Climatic Change*. New York: Cambridge University Press, 1978.)

found, and perhaps even disastrous, consequences.

The outlook does not appear too inviting. A significant cooling would bring a glacial advance and pronounced thermal cold stress conditions to regions in higher latitudes. A warmer climate would lead to melting glaciers and rising ocean levels, and climatic zones that are now optimal for many agricultural activities may well be shifted to less favorable and productive soils. A third possibility is that the climate would become more variable, making planning for food growth and a stable economy very difficult indeed. It is only the fourth scenario—that of a climate and climatic pattern not significantly different from that of the present—that is really desirable, because so many aspects of life (agriculture, dwellings, and transport among other) are geared to our existing climate.

**SUGGESTED READING**

BRYSON, R. "A Perspective on Climatic Change." *Science* 184 (1974): 753–60.

GRIBBIN, J., ed. *Climatic Change*. New York: Cambridge University Press, 1978.

HAYS, J. D.; J. IMBRIE; and N. J. SHACKLETON. "Variation in the Earth's Orbit: Pacemaker of the Ice Ages." *Science* 194(1976): 1121–32.

LANDSBERG, H. E. "Whence Global Climate: Hot or Cold? An Essay Review." *Bulletin of the American Meteorological Society* 57 (1976): 441–43.

LE ROY LADURIE, E. *Times of Feast, Times of Famine*. London: George Allen and Unwin Ltd., 1971.

# 12 SYNOPTIC CLIMATOLOGY

**INTRODUCTION**
**SOME ASPECTS OF SYNOPTIC METEOROLOGY**
**THE SYNTHESIS OF SYNOPTIC WEATHER MAPS**
Extratropical Latitudes
The Tropics
Upper-Air Patterns
**APPLICATIONS**

## 12.1
### INTRODUCTION

The most common way to describe climate is through the use of numbers, that is, as means of the climatic elements. This method is employed throughout this text, particularly in the chapter on regional climates. In the application of meteorology and climatology to agriculture, commerce, transportation, and certain biological sciences, the mean values of a particular element, and perhaps some indication of that element's variability, usually are sufficient.

There is a disadvantage to this way of describing climate. There is nothing in the numbers which indicates the genesis of climate—the fundamental physical factors that produce and govern it. In fact, as was shown in Chapter 8, most of the climate classification systems are empirical (deduced from observation) rather than genetic (based on the origin of climate). One way of differentiating space-time variations of climate genetically is through deduction, either by instrumental observations or physical theory, of the components of the energy balance equations (see Chapter 2). However, even this approach has its shortcomings, since only vertical exchanges of energy are considered.

In order to incorporate the horizontal movement (advection) of heat, moisture, and mass into a derivation of the physical basis of weather, it is necessary to characterize atmospheric flow patterns. This is accomplished by analyzing representations of the state of the atmosphere at an instant of time, either as maps or as cross sections. Such maps show the geographical variation of pressure—from which wind speed and direction may be inferred—and the location of fronts for the surface, and the geographical variation of the height of a constant pressure surface and the temperature of that surface for the upper air, usually at standard levels such as 850, 700, 500, 300, and 200 mb. An example of such surface maps is Figure 4.22, and Figures 4.36 and 4.37 show the mean height of the 500-mb surface. Cross sections are a "sideways" picture of the atmosphere, with height and a transect such as north-south or northwest-southeast as the axes. They usually show some combination of pressure, winds, temperature, and humidity. Figure 4.42 is a climatological cross section. These representations are indispensable to the diagnosis and prediction of weather, and therefore for **synoptic meteorology**. When they are synthesized over periods of time such as a month or year, they constitute **synoptic climatology**.

These long-term depictions of atmospheric circulation may be statistical abstracts (e.g., simple arithmetic averages of many instantaneous maps), or a more subjective approach may be used. By inspecting a long series of "snapshots" of atmospheric circulation, the synoptic climatologist notices differences, similarities, and preferred circulation patterns (those that occur most frequently and for the longest time) and synthesizes this information in a relatively few "model" maps. Next, he or she assesses the weather elements in relation to these circulation categories. Thus, synoptic climatology characterizes local or regional climates by examining the relationships of weather elements, singly or collectively, to atmospheric circulation patterns and processes synthesized over relatively long time periods.

In this chapter we show how an added dimension to the description and analysis of climate can be obtained by the synoptic method. Some applications to climate-related phenomena also will be described. So that we may progress from the simple to the complex, we consider first the relationship between surface weather variables and their accompanying synoptic patterns at an instant of time.

## 12.2
### SOME ASPECTS OF SYNOPTIC METEOROLOGY

In Chapter 4 transient disturbances, a significant aspect of the weather of the middle and high latitudes, were illustrated in Figures 4.5 and 4.22. We learned that by noting the location of a particular place with respect to the features of the transient disturbances, we could infer something about the weather occurring there. For example, the weather ahead of a cold front generally features falling pressure and an increase in cloudiness, dewpoint, and temperature as warm and moist air is advected from the south (Northern Hemisphere). After the cold front passes, the wind shifts to a direction somewhere between northwest and northeast, pressure rises, and decreases in cloudiness, absolute humidity, and temperature are the rule. Figure 4.23 is an example of these changes for a specific location.

More generally, we can infer the various elements of weather from any synoptic map, given the date and time of that map. However, the ease and accuracy with which this is done depend on the synoptician's experience and whether or not he or she is familiar with prior maps. Another factor is the closeness of the location to weather-producing features such as fronts, lows, and highs. In Figure 4.5, for example, the weather for the southeastern United States is more easily inferred than that for California, for example, where the pressure pattern is rather diffuse. Another very important qualification is that local (meso- and microscale) conditions often modify the large (synoptic and large macroscale) influences. In some circumstances the weather at a location can be quite different from that deduced from weather maps, especially at times of light winds or calm and where there are pronounced topographic differences (see Chapter 10).

Some other examples of how we may infer the weather from synoptic representations are shown in Figures 12.1, 12.2, and 12.3. In Figure 12.1 an enormous high has moved into the United States from central Canada. To the east of the ridge line, winds are nearly everywhere westerly to northerly, bringing cold and dry air to these locations. An exception is the area from the western Great Lakes to the Appalachians to the Maritime Provinces of Canada, where precipitation is occurring in the "backwash" of the low (compare with Figure 5.17). Notice the area of clear skies that results from subsidence in and around the high. Such outbreaks of Canadian air are a frequent occurrence in winter and may threaten the citrus-growing areas of southern Texas and, somewhat later, Florida with subfreezing temperatures.

Figure 12.2 shows the synoptic conditions associated with very hot, dry weather in southern California. The eastern portion of the Pacific subtropical high has strengthened, moved into the western United States, and has been cut off from the main cell, while a trough of low pressure extends from Baja California northward along the west coast (dashed line). Already warm, dry air moves into southern California from the east, and its heat and low humidity are accentuated by downslope movement to lower elevations. This phenomenon is known locally as the Santa Ana (see Section 10.2.2). Temperatures in this instance reached as high as 41°C (106°F) in the Los Angeles area.

Heat wave conditions in the central and eastern United States develop under a synoptic pattern such as that shown in Figure 12.3. The western extension of the Atlantic subtropical high—sometimes referred to as the Bermuda high—engulfs virtually all of the country east of the Rockies. This system pumps initially warm, moist air from the Atlantic and Gulf of Mexico inland, where it is heated from beneath and thus made unstable. Scattered showers are likely to break out in highly localized areas and offer some relief to the oppressive heat and humidity, but this area in general is very hot. Only north of the fronts—and their location is typical for the very hot days of summer—are there cooler temperatures.

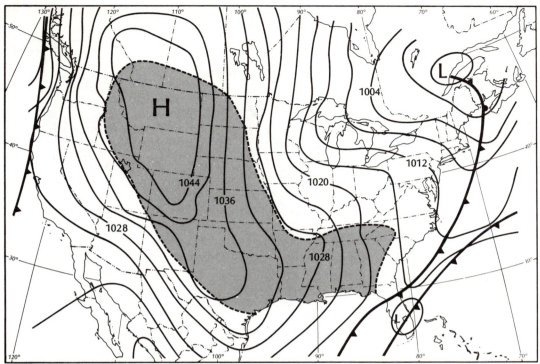

**FIGURE 12.1**
A Canadian air mass moves southward into the United States, bringing clear, cold weather. Pressure is measured in millibars. The dashed line encloses the area of clear skies associated with the high. (1 A.M. EST, November 29, 1979.)

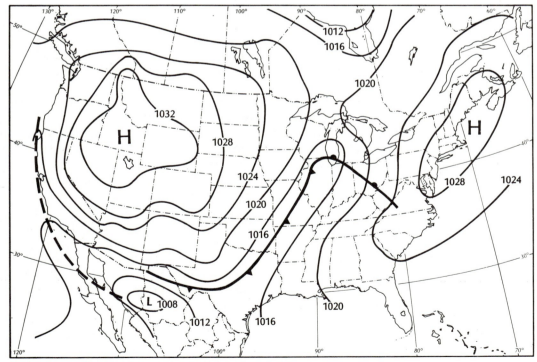

**FIGURE 12.2**
The synoptic situation associated with very hot, dry weather in southern California. Notice that surface winds are moving strongly from the east in this area. The dashed line is a pressure trough. (1 P.M. EST, October 15, 1967.)

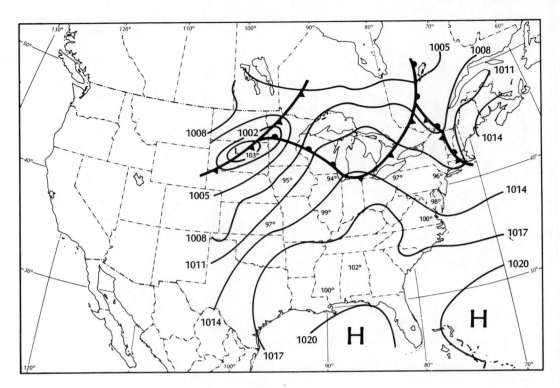

**FIGURE 12.3**
The surface synoptic situation often associated with very hot, humid summer weather in the United States east of the Rockies. Temperatures (°F) are the maxima for the preceding 12 hours. (From *An Introduction to Climate*, 4th ed., by G. T. Trewartha. Copyright © 1968 by McGraw-Hill Book Company. Used with the permission of McGraw-Hill Book Company.)

It must be stressed again that the correct interpretation of synoptic patterns is possible only when the calendar date and time and prior synoptic patterns are known. A situation such as that shown in Figure 12.3 could develop in other seasons (although this is unlikely), and the very high temperatures shown require that this pattern be established for at least a couple of days so that temperatures could increase to these very high values.

## 12.3
### THE SYNTHESIS OF SYNOPTIC WEATHER MAPS

#### 12.3.1 EXTRATROPICAL LATITUDES

Since climate is long-term weather, we need to show how "snapshots" of the surface circulation can be combined, or synthesized, to represent long periods of time. This is not an easy task. In extratropical latitudes, especially in winter, transient disturbances develop, mature, and dissipate so rapidly that simply averaging pressure for periods longer than about a week results in an obliteration of all significant features. This effect was described in Section 4.5 when filtering was discussed. However, there are other approaches to synthesizing weather maps, and we will now examine those most frequently used. The studies to be cited are not meant to be comprehensive; they show only some of the ways in which instantaneous representations can be combined to produce a synoptic climatology.

If our interest is in the frequency of occurrence of fronts for particular seasons, then illustrations such as Figures 4.29 and 4.30 are helpful. Such maps are prepared by taking unit areas of the surface (e.g., every 5° of latitude and longitude) and noting the presence or absence of a front in such an area at an instant of time. Their occurrence over many such instants of time indicates the geographical variation of frontal frequency.

Another way of abstracting the salient features of synoptic weather maps involves tabulating the frequency of occurrence of cyclones and anticyclones over a long period of time for particular geographical areas. Figures 12.4 to 12.7 show how often closed cells of low and high pressure occur over the Northern Hemisphere in winter and summer. One misleading

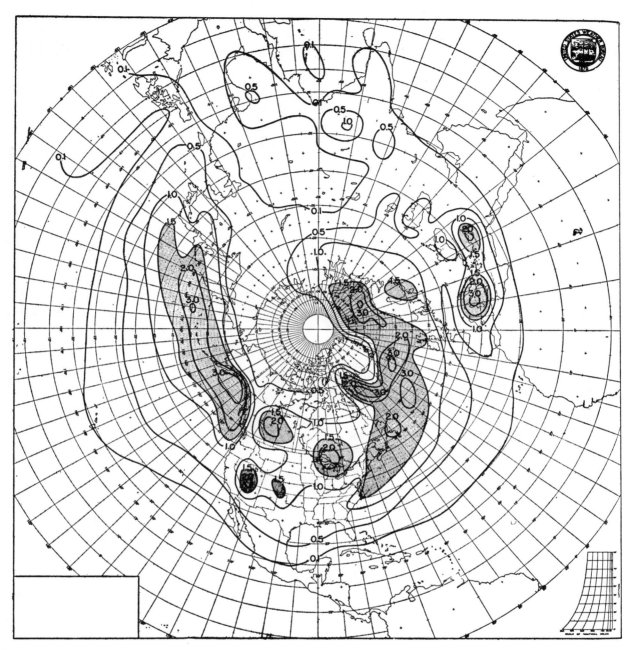

**FIGURE 12.4**
Percentage frequency of cyclone centers in squares of 100,000 km² in winter. (From H. C. Willett and F. Sanders, *Descriptive Meteorology*, 2d ed. New York: Academic Press, 1959.)

aspect of this manner of representation is that the semipermanent features of the surface circulation appear as markedly as do areas of frequent occurrence of transient systems. The latter are represented by the relatively high cyclone frequency centered just north of the Great Lakes, for example (Figure 12.4). A similarly high frequency of cyclones occurs in the southwestern United States in summer (Figure 12.5) because a low becomes established there in early summer and persists at that location until fall. Compare these representations with Figures 4.12 and 4.13.

It also may be of interest to know where cyclones begin. Are there preferred regions for cyclogenesis? And, once begun, do these lows follow preferred tracks? Figures 12.8 and 12.9 (pp. 279-80) show frequency of occurrence of cyclogenesis in winter and summer in the Northern Hemisphere, and Figure

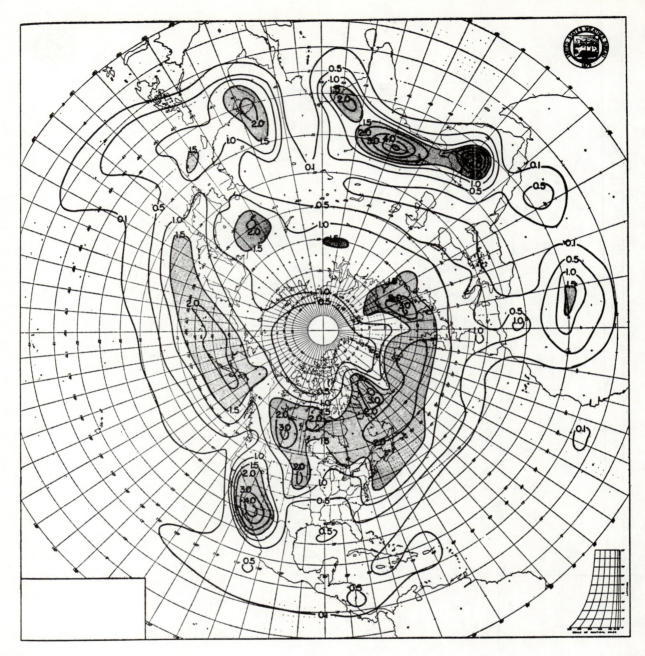

**FIGURE 12.5**
Percentage frequency of cyclone centers in squares of 100,000 km² in summer. (From H. C. Willett and F. Sanders, *Descriptive Meteorology*, 2d ed. New York: Academic Press, 1959.)

12.10 (p. 281) shows primary cyclone tracks for North America for four months of the year. Notice the correspondence between the start of these tracks and preferred regions for cyclogenesis in January (winter) and July (summer). There is also a predominant eastward or northeastward movement of these lows, and favored breeding grounds include areas in which there is pronounced horizontal temperature contrast, as from the east coast of the United States to the western Atlantic Ocean in all months, and in the northern Gulf of Mexico in winter.

So far we have emphasized characteristics of the surface synoptic map, such as the place of origin and direction of movement of surface pressure features (lows and highs) and fronts, and the geographical variation of their occurrence. We also may want to specify the circulation types that affect particular areas by constructing a catalog of the types most frequently

### SEC. 12.3 | THE SYNTHESIS OF SYNOPTIC WEATHER MAPS

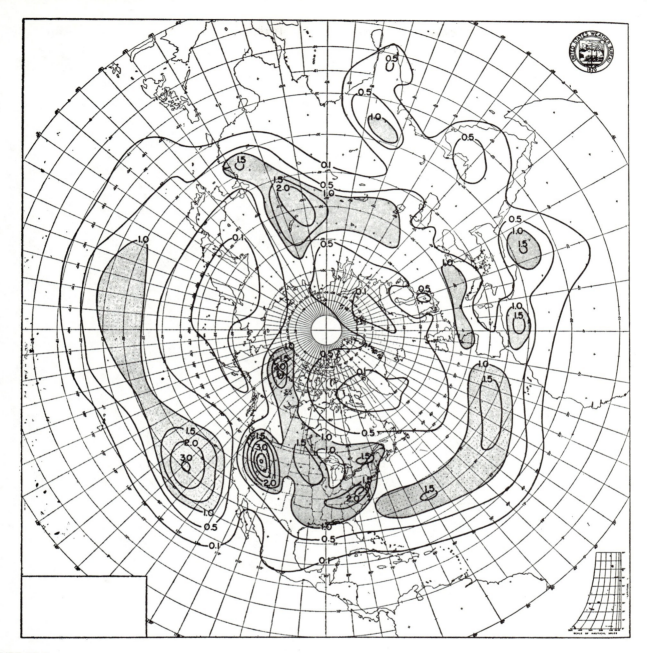

**FIGURE 12.6**
Percentage frequency of centers of anticyclones in squares of 100,000 km² in winter. (From H. C. Willett and F. Sanders, *Descriptive Meteorology*, 2d ed. New York: Academic Press, 1959.)

observed there. The area of concern may range from entire continents to individual locations. Such a catalog for North America is shown in Figure 12.11 (p. 282) and Table 12.1 (p. 283). The major circulation features used are the Aleutian low, the North Pacific anticyclone, the trajectory of polar air outbreaks, and the movement of cyclones. This classification was developed for winter, but it could apply to summer as well.

In summer polar air outbreaks would be less frequent and restricted to higher latitudes, and the North Atlantic anticyclone (evident in Figure 12.11 only in the upper left map) would be more conspicuous, to name just two differences.

Each circulation type can of course be associated with weather at any place on the continent. Notice that type $B_{n-c}$ is very much like the synoptic situation

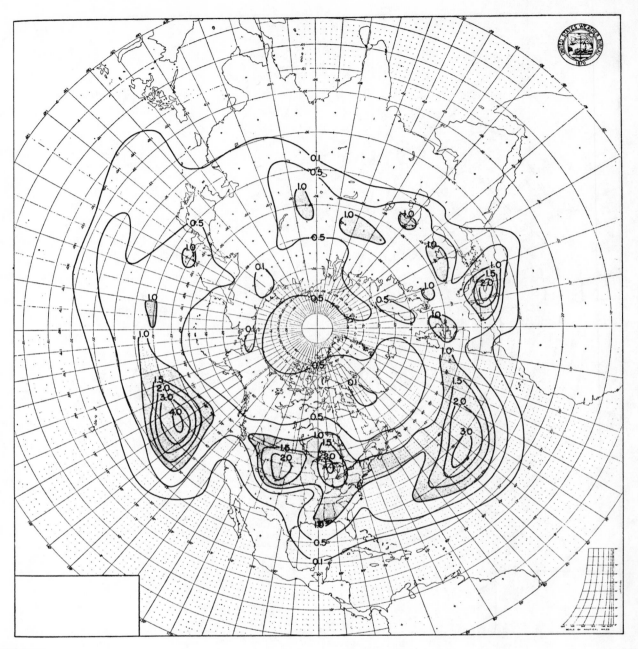

**FIGURE 12.7**
Percentage frequency of centers of anticyclones in squares of 100,000 km² in summer. (From H. C. Willett and F. Sanders, *Descriptive Meteorology*, 2d ed. New York: Academic Press, 1959.)

shown in Figure 12.1, which showed the weather over much of the United States. Each of these types also can be given a written description which more explicitly describes that circulation pattern (Table 12.1). The types are first distinguished according to whether the flow is predominantly meridional (marked north-south components to the flow at the surface and upper level) or zonal (predominant east-west flow). For the former the positions of the western ridge and eastern trough are important, while for the latter the latitude of cyclone tracks across the Rockies is significant in further distinguishing this subtype.

Associations between synoptic patterns and weather elements also can be made for areas smaller than continents. In this case synoptic-scale features become very important, and large-macroscale features less so (see Table 4.1). Two regional studies—one for the British Isles and the second for the Japa-

### SEC. 12.3 | THE SYNTHESIS OF SYNOPTIC WEATHER MAPS

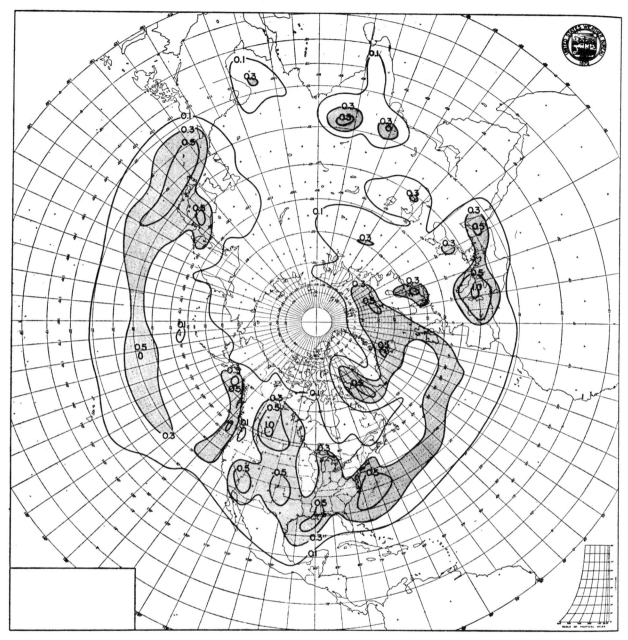

**FIGURE 12.8**
Percentage frequency of occurrence of cyclogenesis in squares of 100,000 km² in winter.
(From H. C. Willett and F. Sanders, *Descriptive Meteorology*, 2d ed. New York: Academic Press, 1959.)

nese island of Hokkaido—are good examples. In both locations the element of concern is precipitation. Figure 12.12 (p. 284) shows that different isohyetal patterns (an isohyet is a line connecting points of equal precipitation) can be associated with the tracks of cyclone centers. In general, precipitation amounts are greatest at and near these tracks.

Figure 12.13 (p. 285) shows how precipitation, expressed as the occurrence of rain in a 6-hour period, can be stratified according to the prevailing wind direction as deduced from synoptic maps. For example, for the southerly flow situation, the southeast-facing part of the island has precipitation (greater than 1 mm accumulation) in the 6 hours prior to 1800 on over 50 percent of the days in winter. This areal differentiation is not due to geographic differences in the occurrence of precipitation-producing systems (e.g., low centers and fronts) because the size of such sys-

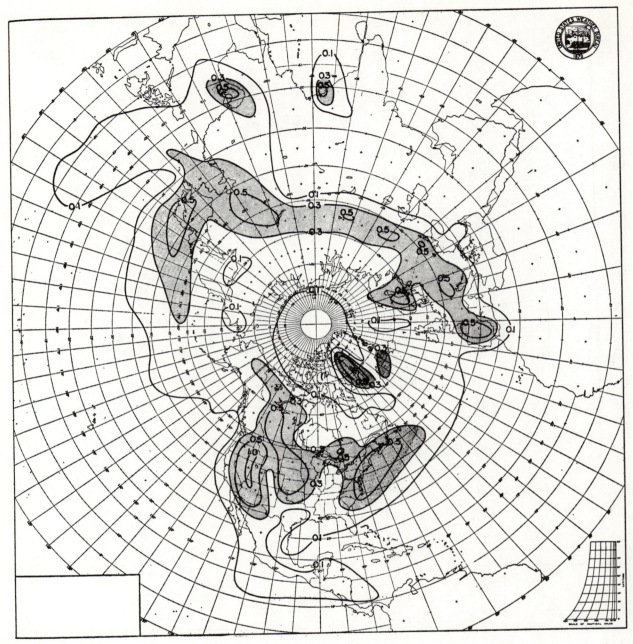

**FIGURE 12.9**
Percentage frequency of occurrence of cyclogenesis in squares of 100,000 km² in summer. (From H. C. Willett and F. Sanders, *Descriptive Meteorology*, 2d ed. New York: Academic Press, 1959.)

tems is larger than the island itself. Instead, it is due to orographic effects. Notice that the windward areas of the island have the most precipitation occurrences.

When synoptic patterns are viewed from the perspective of a specific location, as opposed to that of a continent or region, the concurrent weather at that location can be specified in great detail. Daily surface synoptic charts at 0600 for a 4-year period were examined for New Orleans, Louisiana, and each situation placed in one of eight types. Figure 12.14 (p. 286) shows these types, each for the date on which it was best represented. To a lesser extent these types could be associated with other locations in the analysis area (the 48 States and southern Canada), although the associations are not as well developed with increasing distance from New Orleans.

Some of the weather characteristics of the three most frequent types in the New Orleans study are

### SEC. 12.3 | THE SYNTHESIS OF SYNOPTIC WEATHER MAPS

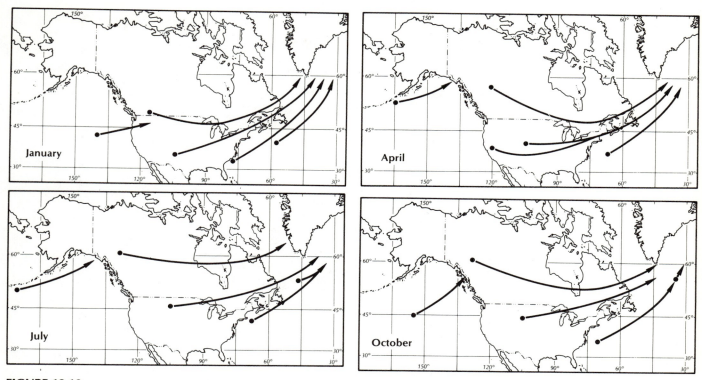

**FIGURE 12.10**
Primary cyclone tracks over North America during different seasons of the year. Each of the X's identifies a region where cyclone formation is relatively frequent (compare with Figures 12.8 and 12.9). [From C. H. Reitan, "Frequencies of Cyclones and Cyclogenesis for North America, 1951–1970," *Bulletin of the American Meteorological Society* 102 (1974): 861–88.]

shown in Figure 12.15 (p. 287). Notice how these characteristics are distinguished by the types. Summer temperatures and absolute humidity, as indicated by dewpoint temperature, are much the same, but these elements for winter are greatly different. The first two types, continental high and frontal overrunning (see discussion of overrunning in Section 5.8.1), have north winds but are distinguished by cloudiness and winter humidity. In the Gulf return type winds are south to southeasterly. The first and third types are at least fairly frequent throughout the year, but frontal overrunning seldom occurs in summer.

Calendars of synoptic types can be prepared from studies of this kind. The average duration of types and their seasonal variations also can be determined. Changes in the frequency of occurrence of types can be associated with trends in climatic variables such as temperature and precipitation.

The results of abstracting composites of weather characteristics for synoptic types are very similar to those obtained when the same method is applied to air masses (see Chapter 7). Some correspondence can be expected; it is likely, for example, that the air mass that affects New Orleans during the continental high type is cP (continental polar), while that during Gulf return is mT (maritime tropical). In other synoptic situations, however, more than one air mass may be present. And, a given air mass may occur under varying synoptic conditions.

### 12.3.2 THE TROPICS

Synoptic patterns in extratropical latitudes consist in part of transient cyclones and anticyclones, and our climatology of synoptic weather there has been based on syntheses of these. In the tropics such transient features are largely absent; tropical storms and hurricanes are a marked exception, but they are very infrequent compared with their counterparts in higher latitudes. Another important difference between these areas is that instantaneous pressure patterns are generally diffuse and isobars widely spaced in the tropics. It is correspondingly more difficult to associate wind speed and direction with isobar configuration and spacing, as was done by applying the geostrophic or friction wind approximation (see Section 4.3.3). These approximations are not good indicators of the relationship of wind to pressure in low latitudes.

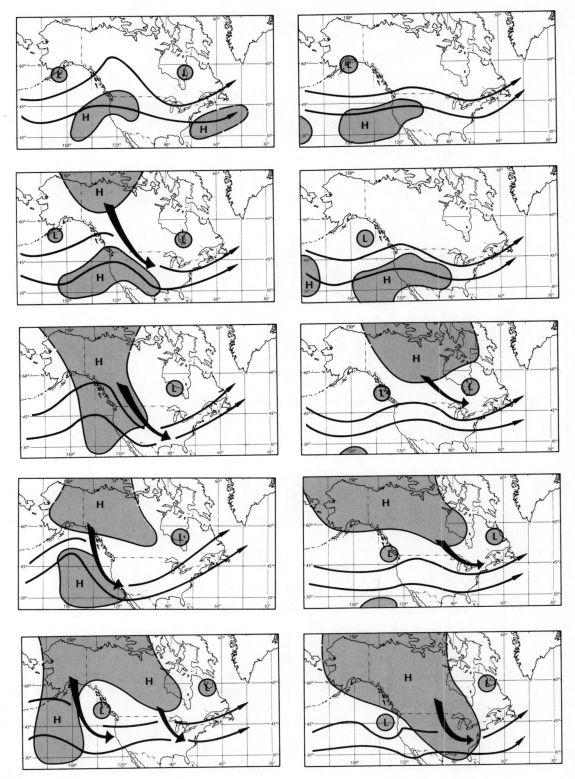

**FIGURE 12.11**
Principal circulation types for North America. Heavy lines indicate upper level mean flow. Stippled areas indicate quasi-stationary low pressure centers; hatched areas indicate persistent surface anticyclones. Open arrows show polar air outbreaks. (From R. D. Elliot, "Extended-range Forecasting by Weather Types," in *Compendium of Meteorology*, ed. T. F. Malone, p. 836. Boston: American Meteorological Society, 1951.)

**TABLE 12.1** North American circulation pattern types.

| Type | Characteristic Features | | |
|---|---|---|---|
| Meridional | Position of Western Ridge | Position of Eastern Trough | Remarks |
| $B_{n-a}$ | 115°–120° W | 90° W | Trough over central North America; strong Great Basin high. |
| $B_{n-b}$ | 115°–120° W | 90° W | Similar to $B_{n-a}$, but stronger frontal contrasts. |
| $B_{n-c}$ | 135° W | 100° W | Strong western ridge, deep eastern trough; intense polar outbreaks over Midwest. |
| A | 145°–150° W | 110° W | Strong Pacific high extending well north; lows move southeastward across Pacific northwest U.S. |
| D | 160° W | 130°–135° W | Extreme meridional pattern, west coast trough. |
| Zonal | Latitude of Cyclone Tracks across Rockies | | Remarks |
| B | 59° N | | Storm track displaced to north. |
| $B_s$ | 55° N | | Similar to $B_{n-a}$, but weaker western ridge. |
| $E_L$ | 47° N | | Westerlies displaced to south; high over northern Canada. |
| $E_M$ | 40° N | | Cyclones much farther south across continent. |
| $E_H$ | 34° N | | Extreme type; strong high over Alaska and northwest Canada extending southeastward. |

For these reasons another approach is used to portray climatological wind fields in the tropics. To explain this approach, we must digress briefly and introduce two new concepts. We first define **mean resultant wind** as the vector average of wind for a specified period of time at a location. (Recall that winds are vector quantities since they have both speed and direction.) The mean resultant wind is obtained by dividing the actual wind—whose direction and speed are given by the orientation of a vector arrow and its length, respectively—into two components, commonly called **u** (east-west) and **v** (north-south) components (Figure 12.16, p. 287). This can be done for one wind observation, or for many when the **u** and **v** components are each averaged, and then combined to produce the mean resultant wind. The same result may be obtained by successively placing the tail of one arrow at the head of the next.

The circulation of the atmosphere at a specified level is explicitly indicated by the use of **streamlines.** Streamlines are lines drawn parallel to the instantaneous motion at a location, as represented there by the orientation of the wind arrow. The number of streamlines drawn is a matter of choice. Figure 12.17 (p. 288) shows the streamline pattern when hurricane Camille was in the Gulf of Mexico. Notice how clearly areas of cyclonic motion—the hurricane itself, almost centered in the Gulf—and anticyclonic motion, as in the western Atlantic and at 38° N 95° W, are delineated by streamlines.

The average (climatological) circulation can be represented by streamline analyses of mean resultant winds. This is especially useful in the tropics, where, because of the comparative absence of transient systems, the mean flow is meaningful. However, this approach indicates only the direction of the wind, not its speed. For either instantaneous or averaged representations, therefore, wind speed is indicated by its isopleths, called **isotachs,** superimposed on the streamlines.

Figure 12.18 (p. 289) is such a representation for a part of the tropics in January and July. Notice how clearly centers of anticyclonic and cyclonic circulation are depicted. (Keep in mind that the direction of these motions is reversed in the two hemispheres.) This is the circulation at the **gradient level,** or the lowest level at which predominantly friction-free flow occurs. Over most of this region the gradient level is about 900 m (3000 ft) above sea level.

In the model of global wind belts in Figures 4.9 and 4.11 we noted that the circumequatorial trough of low pressure, or the ITCZ, acts as a convergence

**FIGURE 12.12**
Rainfall and cyclone tracks over the British Isles. The isohyets show average amounts (mm) for the number of cyclones indicated. (From J. S. Sawyer, "Rainfall of Depressions Which Pass Eastward Over or Near the British Isles." Professional Note No. 118. Norwich, England: British Meteorological Office, 1956. Used with the permission of Her Majesty's Stationery Office.)

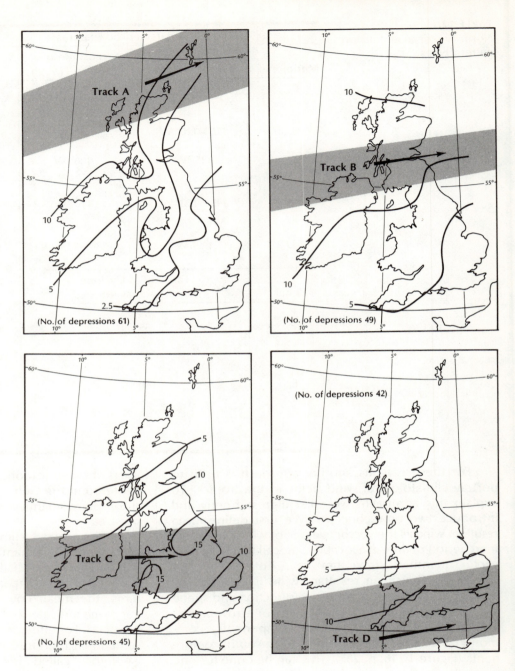

zone. Northeast and southeast trade winds converge at and near the equator, producing rising air, which, coupled with the convective mechanism, results in a latitudinally shifting belt of rainfall. This characterization applies better over oceans than over land, as is strikingly shown in Figure 4.16. But even over the oceans there are exceptions to this idealization, particularly when the ITCZ is displaced more than about 5° from the equator (Figure 12.19, p. 290) and in some other circumstances. Equatorial westerlies are the result of this displacement, as explained in Section 4.6.2 and illustrated by Figure 4.17.

Over tropical Africa the correspondence between the positions of the ITCZ and zones of maximum rainfall is not what would be expected from this idealization. In July the ITCZ extends east-west through the Sahara Desert of North Africa (see Figures 4.13 and 4.15), which is certainly not a rainy area. To understand why this is so and to be able to associate synoptic features with rainfall distribution in tropical Africa, recall the point made in Section 7.3.1. We learned that in order for significant precipitation to occur convergence must extend through at least a moderate depth of the atmosphere. And, since precipitation forms in

### SEC. 12.3 | THE SYNTHESIS OF SYNOPTIC WEATHER MAPS

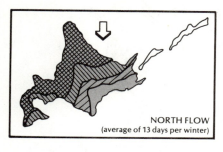

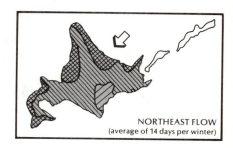
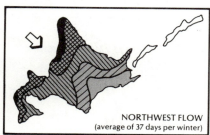

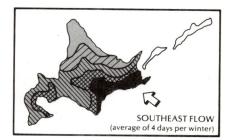
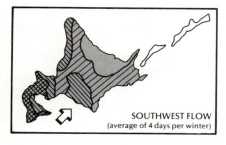
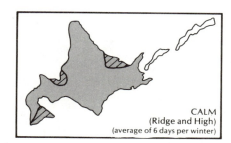
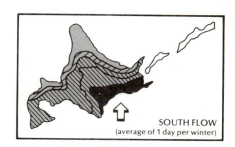

**FIGURE 12.13**
Percentage occurrence of days with precipitation on Hokkaido, Japan, stratified by direction of wind flow, for winter (December–February). The days included comprise 86 percent of all winter days. (From W. Jacobs, *Wartime Developments in Applied Climatology*. Meteorological Monograph No. 1. Boston: American Meteorological Society, 1947.)

the lower and mid-troposphere, it is particularly important that convergence be present at these levels.

Figure 12.20 (p. 291) shows the mean height of the 500-, 700-, and 850-mb surfaces and mean sea level pressure along the 0° meridian (see Figure 4.12 for orientation) in July. Although surface pressure is lowest at about 23° N, pressure is relatively high (high heights) at the three upper levels, and the divergence there is of course inimical to precipitation. The eastern sector of the North Atlantic subtropical high therefore extends farther eastward at upper levels; only at the surface is there relatively low pressure. The zone of maximum rainfall occurs south of the ITCZ, where there is upper-level convergence, as shown by the trough axis. From analyses such as these it has been shown that the maximum rainfall zone is located about 10° south of the ITCZ in the months of January, April, July, and October. This relationship probably holds also for the remaining months.

Furthermore, recall that one of the implications of the hypsometric relationship (Figure 4.33 and the discussion related to it) is that warm-core lows, such as the Saharan low in July, will tend to weaken with height and be replaced by high pressure at upper levels. Figure 12.21 (p. 291) shows this relationship for both a warm-core low and a cold-core high. From such analyses as are represented by Figure 12.20 we have observational evidence that this is indeed the case.

#### 12.3.3 UPPER-AIR PATTERNS

The surface synoptic pattern is of primary interest in inferring the nature of surface weather elements. As stated throughout this chapter, the direction of wind flow, areas of convergence and divergence, and the presence of fronts—all deduced from customary synoptic representation—enable us to infer the accompanying surface weather. But surface patterns indi-

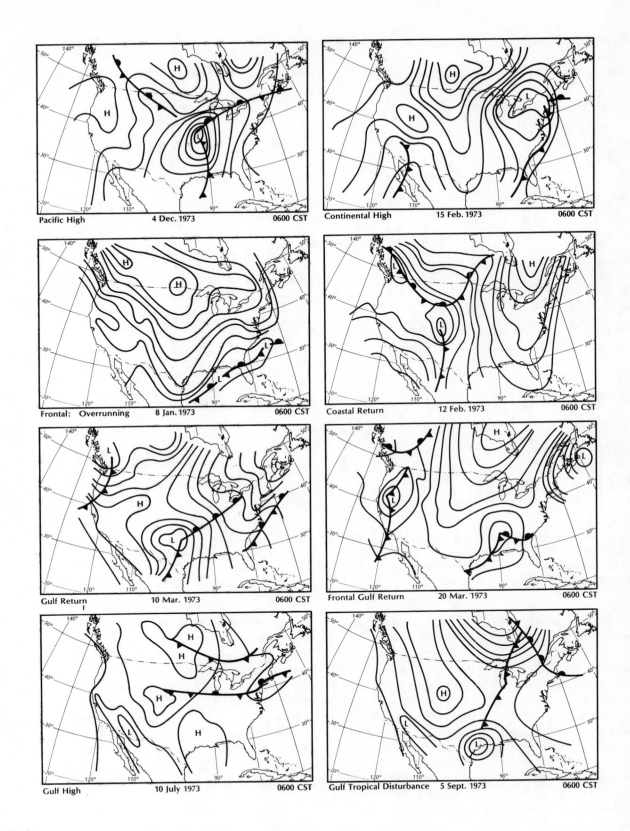

**FIGURE 12.14**
Representative examples of eight synoptic weather types for New Orleans, Louisiana.
(From R. Muller and C. Wax, *Geoscience and Man*. Baton Rouge: Louisiana State University Press, 1977.)

## SEC. 12.3 | THE SYNTHESIS OF SYNOPTIC WEATHER MAPS

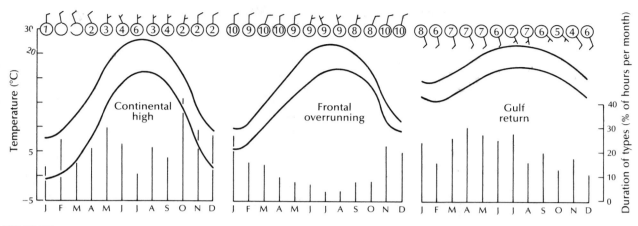

**FIGURE 12.15**
Characteristics of three of the most frequent synoptic types in Figure 12.14. Values shown are an average of observations at 0600 and 1500 CST for each of the twelve months. The upper curve is temperature; the lower, dewpoint temperature. The numbers in circles show cloudiness in tenths; wind arrows show speed and direction. The vertical bars indicate the percentage of monthly hours during which the type occurs. [Data from R. Muller, "A Synoptic Climatology for Environmental Baseline Analysis: New Orleans," *Journal of Applied Meteorology* 16 (1977): 20–33.]

cate only a limited part of the atmosphere. Inferences about temperature and especially about precipitation, which forms at higher levels, are more accurate when the entire troposphere, or at least that portion of it up to 500 mb, is examined. The explanation in Section 12.3.2 for the absence of precipitation around the ITCZ when it is displaced northward into the Sahara is a case in point.

There is yet another reason for examining upper-air conditions. We have seen that surface pressure patterns, especially those of the extratropical latitudes in winter, are extremely variable. This variability is in

**FIGURE 12.16**
Derivation of the mean resultant wind. Each wind observation is represented by a single arrow (vector), **w**. This is divided into **u** and **v** components as shown (notice that $\mathbf{w} = \sqrt{\mathbf{u}^2 + \mathbf{v}^2}$). The means of the **u** and **v** values are then combined to produce the mean resultant wind.

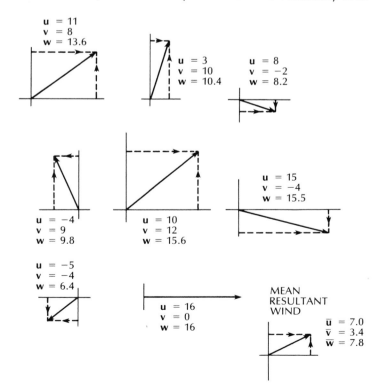

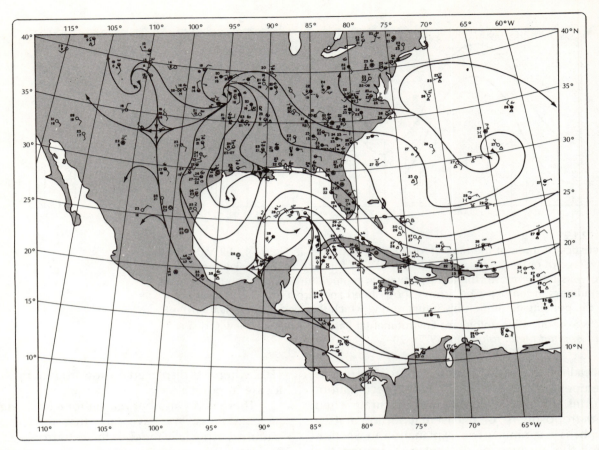

**FIGURE 12.17**
Surface streamline analysis of 7 A.M. EST, August 16, 1969, showing hurricane Camille in the Gulf of Mexico. (Courtesy of William Shenk, NASA.)

part due to pronounced horizontal temperature gradients and topographic variations. At levels above the surface these forcing mechanisms weaken and, in the case of temperature gradients, become more systematically north-south oriented. The result is that mid-tropospheric circulation patterns (700 to 400 mb) are less detailed, and thus smoother, and tend to change more slowly than those at the surface. There are fewer closed systems at upper levels (i.e., troughs and ridges rather than cyclones and anticyclones), and, as described in Section 4.7.2, the flow is circumpolar, with pressure (or height) gradients strengthening and extending farther equatorward in winter in response to a corresponding increase in temperature gradients. Troughs and ridges comprise long waves, and although these may show marked amplitude over short periods, these amplitudes are reduced on maps averaged for periods of a month or more. In Figure 4.38, then, while daily maps may show patterns (a) through (d), the result of averaging is usually the flow pattern represented by (b).

Also, as explained in Figure 4.38, troughs and ridges tend to transport cold air equatorward and warm air poleward, respectively. Suppose that the mean pattern for a particular month showed a greater amplitude than the long-term mean for the same month for a long wave with a ridge to the west and a trough to the east. This implies that areas to the east and west of a north-south line midway between the axes of the ridge and trough experienced colder and warmer weather than average, respectively, and this is usually the case. Similarly, a flatter pattern (less amplitude) implies the reverse of these conditions (Figure 12.22, p. 292). Also, if the long wave for a particular month was displaced east or west of its long-term mean position, there would again be departures from average temperatures.

Precipitation departures from the average also are associated with mid-tropospheric pressure patterns. Climatologists have verified this correspondence for the height of the 700-mb surface. Over most of the United States heavy winter precipitation occurs

## SEC. 12.3 | THE SYNTHESIS OF SYNOPTIC WEATHER MAPS

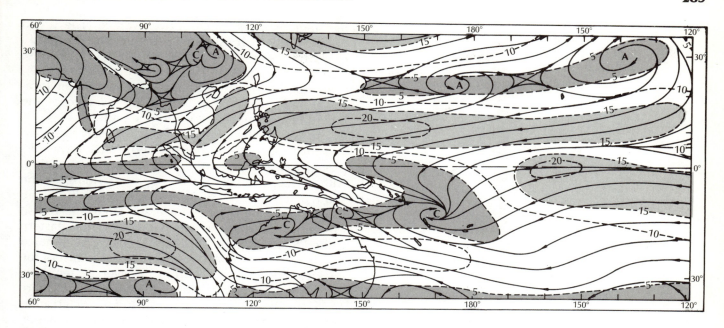

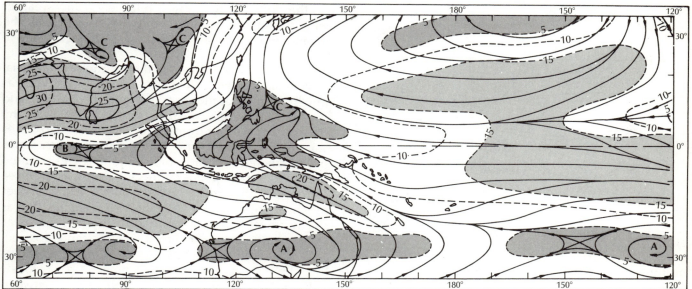

**FIGURE 12.18**
Streamline analysis of mean resultant winds in the western Pacific in (a) January and (b) July. The isopleths are isotachs (lines of constant wind speed), in knots. (From G. Atkinson, *Forecaster's Guide to Tropical Meteorology*. U. S. Air Weather Service, 1971.)

with below average heights, and drier (than normal) weather with above average heights.

In addition to relating temperature and precipitation departures from the average to anomalous pressure fields and heights, synoptic climatologists have shown that there are preferred locations for such departures with respect to long waves. When mean heights for the 700-mb surface are obtained for 5-day periods (although the models to be shown are generally applicable on any time scale from daily maps to 30-day means), areas of above- and below-average temperature and precipitation can be fixed with respect to troughs and ridges. As Figures 12.23 and 12.24 (p. 292) show, these areas vary somewhat in different parts of the United States during winter. Areas of heavy precipitation range from 1930 km (1200 mi) to the east of the trough axis in the East and the Ohio valley to the axis itself in the Rocky Mountain states and northern plains. The areas of coldest temperatures in winter occur slightly to the east of the trough

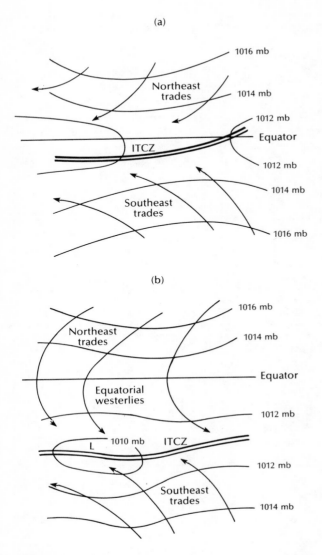

**FIGURE 12.19**
Wind flow and pressure patterns when the ITCZ is (a) near the equator and (b) displaced more than 5° from the equator. (From J. S. Sawyer, "Memorandum on the Intertropical Front." Meteorological Report No. 10. Norwich, England: British Meteorological Office, 1952. Used with the permission of Her Majesty's Stationery Office.)

axis in the Great Plains, but somewhat to the west of it in the remainder of the country.

A specific example of how departures from the average flow at 700 mb can affect surface air temperatures is shown in Figure 12.25 (p. 294). January 1977 was an extremely cold month over virtually all of the conterminous United States, and the 1977 winter (December, January, February) was the first of three consecutive severe winters. In contrast, February 1976 had temperatures much above average. The broad outlines of why these departures occurred are indicated by mean monthly 700-mb patterns for these months. The associations involve the patterns (i.e., the amplitude of troughs and ridges and their locations) and the departures from the average of the height fields which result from the patterns.

In January the western ridge was greatly amplified and the low center in northeastern Canada was displaced well to the southeast of its average position, with a deeper than usual trough extending to the south of it [Figure 12.25(a) and (b)]. As a result, 700-mb heights were much above average along the British Columbia coast and much below it in the eastern United States [Figure 12.25(c)]. Winds were funneled out of the Canadian Arctic toward the area east of the Rockies, and departures from the average temperature were especially striking in the Ohio valley [Figure 12.25(d)]. Only the far Southwest, and restricted areas in the Northwest, were warmer than average. Alaska, on the other hand, enjoyed relatively balmy weather with southwesterly flow from the Pacific, and the mean temperature at Fairbanks was 12C° (22F°) above average!

In marked contrast to the strongly meridional flow of January 1977, the pattern of February 1976 was much "flatter," or more zonal [Figure 12.25(e), (f), and (g)]. Fast, generally westerly flow was the rule, the transport of cold air from Canada to the United States was greatly limited, and warm maritime air masses predominated over most of the country. This resulted in above-average temperatures for almost all of the 48 states and departures greatly above average over much of the Midwest and Great Plains states. It was the warmest February on record for such places as Washington, D.C., Rockford, Illinois, and Roswell, New Mexico [Figure 12.25(h)].

The two months just examined are obviously extreme situations, and the correlations between upper-air patterns and surface air temperatures are invariably highest in such extremes. For more moderate situations the associations are, as a rule, not as direct. It also happens, in some limited circumstances, that departures from the usual mean monthly flow patterns and the occurrence of above- and below-average precipitation are related, as Figure 12.24 suggests. In these cases the tracks of major storm systems may be inferred from the 700-mb flow. However, these associations are even less apparent than associations with temperature and involve diagnostic procedures that go beyond this text.

The movement of surface features such as lows, highs, and fronts is difficult to predict more than a few

## SEC. 12.3 | THE SYNTHESIS OF SYNOPTIC WEATHER MAPS

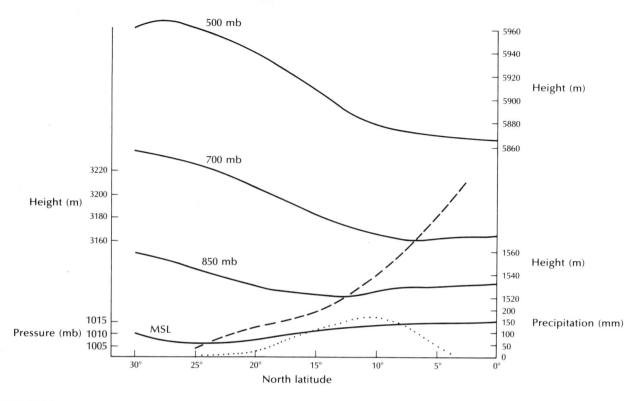

**FIGURE 12.20**
The relationship between the heights of the 500-, 700-, and 850-mb surfaces and sea level pressure (MSL), and July precipitation (dotted line), along 0° longitude for latitudes 4° to 30° N (northwest Africa). The dashed line is the trough axis. There are separate scales for each of the three surfaces, but the height interval is constant. The precipitation scale (mm) is in the lower right. (Based on data from H. van de Boogaard, *The Mean Circulation of the Tropical and Subtropical Atmosphere—July*, NCAR Technical Note, Boulder, Colo.: National Center for Atmospheric Research, 1977, and F. L. Wernstedt, "World Climatic Data—Africa," Pennsylvania State University Department of Geography.)

days in advance. But because upper-air features change more slowly and tend to persist for longer periods of time, climatologists have had some success in predicting the mean state of the mid-troposphere for coming months. Since it is possible to correlate this mean state with departures from average surface

**FIGURE 12.21**
The hyposometric relationship requires that the vertical decrease in pressure be greater in cold than in warm air. Therefore, warm-core lows and cold-core highs tend to weaken with height and be replaced by the opposite pressure feature at upper levels. By the same reasoning, warm-core highs and cold-core lows tend to strengthen with height. Here the terms warm, cool, and cold refer to temperatures through the entire columns of air shown.

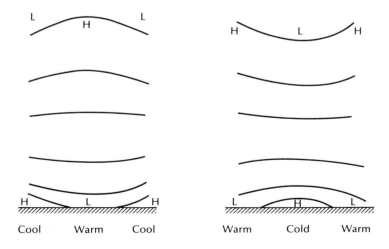

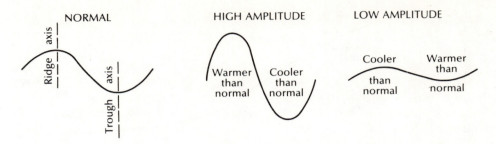

**FIGURE 12.22**
The amplitudes of long waves in the mid-tropospheric westerlies influence surface temperatures. The schematic high and low amplitude patterns are an average for a month, while the normal pattern is an average for many such months.

temperatures—and to a lesser extent with precipitation—it is possible in turn for forecasters to make reasonable estimates of how temperature and precipitation will differ from the average in the month or season to come.

The causes of variations in mid-tropospheric flow are not well understood, but an interesting theory relates the positions and intensities of troughs and ridges over North America to sea-surface temperatures in the North Pacific Ocean. Departures from the average of these temperatures have been shown in some circumstances to correlate with upper-air features, and this linkage may have predictive value. Such correlations between weather conditions in one part of the globe and those that are occurring or have occurred elsewhere are called **teleconnections**. The study of teleconnections, an important aspect of synoptic climatology, has received much attention in recent years.

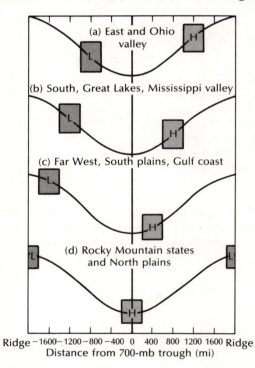

**FIGURE 12.23**
Optimum regions for heavy (H) and light (L) precipitation in winter in different parts of the United States relative to a sinusoidal 700-mb contour. Distance from the 700-mb trough given along the abscissa (X-axis) is assumed to be 2000 mi to the ridges up- and downstream, with positive values when the trough is west of the reference area and negative values when the trough is east of the area. [From W. H. Klein, "Synoptic Climatological Models of the United States," *Weatherwise* 18 (1965): 252.]

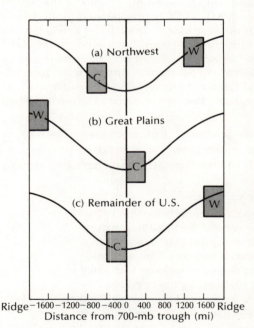

**FIGURE 12.24**
Optimum regions for cold (C) and warm (W) temperatures in winter in different parts of the United States relative to a sinusoidal 700-mb contour. (See Figure 12.23.) [From W. H. Klein, "Synoptic Climatological Models of the United States," *Weatherwise* 18 (1965): 252.]

## 12.4 APPLICATIONS

Mean values of weather elements can be used not only to describe climate, but as correlates with both natural and man-made phenomena that are influenced by weather. (These relationships are discussed in Chapters 13, 14 and 15.) In the case of quantifiable phenomena, one of the most practical and straightforward ways of finding and clarifying such relationships is to use the mathematical methods of regression and correlation (see Chapter 6). Examples include the relationship between heating degree-days and fuel oil consumption; growing season precipitation and the yields (and other characteristics) of grains such as wheat, corn, and rice; and the number of deaths expected in large metropolitan areas during times of excessive heat and humidity.

When the weather-influenced phenomena are not quantifiable, some kind of stratification by weather element or elements is helpful in finding and clarifying relationships. One example would be the relationship between the effects of snow accumulation over a few hours or days and the extent of disruption of commerce, transportation, and industry in large cities. Another example includes the relationships between tides, currents, and sea states, and the speed and location of the jet stream, on the planning of ship and aircraft routes, respectively.

There are two shortcomings to these approaches. First, when we use ordinary correlation and regression procedures, we assume that dependent and independent variables (e.g., wheat yields and growing season precipitation, respectively) are linearly related. That is, for a given increase in the weather element, one expects a proportional increase (or decrease) in the weather-influenced variable over all observed ranges of the former. Although this is often reasonable as a first assumption, there are many cases of nonlinear interaction, or the issue of timing may be critical. Thus, it has been shown that mortality during heat waves increases exponentially for temperatures above a threshold value, and that grain yields depend as much on the timing as the amount of precipitation. Also, in some applications the extremes of a weather element are more significant than its mean.

The second shortcoming is that the causative weather element may only be suspected or may not even be known. Or, there may be evidence that two or more elements are causative, but the relative importance of each may be in question. Given the importance of light, moisture, and heat to the full development of stands of deciduous trees, how important is each, and how do these influences vary during both the growing and dormant seasons?

The methods of synoptic climatology afford a partial solution to these shortcomings. No mathematical restrictions are imposed, and no assumptions need to be made about which weather elements are involved. Furthermore, viewing weather from a synoptic perspective implies something about its genesis, an attribute that is lacking when single or multiple independent variables are associated with a dependent variable, as in mathematical techniques. Also, sequences of weather events—an ordering in time of the synoptic conditions that govern corresponding changes in weather elements—are explicit in the synoptic method and can help resolve problems related to the timing and duration of conditions suspected or known to be causative.

In the remainder of this chapter examples are given of how the synoptic method has been employed to elucidate relationships between weather map patterns and phenomena ranging from biological response to other meteorological characteristics. Some of the studies to be cited involve periods too short to constitute a climatology, and thus they might better be called applications of synoptic meteorology. Presumably, however, these could be extended to longer periods.

Some interesting studies relating the incidence of disease, or of human health and well-being in general, to phases of an idealized wave cyclone have been conducted by what has become known as the German school of meteoropathology (roughly, weather-sickness relationships). As explained in Section 4.6.3, there is a characteristic ensemble of weather elements accompanying each of the various areas in and around an extratropical wave cyclone. Thus, a transect through such a model includes a number of sectors, each of which can be associated with the initiation or exacerbation of certain diseases (Figures 12.26 and 12.27, p. 296). The model (Figure 12.26) was prepared for the British Isles and western Europe, and the epidemiological data are from clinics and hospitals in Germany. Notice that in general prefrontal and postfrontal weather are biologically unfavorable and favorable, respectively.

Over the last several decades air pollution has become a serious problem in many of the technologically advanced nations. The meteorological conditions most favorable for high concentrations of pollutants occur when winds are light, and this often happens at or near the centers of anticyclones. The situation worsens when anticyclones move very

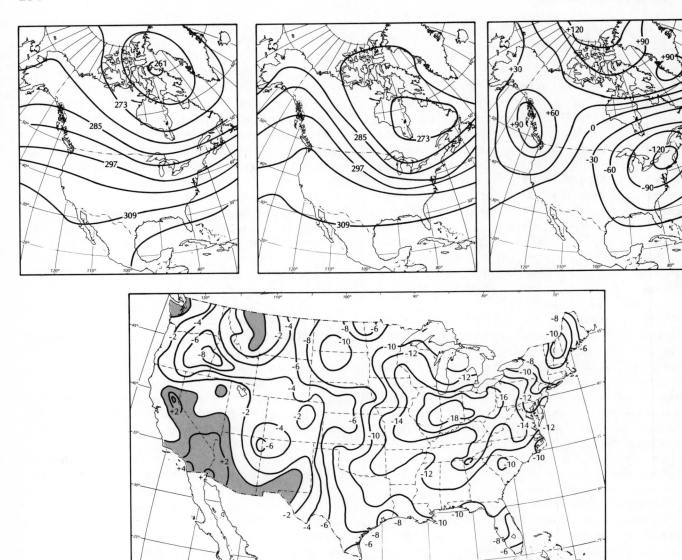

**FIGURE 12.25**
Associations between 700-mb patterns and surface air temperatures for January 1977 and February 1976. (From U. S. National Weather Service, NOAA.)

slowly or stagnate. Of course, serious episodes can only occur when the lack of circulation is accompanied by the emission of pollutants from local sources. Thus, we can say that certain atmospheric conditions create a high pollution potential.

A climatology of stagnating anticyclones has been prepared for the eastern United States. The number of such cases from 1936 to 1965 is shown in Figure 12.28. This map was prepared by assuming that the surface pressure gradient is a more accurate indication of horizontal wind speed than surface wind measurements, and by determining from surface synoptic maps for this period those occasions and areas where the gradient was such as to produce a wind speed less than 7.5 knots (8.7 mi h$^{-1}$, or about 4 m s$^{-1}$). In northern Florida, for example, there were over ten occasions in these 30 years when this wind speed or less was inferred for seven or more consecutive days. The maxima in the southeast are due to the western extension of the Atlantic subtropical high and to transient anticyclones originating in Canada, which often slow or become stationary along the east coast. The point made earlier in this chapter about the influence of local conditions is particularly applicable in this exam-

# SEC. 12.4 | APPLICATIONS

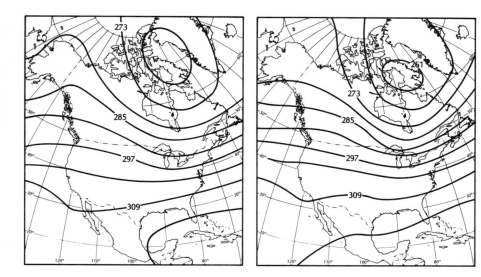
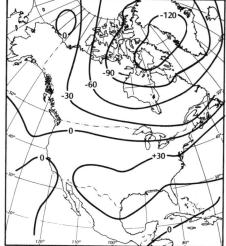
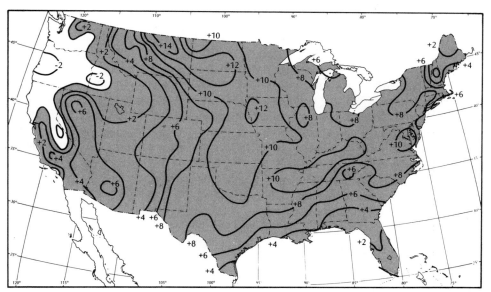

ple. It is likely that where local topography inhibits circulation, as occurs in valleys, conditions would be more extreme than indicated in Figure 12.28.

*Convective activity* is the general term for manifestations of convection in the atmosphere. Its principal features are showers and thunderstorms, including severe weather events such as hail and tornadoes, although its effects are seen also in cumulus clouds from which no rain falls. Meteorologists have long recognized that the likelihood of convective activity in middle latitudes depends on features discernible on the synoptic scale. Fronts act as lifting mechanisms and thereby initiate or augment convection. The extent of vertical motion in the mid-troposphere is related to the position of troughs and ridges at this level, as implied by Figure 12.23.

The occurrence of convective activity in western Texas was studied in relation to surface and upper-air (500 mb) synoptic features. The particular measure used to indicate this occurrence was not precipitation received at the ground, as is usually the case, but rather the number of rain cells per day. The number of rain cells was determined from photographs of the Plan Position Indicator (PPI) scope of a weather radar.

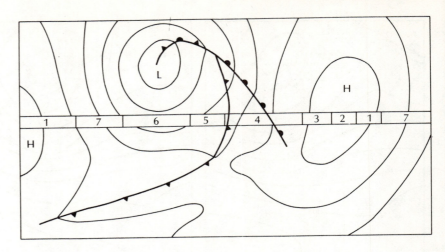

**FIGURE 12.26**
Sectors through a model wave cyclone (transient disturbance) whose weather characteristics are related to disease manifestations (see also Figure 12.27). (After H. Brezowsky, "Morbidity and Weather." In *Medical Climatology,* ed. S. Licht, pp. 358–99. New Haven, Conn.: E. Licht Publishers, 1964.)

PPI representations are familiar to television viewers; many stations have their own radars and televise a picture of the PPI scope when there is precipitation within the viewing area.

In this study the average daily number of initial radar "echoes," or rain cells (those that originate within the area covered by the radar rather than those that originate elsewhere and move into it), was associated with surface and 500-mb patterns. Some of the results are shown in Figure 12.29. The highest average number of echoes occurred when a front was approaching western Texas and was accompanied by a trough at upper levels [Figure 12.29(a)]. The next most numerous echo-producing situations—both virtually the same in this respect—were the surface pattern of (a) coupled with the upper-air pattern of (b), and the surface pattern (b) with upper-air pattern (a). The (b) situation was fourth in this ranking. Notice that (b) shows a very typical summer surface pattern for this area: the westward extension of the Atlantic subtropical high covers all of it. Also, the upper-air patterns are different because the trough axis is farther west in (b) than in (a). Compare these patterns with the models for 700 mb in Figure 12.23.

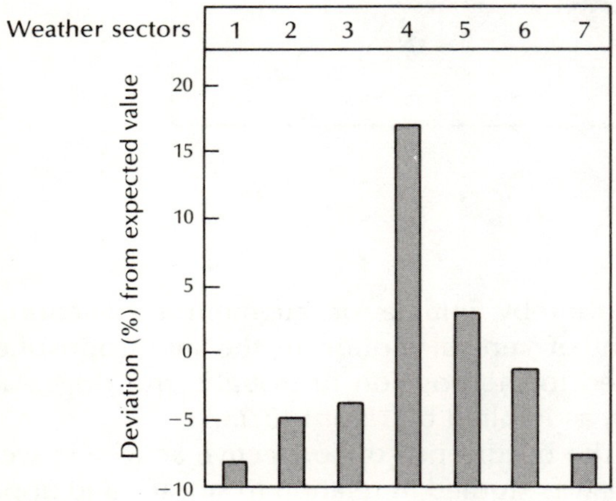

It may seem surprising that thunderstorms are likely to occur in surface pattern (b), in which no fronts are shown. But remember that the circulation phenomena which produce convective activity occur on meso- as well as synoptic scales (see Table 4.1) and include squall lines and other systems which are not apparent in synoptic analyses. The observing stations from which analyses such as Figure 12.29 are prepared are spaced too far apart to enable the analyst to provide the detail sufficient to characterize such mesoscale systems, and these fall "within the cracks," so to speak. This points out another deficiency of synoptic-scale representations (the other being local influences). Finer-scale analyses are necessary if associations are to be made between severe storms, as just one example, and atmospheric circulation patterns.

As a final example of the application of synoptic methods, we offer the findings of joint research between meteorologists, entomologists, and plant pathologists. Biological scientists recognize that atmospheric conditions are instrumental in the development and spread of living organisms such as spores, pollen, and insects. The dissemination of such organisms is strongly controlled by wind patterns, as is the movement of air pollutants. The routes taken by plant pathogens obviously depend on the wind, and often it is possible to trace a substance back to its

**FIGURE 12.27**
Frequency of occurrence of various disease manifestations, principally those of the respiratory and cardiovascular systems, in relation to the weather sectors of Figure 12.26. (After H. Brezowsky, "Morbidity and weather." In *Medical Climatology,* ed. S. Licht, pp. 358–99. New Haven, Conn.: E. Licht Publishers, 1964.)

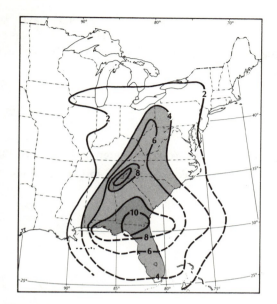

**FIGURE 12.28**
Air pollution potential in the eastern United States: total number of stagnation cases (seven or more days) from 1936 to 1965. (From J. Korshover, Public Health Service Publication No. 999–AP–34, 1967.)

ment of air has been upwind of its present position over the last several hours or days, its former location can be approximated on a series of synoptic maps. The surface map is usually of most interest, although upper-air maps can be used as well. Using the friction wind approximation for both speed and direction, or the geostrophic approximation for above-surface winds, we can follow an air element upwind and determine where that element was at a time halfway between the present map and the next most recent, retaining the friction (or geostrophic) relationship to each successive isobar intersected. The position of the element is then plotted on the second map and the same procedure applied. By successive plots it is thus possible to trace the pathogen, pollutant, or even insect backward in time. This method works best when pressure systems move or alter slowly, and departures between actual and approximated positions increase the farther back in time we go. Also, to approximate the vertical displacement, different techniques must be used.

Analyses of synoptic maps using the trajectory method and more advanced techniques have been helpful in tracing the transport of plant pathogens such as black stem wheat rust, peanut rust, and potato blight and of the viruses leading to diseases in fowl and foot-and-mouth disease in cattle.

The dispersal patterns of insects also depend on atmospheric circulation, although insects are not passive, as are spores, pollens, and viruses. Synoptic analyses have been used to determine the origin of moths, the caterpillar stage of which is damaging to

origin, or, alternatively, to determine where it is likely to go when released. The method of **trajectory analysis** is used in such studies, and we will digress for a moment to explain it.

Trajectories are tracks of air elements over the earth's surface and are distinguished from both streamlines (see Section 12.3.1) and isobars. If an ele-

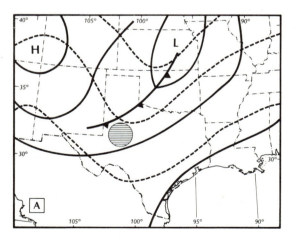 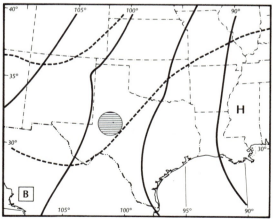

**FIGURE 12.29**
The synoptic patterns (surface and 500-mb) most likely to be associated with convective activity in western Texas. Dashed lines are 500-mb contours. The circular area is that for which the study was made.

crops; grasshoppers; and locusts. The movement and distribution of locust swarms are largely determined by the low-level wind field, in which they are carried along at a speed somewhat less than that of the wind. Concentrations thus occur in convergence zones, and the largest locust swarms in Africa have been observed in and around the ITCZ. The subject of locusts and the weather is discussed further in Section 13.8.

**SUGGESTED READING**

BARRETT, E. C. *Climatology from Satellites*. New York: Methuen and Co. Ltd., 1974.

BARRY, R. G. and A. H. PERRY. *Synoptic Climatology: Methods and Applications*. New York: Methuen and Co. Ltd., 1973.

# 13
# CLIMATE, AGRICULTURE, AND FORESTRY

**INTRODUCTION**
**RADIATION AND VEGETATION**
**TEMPERATURE AND VEGETATION**
**WATER AND VEGETATION**
Evaporation and Evapotranspiration
Precipitation and Drought
**PHENOLOGY**
**CLIMATE AND CROP YIELD**
**ATMOSPHERIC MODIFICIATION**
Irrigation
Mulches
Shelterbelts
Freeze Protection
Reduction of Evaporation
Cloud Seeding
**CLIMATE, PLANT DISEASES, AND PESTS**
**CLIMATE AND LIVESTOCK**
**CLIMATE AND FORESTRY**

## 13.1
### INTRODUCTION

In the final three chapters we are concerned with applied climatology. Recently the statement, "If you eat, you are involved in agriculture," has become a popular phrase. Perhaps one could also add another: "If you grow food, you are involved in meteorology"! Although plant and animal breeding have advanced rapidly, the organism still has to contend with the atmosphere. Farmers constantly are hoping to get just the weather they want over their fields; unfortunately, this may be different from that required by a neighbor. Although there have been many forms of weather modifications, such as irrigation, shelterbelts, mulches, and freeze protection, on the large scale the farmer is still at the mercy of the weather.

The role of climatology comes early in the process of farming, for not only must it be decided what crops can be grown, but also which ones will grow economically in the area. Occasionally this fact is overlooked as the limits of some crops are pushed farther into climatically inhospitable land. For many hundreds of years attention has been given to the cycle of plant development as it relates to the weather. Some of the stages of plant growth have been recorded and then analyzed with respect to the weather. This is the field of **phenology** (see Section 13.5), which preceded the present emphasis on more scientific approaches. To understand the wide role of the atmosphere in agriculture and forestry, we will begin by surveying the impact of some important elements on vegetation.

## 13.2
### RADIATION AND VEGETATION

Although the right amount of solar energy is important to plant development, it is the intensities within certain very specific wavebands that are fundamental. For example, ultraviolet radiation (below about 0.30 $\mu$m) can be harmful, even lethal, to plant growth.

There are three main radiation-induced plant responses, each having its own triggering spectral bands. The process of **photosynthesis,** the formation of carbohydrates from water and carbon dioxide in chlorophyll-containing tissues exposed to radiation, is related mainly to the intensity in bands centered around 0.45 and 0.65 $\mu$m. **Photoperiodism,** a plant's

**FIGURE 13.1**
(a) The spectral absorptance and reflectance of the stems and leaves of desert plants.
(b) The spectral absorptance, reflectance, and transmittance of the thin, light green leaves of *Mimulus cardinalis* (figwort).

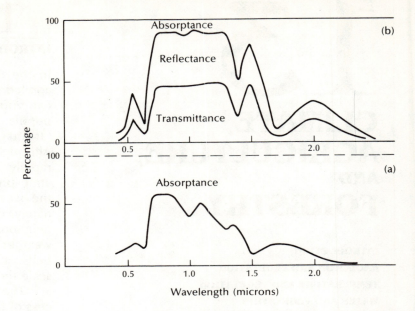

response to the 24-hour pattern of light and dark, has a critical spectral band from about 0.7 to 1.0 μm. The third response is **phototropism**, where the direction of plant growth is determined by radiation in the wavelengths between 0.43 to 0.48 μm. Because of these three responses the visible and near infrared bands have the most important implications for vegetation responses and may be a more meaningful element to measure than the total short-wave radiation.

Solar radiation ($Q$) that impinges upon a leaf is reflected ($Q_R$), absorbed ($Q_A$), and transmitted ($Q_T$), so that

$$Q = Q_R + Q_A + Q_T$$

If $r$ is the albedo (often called the reflectivity) of the vegetation, then $Q_R = aQ$. For most field crops $a$ is about 0.10 to 0.25. The **transmissivity** of leaves is approximately 5 percent, or 0.05, although it can vary from 0.02 for some evergreen plants to 0.10 for certain deciduous trees and grasses. Therefore, the absorbed percentage will be about 70 to 85 percent of $Q$. However, it has been shown that leaf absorptance is definitely a function of the wavelength of the impinging radiation and varies with different plants. Figure 13.1(a) shows the general (smoothed) pattern of absorptance and reflectance. For this type of plant (prickly pear cactus, cushion cactus, and similar plants), the transmissivity is essentially zero. The important feature is that there is low absorptance in the region where there is much incoming solar energy—a fact that assists desert species to survive in an environment of high temperature and high insolation. In Figure 13.1(b) the variations with wavelength of all three components is given for *Mimulus cardinalis* (a type of figwort). Notice how the transmissivity and reflectance parallel each other reasonably well (except around 1.5 to 2.0 μm) and that both vary from very small percentages (around 5 percent or less) to as much as 40 percent in the near infrared.

The percentage of incident radiation used in the photosynthetic process must be small, as we can see in the following calculation. Only some 40 percent of the energy received is in the visible range; of this about one fifth is reflected. In addition, it has been shown that the efficiency of the basic photosynthesis process is around 20 percent. Therefore, the maximum utilization of the incident radiation is only some 6.4 percent ($0.4 \times 0.8 \times 0.2 = 0.064$). In practice the maximum is closer to 5 percent, with a long-term mean of only 1 or 2 percent.

Light rays (0.4 to 0.7 μm) are essential for photosynthesis, but they do not penetrate well into dense stands of vegetation. This means that the lower leaves suffer from light shortage and do not grow well. As far as crop plants are concerned, it would be advantageous for insolation to penetrate the stand and thereby improve the weight of dry matter produced, which is related to the amount of photosynthesis. The ideal arrangement for radiation penetration is when the upper leaves are vertical and the lower ones horizontal. Such a leaf orientation is often bred; the so-called wonder rice is of this type.

The amount of sunlight penetration also is affected by the solar altitude and the orientation of the rows (north-south or east-west). The sunlight that penetrates deep within the vegetation is referred to as a **sunfleck.** For vertical crops, such as cereals, the quantity of sunflecks is appreciable and can lead to the growth of weeds. However, this feature also can be used to undersow a second crop, such as red clover or alfalfa, that develops after the other crop is harvested.

When the light intensity reaches a certain threshold, the leaf becomes light-saturated and the photosynthetic rate becomes dependent on temperature. Plants with a high value of light saturation are called **sun species** (such as rice, sugar cane, and most field crops), while those with a low value are called **shade species** (such as oxalis and philodendron). Tea and cocoa have traditionally been grown with some degree of shading, but recent experiments suggest they grow better in the open. Moderate shading is preferred by coffee and sunflowers.

Photoperiodism, the response of the plant to the length of daylight, was first identified in 1920. Plants are usually subdivided into three classes according to their response:

1. *Short day* (long night) plants flower or respond when daylight is less than 12 hours; examples are some orchids, soybeans, sweet potato, and winter rice.

2. *Neutral* plants are independent of daylight; examples are carrot, pansy, squash, summer rice, and tomato.

3. *Long day* (short night) plants flower or respond when daylight is greater than 12 hours; examples are barley, larkspur, radish, spinach, and winter wheat.

Classification is complicated because some plants change their response to day length with stage of development. For instance, the strawberry needs short days for floral initiation but long days for fruit formation. In addition, the influence of day length is modified by other elements of the environment, especially temperature. In the tropics, where day length varies little through the year, some plants respond to differences as small as 15 minutes. Table 13.1 shows the day length for selected latitudes and days. Since this table includes the effect of refraction it is seen that the day length at, for instance, the equator is increased by some minutes over the simple geometric calculations (Chapter 2).

# 13.3
## TEMPERATURE AND VEGETATION

Even if radiation and moisture conditions are not limiting, each plant will have certain threshold temperatures above ($T_u$) and below ($T_l$) which it will not grow. Additionally, it will exhibit the greatest growth rate at an optimal temperature ($T_o$). These three values are known as the **cardinal temperatures** and generally vary both with the type of plant (Table 13.2) and its developmental stage.

Respiration is a process in which carbon dioxide is produced from oxygen. The respiration rate is increased by high night temperatures. High day temperatures increase the photosynthetic rate, but above a threshold that depends upon the plant (generally around 30° to 37°C, or 86° to 98°F) the rate decreases. The **compensation point** is reached when the respiration and photosynthetic rates are equal and is a function of temperature and light intensity.

**Thermoperiodicity** is a plant's response to the diurnal variation of temperature. For some plants

**TABLE 13.1**
Day length on 21st of each month at various latitudes.

| Month | Latitude (°N) | | | | | | | | | |
|---|---|---|---|---|---|---|---|---|---|---|
| | 0° | 10° | 20° | 30° | 40° | 50° | 60° | 65° | 70° | 80° |
| J | 12.07 | 11.39 | 11.07 | 10.33 | 9.49 | 8.48 | 7.08 | 5.39 | 2.00 | 0 |
| F | 12.07 | 11.52 | 11.36 | 11.18 | 10.58 | 10.28 | 9.44 | 9.09 | 6.30 | 0 |
| M | 12.07 | 12.07 | 12.07 | 12.09 | 12.11 | 12.13 | 12.18 | 12.20 | 12.30 | 13.15 |
| A | 12.07 | 12.24 | 12.42 | 13.04 | 13.30 | 14.07 | 15.05 | 15.52 | 19.00 | 24.00 |
| M | 12.07 | 12.37 | 13.09 | 13.47 | 14.34 | 15.40 | 17.35 | 19.26 | 24.00 | 24.00 |
| J | 12.07 | 12.43 | 13.21 | 14.07 | 15.01 | 16.23 | 18.53 | 22.03 | 24.00 | 24.00 |
| J | 12.07 | 12.37 | 13.10 | 13.48 | 14.36 | 15.44 | 17.41 | 19.34 | 24.00 | 24.00 |
| A | 12.06 | 12.24 | 12.42 | 13.04 | 13.32 | 14.09 | 15.09 | 15.56 | 20.00 | 24.00 |
| S | 12.06 | 12.08 | 12.08 | 12.10 | 12.13 | 12.17 | 12.23 | 12.27 | 12.45 | 13.15 |
| O | 12.07 | 11.51 | 11.35 | 11.17 | 10.55 | 10.26 | 9.41 | 9.06 | 7.30 | 2.00 |
| N | 12.07 | 11.38 | 11.07 | 10.32 | 9.48 | 8.47 | 7.07 | 5.37 | 2.30 | 0 |
| D | 12.07 | 11.32 | 10.55 | 10.12 | 9.20 | 8.04 | 5.52 | 3.34 | 0 | 0 |

**TABLE 13.2** Cardinal temperatures for selected crops.

| Crop | $T_u$ | $T_o$ | $T_l$ |
|---|---|---|---|
| Sorghum | 41°–50°C (106°–122°F) | 31°–37°C (88°–99°F) | 15°–18°C (59°–64°F) |
| Barley, oats, wheat | 31°–37°C (88°–99°F) | 25°–31°C (77°–88°F) | 0°–5°C (32°–41°F) |

(e.g., potatoes and tobacco) it is the night temperature that determines productivity and flavor; for others (e.g., peas and strawberries) it is the day temperature that dominates. The quality of seeds and fruits also is affected by thermoperiodicity. For example, the sucrose concentration in both sugar beet and sugar cane increases with the lowering of night temperatures.

A concept that has been used for over 200 years is the index of the number of heat units ($H$), or degree-days, and its relationship to plant growth. The value of $H$ is given by

$$H = \sum_{i=1}^{N} (\overline{T}_i - T_{th})$$

where $N$ is the number of days in the period (week, month, season, or year) under consideration, $\overline{T}_i$ is the mean temperature on day $i$, and $T_{th}$ is a threshold temperature selected by experimentation. If $\overline{T}_i < T_{th}$, $H$ is zero.

For example, if $T_{th} = 10°C$ (50°F) and $\overline{T}_1 = 20°C$ (68°F), then the first day will contribute 10 units to the value of $H$ (or 18 units if °F are used). The value of $T_{th}$ varies from crop to crop and occasionally changes with the stage of development of the plant. For example, $T_{th}$ for peas is often taken as 4°C (40°F), while for citrus 10°C (50°F) is used. An average value for many temperate zone crops is around 6°C (43°F). Sometimes the value of $H$ is called the **growing degree-days** when $T_{th}$ is 10°C (50°F), but recently the U.S. National Weather Service has introduced a modified expression to indicate that temperatures above 30°C (86°F) do not contribute to crop development.

Using growing degree-days can be quite successful in determining when a crop reaches a certain stage of development, but it is not very helpful in assessing or forecasting crop yields. There are many disadvantages to the use of this form of relationship, but there are two of prime importance. First, a large value is given when $\overline{T}_i$ is high, when in fact the effect could be detrimental. Second, the diurnal range is ignored in spite of the fact that thermoperiodicity is known to be important. In recent years the concepts of photothermal units (temperature multiplied by the day length) and radiation units have been tried in some studies of crop development and yield.

# 13.4
## WATER AND VEGETATION

Vegetation needs to have sufficient water at all stages to sustain optimal growth, assuming other elements are present in their correct quantities. Insufficient water can cause wilting, while too much water can result in waterlogging of the soil and a lack of available oxygen for the plant. These limits mean that all aspects of the water budget should be considered. Very often, however, more attention is paid to the water gain, in the form of precipitation, than to the water loss.

Most of the water gain to soil or plant is due to precipitation, although a small amount comes from condensation occurring directly from the air (called dewfall, to distinguish it from the dew that results from moisture in the ground or plant). In some areas flooding is also an important source of water, particularly in regions where **recession agriculture** (planting in the wet areas as the water recedes) is practiced.

Before proceeding a few terms must be defined. Near the surface there is a **zone of aeration** in which the pore spaces contain both air and water. This zone can be of almost zero thickness in swamps to hundreds of feet in some arid areas. Water in this zone is called **soil moisture**. Below the zone of aeration there is the zone of saturation, or ground water zone. Soil moisture can be present as water vapor, hygroscopic moisture adhering to soil particles, capillary water in the smaller pores and gravity water in transit to the large pore spaces, or interstices. **Infiltration** occurs with the movement of water through the soil surface and is distinguished from **percolation** which is the movement of water through the soil (the gravity and capillary water in the zone of aeration).

Water loss at the surface can occur in three ways: infiltration, evaporation and surface runoff. Evaporation is the process in which a liquid (water, in this case) transforms to its gaseous state. When vegetation is involved in the system, transpiration also occurs and the composite word *evapotranspiration* is used (see Section 5.6). Because of the importance of evaporation and evapotranspiration in agriculture, a small section is included here. For a more general discussion see Section 5.6.

## 13.4.1 EVAPORATION AND EVAPOTRANSPIRATION

There are seven possible types of evaporation/evapotranspiration components. Divided according to the surface under consideration and whether the water loss is actual or potential, the types are

1. Loss from an open water surface (evaporation, $E_o$).
2. Potential loss from a soil surface in which there is no water shortage (potential evaporation, $PE_s$).
3. Loss from a soil surface under natural conditions (actual evaporation, $AE_s$).
4. Potential loss from vegetation that has all the water needed for full growth (potential transpiration, $PT$).
5. Loss from vegetation under natural conditions (actual transpiration, $AT$).
6. Potential loss from a soil/vegetation complex that is never short of water (potential evapotranspiration, $PET$).
7. Loss from a soil/vegetation complex under natural conditions (actual evapotranspiration, $AET$).

Potential evapotranspiration ($PET$) is usually considered as the evaporation from a short, green crop that fully shades the ground and is always well supplied with water. Under the same weather conditions the mean $PET$ over a long period does not exceed $E_o$. The concept of a potential value is introduced to differentiate from the real or actual evapotranspiration. In a real situation there is rarely a full cover of short, green vegetation with no deficit of water.

Generally, the components of major interest are $E_o$, $PET$, and $AET$. The evaporation from an open water surface, $E_o$, is usually measured by means of an evaporation pan, although some empirical relationships have been suggested. Evaporation pans present some special problems because they are not truly representative of the open water conditions, such as a pond or lake. Insolation is absorbed by the metal pan so that the water temperature is raised. During precipitation much splashing occurs, and air flow across the pan, which is generally raised above ground, is not typical of that over a lake.

The most accurate values of $PET$ have been obtained by studying special catchment areas or by use of lysimeters. A **lysimeter** is a buried tank that contains soil and the vegetation being investigated. The lysimeter can be weighed with extreme accuracy to assess the water balance, or careful measurements are made of water input and water drainage. Because of the complexity and cost of such field studies, many empirical expressions for $PET$ have been suggested. The most frequently used is Thornthwaite's expression (see Section 5.6). His relationship works reasonably well over long periods of time (e.g., using mean monthly data) in the temperate continental regions where temperature and radiation are correlated. His equation does not consider wind movement or warm/cool air advection, however.

Another method uses the equation

$$u = ktp$$

where $u$ is the monthly water consumption (in.), $k$ is the crop constant, $t$ is the mean monthly air temperature (°F), and $p$ is the mean monthly percentage of daytime hours. This approach recognizes that the opportunity for transpiration is limited by the day length and that different crops will have different water use. For instance, in irrigated crops in the western United States the crop constant for rice is about 1.0; tomatoes, 0.7; cotton, 0.6; and citrus orchards, 0.5 to 0.65. Recent work suggests that this formula may be improved by including relative humidity and wind speed.

Methods of calculating evapotranspiration or evaporation fall into three classes: the empirical method, the aerodynamic method, and the energy balance approach. The empirical methods, such as Thornthwaite's and $ktp$, have just been described. The aerodynamic (also called Dalton or mass transfer) method uses an equation that expresses $E$ as proportional to the product of the saturation deficit, a function of wind speed, and a value dependent on the type of surface (see Section 5.6). Generally, the wind speed function is assumed to be linear. In this method the radiation component is ignored.

A better method is the energy balance approach, which is explained in Section 5.6. Using this method, and some acceptable simplifying procedures, H. L. Penman, a leader in the study of the physical processes of agricultural meteorology, derived the expression

$$E = \frac{(cR + E_A)}{(c + 1)}$$

where $E$ is the transport of water vapor (evapotranspiration in units of flux density); $c$ is a term related to the slope of the saturation water vapor pressure curve at the mean air temperature (see Section 5.3); $R$ is the heat budget term (see Section 2.3.3); and $E_A$ is an aerodynamic expression (just described).

For an open water surface $E = E_o$, while for a green crop (water is not limiting) $E = PET$. The ratio $PET/E_o$ approximates 0.7, ranging from 0.6 with short day length to 0.8 with long day length. These values are suggested only for long period estimates, not day-to-day studies.

## 13.4.2 PRECIPITATION AND DROUGHT

When considering the moisture gain of soil and plants, it is not realistic to count all of the precipitation received. Some loss occurs because of surface runoff, infiltration, and percolation (the downward movement of water through the soil), so the plant does not benefit from all the precipitation received. Many attempts have been made to estimate "effective precipitation," but perhaps the best is the simple approach suggested by the U.S. Department of Agriculture. From the monthly precipitation amount, the effective precipitation is calculated by a formula, shown graphically in Figure 13.2. As monthly amounts exceed about 150 mm (6 in.), there is comparatively little increase in effectiveness, and only 5 percent of amounts above that threshold are effective.

**Drought** means different things to different people. The drought of the meteorologist, the farmer, the hydrologist, and the economist do not necessarily coincide. In many meteorological services drought is said to exist when there is a significant decrease of precipitation from the climatologically expected mean; for the agriculturist, a drought occurs when soil moisture and rainfall are inadequate to support healthy crop growth to maturity. Each definition has weak points because of undefined words.

A number of measures of drought have been proposed. The one in general use by the U.S. National Weather Service is the **Palmer Drought Index** (see Chapter 5), which is really a form of soil moisture index. Defining or assessing drought is further complicated by the effects of water stress, which will give differing reductions in crop yield depending upon the developmental stage in which it occurs. For example, a water stress at the heading or blooming stage of wheat will be much more severe than at the tillering or mature stage.

There is a common belief that droughts occur in a cyclic, and thus forecastable, pattern. If this were the case, meteorologists would have been able for many years to give warnings of droughts. Some studies have indicated that in certain very large areas the probability of a drought occurring somewhere within the region does show somewhat regular patterns. The most well known of these is the 22-year cycle suggested to occur in the plains region of the United States. However, studies of small areas do not show this feature. It is likely that in the near future drought forecasts will be of a probabilistic nature, similar to present-day precipitation forecasts.

# 13.5

## PHENOLOGY

**Phenology** is the relationship of climate to the periodic biologic activity of plants and animals. The study of phenology is essential if we are to improve our understanding of the climate-crop relationships. In Section 13.3 we described the degree-day, or heat unit. This concept was developed over 240 years ago by R. A. F. de Réaumur, who believed that the sum of daily temperatures from one stage of development to the next should be nearly constant for the same plant from year to year.

The importance of climate-crop relationships was ably demonstrated in New Jersey by Thornthwaite nearly 50 years ago. He used the growth unit (degree-day) concept to determine when peas should be planted so that they could be harvested on a certain day. Because surpluses or shortages of peas were avoided, this system was of great economic benefit. Day length is often incorporated in these numerical techniques.

In 1938 **bioclimatics** was defined by A. D. Hopkins as the science concerned with relationships between life and what he called the three aspects: climate, seasons, and geographic distribution. In his study of these relationships Hopkins wrote, "Phenomena of life and climate as modified by terrestrial influences should be equal under equal influences at the same level across the continents along lines (isophanes) which depart from the parallels of latitude at the

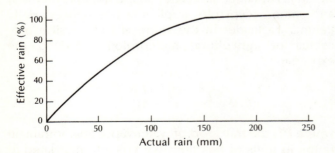

**FIGURE 13.2**
Effective vs. actual precipitation.

assumed constant rate of 1° of latitude to 5° of longitude."* The latter part of the quote is an assumption based on observations of blossoming times of such plants and trees as almond, cherry, foxglove, and lilac. Hopkins was suggesting that these isophanes did not run along the latitudes, but that, for example, the same isophane would go through the points (35° N, 80° W), (40° N, 105° W) and (43° N, 120° W)—a set of coordinates in which latitude increases by 1° as longitude increases by 5°. He also proposed that a 120-m (400-ft) increase in altitude is equivalent to a 1° poleward shift in latitude. Although he makes some sweeping statements, his paper is worth studying for its many examples and concepts.

## 13.6
### CLIMATE AND CROP YIELD

Even though the success of using climatic parameters to predict various stages of crop development has been limited, it has given great impetus to the more important facet of crop yield. Numerous studies of crop yield have been undertaken. Such studies use two basic approaches: information can be gathered over small areas, such as a research station, or over a large area, such as all or part of a state or country. In the first case there is greater control of the variables and more information than otherwise could be obtained, such as planting dates and times and amounts of fertilizer and pesticide application. Such studies generally result in good correlations between yield and climate and other variables.

The first approach cannot be applied to large areas because environmental conditions, especially meteorological, cannot be determined with the same accuracy as for small plots. Also, the excellent agronomic practices used on relatively small plots at research or experiment stations are not feasible for the farmer concerned with crops covering vast areas. For these reasons, the expressions for calculating yields can hardly be extrapolated from small plots to large acreages.

In practice, where crop yield estimates for large areas must be made, the method used relates, often linearly, some climatic variables to the yield. This technique has many problems, as the following list indicates.

*A. D. Hopkins, *Bioclimatics, A Science of Life and Climate Relations.* U.S. Department of Agriculture Publication 280 (1938), pp. 3-9.

1. Is the yield figure representative? It may not be if it has been obtained from small samples. Does it combine irrigation and nonirrigated areas? Is the planted area differentiated from the harvested area?

2. Are the climatic data representative? A single station's values cannot truly represent those of a large area, especially in such elements as rainfall, which is spatially so variable.

3. How can the effects of management, fertilizers, pesticides, and crop varieties be taken into account? The normal practice is to assume that some technological advance has affected yield from year to year, either in a linear or a sigmoid (s-shaped) form, and then to correlate the yield deviations from this with climatic variables. Although this is accepted procedure, the true technological advance is unknown or can only be approximated.

4. What time interval is important? Generally monthly data are used in the regression models, but these can mask completely the short-period variations in rainfall that are so important in determining yield. For example, wheat is much more susceptible to water deficit at heading and blooming than at other times, and soil moisture variations should be included.

5. Is the derived equation a good predictor? Simply because a relationship has a good correlation coefficient does not necessarily mean it is a good forecasting tool.

6. Are the estimates of planting and harvesting time realistic? Over large areas average values are used, and yet variations of ±2 to 3 weeks are likely. These variations mean that at any selected time the crop can be in various developmental stages in different, but neighboring, areas. Since a crop generally has responses to climatic stress that depend upon its developmental stage the averaging technique can introduce considerable error.

In Canada studies have used soil moisture estimates in six zones of soil depth to estimate wheat yield, with the model incorporating five periods of development. Some models take account of soil types, interrelationship between variables, nonlinear terms, and variations in the time of planting. At present most models are too coarse to be used as good predictors. However, over very large areas such

**TABLE 13.3**
Some climatic requirements of a selection of crops.

| Crop | Temperature | Rainfall | Relative Humidity | Wind Speed | Sunshine | Elevation |
|---|---|---|---|---|---|---|
| Apples | Require winter cold for successful cultivation (about 1000 hr below 7°C, or 45°F); roots can be damaged below −7°C (20°F), but snow can protect them to withstand −34°C (−30°F); sharp severe freezes after moderate temperatures are hazardous; optimal mean summer temperatures are 18°–24°C (65°–75°F). | Need good soil moisture in growing season. | | | | |
| Bananas | Need monthly mean over 21°C (70°F), 27°C (80°F) is optimal; a cool season reduces leaf area so that bunches do not fill and sun scald can result; below 13°C (55°F) is bad, freeze is deadly. | 5 cm (2 in.) monthly is essential; best is 2.5–5 cm (1–2 in.) weekly. | | When fruiting, winds of over 24 km h$^{-1}$ (15 mi h$^{-1}$) can cause blowdowns, so thunderstorms are bad. | | |
| Cabbage | Will survive −18°C (0°F) if in right stage of development; susceptible to unseasonal heat or cold; best monthly means are 16°–21°C (60°–70°F)—above this, growth is slow and quality is poor. | Annual amount needed 75–100 cm (30–40 in.) or irrigated; optimum is 2.5 cm (1 in.) per week in growing season. | Low values interfere with normal development, but reduce some diseases. | | | |
| Citrus Fruits | Grow over 13°C (55°F) and below 38°C (100°F); low temperatures (below freezing) are acceptable if of short duration; below −7°C (20°F) can damage or kill trees; sweet oranges and grapefruit most cold tolerant. | About 90 cm (35 in.) needed annually; irrigation usual; fruit-bearing tree requires more moisture. | High values give thinner skins and juicy fruit with better quality. | | | Basically subtropical fruit; in tropics oranges and tangerines grown at about 1000 m |

| | | | | |
|---|---|---|---|---|
| | | | | (3000 ft); lower elevations needed for grapefruit, pomelo, and lime. |
| Cotton | Needs 180–200 consecutive freeze-free days; average hot season is 25°C (77°F); continuous growth and development at 32°C (90°F); over 37°C (98°F) checks growth of main stem. | Requires between 50 and 150 cm (20 and 60 in.) annually; moisture preceding bloom and boll setting is most important; dry season needed. | | Over 50% possible needed. |
| Rice | Growing period of 120–150 days; temperatures above 15°C (59°F) needed, but not above 35°–38°C (95°–100°F); optimum is around 20°C (68°F) at night and 30°C (86°F) during day; water temperature important in germinating and seedling stages. | Needs abundant water, especially in late part of vegetative period; values of 4 to 5 cm (1.5 to 2 in.) daily are often quoted. | | Growing period to have less than 14 hours photoperiod; abundant sunshine. |
| Spring Wheat | Spring of 21°C (70°F) is optimal; 90 freeze-free days; optimal temperatures 5°–20°C (41°–68°F). | Annual rainfall 25–75 cm (10–30 in.); low rainfall for hard variety; high rainfall for soft variety and Durum; drought at jointing, flowering, and dough stages reduces yields. | | |
| Winter Wheat | Winter not too severe; below 5°C (41°F), growth ceases. | Hard variety needs drier conditions; soft variety accepts wetter conditions; less susceptible to drought periods than spring wheat. | | |

**TABLE 13.4**
A thermal classification of vegetables.

| Type | Subtypes | Vegetables |
|---|---|---|
| Cool region [16–18°C (60°–65°F)], intolerant of monthly means above 24°C (75°F) | Hardy, freezing not injurious | Beets, cabbage, parsnips, spinach |
| | Cool season type, usually freeze damaged | Carrots, cauliflower, celery, lettuce, potatoes |
| Tolerant of wide temperature range | Adapted to monthly means of 13°–24°C (55°–75°F); tolerant to freeze under certain conditions | Garlic, leek, onion, shallot |
| | Adapted to monthly means of 18°–26°C (65°–80°F); not tolerant to freeze | Beans, cucumber, peppers (some), pumpkins, sweet corn, tomato |
| Warm region, long season; do not thrive below 21°C (70°F) | | Eggplant, okra, sweet potato, watermelon |
| Perennial (living from year to year) | | Artichoke (globe), asparagus, rhubarb |

as the U.S. plains or the Ukraine-Kazakhstan area of the USSR they can prove useful as gross indicators.

Table 13.3 lists some of the climatic requirements for achieving a good yield for a selection of crops. Some of the crops have many varieties, for which the requirements vary greatly; Table 13.3 does not attempt to identify all these variations, but gives general guidelines. Table 13.4 shows the thermal needs of a selection of common vegetables. Detailed information can be obtained from specialist organizations such as Agricultural Extension agents or the U.S. Department of Agriculture. These sources also can give information on the influences of soil type on yield.

# 13.7
## ATMOSPHERIC MODIFICATION

To the average person the phrase *atmospheric modification* means one thing only—cloud seeding and enhancement of rain. However, there are six important ways in which local atmospheric conditions can be changed. Some of these alter conditions only in the lower layer of the atmosphere, and in very limited areas, and then often inadvertently. However, they do alter the conditions.

### 13.7.1 IRRIGATION

It is possible that irrigation was the earliest way in which people altered the microhabitat, after they constructed shelters for comfort. In many parts of the world the growing season is warm and relatively dry, so irrigation is necessary for a successful crop. Although crops that are partially drought resistant are usually considered first, they may not be profitable unless irrigation water is available during drought periods. To conserve water, irrigation is generally carefully planned; many planning systems, using climatic variables, have been devised to assist farmers in scheduling their times of irrigation. These systems use estimates of soil moisture content made on the site, as these are more reliable than the empirically derived formulas for soil moisture.

Irrigation water is applied in many ways: via furrows, through grated or holed pipes, with drip systems, and by overhead sprinklers. The drip method, where water is applied close to the root system, is very efficient but is difficult and costly to install. In large-scale projects a system revolving around a central point can irrigate tens of acres at a time.

An irrigated area has a lower temperature than its surroundings due to the increased evaporation, which also causes a higher relative humidity. The most extreme changes are found in the oases of the hot desert areas. In southern Arabia one of the authors measured a change from 48°C (118°F) and 12 percent relative humidity in the desert to 39°C (103°F) and 31 percent relative humidity at the middle of a small oasis only some 50 to 70 m (55 to 75 yd) in radius.

### 13.7.2 MULCHES

Mulches are any material spread over the soil surface for the purposes of improving natural conditions,

## TABLE 13.5
The thermal effects of some mulches.

|  | Black Film | Translucent Film | Aluminum Foil | Kraft Paper | Hay |
|---|---|---|---|---|---|
| Ground heat storage | − | + | − − | − | − − |
| Midday soil surface temp. | 0 | + + | − | − | − − |
| Night soil surface temp. | + | + | + | 0 | + |
| Mean soil temp. (−3 cm) | 0 | + | 0 | − | − |
| Diurnal range temp. (−3 cm) | − | + | − − − | − | − − |
| Soil moisture conservation | + + | + + | + + | + | + |

NOTE: + indicates increase; − indicates decrease; 0 indicates no change. SOURCE: Reproduced with permission of Connecticut Agricultural Experiment Station, New Haven, from Waggoner, Miller, and DeRoo, 1960, p. 30.

chiefly temperature and moisture. Mulches change the physical and chemical characteristics of the interface layer between the soil and the air and thereby alter the radiation absorption and air flow, with consequent modifications of temperature and soil moisture. Originally mulches consisted of straw and natural materials, but now they include paper, plastic or aluminum sheets, rocks, sawdust, wood chips, and peat. In a study in the USSR an application of coal dust resulted in the cotton crop developing one month in advance of the untreated plots. A summary of the effects of some mulches is given in Table 13.5.

### 13.7.3 SHELTERBELTS

Strictly, shelterbelts and windbreaks are not identical. The former applies to a long barrier protecting fields, the latter to a protective planting around a house, garden, or orchard. Windbreaks may be the first way in which people made a deliberate attempt to alter the weather. The aborigines of Australia and the bushmen of the Kalahari Desert of Southern Africa use screens to protect themselves from the wind. In agriculture shelterbelts are used to control soil erosion and snow drift and reduce wind pressure on or damage to objects in their lee. Naturally, a shelterbelt is most effective when the "undesirable" wind comes from a single direction.

A solid screen, while very effective in reducing wind speed in the immediate lee of the shelter, does not affect conditions significantly farther than about 15 times its height (Figure 13.3). The most efficient structure is one in which the shelter is about 50 percent open. In this case the shelter's effect extends downwind to a distance of about 25 times the height of the barrier (Figure 13.3). An efficient shelterbelt or windbreak should use both deciduous and coniferous trees to ensure good density in all seasons.

A shelterbelt's influence is not confined to wind movement. Solar radiation and net radiation are altered in the immediate neighborhood of a shelterbelt. This change can be useful as a protection in very sunny areas and may give some degree of frost protection along a boundary region. Evapotranspiration and loss of soil moisture is normally reduced in the lee of a shelterbelt because of the lower wind speed. Air temperatures are generally little affected, but an increase in absolute and relative humidity and dewfall at night is often reported. Evidence suggests that the vapor pressure within the shelter area can be increased appreciably during the daytime but that there is a smaller difference at night. In the immediate vicinity of the shelterbelt the distribution of precipitation is greatly affected, but no appreciable increase in the total amount of precipitation in the sheltered area has been identified. The same effects apply to the concentration of carbon dioxide. The combined effects of shelterbelts are shown in Figure 13.4. In the study from which Figure 13.4 was taken, it was reported that in general crop yields were increased downwind in the region from 1½ to 12 times the height of the barrier, maximizing at a distance of about 4 to 8 times. Adjacent to the barrier the yields were less, due in part to the use of some soil moisture by the trees.

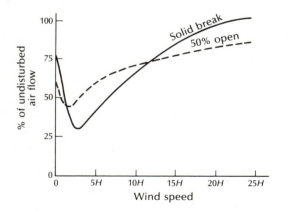

**FIGURE 13.3**
Effect of a shelterbelt of height H on wind speed.

**FIGURE 13.4**
Local climate of open fields compared with fields protected by shelterbelts. (From R. A. Read, "Windbreaks for the Central Great Plains: How to Use Trees to Protect Land and Crops." Paper for U.S. Department of Agriculture, Forest Service, Rocky Mountain Forest and Range Experiment Station, Lincoln, Nebraska, 1965.)

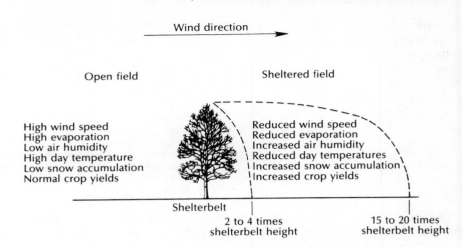

### 13.7.4 FREEZE PROTECTION

Often the words *freeze* and *frost* are used interchangeably, but it is advisable to use the accepted definitions. A **freeze** occurs when the air temperature goes below 0°C (32°F), but a **frost** condition exists when the temperature of the earth's surface and earth-land objects (such as vegetation) falls below 0°C (32°F). Freeze conditions can occur in one of three ways: cold air can be produced locally by radiative heat loss; cold air can be advected into the area; or the two conditions can be combined. When there is advection of cold air, the techniques of freeze protection are usually inadequate for large areas. This is particularly true when high wind speeds necessitate the continuous supply of a large quantity of energy to keep the air temperature around the plant above the critical temperature.

There are five basic methods of freeze protection. A simple technique is to cover the plant with a material opaque to long-wave radiation. Straw, soil, glass covers (cloches), and plastics can be used. Recently a nontoxic, protein-based foam has been tried on row crops such as strawberries and tomatoes. The protection usually is effective in preventing freeze damage with air temperatures as low as −4°C (25°F). The foam either disperses spontaneously or can be dissipated easily by water.

The second method uses smoke from oil-burning smudge pots or from burning old rubber tires or other materials. The clouds of smoke act as a screen that intercepts, in part, the thermal (long-wave) radiation from the surfaces of leaves, fruit, and soil. However, the method is not very efficient, and wind movement can transport the smoke screen from the zone. The combination of the heat of combustion plus the partial moderation of the long-wave radiation loss can be effective, under optimal conditions, against temperatures of as low as −3°C (27°F).

Wind machines or fans have proved to be successful in freeze protection in some areas. When radiative cooling occurs, there is a shallow inversion layer in which the air near the surface may be 8C° (15F°) cooler than air 15 m (50 ft) above. If the fan location and size are correct, mixing the stable air then can bring about a slight warming of the lower levels.

In the sprinkling technique, the latent heat of fusion as the water turns from liquid to ice is used to keep the leaf or fruit from freezing. The rate of sprinkling is very critical; with too little the vegetation freezes, while too much may lead to an ice buildup that cannot be supported. When wind speeds are high or relative humidity is low (below 60 percent), this method is not very efficient.

The most common method of freeze protection is to use some form of heating—burning coal, oil, wood, or solid fuels. A large number of small sources of heat are preferred to a few large heaters, because in the latter case the ascending warm currents can break through the inversion, draw in cold air, and make matters worse. Studies indicate that three or four solid-fuel heaters per tree can maintain a 3 to 7C° (5 to 13F°) increase and as much as 4C° (7F°), even in winds of 16 km h$^{-1}$ (10 mi h$^{-1}$). Clearly, in all of these methods a precise forecast is essential so that the farmer does not commence the costly protection procedure needlessly.

### 13.7.5 REDUCTION OF EVAPORATION

For nearly 40 years experiments have been conducted to try to reduce the evaporative loss from both open water surfaces and plants. Use of a monomolecular layer of a harmless chemical (generally cetyl alcohol, a mixture of duo- and hexadecanol hydrocarbons) to cover the water surface has been popular, but because the film has a very low surface tension, a slight wind will break the cover extremely easily. Maintaining coverage over a lake or reservoir is therefore difficult, and at present the method is not really economic. For plants, antitranspirants that partially close the stomates have been tried, but they are effective generally only in greenhouses and not under field conditions.

### 13.7.6 CLOUD SEEDING

It was demonstrated some 30 years ago that introducing artificial nuclei into suitable clouds may induce precipitation. Since then many experiments have been tried around the world, with varying results. The usual seeding agents are solid carbon dioxide, silver iodide, water droplets, or other large hygroscopic nuclei (such as salt). The latter two are suitable for use with clouds at a temperature above freezing.

One of the major difficulties has been developing a method of getting the nuclei into the cloud so that they are well dispersed. Techniques have included using aircraft, which is expensive; using rockets timed to explode at or near the cloud base, which can be dangerous due to falling debris and are often inaccurate; and using ground-based burners and depending on the wind to carry the particles upward, which also is very inefficient and does not ensure that the nuclei enter the clouds.

Tests to determine the efficacy of cloud seeding must be designed to ensure that the seeding periods are chosen randomly. Otherwise it is impossible to tell whether the precipitation was induced artificially or by nature. Such experiments have to cover a long period of time to increase the significance of the results. The few well-designed tests conducted tend to indicate a slight (5 to 10 percent) precipitation increase due to cloud seeding, but some experiments in other areas suggest that precipitation can be reduced due to overseeding.

Cloud seeding also has been used to try to suppress or reduce the incidence of hail. Initially it was thought that by introducing many nuclei or embryos into the atmosphere a cloud would become overseeded and dissipate. However, a more recent interpretation is that the increased numbers of nuclei may lead to many smaller droplets or hailstones and thereby reduce the damage. For the most part this method has not been used in scientifically designed experiments, and its efficacy is not determined completely.

## 13.8 CLIMATE, PLANT DISEASES, AND PESTS

Almost all aspects of plant diseases are affected by atmospheric features. As in the case of crop yield studies, most investigations of crop diseases and pests have been simplified due to the complexity of their relationship to climate. The relationship between weather and disease is the basis for meteoropathological forecasting. Plant diseases cause an amazing annual loss in world food production—estimated to total over $3 billion damage in the United States alone.

Many diseases necessitate the use of sprays and dusts. For these to be effective as well as economical, current and future weather conditions must be considered carefully because the air flow will affect greatly their application. When spraying from aircraft, it is optimal to have low wind speeds of 8 km h$^{-1}$ at 1.5 m (below 5 mi h$^{-1}$ at 5 ft height). Above twice this value there are appreciable drifting effects, and at speeds of over 50 km h$^{-1}$ (30 mi h$^{-1}$) good application is impossible. When convection has developed, turbulence increases and nearby susceptible crops may be affected; for this reason application at night or in the early morning hours is generally best. If rain occurs within a couple of hours of spraying, it acts as a washing agent and can remove the chemical. However, dew on the vegetation will act as a sticking agent and improve the efficiency.

The major causes of epidemic or sporadic plant diseases are bacteria, fungi, nematodes, and viruses. Bacteria which are responsible for wilting, leaf spots, galls, and blight enter plants through stomata or wounds or are deposited by insects. A film of water on a plant surface often provides a fluid medium for transport of the bacteria into the plant. Fungi enter the leaves through the stomata and can cause wheat rust, leaf spots, and leaf curl. The proper combination of moisture and temperature results in rapid production of spores, which are spread by wind flow. High temperatures reduce the time between inoculation and spore production. Low light intensity favors the development of powdery mildews and wilts, while high intensities assist cereal rust diseases. Nematodes, wormlike pathogens, generally live in the soil and are often restricted in number during times of drought. Viruses infect a plant through insect feeding or mechanical inoculation. A very expensive virus to control is the tobacco mosaic. Insects play an important role in the transmission of the viruses, so prediction of the weather suitable for insect movement is essential.

Weeds are also an agricultural pest. Some, such as the field bindweeds, are tolerant of a wide temperature range but need water. Others require cool temperatures and are tolerant of water (Canada thistle), or need high moisture and temperature (purple nutsedge and coffeeweed).

We will discuss two specific examples to illustrate the wide diversity of plant diseases and pests. Because

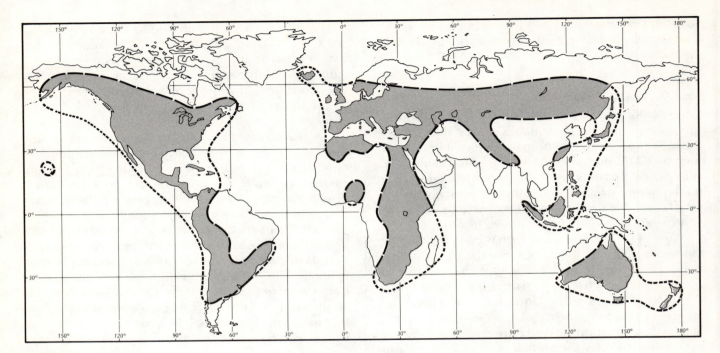

**FIGURE 13.5**
World distribution of potato blight.

of its economic and wide-ranging importance, the problem of the potato late blight (*Phytophthera infestans*) has been the topic of many investigations for over 50 years (Figure 13.5). Early works indicated that there are four climatic conditions, all of which must be present, under which the fungi thrive and spread. These conditions are at least 4 hours of dew at night; night minimum temperature above 10°C (50°F); the next day quite cloudy (8/10 or more); and more than 0.1 mm (0.004 in.) of rainfall. Subsequent studies in Britain modified these conditions, stating that the first outbreak of potato blight follows in one to three weeks after a period of at least two days in which the air temperature exceeds 10°C (50°F) and relative humidity is over 75 percent. It appears that the threshold values for temperature, humidity, and precipitation vary with region and type of potato, but all three elements play important roles. In areas of the United States that experience warm summers, the temperature threshold is usually higher than the 10°C (50°F) just stated.

Detailed studies have shown that the blight spores need water on the leaves and that the potato is particularly sensitive in the early tuber and late maturity stages. The susceptibility to this particular disease is related to percentage of carbohydrate in the plant. This finding provides a clue to a relationship that likely holds in many climate–plant disease problems: Both the weather and the condition of the plant are fundamental and interrelated.

Our second example involves the insect pest *Schistocerca gregaria*, the desert locust. At times of maximum outbreak (plagues) the swarms ravage some 28 million km² (11 million mi²) in Africa and southwest Asia (Figure 13.6). The life cycle of these insects is interwoven with climatic conditions. Although they often breed in the desert regions, the greatest population increases occur with the advent of unusual rains in the area. When these occur the moist soil increases both the ease of egg laying and the viability of the eggs. By the time the hoppers emerge, and through their five developmental phases, there is

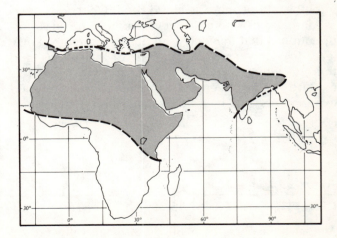

**FIGURE 13.6**
The main geographical distribution of the desert locust.

sufficient vegetation to support a large percentage that would otherwise perish from heat or from their cannibalistic companions. When they can fly, the insects are carried along by winds in excess of about 16 km h$^{-1}$ (10 mi h$^{-1}$) and are brought into zones of convergence, which causes more aggregation. In addition, the convergence of the air can lead to ascending currents, precipitation, and, for these semiarid areas, the rapid emergence of relatively lush vegetation, which provides the food to help enlarge the swarm. Swarms can contain billions of locusts, which eat their own weight in food every day.

## 13.9
### CLIMATE AND LIVESTOCK

Because of their dominant position in animal production, more studies have concentrated on cattle than on all other domesticated animals combined. Each type of domesticated animal has its own zone of thermal tolerance (Table 13.6), a zone that varies with its stage of development, as was the case with plants. Generally reproduction and productivity are affected more by heat than cold, but efficiency and health are harmed more by cold.

**TABLE 13.6**
Zones of thermal tolerance of newborn and adult domestic animals.

| Animal | Zone of Thermal Tolerance | |
|---|---|---|
| | Adult | Newborn |
| Sheep | −3° to 20°C(27° to 68°F) | 29° to 30°C(84° to 86°F) |
| Cow | 0° to 16°C(32° to 61°F) | 13° to 25°C(55° to 77°F) |
| Pig | 0° to 15°C(32° to 59°F) | 33° to 34°C(91° to 93°F) |
| Hen | 18° to 28°C(64° to 82°F) | 34° to 35°C(93° to 95°F) |

Beef cattle are very hardy and, except in severe conditions (blizzards, floods, or prolonged periods of very low temperatures), are climatically affected most through their pasture and water needs. Dairy cattle are more climate sensitive; extreme heat is a severe stress that leads to decreased milk production and a decrease in fertility. Some form of shading that reduces the radiation load but allows good ventilation is desirable, and trees are the optimal solution. Tropical breeds, zebu and brahma, have a greater tolerance to heat than do the European breeds and apparently have the ability to diffuse more water through the skin (whether it is true sweating is still being investigated) to obtain evaporative cooling. Figure 13.7 shows the interrelationships between temperature, humidity, and wind, and the comfort zone for some types of cattle.

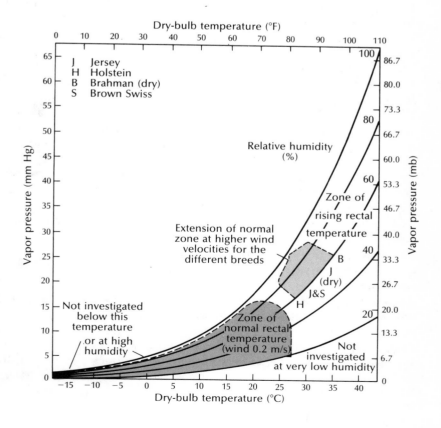

**FIGURE 13.7**
Comfort zones for cattle. (From H. H. Kibler and S. Brody, Bulletin 552. Columbia: Missouri Agricultural Experiment Station, 1954.)

The effects of the atmosphere on sheep have been studied, but not as extensively as those on cattle. The insulating properties of the fleece are immense. Sheep can sustain themselves even at −46°C (−50°F) with minimal sheltering because the fleece traps a thick layer of air near the skin. However, if snow packs into the fleece (as happens when the animals crowd together) or if high winds or rain reduce the insulation value, the sheep become very susceptible to cold. Newly shorn animals, especially the lambs, are particularly affected by cold and wet conditions.

Pigs are the least tolerant of the productive domestic animals and need to have protection from high insolation plus water for wallowing. Very calm conditions and high temperatures lead to a decrease in weight gain, but a wind flow of around 1.5 km h$^{-1}$ (1 mi h$^{-1}$) was shown to reverse this trend. The heavier the animal, the lower is the upper tolerance temperature.

Although the domestic fowl has a high body temperature [42°C (107°F)], temperatures above 38°C (100°F) can cause heat prostration. Newly hatched chicks are particularly vulnerable. Above 32°C (90°F) production can suffer, and when high temperatures combine with high relative humidity, eggshell thickness is reduced. A degree of night cooling (to 16°C or 60°F) in hot seasons is desirable to maintain egg production. The length of light time (day or artificial) is very important in egg production and is one of the triggering mechanisms in egg laying.

# 13.10
## CLIMATE AND FORESTRY

In this section we are concerned not with the climate around trees, as in Section 10.3.3, but with trees as a crop. As with other agricultural crops, trees are greatly influenced by atmospheric conditions (see Figure 11.6). Because of their long maturing time, trees are more likely to experience an extreme season that may reduce drastically their productivity than are other crops.

The young seedlings and saplings are very climate sensitive, so favorable microhabitats, usually in a sheltered area, must be available. The single tree left alone in a large open expanse is unlikely to produce a forest. The seedlings are best grown under a protective canopy, not necessarily complete, of other trees so as to reduce temperature extremes and wind damage. However, very dense stands tend to produce trees of small diameter because the trees grow upward rapidly seeking light. Branches can be snapped off by ice accretion, and in windswept coastal regions saltburn or browning can cause damage and stunting of growth. Severe gales can lead to blowdown, the percentage increasing with tree height and density and affected by root system characteristics. In 1968 a January gale in Scotland destroyed a volume of timber equivalent to twice the country's annual cutting.

Weather also plays a role in transporting timber. The logging roads can become quagmires in heavy rains, bogging down the heavy trucks, while even the logs floated downriver can be jammed or damaged by storm conditions.

The total productivity of forest areas has been investigated by Holdridge, who developed the life zone concept (see Section 8.8). He related productivity to the actual life zone in which the forest grew by using a formula including tree height, diameter, and density. For example, in stands where no strongly restrictive growth factors are present, if a tropical rainforest is ranked as 100 percent productive, then a tropical dry forest would be 67 percent; a subtropical wet forest, 45 percent; a cool temperate wet forest, 22 percent; and a boreal nothern rainforest, 4 percent.

Special studies have related the growth of beech and sugar maples to maximum temperature fluctuations, indicating growth decreased above 27°C (80°F). Another investigation showed how the growth of Ponderosa pine depended upon the departure of rainfall from the average over a 2-year period. Productivity is also related to insect infestations, which are greatly affected by climate and weather.

The most significant hazard for forests is fire, which causes millions of dollars of damage. Hot weather and dry conditions following a time of rapid vegetation growth are particularly conducive to fire. When the undergrowth and forest litter become very dry, the fire risk is great. Some forestry organizations assess the water content of wooden sticks freely exposed to the weather to give information on the relative danger of fires. The Forest Fire Danger Meter combines the wind speed, the amount of the last rainfall, the time elapsed since it fell, and the fuel moisture to rate forest conditions. Although human action causes many fires, it has been estimated that lightning starts 25 to 50 percent of them in the northwest United States and southwest British Columbia. High wind is an extra hazard as it leads to the creation of new fires by flying embers and sparks. It is possible that in some areas, under favorable conditions, cloud seeding may help in fire control.

## SUGGESTED READING

CHANG, J-H. *Climate and Agriculture*. Chicago: Aldine, 1968.

HOPKINS, A. D. *Bioclimatics, A Science of Life and Climate Relations*. U.S. Department of Agriculture Publication 280. Washington, D.C., 1938.

WORLD METEOROLOGICAL ORGANIZATION. *The Forecasting from Weather Data of Potato Blight and Other Plant Diseases and Pests*. Technical Note No. 10. Geneva, 1955.

──────. *Meteorology and the Desert Locust*. Technical Note No. 69. Geneva, 1965.

──────. *Weather and Animal Diseases*. Technical Note No. 113. Geneva, 1970.

──────. *Drought and Agriculture*. Technical Note No. 138. Geneva, 1975.

# 14

# CLIMATE, PEOPLE, AND ARCHITECTURE

**INTRODUCTION**
**HUMAN BIOMETEOROLOGY**
Direct Effects of the Atmosphere on People
The Body's Heat Balance
Indices of Comfort
Clothing
Health
Food and Diet
**BUILDINGS**
Climatic Variables and Building Design
Urban Climates

## 14.1
### INTRODUCTION

This chapter continues our discussion of applied climatology, a subject linked with many facets of living—agriculture, human physiology, architecture and design, urban planning, engineering and hydrology, transport and communications, wildlife and fisheries, recreation and parks, industry and commerce, to name a few. Even politics appears not to be immune. In 1979 it was the inability of Chicago's mayor Michael Bilandic to cope with an unseasonal snowfall that contributed greatly to his defeat at the polls.

The term *biometeorology* refers to studies of the relationships between organisms and their atmospheric environment. Therefore, biometeorology has many branches, as it is concerned with animals, insects, and human beings, among other organisms.

## 14.2
### HUMAN BIOMETEOROLOGY

People are subject to the impact of the weather elements. Until clothing was fashioned, crude shelters developed, and fire controlled, human settlement had to be in regions of benign climate where temperatures were not too extreme. Now we are able to modify our immediate microenvironment so that anywhere, even the moon, can be made habitable for a short period of time. Nevertheless, in our day-to-day living the atmosphere plays a most important role.

#### 14.2.1 DIRECT EFFECTS OF THE ATMOSPHERE ON PEOPLE

People are really quite tolerant to atmospheric conditions over a wide range of values, but often there are some upper or lower thresholds beyond which detrimental effects occur. Solar radiation, particularly in the ultraviolet range ($<0.3$ μm), causes tanning of fair-skinned people because of the pigment melanin in their skin. As the intensity or length of exposure increases, skin blistering and, in some cases, skin cancer can occur. High intensities in the ultraviolet wave band are more usual at elevated sites, where there is less of the atmosphere to deplete solar radiation. The eyes are particularly sensitive in this range and can suffer solar conjunctivitis or even cataracts. At greater wavelengths (0.32 to 0.65 μm) insolation can cause sunburn, even on overcast days or behind glass, so a gradual increase in exposure time and other safety measures are needed. On the beneficial side, ultraviolet radiation devitalizes some germs and bacteria and develops antirickettsial compounds (rickets is a bone-softening disease of children). After long periods of dark or overcast conditions it is thought that sunlight can have a psychological benefit.

Our tolerance to temperature alone is very great. It is when extreme temperatures are combined with other elements, such as humidity and wind, that suffering is greatest. People in Australia and South America have been found living almost naked in freezing temperatures; others, clad in short sarongs, can cross deserts where afternoon temperatures reach above 49°C (120°F). People will also voluntarily take sauna baths at temperatures above the boiling point of water and then sometimes plunge into snow or icy water. Of course, people can adapt or acclimatize to a particular temperature. The indigenous people of the hot deserts can walk on sand at 71°C (160°F), and those in some intensely cold regions walk barefoot across snow and ice. However, at very high temperatures heat stroke and desiccation can occur, and intense cold can lead to frostbite of the extremities and strain on the heart as the lungs are cooled by the inhalation of icy air. Pneumonia and other pulmonary diseases are common on the Andean altiplano, where the cold season is called "the harvest of death." When wind movement and low temperature are combined, the cooling effect is magnified, as expressed by the wind-chill equivalent temperature (see Section 14.2.3). Also, when high temperature and high water vapor pressure occur together, people can suffer the irritating rash known as prickly heat or, under intense conditions, heat stroke.

At high altitudes the reduced partial pressure of the oxygen in the air is insufficient for maintaining optimal operation of the body. It is well established that people cannot live permanently above about 5240 m (17,200 ft), even though some may be accustomed to high altitudes. The city at greatest elevation is Wenchuan, China, at 5150 m (16,900 ft); the highest observatory is Chacaltaya, Bolivia, at 5400 m (17,700 ft); and the highest settlement discovered is some ruins at 6400 m (21,000 ft), also in the Andes. It is possible, through genetic adaptation developed over generations, to develop physiological mechanisms which compensate for the reduced partial pressure of oxygen which occurs at such heights.

Weather conditions play a major role in the transport and concentration or dispersion of many undesirable substances—aerosols, pollens and other allergens, insects, and smog, to name a few. Photochemical smog develops when certain chemicals, introduced into the atmosphere by people, are acted upon by intense insolation.

Air pollution events with serious effects have occurred from time to time. The most disastrous event resulted in over 4000 deaths in London, England, in December 1952. The deaths were mostly caused by pulmonary diseases in the very young and the elderly.

### 14.2.2 THE BODY'S HEAT BALANCE

As homeothermic animals, human beings need to maintain an almost constant body or core temperature to survive. This is in contrast to the poikilothermic (coldblooded) creatures whose body temperatures can fluctuate within very wide limits. The body's internal thermal state can vary between only narrow limits if the essential organs are to function efficiently. The average body temperature, $T_B$, is 36.1° to 37.2°C (97° to 99°F); and beyond the interval 35.6° to 38.9°C (96° to 102°F) most people begin to exhibit signs of physical stress. It is clear that to preserve the constancy of $T_B$ the heat gain to the body must equal the heat loss. In Figure 14.1 the components of the body's heat balance are shown. The balance of heat gain and loss is represented by the following equation:

$$M + R + C + P + E = 0$$

where $M$ is metabolic heat, $R$ is the radiation term, $C$ is the convection term, $P$ is the conduction gain, and $E$ is the evaporative loss (negative).

The heat generation by an average person when resting is about 50 kcal m$^{-2}$ h$^{-1}$, a unit called a **MET** (one MET = 0.08 ly min$^{-1}$). The change of $M$ with various activities is given in Table 14.1. The basal metabolic rate of one MET is enough to raise the body temperature by about 1C° (2F°) per hour unless the heat is dissipated in some manner. Food is the main source of this energy, but 80 percent is used for body repair, growth, and heat production; only 20 percent is used in daily activities.

The radiation term has two components: $R_+$ (gain) and $R_-$ (loss). The radiation gain, $R_+$, is composed of short-wave radiation from the sun—direct, diffuse, and reflected—and long-wave radiation from the atmosphere, ground, and all surrounding surfaces. During daylight hours the short-wave component usually dominates, with the direct input being the most important, especially on clear or partly cloudy days. Of course, under certain conditions the diffuse and/or reflected components may become

**FIGURE 14.1**
The body's heat balance.

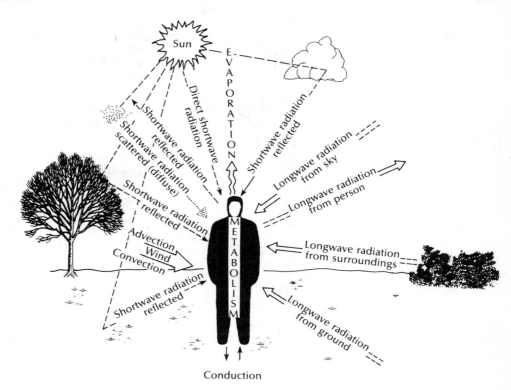

very important. Figure 14.2 shows the hourly changes in the direct radiation falling on a standing person of average size on selected days at latitudes 0°, 40° N, and 65° N. As the sun gets close to the zenith, the direct radiation component is less than when the sun is at medium altitudes. This occurs because only a small surface area is exposed to the direct beam at the time of high sun—a point illustrated by the size of one's shadow. The pattern of the graphs would be altered slightly if the clear-day diffuse radiation were added, which could be done if certain assumptions were made concerning its distribution.

The radiative heat loss, $R_-$, is due to long-wave radiation and is proportional to $T_S^4$, where $T_S$ is expressed in degrees Kelvin (K). (Remember from Section 3.1 that K = 273 + °C.) At night $R_-$ can be an appreciable percentage of the total heat loss.

The convective term, $C$, can be positive or negative. A heat gain, $C_+$, occurs when the temperature of advected and/or convected air, $T_A$, exceeds the skin temperature, $T_S$. This gain is proportional to $(T_A - T_S) V^{0.3}$, where $V$ is the wind speed. Skin temperature varies with the region of the body, with hands and feet usually being cool, but a reasonable average is 33°C (92°F). If $T_A$ is less than $T_S$, there is heat loss, expressed by $C_-$.

The conductive term, $P$, can also be positive or negative. If a conductive heat gain, $P_+$, is to occur, part of the body must be in actual physical contact with a warmer surface. Under normal conditions this is a small value, as is $P_-$, which occurs if the surface is cooler than the body.

The evaporative heat loss occurs when moisture (water) changes state to water vapor (see Section 5.2),

**TABLE 14.1**
Metabolic rate for various activities.

| Activity | MET Units Required |
| --- | --- |
| Sitting at rest | 1.0 |
| Typing | 1.3 |
| Driving | 1.6 |
| Walking (2 mi h$^{-1}$) | 2.0 |
| Walking (downstairs) | 3.8 |
| Walking (4 mi h$^{-1}$) | 4.0 |
| Walking (4 mi h$^{-1}$, against wind of 20 mi h$^{-1}$) | 6.0 |
| Walking (upstairs) | 11.0 |
| Sprinting (22 mi h$^{-1}$) | 40.0 |

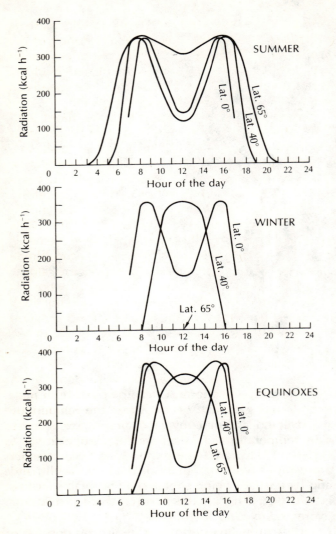

**FIGURE 14.2**
Direct radiation load on a standing person in latitudes 0°, 40°, and 65° at solstices and equinoxes. (Reproduced by permission from *The Annals of the Association of American Geographers,* Volume 61, 1971, p. 486, fig. 1, W.H. Terjung and S.S. Louie.)

because when 1 g of water is evaporated, 580 cal of heat are used. This process happens in two main regions, the skin and the lungs/upper respiratory tract. Very approximately, at an air temperature of about 10°C (50°F) the quantity $(R + C)$ is some 90 percent of the heat loss and $E$ is relatively small. But at 21°C (70°F) this is reduced to 75 percent ($E$ increases to about 25 percent), and at 30°C (86°F) $E$ exceeds $(R_- + C_-)$.

Heat flows from the inner body, at temperature $T_B$, to the skin because there is a gradient of temperature. The quantity of heat flow is proportional to $(T_B - T_S)$ and is inversely proportional to the insulation between the two surfaces. There is also a heat flow from the skin to the clothing and from clothing to the air, when the air is at a lower temperature than the clothing and the clothing is cooler than the skin. Heat flow is always from the warmer to the cooler surface or area. The air itself has an insulation value, which is at a maximum under calm conditions and reduces rapidly as the wind speed increases. Thus a combination of cold air and high wind feels colder than cold air with a low wind movement, because more heat is flowing away from the body. This interrelationship between climatic elements is the basis for studying the impact of the atmosphere on human comfort.

## 14.2.3 INDICES OF COMFORT

Four main weather elements contribute to our physical well-being: temperature, humidity, wind speed, and radiation. Each of these four appears in some form in the heat balance equation (see Section 14.2.2). There is no single atmospheric index that combines all four elements to represent the overall impact of the weather on people under all conditions, but simplified expressions have been proposed for two special climatic regimes—the cold stress region and the heat stress region.

A cooling sensation is felt whenever air at a temperature below $T_S$ flows across bare skin, due to the advection of cool air. Such a sensation can therefore occur even with temperatures of 21° to 32°C (70° to 90°F), but here we will be concerned only with temperatures below about 4°C (40°F). In the range from 4° to 18°C (40° to 65°F) a suitable index most likely would have to include some parameter representing atmospheric water. The generally accepted approach for expressing the sensation of the combination of low temperature and wind speed is by means of the **windchill equivalent temperature,** or, as it is more loosely called, the windchill factor. This method, developed in the late 1930s and early 1940s, applies only to bare skin, however, and in extremely cold, windy conditions no sensible person will be exposing much uncovered flesh. In the early 1970s an index that applies to clothed persons was proposed.

The windchill equivalent temperature, $T_{wc}$, is calculated by equating the combination of air temperature, $T_A$, and wind speed, $V$, to $T_{wc}$ and low wind speed conditions so that heat loss is the same in both situations. A practical problem occurs in selecting the criterion of low wind speed; in actuality, since neither the atmosphere nor a person under intense cold stress is completely at rest, values of 5 to 8 km h$^{-1}$ (3 to 5 mi h$^{-1}$) have been used. Notice that the lower the accepted threshold, the lower will be $T_{wc}$. Figure 14.3, a graph for determining $T_{wc}$ from $T_A$ and $V$, uses the

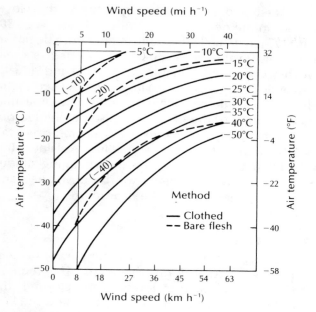

**FIGURE 14.3**
Windchill nomogram using the clothed and bare flesh approaches. (From J. F. Griffiths, *Climate and the Environment*. London: Elek Books Limited/Granada Publishing Limited, 1976.)

threshold value of 8 km h$^{-1}$ (5 mi h$^{-1}$). It can be seen that for $T_A = 0°C$ (32°F) and $V = 26$ km h$^{-1}$ (16 mi h$^{-1}$), $T_{wc}$ is about −5°C (23°F). Using the earlier method (bare flesh), the windchill equivalent temperature would be around −10°C (14°F). The initial formula (bare flesh) always gives lower windchill equivalent temperatures than the more realistic formula under the same conditions. For $T_{wc}$ below about −30°C (−22°F), conditions are bitterly cold, while at −34°C (−30°F) exposed flesh will freeze. If there is insolation during the cold stress periods, the environmental temperature will be higher than if there is none, but often in extremely cold situations this increase is insignificant. However, due to the absorption of the insolation by the body and clothing the effect can be considerable. A special case of self-inflicted windchill stress conditions is that of the skier whose speed can make −10°C (14°F) feel like −25°C (−14°F).

When the air temperature exceeds about 22°C (72°F), the water vapor content of the air becomes important because of the evaporative heat loss needed to retain the body's heat balance (see Section 14.2.2). If the atmospheric humidity is high, the evaporative heat loss is hindered or reduced. Therefore, the **Temperature-Humidity Index** (THI), originally called the Discomfort Index, is often used. The THI is an empirical method of combining the two elements to give a single value that represents atmospheric stress on people better than temperature alone. Notice that wind speed and radiation are not included. Generally, discomfort from heat and humidity decreases as air movement increases, but insolation will add to the unpleasantness experienced. In the extreme condition when the dewpoint exceeds 33°C (92°F), no effective evaporation occurs.

Figure 14.4 shows a graph from which the THI can be calculated, given $T_A$ and either the wet-bulb temperature, $T_{wb}$ as an indication of atmospheric humidity (see Section 5.3), or the relative humidity, $RH$. The use of $T_{wb}$ is preferred since the perspiring skin acts much like a wet-bulb thermometer. As we will see, the THI values for $T_A$ below about 35°C (95°F) are closely represented by $0.5(T_A + T_{wb})$. It has been found that with a THI below 21°C (70°F) no discomfort is felt, but 50 percent of people are uncomfortable at a THI of 24°C (75°F). At values above about 29°C (85°F) there is a distinct stress. A record THI value of 37°C (98.2°F) was experienced at Death Valley, California, in July 1966 and August 1978. Since this is a sensation index, based on the responses of people with varying back-

**FIGURE 14.4**
Temperature-Humidity Index (THI) chart.

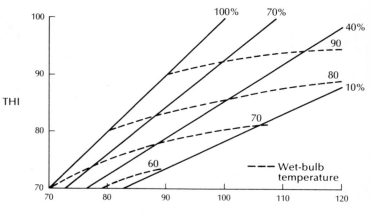

grounds, cultures, and perceptions of heat, it may not apply equally to, for example, an Eskimo or an Indonesian.

No discussion of comfort indices would be complete without reference to the **effective temperature (ET) index** developed over 50 years ago. Effective temperature is the temperature of virtually still (0.2 m s$^{-1}$, or 0.5 mi h$^{-1}$) saturated air which produces the same sensation of comfort or discomfort as the actual conditions of temperature, humidity, and wind speed. The technique used to obtain the ET has been the subject of much discussion. In Figure 14.5 a graph for obtaining the ET is given. The example shows how, with a temperature of 35°C (95°F) and a wet-bulb temperature of 21°C (70°F), the ET varies from 27°C (80°F) with a wind speed of 0.1 m s$^{-1}$ (0.25 mi h$^{-1}$) to 24°C (75°F) at a speed of 7.5 m s$^{-1}$ (16 mi h$^{-1}$). The range of ET values within which an individual feels comfortable is affected by the metabolic rate and the clothing worn (assumed as 1 clo in Figure 14.5; the clo is defined in the next section). With an ET between 30° and 31°C (86° to 88°F) the body temperature begins to rise, while a value of 35°C (95°F) is accepted as a reasonable upper limit of tolerance. If an individual's metabolic rate is high, these limits are reduced. Actually, the THI is derived from a simplified approach to the estimation of the ET. It must also be pointed out that sometimes the term *effective temperature* is used for other indices than that the one just explained. It is hoped that a satisfactory index incorporating all four important climatic elements will be developed and come into general use in the near future.

### 14.2.4 CLOTHING

In order to be comfortable at low temperatures, people must wear some form of clothing. This requirement is usually stated in terms of the clo unit, the insulation value that would permit a heat flow of 1 kcal m$^{-2}$ h$^{-1}$ with a temperature gradient of 0.18C° across the material. One clo is roughly equivalent to the insulation value of a medium-weight wool business suit.

Under light wind flow an unclothed individual is comfortable at rest with an air temperature of 30°C (86°F). The temperature can be reduced to 21°C (70°F) if a 1-clo insulation is worn. At 12°C (54°F) an insulation of 2 clo is necessary to achieve comfort. A realistic maximum of 4 to 5 clo units is normally accepted as the clothing form that still permits mobility.

In hot, dry climates clothing should assist in reducing both the radiation load and the evaporative loss; light-colored, closely woven, and loose fitting outer garments with some underclothing are ideal. In hot, moist climates thin, open-weave material with only minimum underclothes is optimal. In regions of cold stress, layers of clothing are advised because they not only trap air to act as an insulator but can be easily removed or donned as metabolic activity increases or decreases, respectively. Water can cause a reduction in the insulation of some materials, such as wool, by substituting water for air—the former being the better heat conductor. Some investigators suggest it is evaporative heat loss from the wet clothing that reduces its insulating quality.

### 14.2.5 HEALTH

Many common diseases show a seasonal and a climatic variation, although a cause-and-effect connection may be hard to identify. In general, respiratory diseases occur more frequently in winter than in summer, while the opposite is true of infective and parasitic diseases. In some areas, such as New York and Japan, studies have shown that as better protection is

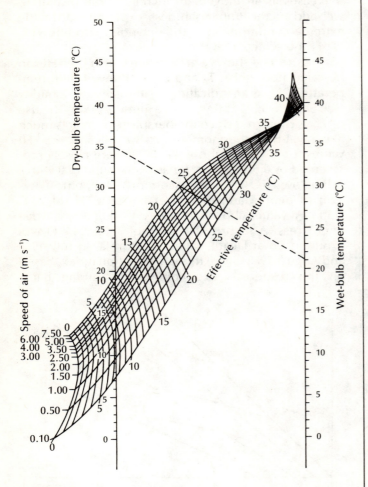

**FIGURE 14.5**
Chart of the effective temperature (ET) index. [From F. P. Ellis, R. E. Smith, and J. D. Walters, "Measurement of Environmental Warmth in SI Units," *British Journal of Industrial Medicine* 29 (1972): 361–77.]

taken against the diseases (improved sanitation and medical facilities, or homes more effectively insulated from the cold), the seasonal differences have declined or even vanished. A special aspect of the respiratory disease–climate relationship involves air pollution. Serious pollution episodes have occurred in London, England, in 1880 (1000 deaths), 1952 (4000), 1956 (1000), 1957 (700), and 1962 (700), while New York runs second and far behind in such statistics.

The impact of weather and climate on people is not limited to physiological responses; there are psychological aspects as well. Spells of overcast weather often cause a large percentage of people to become depressed and moody—a condition which disappears with the first sunny day. Similar responses to rainy periods have been suggested. However, because psychological and physiological responses and effects are interrelated, it is difficult to determine which is dominant. Some investigators suggest it is the change in air ions in the atmosphere that triggers a psychological response in humans, but the small number of controlled experiments reported do indicate that some physiological responses, such as changes in blood pressure and globulins, occur also.

Illnesses triggered by meteorological factors are referred to as **meteorotropic diseases.** Many attempts, especially in Europe, have been made to discover these complex interrelationships. Identifying weather types and their effects on such diseases as bronchial asthma, certain rheumatic complaints, and heart diseases has been a popular line of study. These studies often have found that asthma frequency is not consistently related to pollen or spore count. In foggy conditions asthmatics find some relief, but bronchitic patients suffer badly. For rheumatic sufferers, cold and damp environments are not advised, while hot and humid conditions put extra strain on persons susceptible to some heart problems. Dry heat exhibits no such effects.

Another complex problem concerns the influence of climate on drug action in humans. Within the normal range of environments (suggested as 0° to 40°C, or 32° to 104°F) the effect is small, but modern surgical techniques create special hypothermic (less heat than normal) microscale climates. Research on test animals has suggested the lower toxicity of caffeine at high temperatures, a decrease of tolerance to alcohol in hot climates, and an increase in morphine's toxicity and mortality when the body temperature decreases.

## 14.2.6 FOOD AND DIET

Climate and weather play a most important indirect role in human well-being through their impact upon food and dietary habits. In this section we will examine food preparation and cooking rather than the growing of food, which was considered in Chapter 13.

Rises in temperature, by increasing the activity of microorganisms and enzymes, can cause rapid spoilage. High temperatures also cause fats to liquefy and separate—a very real problem for pastry makers. Optimal growth of yeasts occurs around 16° to 21°C (60° to 70°F), so this is a good range for bread and wine making. Very high temperatures (70°C, or 160°F) are needed to kill most varieties of molds, but low temperatures slow the growth rate of microorganisms, and yeast growth stops below about 10°C (50°F). Even extremely cold temperatures will not destroy some bacteria, as they simply remain dormant.

With high temperature and low humidity, moisture is lost from some foodstuffs. Thus bread and cakes become dry and stale unless protected in airtight containers. Of course, this same combination of conditions can be used for drying or preserving some foodstuffs, such as coffee beans, fish, fruits, and tea. When high temperature occurs with high humidity, bacteria develop rapidly and cause decay of many perishable foods. Cases of serious food poisoning have resulted from eating foods contaminated in this manner.

High atmospheric humidity can affect many dry foods. Cereals may deteriorate rapidly, salt and sugar tend to lump together, and cookies become soggy. On the other hand, in hot, arid areas water housed in porous pots can be cooled to around the wet-bulb temperature by evaporation into the very dry air.

Pressure shows a decrease with altitude of about 1 mb per 9 m (1 mb per 30 ft) up to a height of around 3000 m (10,000 ft). At such reduced pressures water boils at lower temperatures than the 100°C (212°F) needed at sea level. The decrease in boiling point is about 1C° per 300 m (2F° per 1000 ft), a distinct effect on some cooking processes. For example, because tea needs boiling water for best infusion, it is not likely that a tea connoisseur will be satisfied with a pot of tea at above 1000 m (3000 ft). Coffee, which ideally needs water at a temperature of 82°C (180°F) to dissolve the flavor-bearing oils and leave the alkaloids in the dregs, does not suffer from this problem. Most recipes need no adjustments up to elevations of around 1000 m (3000 ft), but above this height the amounts of baking powder or yeast advised should be reduced, pressure cooker settings should be increased, and flour mixtures need special treatment if they are raised by air or steam.

Sunlight acts to destroy vitamin C and can affect bottled fruit, fruit juices, and milk if the containers permit penetration of rays. Green vegetables can wilt and yellow on exposure to intense sunlight.

People living in hot climates should have a diet with reduced fat and carbohydrates, many cold dishes, and uncooked fruits and vegetables. Liquids are normally drunk more frequently than in the cool climates to replace the loss due to evaporation and prevent an imbalance of the body salts. In cold climates the body has a high calorific need, so relatively large quantities of fat and carbohydrates should be consumed. Although outdoor cooking is advised in very hot weather, cooking indoors is desirable during cold periods because the waste heat helps warm the home.

# 14.3
## BUILDINGS

Because most areas of the world have a sufficiently inhospitable climate at some period of the year, shelter from the elements is needed. In its most primitive form, shelter takes the form of a simple lean-to, which is the type used by a few aboriginal tribes in Australia and in the Kalahari desert of Southern Africa. However, generally people need and aspire to something more substantial. In much of the United States and western Europe during the 1950s and 1960s it became almost fashionable to build a box-type house in which the interior climate, or cryptoclimate, was almost isolated from the external climate and was controlled to the inhabitants' liking by means of a thermostat. With the rather tardy realization of the need to conserve energy, we are beginning to design with the climate and not in spite of it.

### 14.3.1 CLIMATIC VARIABLES AND BUILDING DESIGN

The objective of building design should be to achieve a cryptoclimate that is comfortable for the activity to

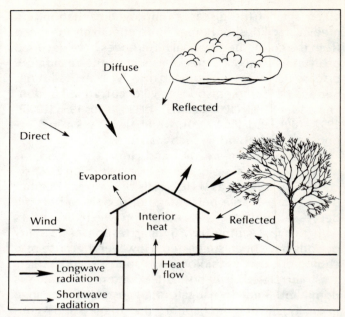

**FIGURE 14.6**
The energy balance components of a building.

be undertaken there, with the minimum expenditure of energy for heating, cooling, and lighting. When a building is erected, there are two aspects to be considered: how climate affects the structure and how the structure affects the climate. Figure 14.6, which shows the energy balance components for a structure, is similar to Figure 14.1, which shows the body's heat balance. In the case of buildings, however, evaporation generally plays a very small role and the conductive heat flow (Section 14.2.2) is quite appreciable, especially for slab constructions. The heat generated in the building interior can come from appliances (refrigerators, cookers, and so on), lighting, and its occupants. Certain general features of the building may be altered to change the resultant cryptoclimate; these are the designer's options, although

**TABLE 14.2**
Building design options and their general effects.

| Unit Variable | Remarks |
|---|---|
| Site selection | Determines climate that will impact upon structure. |
| Orientation and spacing of units | Affects radiation patterns and air flow in the microenvironment. |
| Landscape features | Affect mainly elements of temperature, humidity, wind flow, and radiation on microscale. |
| Outer features (design and materials) | Affect heat balance at interface; help determine cryptoclimate. |
| Floor plan (room location) | Important in determining variation of cryptoclimate within a room. |
| Mechanical equipment | Last resort to obtain comfortable cryptoclimate. |

## SEC. 14.3 | BUILDINGS

**TABLE 14.3**
Some methods of modifying the climatic variables.

| Climatic Variable | Remarks |
|---|---|
| Insolation | Increased by use of dark, heat-absorbing materials and nearby reflecting surfaces. Decreased by use of light-colored materials, shade trees, and overhangs. |
| Air flow | Increased by channeling. Decreased by fences, shrubs, and trees. |
| Temperature | Increased by use of dark, heat-absorbing materials, nearby reflecting surfaces, and protection from cold winds. Decreased by use of light-colored materials, shade trees, and overhangs and by keeping structure open to cooling breezes. |
| Humidity | Increased by surrounding vegetation, pools, and fountains. Decreased by significant amount only by mechanical equipment. |
| Precipitation | Protect walkways by overhangs and guttering of eaves by entrances; use steeply sloping roofs if snow buildup large. |

not all may apply in every instance. These features are called the *unit variables* (Table 14.2).

The climatic variables of greatest significance in the building design are long- and short-wave radiation, air flow, temperature, humidity, light (natural illumination), and precipitation. Generally, by carefully selecting the right pattern of the unit variable, the climatic element can be changed so that comfort is increased. Suggested ways of modifying the climatic variables are given in Table 14.3, which is not intended to be comprehensive.

The insolation on a house is appreciable, with most being absorbed by the roof; thus good ventilation of the attic space and excellent insulation is essential to keep the interior comfortable in warm or hot weather. For a rectangular building, to ensure maximum winter insolation and minimum summer insolation load on the building the best orientation is with the main axis running east-west. However, the improvement over a north-south axis is only about 5 to 10 percent, and usually it is preferable to orientate the building to make the best use of prevailing winds. A very good insolation and light control is deciduous trees, because they are in leaf when insolation needs to be reduced yet allow the radiation through in the cold season.

The positioning of trees, fences, or shrubs around a building makes a big difference to the air flow. Although the number of variations is infinite, Figure 14.7 gives some idea of what can occur. Landscaping in the vicinity of a building can have pronounced and significant effects on air flow around and through the building. Figure 14.7 shows modifications in the air flow patterns determined from wind tunnel studies. The top series shows how a counterclockwise cell can develop in the lee of a thick hedge. The second series shows the effect of a very tall tree on the air movement; when the tree is very close to the building, there is little flow across the roof or within the building. Some of the numerous combinations of tree and hedge appear in the third series; this illustrates the counterflow that can be caused (section A). The last series depicts how various channeling patterns within the house can be induced by positioning a high hedge.

The importance of wall materials is shown in Figure 14.8 (p. 327), where the internal and external temperatures for two types of structures are compared. We can see that a structure of suitable material can reduce the indoor daily temperature range compared with that outdoors. Also, we can see that the time of maximum temperature indoors is 4 hours after the time of maximum temperature outdoors. To get a time lag of 12 hours in temperature extremes from indoors to outdoors, a wall thickness of about 22 cm (9 in.) of wood, 34 cm (14 in.) of brick, or 39 cm (16 in.) of concrete is needed. These are unreasonably large, and in practice the time lag is often no more than 6 hours.

Variations in the Temperature-Humidity Index (see Section 14.2.3) in two houses in Singapore (Figure 14.9 p. 327) illustrate the superiority of the older house over the modern one. This is often the case because the older houses were constructed when very efficient heating was difficult and air conditioning was not available, so more attention was paid to the effects of climate.

Precipitation and roofs can also combine to create special problems. Perhaps the most serious of these is snow buildup in winter in higher latitudes. Many Canadian studies on this subject suggest calculating, for design purposes, a maximum load using a snow density of 0.2 and a roof buildup of 80 percent of the maximum snow load on the ground and assuming a 30-year return period. With these criteria, Edmonton and Toronto have maximum roof loads of 150 kg m$^{-2}$ (30 lb ft$^{-2}$); Quebec, 320 kg m$^{-2}$ (66 lb ft$^{-2}$); and the Gaspé Peninsula, 425 kg m$^{-2}$ (88 lb ft$^{-2}$). When rain falls, roof guttering can channel a large quantity of water to just a few downspouts, leading to soil ero-

**FIGURE 14.7**
How different types of landscaping modify air flow patterns. (From R. F. White, *Effects of Landscape Development on the Natural Ventilation of Buildings and Their Adjacent Areas*. Research Report No. 45. College Station: Texas Engineering Experiment Station, 1954.)

Low hedge (less than 3 ft high)

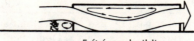
5 ft from building

20 ft from building

High hedge (8–10 ft high)

5 ft from building

20 ft from building

Trees (30 ft high)

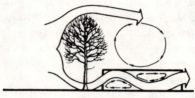
5 ft from building

30 ft from building

Tree–low hedge combination

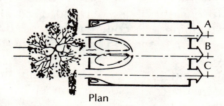
Plan
Hedge: 5 ft from building
Tree: 10 ft from building

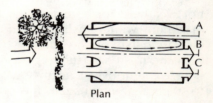
Plan
Hedge: 10 ft from building
Tree: 20 ft from building

Building–hedge combination

Building turned 90° into the breeze with no planting

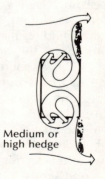
Medium or high hedge

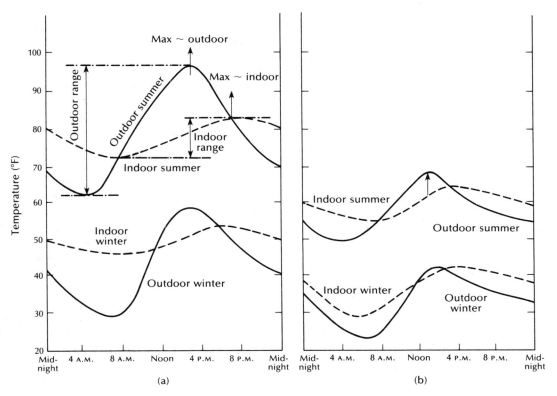

**FIGURE 14.8**
Comparison of typical internal and external temperatures in (a) an adobe in hot, dry climates and (b) a wood structure in the mid-latitudes. (From J. E. Oliver, *Climate and Man's Environment: An Introduction to Applied Climatology.* New York: John Wiley & Sons, 1973.)

**FIGURE 14.9**
Variations in the Temperature-Humidity Index within an old colonial-style house and a modern suburban house in Singapore during part of June 1967. (From P. G. Greenwood and R. D. Hill, "Buildings and Climate in Singapore," *Journal of Tropical Geography* 26(1968): 37–47.)

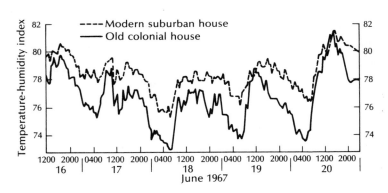

sion and waterlogging. The gravity of this problem is illustrated by the fact that 2.5 cm (1 in.) of rain on a 240 m² (2400 ft²) roof weighs around 5 tons.

## 14.3.2 URBAN CLIMATES

We have seen how even a single building can alter the climate of its immediate vicinity. When there is a complex or community of buildings, it is clear that a greater change will develop. The major alterations can be classified as thermal, hydrologic, and aerodynamic, and those pertaining to air quality.

In the more polluted city air less insolation—both direct and diffuse—penetrates to the street level. The reduction is emphasized during winter when the solar altitude is small. However, because of the complex of horizontal and vertical surfaces in the city, more insolation is trapped during the reflection process than occurs in the countryside. The stone and rock materials of a city usually have a greater heat capacity than that of soil, so they accept and store more heat energy in a shorter time. The greatest temperature difference between the urban and rural areas generally occurs a few hours after sunset on clear, calm, and dry days when heat is still retained in the buildings but the countryside is comparatively cool.

The major hydrologic changes in a city are due to the way in which the precipitation is redistributed by roofs, pavement, gutters, and other drainage systems. Much of the water is soon removed to underground channels, while in rural areas the precipitation is retained more evenly and is available for evaporation, thereby cooling the air. City air soon becomes relatively dry, allowing dust and other particulates to be lifted on the convection cells developed over the urban regions. Thus these particles can act as condensation nuclei for the formation of clouds and, possibly, precipitation. Because of the time taken between the ascent of the nuclei and the precipitation reaching the ground, prevailing winds often carry the rain beyond the city limits. Because cities are, in general, warmer than their environs (the "heat island" effect), light snowfalls sometimes melt before reaching the ground, but heavy falls are affected little. Snow is usually retained longer in the rural regions. Table 14.4 summarizes the urban-rural differences as measured in a number of mid-latitude regions.

A city's numerous buildings, often of varying heights, cause a change in the air flow pattern within the city because they increase the frictional drag and turbulence while simultaneously bringing about channeling of air. The extra drag will, on the average, tend to reduce the air speed in the city; but in certain locations the extra drag will, in concert with the funnel effect, give rise to "windy corners" and "calm pockets," which change in location with wind direction and gustiness.

The city air carries a large load of contaminants, some of which are natural particulates from the breakdown of materials such as bricks and concrete. Because precipitation is soon removed, the dry particles are readily transported into the atmosphere. In addition, cities are often the areas in which industries feed their effluents into the air. Even in medieval times the noxious smells of the towns prompted laws to control certain sources, such as the burning of coal. A man in London was executed in 1307 for breaking the law—a drastic penalty, far removed from the ease with which industrial polluters can circumvent regulations today. The particulates supply condensation nuclei for hydrometeors, but any increase in rainfall in cities is probably due to other factors since additional condensation nuclei probably reduce precipitation amounts. Air flow can cause the maximum precipitation increase to be located far away from the city center. Studies of LaPorte, Indiana, some 50 km (30 mi) downwind of the steel mills of Chicago and Gary, suggest that the city receives 31 percent excess rain, 38 percent more thunderstorms, and 246 percent more hail than the countryside around the city, but more recent analyses have not shown such convincing evidence.

**TABLE 14.4** Average changes in climatic elements caused by urbanization.

| Element | Parameter | Urban Compared with Rural Areas |
|---|---|---|
| Radiation | On horizontal surface | −15% |
|  | Ultraviolet | −30% (winter); −5% (summer) |
| Temperature | Annual mean | +0.7°C |
|  | Winter maximum | +1.5°C |
|  | Length of freeze-free season | +2 to 3 weeks (possible) |
| Wind speed | Annual mean | −20 to −30% |
|  | Extreme gusts | −10 to −20% |
|  | Frequency of calms | +5 to +20% |
| Humidity | Relative—Annual mean | −6% |
|  | Seasonal mean | −2% (winter); −8% (summer) |
| Cloudiness | Cloud frequency and amount | +5 to 10% |
|  | Fogs | +100% (winter); +30% (summer) |
| Precipitation | Total annual | +5 to 10% |
|  | Days (with less than 0.2 in.) | +10% |
|  | Snow days | −14% |

SOURCE: From H.E. Landsberg, "Climates and Urban Planning." In *Urban Climates*, Technical Note 108, p. 372. Geneva: World Meteorological Organization, 1970.)

**SUGGESTED READING**

LANDSBERG, H. E. *The Assessment of Human Bioclimate*. Technical Note 123. Geneva: World Meteorological Organization, 1972.

LICHT, S., ed. *Medical Climatology*. New Haven, Conn.: E. Licht Publishers, 1964.

MACHTA, L., J. K. ANGELL, J. KORSHOVER, and J. M. MITCHELL. "Demise of the La Porte Precipitation Anomaly." *Bulletin of the American Meteorlogical Society* 58 (1977): pp. 106–9

OLGYAY, V. *Design with Climate*. Princeton, N.J.: Princeton University Press, 1973.

WORLD METEOROLOGICAL ORGANIZATION. *A Survey of Human Biometeorology*. Technical Note 65. Geneva, 1964.
———. *Urban Climates*. Technical Note 108. Geneva, 1970.
———. *Building Climatology*. Technical Note 109. Geneva, 1970.

# 15
# CLIMATE AND TRANSPORT, ENERGY, BUSINESS, AND LEISURE ACTIVITIES

**INTRODUCTION**
**TRAVEL AND TRANSPORT**
Water
Road
Rail
Air
**POWER GENERATION AND TRANSMISSION**
**UTILITIES**
**INDUSTRY**
**COMMERCE**
**COMMUNICATIONS**
**TOURISM**
**RECREATION AND ENTERTAINMENT**

## 15.1
### INTRODUCTION

This chapter rounds out our discussion of applied climatology. With the broad spectrum of topics presented, this text has not been an in-depth survey, but it is hoped that the many fascinating, challenging, and important problems described herein will demonstrate climate's wide-ranging influences.

## 15.2
### TRAVEL AND TRANSPORT

Climate and weather have played fundamental roles in people's movements since they began to travel. It is likely that early peoples were prevented from making some hunting forays because of inclement conditions and perhaps from fear and superstition when thunder, lightning, and unusual optical phenomena occurred. In this section, because of the many roles of atmospheric conditions, we will discuss water, road, rail, and air problems separately.

#### 15.2.1 WATER

Although this section emphasizes travel by open water, many of the comments also apply to movements on rivers. If sea travel is to be accomplished successfully, safe and protected harbors must be available. Perhaps the first requirement is for protection from the waves and winds of the open sea. Famous harbors such as Manila, Dar es Salaam, Seattle, and Antwerp all have ideal locations. Wind-driven currents can cause problems with the silting of the approaches, although it is often possible to direct the currents so that the sand and soil are dumped in an offshore deep area. Such a solution works admirably in West Africa, where the Vridi Canal has effected a suitable clearance. Another important consideration in sea travel is the ice-free period. In Norway, influenced by the warm Gulf Stream, the harbors are generally kept open, while those at similar latitudes in Sweden are ice-bound so that ore from the iron mines has to be shipped across the border and the mountain range. The Great Lakes and the St. Lawrence waterway are other examples where ice is an important factor. In the latter case the mean February temperature is a good indicator of the time of opening of the river, with a low value indicating a late opening. In harbors high winds can make docking dangerous, and at loading and unloading times heavy rains can slow or prevent operations as well as necessitating protection for perishable cargoes.

On the open sea the early sailors were often subject to the whims of nature. Many voyages must have ended in disaster due to unfavorable winds and currents. Generally, however, the early seamen kept quite close to shore. About 2000 years ago Greek sailors realized that during the warm season in the Indian Ocean/Arabian Sea winds blow from the southwest toward India, while in the cool season they blow from the northeast. This was an early recognition of the monsoonal and trade wind circulation. They used this knowledge to schedule their sailings and thereby avoid making the arduous and costly trek across the Arabian peninsula. It was this reduction in trade and revenues that contributed significantly to the decline of the Sabean (Sheba) Empire of southwestern Arabia. During the centuries of the sailing ships knowledge of wind patterns was essential for safety and for beating rivals with similar cargoes to the unloading ports. The great races of the tea clippers during the nineteenth century, with financial gain as the main prize, were won or lost through a combination of wind flow, nautical knowledge, and good fortune. A system or scale of estimating winds at sea developed by Admiral Sir James Beaufort in 1806 (Table 15.1) is still, with some modification, in use today.

Ships and their crews are especially at the mercy of weather extremes. The terrible, almost continuous, gale force winds around Cape Horn spelled disaster to many ships as they were driven onto the rocks. Vessels caught in hurricanes or typhoons are tossed around helplessly, and survivors, if any, tell of hair-raising experiences. Ice and fog also cause severe problems at times. Icebergs breaking from the main ice sheets, sometimes due to relatively warm temperatures, are a hazard. For example, 1513 persons were

**TABLE 15.1** The Beaufort scale.

| Beaufort Number | Wind Speed (mi h$^{-1}$) | U.S. National Weather Service Term | Effects Observed on Land |
|---|---|---|---|
| 0 | Under 1 | Calm | Calm; smoke rises vertically. |
| 1 | 1–3 | Light | Smoke drift indicates wind direction. |
| 2 | 4–7 | Light | Leaves rustle; wind vanes move; wind felt on face. |
| 3 | 8–12 | Gentle | Leaves, small twigs in constant motion; hair is disturbed. |
| 4 | 13–18 | Moderate | Dust, leaves, and loose paper raised from ground; small branches move. |
| 5 | 19–24 | Fresh | Small trees in leaf begin to sway; force of wind felt on body. |
| 6 | 25–31 | Strong | Larger tree branches in motion; whistling heard in wires; umbrellas hard to use. |
| 7 | 32–38 | Strong | Whole trees in motion; difficulty in walking. |
| 8 | 39–46 | Gale | Twigs and small branches broken off trees; progress in walking impeded. |
| 9 | 47–54 | Gale | Slight damage to structures; slate blown from roofs; people can be blown over. |
| 10 | 55–63 | Whole gale | Trees broken or uprooted; considerable damage to structures. |
| 11 | 64–72 | Whole gale | Usually widespread damage. |
| 12 | Over 72 | Hurricane | Widespread damage. |

lost as the *Titanic,* a huge (46,300-ton) luxury liner, sunk on its maiden voyage from Southampton to New York in 1912 only minutes after hitting an iceberg. Today radar devices have reduced the probability of similar disasters. Ice can also build up around shores and cause damage to propellers and steering mechanisms; areas around the Great Lakes are particularly bad in this respect. The phenomenon of ice accretion, when ice is deposited steadily and rapidly on exposed surfaces, reduces both stability and freeboard, the distance from the water line to the deck. Ice accretion has caused vessels to become top-heavy and capsize, plunging those aboard into the freezing waters, where survival time is measured in minutes. Fog problems, now not so serious because of radar, are most pronounced when poor visibility occurs in harbor areas. A great deal of time can be lost, and, with the high operational cost of vessels, this can be very expensive. Some fog-prone areas, such as the Newfoundland Banks, parts of the Aleutian Island chain, and the Dogger Bank of the North Sea, are still given a wide berth by many ships.

Cargoes can be very much affected by atmospheric conditions. Items stored on deck can reach very high temperatures when exposed to radiation, making good insulation desirable. Efficient waterproofing is needed as well, for cargoes can be affected by wind, rain, and spume. Cargoes stored in the upper layers of the vessel normally experience temperatures close to that of the air, but if the vessel is of a dark color, absorbed radiation can lead to very high temperatures. Under these conditions refrigeration, although costly, is sometimes essential. In the lower holds the ambient air is generally close to sea temperature. Humidity conditions in the storage areas can be critical for hygroscopic (water-absorbing) cargoes of wood, leather, and wool or paper materials because warping, cracking, and shrinkage can occur. In addition, if condensation takes place, as "ship sweat" or "cargo sweat," corrosion of metals is likely. Many automobiles have become badly corroded on long journeys, and preventative methods are very costly.

As with most transporting activities, time savings are economically expedient. Generally, a great-circle route is the shortest distance, but adverse weather or sea conditions can make deviations from this desirable or essential. A vessel on a great-circle route may be taken too far into an inhospitable region of storms, high seas, or ice conditions. Winds can speed or slow a ship; a stern wind may increase the speed by 1 or 2 percent, while headwinds may effect a slowing of from 3 to 13 percent, dependent upon wind speed as well as the ship's size and shape. High seas, where wave heights exceed about 2 m (6 ft), necessitate a reduction of speed and are avoided as much as possible. The calculation of weather routing for ships is becoming more common because it has many justifications: extra safety, comfort for crew and passengers, protection of cargo, a reduction in insurance premiums, and a saving in voyage time. The time saving is expressible in numerical terms, and studies have shown winter voyages across the Atlantic are reduced by 10 hours on the average by use of weather routing. It is calculated that for British shipping on the Atlantic routes alone a saving of $4 million per year is possible, and this does not take into account that delays sometimes cause perishable goods to spoil. Even savings of a short time are financially desirable; consider that shortening the journey of a 100,000-ton oil tanker by half an hour more than pays the cost of the specialized weather routing forecast. It is sobering to realize that, although they were available to shipping in many areas from the 1920s, weather data have only been acted upon in the last decade.

River traffic is affected by some of the same ice problems that plague sea traffic. Locks, bridges, navigation aids, and port facilities are all influenced by a heavy ice buildup. It is estimated that on navigable inland waters in Canada the annual loss due to ice buildup is about $50 million. On the St. Lawrence waterway the ice incidence is reduced in places by thermal input from nuclear reactors.

A special river problem is variations of the water level. At low flow vessels cannot carry full loads. In the Gambia River in West Africa a very expensive new ferry linking the capital, Banjul, with Senegal cannot be used during periods of low river flow.

### 15.2.2 ROAD

Road surfaces are greatly influenced by weather conditions. Unpaved roads of dirt or gravel particularly are affected by water; heavy rains will cut gullies in the road and reduce traction, while dry spells will lead to clouds of dust being raised as vehicles pass. The **tractionability** of the soil is a function of its type and moisture content. Soils with less than about 15 percent clay, referred to as the nonplastic type, have poor tractionability in dry situations. Their tractionability improves as the soil moisture content increases, but it becomes only fair when the soils are very wet. The plastic soils, containing over 15 percent clay, have good tractionability when dry but deteriorate with increasing moisture content. Surfaced roads often become slick during and after rains, with sometimes appreciable erosion occurring at the shoulders. Water at the edges can then drain back under the surface and lead to road collapse. Blacktopped roads are notorious for bad visibility at night when they are wet and in freezing conditions are susceptible to a frost heave of the surface, one contributor to pot holes. In areas particularly prone to freezing, freezing and thawing degree-days are used. The calculations generally use both mean air and mean surface temperatures and provide information on the depth to which freezing occurs. The degree-days for a single day are the difference between the daily mean temperature ($\overline{T}_i$) and 0°C (32°F). **Freezing degree-days** occur when $\overline{T}_i$ is below 0°C, and **thawing degree-days** occur when $\overline{T}_i$ is above 0°C.

Ice on the roads brings its own hazards, but the spreading of sand and/or salt brings other troubles—pitting of paint and bodywork and corrosion of metal parts, especially underneath the vehicle, for example. Snow, another hazardous condition, can melt during the day and then freeze at night, making the road a sheet of ice. If snowfall is appreciable, snowplows are called out and vast amounts of snow are deposited along or near sidewalks or on parked vehicles. Even in regions where heavy snow is expected in the winter, an unusually early or late snowfall can cause chaos if the plows are not ready and in operating condition. In the United States the fall of the year will usually yield examples of communities caught unprepared. Some roads are protected by fences which, if the wind is from the right direction, lessen the drifting of snow over the road. A study of the disruption caused by snow (Figure 15.1) shows that the maximum disruption occurs in those regions where snow is a relatively unusual occurrence.

Certain climatic features should be taken into account when designing and positioning high-speed arterial roads so that delays from ice and snow build-up, flooding, and fog patches are minimal. Pollution potential from the vehicles should also be considered, for narrow roads with high construction on both sides are conducive to a concentration of gases. Some road features that are affected by certain climatic or weather aspects include underpasses, which can be flooded after heavy rains; low points in which cold air drainage gives rise to fog patches; certain cuts where the channeling of winds causes vehicles to be overturned; and bridges that need to be covered (as in New England and the Alps) to reduce snow loading and the frequency of ice formation. Some roads are particularly prone to the formation of rather disconcerting mirages when the **autoconvective gradient** (when the temperature lapse rate is so great that air density actually increases with height) is exceeded, while others have long stretches or sharp corners that are in line with the setting or rising sun. Under such circumstances, when the sun is 5° to 10° above the horizon, it is very difficult to see ahead.

The volume of traffic on the roads also is subject to weather conditions. In cities the bus companies and parking lots report increased numbers of patrons during wet weather, especially if the day starts rainy and cool. However, nonessential journeys are usually reduced during these conditions. Conversely, good weekend weather normally is accompanied by an increase in traffic volume as people leave the urban areas. Foggy days reduce traffic by about 50 to 75 percent, depending upon the visibility. Icy and snow-covered roads also reduce the traffic volume. Although most of these comments concern automobile traffic, cyclists are even more exposed to the weather and thus are more influenced by its conditions.

Many accidents on the road have been shown to be related to the weather conditions. A study in the United Kingdom determined that accidents increased 25 percent on wet as compared with dry days due to skidding and aquaplaning, while an investigation in St. Louis, Missouri, showed that with a rainfall intensity of over 25 mm (1 in.) per hour there were three times as many accidents compared with lower inten-

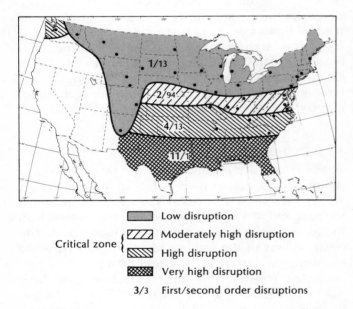

**FIGURE 15.1**
A generalized pattern of snow-caused disruption in the United States. Figures indicate the average number of first- and second-order disruptions per 250 mm (10 in.) of snowfall for the sites within each zone. (From J. F. Rooney, "The Economic and Social Implications of Snow and Ice." In *Water, Earth and Man*, ed. R. J. Chorley, p. 398. London: Methuen & Co. Ltd., 1969.)

sities or dry conditions. A survey in Melbourne, Australia, indicated a 30 percent increase in the number of casualty accidents on rain days over rainless days. At 50 km h$^{-1}$ (30 mi h$^{-1}$) the safe stopping distance on dry concrete is 26 m (84 ft), but this increases to 42 m (140 ft) on wet pavement and 125 m (410 ft) on ice.

Comfort and safety features in vehicles, needed to counteract the effects of climate and weather, can be quite costly. Low temperatures require the use of car heaters or, in severe situations, engine-block heaters, while air conditioners are desirable in hot weather. For the windshield good wipers are essential, but defrosters, defoggers, tinted glass, and sun visors are also usually installed. Certain basic features of the vehicle also are, or should be, dictated by atmospheric conditions; design should reduce air flow resistance, and the coachwork color and paint should absorb or reflect radiation, depending upon whether the vehicle is used mainly in a cold or a warm climate.

Finally, a word should be said on behalf of the pedestrian. Street gutters should not require the prowess of an Olympic athlete to cross after heavy rains, while road surfaces should be such that passing vehicles will not drench the walker with dirty water or slush. Snow and ice are additional hazards, but cities could reduce the number and duration of these dangerous periods by using colored sidewalks (e.g., pink or light green), which would absorb more solar radiation. Wind pressure has a definite effect on people; at speeds of 64 km h$^{-1}$ (40 mi h$^{-1}$) the pedestrian has to lean at a 15° angle to counteract this pressure. The energy expended walking at 1.6 km h$^{-1}$ (1 mi h$^{-1}$) against such a wind is about the same as that used in walking at 6.4 km h$^{-1}$ (4 mi h$^{-1}$) in calm conditions.

### 15.2.3 RAIL

Many of the weather-related problems affecting railroad operations are listed in Table 15.2. The table is self-explanatory, but a few aspects are not covered.

Service on railroads usually is affected by fog, heavy precipitation, and high winds since all of these require a reduction in operating speed. In the United Kingdom alone some $50 million is spent annually on safety precautions and operations during fog. Passenger density is weather dependent and must be included in planning schedules and passenger car availability. Most regions have a known seasonal pattern of passenger use, but many studies have shown greater use of railroads during both very good and very bad weather. Ice on conductor rails reduces the voltage reaching train motors, which was the main reason that a single commuter line in London during January 1970 was halted, delaying over 4000 passengers! On electrically powered equipment, snow and ice on the overhead pantographs (current-collecting frameworks) can cause them to sag, preventing any, or efficient, contact with the power lines. With respect to freight, long delays can cause deterioration of perishable goods. Also, freight trains need not be direct-

**TABLE 15.2**
Weather effects on railroad operations.

1. Roadway and track
   (a) Large temperature changes: cause excessive rail expansion or contraction; "sun kinks" which can bend rails out of shape, keep drawbridges from closing.
   (b) Dry weather: increases danger from fires.
   (c) Low temperature: frost heaving of track and roadbed. Ice and freezing may damage bridge piers. Steel rails become brittle at very low temperatures. Switches may freeze.
   (d) Snow: switches, joints, crossings, etc., can become impacted with snow and rendered inoperative. Drifting of snow or snow slides can block open tracks especially in defiles.
   (e) Wind: only significant when it results in blowing snow or sand, or windchill effect (except for hurricane or tornado velocities).
   (f) Moisture: excessive humidity can cause rusting and corrosion. High rainfall can cause soft spots in roadway or slippage of subgrade materials, landslides. Floods and washouts can occur.

2. Signals, communications
   (a) Temperature: large range can cause breaking of wires with repeated expansion and contraction. Moving parts can freeze.
   (b) Snow: signals covered by drifting snow, lines destroyed by ice and snow coating. Snow slides can carry away signals or wires. Signals and lines especially susceptible to ice storms; communications are disrupted by downing of lines and poles.

3. Yards and terminals
   (a) Temperature: cars roll more freely in hump yards with high temperatures, more slowly with low temperature. May require two different humps. Coal may freeze with low temperature. Diesel fuel oil needs heating when below −18°C to flow more readily.
   (b) Snow: heavy falls as well as drifting can disrupt yard just as it stops movement over open track.
   (c) Switches, rails, roadbed have same weather problems as listed in 1 above.

4. Locomotives and rolling stock
   (a) High temperature: increased incidence of hotboxes in cars. Perishable materials in shipment must be handled in refrigerator cars. Domes needed in liquid cargo cars so liquids can expand and gases can collect and escape.

**TABLE 15.2 (continued)**

(b) Low temperature: frost or ice on rails causes loss of adhesion (coefficient of friction of 0.15 for slippery rails vs. 0.25 for dry rails would mean 40–50 percent reduction in traction and in hauling capacity). Diesels likely to freeze in very cold temperatures should be housed at −40°C or below. Lubrication less effective, resulting in overheated journals and hotboxes.

| Temp. Range (°C) | Diesel Tonnage Reduction (%) |
|---|---|
| −1 to −18 | 0 |
| −18 to −29 | 15 |
| −29 to −40 | 20 |
| −40 to −51 | 40 |

(c) Wind: blowing sand may block tracks, clog air filters, increase wear on bearings.

(d) Rainfall: maximum depth of only 75 mm over rails is permitted for diesels in order to keep moisture out of motors.

SOURCE: From W.W. Hay, "Effects of Weather on Railroad Operation, Maintenance and Construction," *Meteorological Monographs* 2 (1957): 351. With permission of the American Meteorological Society.

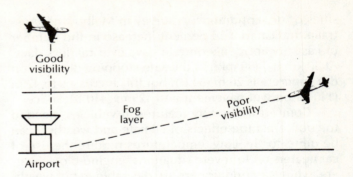

**FIGURE 15.2**
Effect of fog on approach visibility.

ed to ice-bound ports. Many freight items need to be protected from exposure in open sidings or from moisture condensation and related corrosion. In very cold or very hot weather a comfortable cryptoclimate for the passengers and freight can be quite costly to achieve and maintain.

## 15.2.4 AIR

The first impact of the atmosphere upon aircraft operations comes with the selection of the airport location and runway(s) orientation. Because the airport's most important feature is that it should be operational for as much of the time as possible, its site should not be overly prone to fogs; for example, an airport should not be located in a depression or where downslope winds effect an appreciable cooling. Sites downwind of cities where pollution is a problem also should be avoided. Runway orientation is carefully selected after a study of the local wind climatology, and although modern large aircraft are not as troubled by sidewinds as earlier craft, gusty crosswinds are still a major hazard.

Airport operation is influenced by visibility, cloud base height, winds, and runway conditions—all of which involve the atmosphere. As just mentioned, fog can close even a major airport for a day or two. Fog can be rather insidious, even if it is of small depth, because surface features may be visible from above even though the approach visibility may still be below acceptable levels (Figure 15.2). Methods of fog dispersal under cold conditions have used dry ice, silver iodide, and liquefied propane with a reasonable degree of success on the nonadvective types of fog. High and/or gusty winds also pose a problem on takeoff and landing, so detailed knowledge of air flow near the runway is essential. The runway condition itself also is important, with snow, ice, slush, or water on large areas creating a safety hazard of the greatest magnitude. Runway surface temperatures affect takeoff power and the load-carrying capability of the aircraft. Climatology also is used to assist in determining the chances of finding a nearby airport operational if one is closed because of weather conditions, but, of course, current weather is used in this final decision.

Many treatises have been written on weather effects on routing and route conditions, but space precludes mention of all but a few here. The climbing conditions on takeoff are important because the vertical temperature profile greatly affects fuel consumption. This is due to the temperature affecting the air density. Temperature deviations from the standard atmosphere during the cruise phase can alter the flight time; in Atlantic crossings flight time is changed by $\frac{3}{4}T$ minutes, where $T$ is the temperature excess in C°. En route it is desirable to reduce the exposure to headwinds and avoid thunderstorms wherever possible. If an aircraft has to maintain a holding pattern in cold cloud situations while awaiting clearance to land, icing can become a major problem. Clear air turbulence (known as C.A.T.) is another unpleasant, and sometimes dangerous, atmospheric feature caused by wind shears, so identification of these areas is very important. There are, unfortunately, many examples of accidents caused wholly or partly by weather, but these constitute only a very small fraction of a percent of commercial flights. Two recent examples are the Tenerife, Canary Islands, disaster on March 27, 1977, where 582 people died when two 747s collided on a runway of a subsidiary airport when inclement weather closed the regular, well-instrumented airport; and the tragedy at the New York Kennedy International Airport in June 1975, where a commercial flight encountered an intense, small-scale storm on its landing approach and was forced down, with the loss of 114 lives.

Economic impacts of weather on air travel are numerous. As mentioned earlier the payload may have to be reduced due to runway temperature. For supersonic craft it has been calculated that an excess of 11C° (20F°) over the temperature of the standard atmosphere (a not unusual value in many areas) between 7500 and 16,500 m (25,000 and 55,000 ft) means the use of some 2 tons of extra fuel, equivalent to the weight of 20 passengers with their baggage. Delays to a $10 million aircraft, together with the expense of accommodating and feeding the passengers, cause losses to mount. One study estimated that during four winter months of 1962 and 1963 weather effects caused a loss of $37 million to the aircraft industry in the United States because of delay aspects such as those cited here.

# 15.3
## POWER GENERATION AND TRANSMISSION

Both nations and individuals are concerned with the problems related to energy—its generation, transmission, and consumption. In this section the first two aspects are discussed; the third facet is investigated in Section 15.4. There are three different types of power generation that are particularly affected by weather and climate: electrical, solar, and wind.

The effects of climate and weather on equipment and power generation have been well summarized in a study concerned with the north central United States; its major findings are listed in Table 15.3. It is clear that low temperature is the element of most consequence, with snow, sleet, and ice also important. Since these factors are often associated, a period of low temperatures becomes hazardous and usually occurs when the consumer requires above-average energy input. The atmosphere also affects hydroelectric power generation through the catchment conditions. The incidence of snowfall, drought frequency, and the duration of snow cover are some of the important elements that should be considered.

The pollution that can arise from electricity-generating plants presents another problem. This hazard can lead to both air and water pollution, so great care is necessary in choosing a suitable site. It has been suggested that the cooling towers and ponds used to remove the heat generated during power production are instrumental in changing the frequency of clouds, fogs, icing, and precipitation. However, no significant alterations of the atmosphere have been proved.

Transmission of power by overhead lines is seriously affected by severe icing conditions and/or high winds (see Table 15.3). The frequency of occurrence of these conditions has to be considered in the planning stage. Another significant factor in regions subject to great temperature changes, such as occur in areas with continental climates, is the variation in length of the lines from winter to summer due to thermal expansion and contraction.

Recently there has been a surge of interest in solar and wind energy. Their potential power is obviously dependent upon climatic conditions. Generally, those areas with high insolation will need air conditioning—areas of high elevation are exceptions—but solar cooling systems are not yet of great efficiency. The question of the optimal slope and orientation of a flat plate collector is also a climatic one; knowledge of the changes of the solar path with the seasons and the general diurnal variation of cloud amounts and types is needed to answer the query.

Wind power offers a good energy source in some regions, but the location of the generator has to be chosen on the basis of climatic information. Most vanes need wind speeds in excess of about 13 km h$^{-1}$ (8 mi h$^{-1}$) to function efficiently, so seasonal or monthly data concerning the number and distribution of hours with speeds above such a threshold are necessary. Since wind speed increases with height above ground, a rotor should be raised as high as is commensurate with cost, construction, and safety.

# 15.4
## UTILITIES

One of the main problems facing power suppliers, whether they deal in electricity, natural gas, or oil, is the susceptibility of the demand to atmospheric conditions. For long-term variations climatic data are a good guide, but closer to the time of interest these need to be modified by the seasonal or monthly outlooks published by some national weather services.

**TABLE 15.3**
Weather hazards to electric power utilities and distribution lines.

| Type | Low Temperature | High Temp. | Rain | Snow, Sleet, Ice | Humidity | Wind | Lightning |
|---|---|---|---|---|---|---|---|
| Steam | Below 0°C, freezing of water supplies and water intake systems, freezing of coal in railroad cars, freezing of wet conveyor belts, cooling tower problems. Lignite coal freezes easily and creates difficult winter handling problems. | None | None | Slows unloading and handling of coal. | High humidity reduces efficiency of cooling towers. | Might blow water from intake opening. | None |
| Hydro | Temperature 0°C critical because of ice problems, decrease in supply of water runoff. Possible frost damage to hydroplant dams. Winter water supply must be continuous to keep intake equipment free of ice. | None | None | Where snowmelt is source of water, annual changes in snow volume can result in changing surface water supplies. | None | Might blow water from intake opening. | None |
| Diesel | Diesel fuel must be heated prior to motor use. Starting diesel engines with temperatures below −18°C not normally possible because of cold lubricants. | None | None | None | None | None | None |
| Distribution and transmission lines | Maintenance work hindered due to effect on repair crews. | None | None | Lines become coated and poles and lines can be broken by weight. Critical if accompanied by high wind. | None | High winds knock down poles and lines. | Transformers, poles can be damaged if struck. |

SOURCE: From J. W. Waters, "Weather Limitations to Electric Power Utility Operation," *Meteorological Monographs* 2 (1957): 347. With permission of the American Meteorological Society.

Forecasts of conditions from a few hours to a couple of days ahead are necessary in some studies because of their finer detail. Researchers have tried to relate the power demand for heating to the weather elements, specifically temperature, humidity, wind speed, and radiation, and generally their results have been quite consistent. At times of low temperatures, the demand is related significantly ($r^2 = 0.8$ to $0.95$) to the heating degree-days, HDD, where

$$HDD = \sum_{i=1}^{P}(65 - \overline{T}_i)$$

where $\overline{T}_i$ is the mean temperature of day $i$ and $P$ is the period of days under consideration. If $\overline{T}_i$ exceeds 65, then that day's contribution is zero. Some representative annual HDD values are 0 for Honolulu, 250 for Miami, 3000 for San Francisco, 5000 for New York, 6500 for Chicago, and 15,000 for central Alaska (see

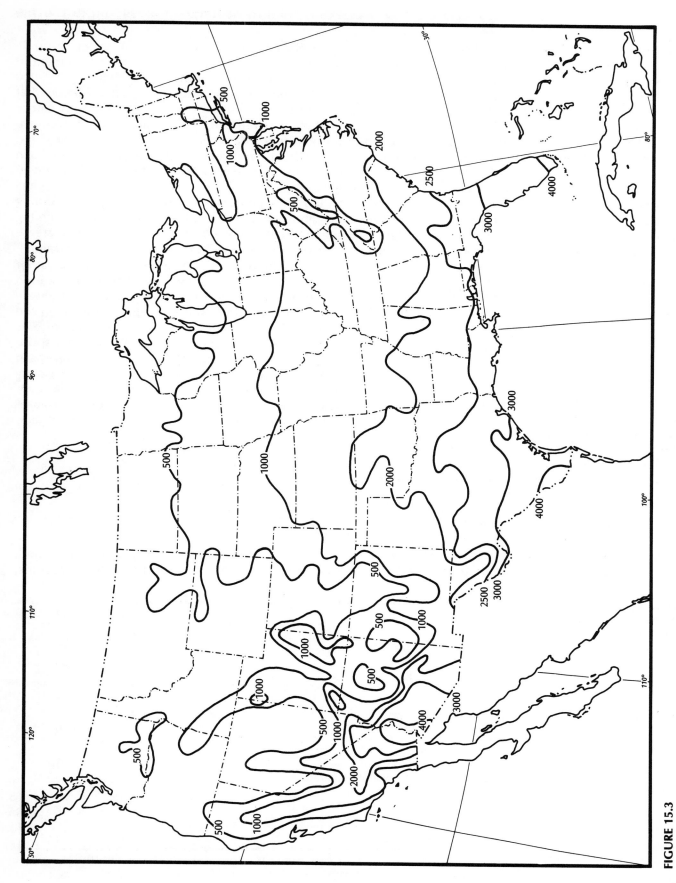

**FIGURE 15.3**
Average seasonal cooling degree-days (base 65°F) from 1941 to 1970. (Courtesy of the U.S. National Weather Service, NOAA.)

**339**

Figure 3.19). A computer-based study showed that a 0.6C° (1F°) decrease in mean annual temperature (or a 0.6C° increase in thermostat setting) would cause a 10 to 12 percent increase in fuel needs in parts of the Gulf States area, decreasing to about 5 percent in the central regions and 1 to 2 percent in the extreme northern United States. Remember that such values are greatly dependent upon the building's location, microscale climate, and construction. Although wind speed (positive correlation) and insolation (negative correlation) play a significant role in the power demands in some areas, their contribution is small compared with that of the air temperature. In Britain, however, a study showed that, compared with a calm day, the effect of a 45 km h$^{-1}$ (28 mi h$^{-1}$) wind was equal to a temperature drop of 2.2 to 2.8C° (4 to 5F°).

On hot days the concept of cooling degree-days (CDD) is used, where

$$CDD = \sum_{i=1}^{P}(\overline{T}_i - 65)$$

When $\overline{T}_i$ is below 65°F, the day's contribution is zero. The value of CDD is related to the air-conditioning load ($r^2$ is about 0.5), but not as closely as the HDD is related to the heating load. In this case insolation and humidity could be expected to have a significant influence, but so far investigators have not discovered any close, and practically important, correlations. Some representative annual CDD values are 50 for central Alaska, 100 for San Francisco, 650 for Chicago, 850 for New York, and 4000 for Honolulu and Miami (Figure 15.3). A uniform increase of 0.6C° (1F°) in mean annual temperature (a thermostat setting decreased by 0.6C°) would increase energy consumption for cooling about 5 to 7 percent along the Gulf States, 10 to 15 percent in the interior and as much as 20 percent in the northern regions of the United States.

A problem associated with such studies is that they include demand variations that are not weather related. It is difficult, if not impossible, to remove these from the data relating to large areas, cities, and towns, but for individual buildings or complexes such correction is possible. Although the examples given generally relate to electrical energy loads, clearly they would be applicable to the heating and cooling demands when gas or oil is used. Studies of lighting requirements showed that heavy clouds increase electrical demand by 350 megawatts in London alone.

Household water consumption is partly weather related, as has been shown by comparing metered versus flat-rate areas. The results of a typical study are shown in Table 15.4; the data apply to areas in the western United States. Average household use is about the same in both cases, but the summer sprinkling is about four times as great in flat-rate areas as in metered areas. This amount for sprinkling is clearly weather related. One specific case, in Victoria, Canada, showed that in the summer the "normal" household's use of water was doubled on sunny and rainless days compared with the average day.

## 15.5 INDUSTRY

Sites for industrial plants are chosen because of both weather-related and non–weather-related reasons.

**TABLE 15.4**
Water use in metered and flat-rate areas (October 1963 to September 1965).

|  | Metered Areas | Flat-rate Areas |
|---|---|---|
| Annual average* |  |  |
| Leakage and waste | 25 | 36 |
| Household | 247 | 236 |
| Sprinkling | 186 | 420 |
| Total | 458 | 692 |
| Maximum day* | 979 | 2354 |
| Peak hour* | 2481 | 5170 |
| Annual† |  |  |
| Sprinkling | 12.2 | 38.7 |
| Potential evapotranspiration | 29.7 | 25.7 |
| Summer† |  |  |
| Sprinkling | 7.4 | 27.3 |
| Potential evapotranspiration | 11.7 | 15.1 |
| Precipitation | 0.15 | 4.18 |

\* Gallons per day per dwelling unit.  † Inches of water.
SOURCE: From F. P. Linaweaver, J. C. Beebe, and F. Skrivan, *Data Report of the Residential Water Use Research Project*. Baltimore: Johns Hopkins University Press, 1966.

Among the latter are the availability of raw materials, water, and labor; the accessibility of the site to transport; and the ease (cost consideration) of assuring comfort for the work force.

Industries can be divided into those in which the activities are mainly indoors or outdoors. The optimal indoor operating ranges for selected industries are given in Table 15.5. While it is possible to ensure that work or storage areas have the required conditions, there is a cost involved in obtaining the best cryptoclimate. If it is possible to locate in regions where the climate already approximates these conditions, there are distinct economic benefits. For example, in California the huge sheds used in aircraft building can be open to the air flow (they need roofs, however) since little heating or cooling is necessary for the workers' comfort. In the case of textiles, it has been suggested that the growth of the cotton industry at the expense of woolens in Lancashire, England, was due to the need for the moist environment of this northwestern county, because cotton thread tends to snap when spun under dry conditions. The iron and steel industry is relatively weather-insensitive, although extremely cold temperatures can lead to increasing brittleness of the products, and a decrease in the water available for processing.

The effect of weather and climate on outdoor industries can also be illustrated by some examples. Table 15.6 shows their influence on the petroleum and chemical industries. Some of these influences are subtle, so different industries must have specific forecasts to achieve maximum efficiency. An interesting example of the climate's effect on an outdoor task occurred in Iceland. After the volcano Heimaey erupted in 1973, heavy ashfall on roofs posed no threats until rain added to the weight and caused roofs to collapse. Because of this the work force was concentrated in areas where rain was forecast.

Weather and the construction industry are intimately related. Table 15.7 shows the critical limits of weather elements that have a significant influence on construction operations, and Table 15.8 summarizes the effects of the most important weather elements. Knowing the likely conditions during the next day allows the construction team to plan operations much more efficiently than if they were unaware of what the weather would do. For example, on a day with temperatures below freezing or with steady, heavy rain there is no need to employ a crew to pour concrete.

Other outdoor industries subject to the vagaries of the weather are shipbuilding and strip mining. In the former, low temperatures and high winds are particularly troublesome, and heavy rains can cause serious problems when strip mining is undertaken. A new industry that is very weather sensitive is offshore drilling. Special weather forecasts are prepared for areas in which rigs are located because some regions are subject to seasons of very severe weather. Lives may be lost and extremely costly platforms and equipment destroyed if such weather is not anticipated.

The fishing industry is particularly susceptible to weather and climate. Changes in the water temperature influence the availability of plankton for food. Warmer water generally will be beneficial for plankton development, but sometimes there are detrimental effects. Along the coast of Ecuador the warm water sometimes spreads southward to the Peruvian coast—a phenomenon called El Niño (meaning the Christ Child) since it occurs around Christmas. This warm water then replaces the cold water upwelling normally experienced, a condition ideal for anchovies and their special plankton food, and thus reduces the catch. In 1972 El Niño reduced the anchovy industry to only 30 percent of the 1970 value. When relatively warm water

**TABLE 15.5**
Optimal indoor operating climatic range for selected industries.

| Industry | Temperature (°F) | Relative Humidity (%) |
|---|---|---|
| Textile industry | | |
|   Cotton | 68–77 | 60 |
|   Wool | 68–77 | 70 |
|   Silk | 71–77 | 75 |
|   Nylon | 85 | 60 |
|   Orlon | 70 | 55–60 |
| Food industry | | |
|   Milling | 65–68 | 60–80 |
|   Flour storage | 60 | 50–60 |
|   Bakery | 77–81 | 60–75 |
|   Candy | 65–68 | 40–50 |
|   Process cheese production | 60 | 90 |
| Miscellaneous industries | | |
|   Paper manufacturing | 68–75 | 65 |
|   Paper storage | 60–70 | 40–50 |
|   Printing | 68 | 50 |
|   Drug manufacturing | 68–75 | 60–75 |
|   Rubber production | 71–76 | 50–70 |
|   Cosmetics manufacturing | 68 | 55–60 |
|   Photographic film manufacturing | 68 | 60 |
|   Cosmetics storage | 50–60 | 50 |
|   Electric equipment manufacturing | 70 | 60–65 |

SOURCE: From W. J. Maunder, *The Value of the Weather*. London: Methuen & Co. Ltd., 1970. After Landsberg, 1960; after Grundke, 1955–56.

**TABLE 15.6**
Effect of climate on operation of petroleum and chemical industries.

| Industry | Low Temperature | High Temperature | Precipitation | Wind |
|---|---|---|---|---|
| Petroleum refineries | Below −7°C critical. Housing of exchangers, pumps, etc., can present safety problem if combustible vapors are present without ventilation. Excessive cooling of pipelines restricts flows, might cause water in gas to freeze or hydrates to form. | Cooling water must be between 15–32°C. Above 32°C cooling towers must be used. Reduces maximum recovery of light oil since increased temperature affects partial pressure of gases. | Causes corrosion and deterioration of units. Old, worn insulation wets easily, contacts give trouble. Oil collectors may become clogged when moisture seeps into unit. | Tall units, cracking, fractionating, cooling towers susceptible to high wind speeds. Gasoline refinery, fires, or fired heaters should be upwind to prevent gas or vapor being blown toward them. |
| Coke chemical | 50% of normal efficiency at 0°C. At −18°C, all weather-sensitive operations must be housed. | High ambient temperatures prevent maintenance of proper oven coking temperature due to reduced upstack draft—requires blowers to supplement draft. | 50-mm rain may shut down all operations. Rain may cool coke ovens and cause shutdown or inefficient operation. | Coal bridge may be damaged by high winds. Wind of 30–40 mi h$^{-1}$ may cool coke ovens too rapidly, affect efficiency of recovery and quality of light oils. Need winds over 4 mi h$^{-1}$ for proper ventilation. |
| Synthetic ammonia | Scrubbing towers subject to freezing. Solubility of scrubbing agent decreases with temperature and so does efficiency. More difficult to remove impurities. | High temperature lowers the efficiency of production; 28% conversion of gas to ammonia at water temperature of 15°C, while at 35°C only 12% conversion occurs. Water scrubbers do not remove as much carbon monoxide. High temperatures cause high vapor pressures which result in loss of stored ammonia. | Stored ammonium sulfate degraded by moisture; complete housing required for protection. | |

SOURCE: From A.W. Booth, "Petroleum Refining and Selected Chemical Industries," *Meteorological Monographs* 2 (1957): 345. With permission of the American Meteorological Society.

enters the English Channel, hake and red mullet are more prolific, but the cold-water fish cod, haddock, and mackerel decrease. Another fishing industry loss occurred in the New York Bight in March 1976, when early warming caused unusual snowmelt and the influx of fresh water. Absence of spring storms and little mixing depleted the benthic oxygen and led to excessive kills of crabs, lobsters, and red hake. Stormy and/or icy conditions often prevent fleets from reaching the fishing grounds and cause great economic loss.

## 15.6 COMMERCE

The retail trade is influenced by many factors—economic, psychological, and sociological influences are foremost. Nevertheless, climate and weather do have a distinct effect. Many shoppers act on impulses to buy, activated by their own wishes or by subtle advertising, and if the weather is inclement, cold, hot, wet, or windy, they may be discouraged from shopping. For this reason closed shopping malls, where people can wander and shop protected from the atmospheric elements, are becoming popular. Even the proximity of parking spaces to the shops is an important consideration.

Many retail trades show a definite seasonal pattern in their sales. In summer certain commodities are in greater demand than at other seasons, such as gardening equipment, pesticides, insecticides, beachwear, sun lotion, outdoor sports equipment, and lightweight clothing. Foods such as salads, cold meats, drinks, ice cream, and picnic goods are popular at this time. Repairs of air conditioners and cars increase as people prepare for holidays and vacations. On hot, humid days, air-conditioned places are frequented.

## TABLE 15.7
Critical limits of weather elements having significant influence on construction operations.

| Operation | Rain | Snow and Sleet | Freezing Rain | Low Temperatures (°F) | High Wind mi h$^{-1}$ | Dense Fog | Ground Freeze | Drying Conditions | Temperature Inversion | Flooding Abnormal |
|---|---|---|---|---|---|---|---|---|---|---|
| Surveying | L* | L | L | 0--10 | 25 | x† | — | — | — | — |
| Demolition and clearing | M | M | L | 0--10 | 15-35 | x | x | — | x | — |
| Temporary site work | M | M | L | 0--10 | 20 | x | x | — | — | — |
| Delivery of materials | M | M | L | 0--10 | 25 | x | — | — | — | — |
| Material stockpiling | L | L | L | 0--10 | 15 | x | — | — | — | — |
| Site grading | M | M | L | 20-32 | 15-25 | x | x | x | — | — |
| Excavation | M | M | L | 20-32 | 35 | x | x | x | x | — |
| Pile driving | M | M | L | 0--10 | 20 | x | x | — | — | x |
| Dredging | M | M | L | 0--10 | 20 | x | x‡ | — | — | x |
| Erection of coffer dams | M | L | L | 32 | 25 | x | x | x | — | x |
| Forming | M | M | L | 0--10 | 25 | — | x | — | — | — |
| Emplacing reinforcing steel | M | M | L | 0--10 | 20 | — | x | — | — | — |
| Quarrying | M | M | L | 32 | 25-35 | x | x | x | x | — |
| Delivery of premixed concrete | M | L | L | 32 | 35 | x | x | — | — | — |
| Pouring concrete | M | L | L | 32 | 35 | — | x | x | — | — |
| Stripping and curing concrete | M | M | L | 32 | 25 | — | x | x | — | — |
| Installing underground plumbing | M | M | L | 32 | 25 | — | x | x | — | — |
| Waterproofing | M | M | L | 32 | 25 | — | x | — | — | — |
| Backfilling | M | M | L | 20-32 | 35 | x | x | x | — | — |
| Erecting structural steel | L | L | L | 10 | 10-15 | x | — | — | — | — |
| Exterior carpentry | L | L | L | 0--10 | 15 | — | — | — | — | — |
| Exterior masonry | L | L | L | 32 | 20 | — | x | x | — | — |
| External cladding | L | L | L | 0--10 | 15 | — | — | — | — | — |
| Installing metal siding | L | L | L | 0--10 | 15 | — | — | — | — | — |
| Fireproofing | L | L | L | 0--10 | 35 | — | — | — | — | — |
| Roofing | L | L | L | 45 | 10-20 | — | — | x | — | — |
| Cutting concrete pavement | M | M | L | 0--10 | 35 | — | x | — | — | — |
| Trenching, installing pipe | M | M | L | 20-32 | 25 | — | x | x | — | — |
| Bituminous concrete pouring | L | L | L | 45 | 35 | x | x | x | — | — |
| Installing windows and doors, glazing | L | L | L | 0--10 | 10-20 | — | — | — | — | — |
| Exterior painting | L | L | L | 45-50 | 15 | x | — | x | — | — |
| Installation of culverts and incidental drainage | M | L | L | 32 | 25 | — | x | x | — | x |
| Landscaping | M | L | L | 20-32 | 15 | x | x | x | — | — |
| Traffic protections | M | M | L | 0--10 | 15-20 | x | x | — | — | — |
| Paving | L | L | L | 32-45 | 35 | x | x | x | — | — |
| Fencing, installing lights, signs, etc. | M | M | L | 0-10 | 20 | x | x | — | — | — |

\* L indicates light; M indicates moderate.
† Indicates operation affected by this condition but critical limit is undeterminable.
‡ Indicates water freeze.
Note 1. All operations are hindered by air temperatures over 32°C (90°F) and THI over 77.
Note 2. The survey data reported in this table are furnished for purposes of illustration only and do not constitute a representation applicable to particular construction or projects.
SOURCE: Environmental Science Services Administration.

During fall fuel sales increase, along with the demand for antifreeze and the services of the furnace or heating specialists. In fall and winter sales of heavier clothing rise, as one study of winter coat sales in New York illustrates (Fig. 15.4). Retailers of snow tires and chains find business on the increase, as, unfortunately, do wreckers during and after ice and snow storms. Pharmacists sell more cold medicines and other cures and palliatives in the winter. In the worst weather local stores tend to do well as shoppers wish to stay closer to home. As spring arrives the sales of beer, carbonated drinks, and iced tea show an upward trend; it has been shown in one study that such sales are directly correlated with temperature. The trade in women's raincoats and umbrellas, shown in Figure 15.5, is also weather dependent.

Income also shows a definite seasonal pattern—a pattern dependent upon the main sources of income of the area (Figure 15.6). For example, Mississippi, which has 11 percent of its total income dependent

**TABLE 15.8**
Effects of weather on construction operations.

| Phenomenon | In Conjunction with | Effect |
|---|---|---|
| Rain | | 1. Affects site access and movement.<br>2. Spoils newly finished surfaces.<br>3. Delays drying out of buildings.<br>4. Damages excavations.<br>5. Delays concreting, bricklaying, and all external trades.<br>6. Damages unprotected materials.<br>7. Causes discomfort to personnel.<br>8. Increases site hazards. |
| | High wind | 1. Increases rain penetration.<br>2. Reduces protection offered by horizontal covers.<br>3. Increases site hazards. |
| High wind | | 1. Makes steel erection, roofing, wall sheeting, scaffolding, and similar operations hazardous.<br>2. Limits or prevents operation of tall cranes and cradles, etc.<br>3. Damages untied walls, partially fixed cladding, and incomplete structures.<br>4. Scatters loose materials and components.<br>5. Endangers temporary enclosures. |
| Low and subzero temperatures | | 1. Damages mortar, concrete, brickwork, etc.<br>2. Slows or stops development of concrete strength.<br>3. Freezes ground and prevents subsequent work in contact with it, e.g., concreting.<br>4. Slows down excavation.<br>5. Delays painting, plastering, etc.<br>6. Causes delay or failure in starting of mechanical plant.<br>7. Freezes unlagged water pipes and may affect other services.<br>8. Freezes material stockpiles.<br>9. Disrupts supplies of materials.<br>10. Increases transportation hazards.<br>11. Creates discomfort and danger for site personnel.<br>12. Deposits frost film on formwork, steel reinforcement, and partially completed structures. |
| | High wind | Increases probability of freezing and aggravates effects of 1–12 above. |
| Snow | | 1. Impedes movement of labor, plant, and material.<br>2. Blankets externally stored materials.<br>3. Increases hazards and discomfort for personnel.<br>4. Impedes all external operations.<br>5. Creates additional weight on horizontal surfaces. |
| | High wind | Causes drifting which may disrupt external communications. |

SOURCE: From Winter Building Advisory Committee, "Winter Building: A Review of Winter Building Techniques." In *Advisory Leaflet 40: Weather and the Builder*. London: Department of the Environment 1971. Used with the permission of Her Majesty's Stationery Office.

upon agriculture, shows much more variation in income than states in which that percentage is down to 3 or 4.

Another market in which weather plays an important role is that of the commodity "futures." In this field, where it is necessary to anticipate the supply of

## SEC. 15.6 | COMMERCE

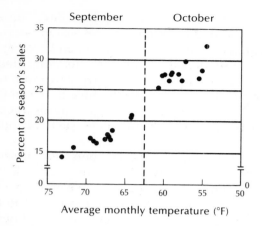

**FIGURE 15.4**
The relationship between the average temperature in September and October and the relative contribution of those months to the total September-to-December sales of women's winter coats. (From F. Linden, "Merchandising Weather", *The Conference Board Business Record* 19 (1962): 15–16.)

as well as the demand for a commodity, there is an obvious weather-related element in the forecasting. Some typical commodities would be cattle, corn, potatoes, sugar, and wheat.

Another weather-related activity is the insurance business. A 1966 study stated that in the United States paid claims for storm damage ranged from $0.5 to $1.0 billion annually over several years. Many types of weather-associated policies are written, especially in the agricultural field, of which the crop-hail insurance is the major type. In 1968 nearly $3 billion worth of such insurance was purchased, with coverage for corn and soybeans constituting some 40 percent of this.

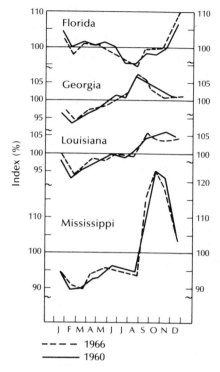

**FIGURE 15.6**
Seasonal pattern of personal income in Florida, Georgia, Louisiana, and Mississippi. (From J. W. McLeary, "Seasonal Income Patterns in the South." *Monthly Review*, Sixth Federal Reserve District, Atlanta, Georgia, November 1968, pp. 150–53.)

Insurance against drought is also a big business. For example, in Montana from 1947 to 1967, 61 percent of the indemnities paid were caused by drought and only 28 percent by hail. Such percentages would show

**FIGURE 15.5**
Daily sales of umbrellas and women's raincoats in a large New York department store. (From F. Linden, "Weather in Business," *The Conference Board Business Record* 16 (1959): 90–94, 101.)

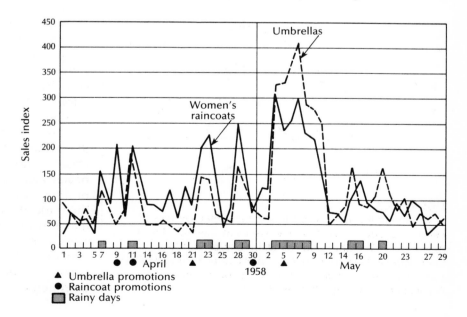

large variations from state to state. Of course, other items than crops are insured against weather hazards. In Britain it was estimated that claims resulting from the effects of severe snowstorms and bitter cold on structures exceeded £20 million (then $48 million).

Other aspects of commerce that are affected by weather and climate include banking, since deposit flow and loan requests are seasonally dependent; law, where cases may involve a weather aspect (slick surfaces, sun reflection dazzle, collapses of buildings); and advertising, which is linked with the retail trading mentioned earlier.

# 15.7
## COMMUNICATIONS

We depend greatly on communications for the rapid dissemination of information, and most of our methods of communication are subject to the whims of the weather. In times of bad weather the communication demand is usually greatest. Icing conditions often result in the breakage of telephone and telegraph wires, while high winds can cause blowdown of the poles and supports. Underground cable, if completely waterproofed, would eliminate this hazard, but it is extremely expensive to install. When transmission is in the microwave frequencies, some of the line problems are eliminated, but there is still an effect due to temperature and humidity. These two elements also affect radio and television because of their influences on wave refraction and propagation. Temperature increases and steep humidity gradients near the surface produce excessive downward refraction and therefore limit the distance of transmission. Another common problem is the poor reception caused by electrical disturbances such as thunderstorms.

The distribution of newspapers can be severely hampered by severe weather. When rain is forecast, the protection of newspapers delivered to homes requires expenditure of time and money. The atmosphere also affects sales or circulation. Most newspapers find that the weather summary or forecast section ranks among the really popular items, and stories of extreme weather, especially when they are local, are read avidly.

# 15.8
## TOURISM

Travel posters generally show clear skies with sand, or snow, and smiling people. Perhaps it is true that persons appear happier on sunny days and more woeful on overcast days, but you must ask the question, "What are my chances of getting a sunny day on my vacation?" If you wish to rest your eyes on the green fields and hills of Ireland, remember that it did not get green by irrigation—there are a lot of rain days in Ireland in any month.

The United Kingdom Meteorological Office reported that by 1970 inquiries about vacation weather outnumbered all other categories of queries. While it is not possible to pick a "best" weather for a holiday, as this depends on one's preferences and activities, the general description of warm, no rain, bright but with some cloud cover, little or no wind, excellent visibility, and average relative humidity is often quoted. Different activities require different types of weather, as shown in Figure 15.7.

The weather impact on a vacation depends upon the planned activities. If theaters and museums are chosen, the weather is relatively unimportant, but a beach vacation requires a specific type of weather. A prolonged period of bad weather will usually harm the trade at a resort, but the odd rainy and cold spell may send people indoors to restaurants, movie theaters, and retail shops. Winter sports resorts are particularly weather-sensitive; too much or too little snow, unexpected thaws, and high winds can all be detrimental.

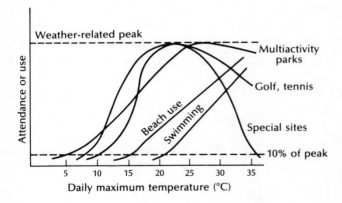

**FIGURE 15.7**
A schematic illustration of the weather-selectiveness of certain outdoor recreational activities in Canada. (From A. H. Paul, "Weather and the Daily Use of Outdoor Recreational Areas in Canada." In *Weather Forecasting for Agriculture and Industry*, ed. J. A. Taylor. Cranbury, N. J.: Associated University Presses, 1972.)

When planning for a vacation, it is important to find out what weather is likely to occur during the visit so that one is prepared, not only with the correct clothing, but psychologically as well. The luggage restrictions in air travel make the selection of optimal clothing imperative. Some questions asked should be, How often does it rain? When does it rain? What is the temperature range? Will it be very dry or humid, very windy or calm? How many hours of sunshine will there be? Some cities take pride in their sunshine amounts. For example, the St. Petersburg, Florida, newspaper *The Sun* offers a free distribution any day the sun fails to shine. During one period of over two years it was not given away once. All of the climatic questions just noted usually can be answered by a visit to the library or by a meteorologist.

Another vacation consideration involves transport (see Section 15.4). Is it going to be easy to reach one's destination? For some winter resorts, passes may be closed. In summer severe storm conditions might delay the journey, especially if it is by air.

## 15.9
### RECREATION AND ENTERTAINMENT

There are numerous examples of the effect of weather on outdoor sporting activities. Often inclement conditions have forced an outdoor event to be curtailed, postponed, or cancelled with a resultant monetary loss. For financial reasons and to create stable or standardized conditions, such structures as the Houston Astrodome, the New Orleans Superdome, and other covered stadiums have been built. The intention is to achieve a comfortable climate—but comfortable for whom? The performers or participants expending energy and generating heat will be comfortable at a lower temperature than spectators sitting in the stands. Perhaps the need to increase the metabolic rate accounts for some of the rather rowdy spectator behavior often witnessed.

Certain weather types are desirable for outdoor sports, while others are detrimental and often dangerous (Table 15.9). Some sports, such as football, soccer, rugby, hockey, baseball, and track events, are affected by both the atmospheric and the ground conditions. In wet or slushy conditions skill becomes less of a factor than adaptation and luck. In these sports either the ends are changed to even the situation (such as kicking against/with the wind) or the participants compete at the same time instead of on consecutive days.

Other sports introduce unequal weather impacts. A perfect example is a cricket match, which, in its first-class form, lasts for three to six days so that the two sides can experience very different conditions. Golf tournaments lasting a few days are another example, as are track events where heats contested over some days could lead to inequities, especially when "fastest losers" can qualify.

Today outdoor concerts attract many tens of thousands of spectators, and heavy thunderstorms may cause their vehicles to become mired in fields. Movies filmed outdoors require special conditions; because of the expense involved in getting stars, extras, crew, and props to a certain location, the director needs to have a reliable forecast to assess his chances of shooting during a certain period. Changes in lighting conditions play havoc with sequential scenes that are filmed days apart. Gardening is also very susceptible to weather conditions. Days of high relative humidity are advised for planting or bedding out, spraying should be avoided on windy days, and hot, windy days with low relative humidity require appreciable watering. Watering during the early morning is preferred, as in the daytime much water is lost by evaporation before it reaches the plant. Fungal susceptibility is increased when saturated air conditions develop during the night. For the do-it-yourselfer outside painting is not advised on hot, dry days because rapid drying will make a uniform finish difficult to achieve. These are the same conditions that make washing or waxing the car unadvisable.

**TABLE 15.9**
Effects of weather on some sporting activities.

| Sport | Detrimental Weather | Beneficial Weather |
|---|---|---|
| Sailing | Sea breezes pushing craft onshore; gusty winds | |
| Gliding | Hail; thunderstorms | Lee waves; thermals |
| Parachuting | Surface winds ($> 7$ m s$^{-1}$, or 16 mi h$^{-1}$) | |
| Skiing | Poor visibility; high winds | Good visibility |
| Hot-air ballooning | High temperatures [$>25°C$ (77°F) lift requires excessive temperatures, which may split fabric] | |

**SUGGESTED READING**

GRIFFITHS, J. F. *Applied Climatology: An Introduction.* London: Oxford University Press, 1978.

MAUNDER, W. J. *The Value of the Weather.* London: Methuen & Co., Ltd., 1970.

# INDEX

Absolute humidity, 109
  mixing and, 123, 125
Absolute (Kelvin) scale, 29, 30
Absorptance, leaf, 300
Adhemar, J. F., 255
Adiabatic processes, 117–120, 123, 125, 238–239
Advection fog, 169
Aerodynamic method of evapotranspiration calculation, 303
Ageostrophic wind, 56
Agriculture, 299
  atmospheric modification, 308–311
  crop yield, climate and, 305
    requirements, 306–308
  forestry, 314
  growing season, 45, 47, 177
  insurance, 345
  light penetration and, 300–301
  livestock, 313–314
  plant pathogens/pests and, 311, 312
  temperature, role of, 302
Air
  climatic change, oceans' effect on, 264
  conductive capacity, 32
  and fog formation, 169
  heat transport in, 16, 18, 20
  maritime and continental, 69, 83, 84, 95, 174
  masses, 82–83, 122–123, 172, 174–176
    climate classification based on, 193
    mountain ranges and, 192
  and mixing, 123, 125
  movement. See Circulation, atmospheric; Vertical motions; Wind
  saturation, attainment of, 116
    ice crystals vs. water, 127
    parcels, 119–120
  temperature, 29, 30, 31, 174
  water vapor in. See Water vapor
Air parcels, 117–120
Air pollution, 318, 323, 328
  potential for, 293–294, 297
Airport/aircraft operations, 336–337
Albedos, 15, 266
  mechanistic model, 268
  of vegetation, 300
Altitude of sun, 6–7, 9
Ammonia industry, climate and, 342
Analysis of variance, 144
Animals, domestic, 313–314
Anthropogenic effects on climatic change, 259, 263, 264, 265
Anticyclones. See High pressure centers
Apples, crop requirements of, 306
Arctic front, 82, 83
Aridity. See Dry climates
Athens, Greece
  Csa climate, 219
  Tower of the Winds, 2
Atmosphere
  air parcels, 117–120
  components, 104, 263–264
  differential transparency, 25
  energy balance, 26
    maintenance of, 26–27, 52
  interface, 12, 15
  See also Energy fluxes: at interface
  solar radiation in, 10, 11, 24–25, 263–264
  temperature

inversions, 41–42, 119
  lapse rates, 40–41
  mean/variability, 39–40
  See also Air; Circulation, atmospheric
Atmospheric counterradiation, augmentation of, 33
Atmospheric effect, 25, 264
Atmospheric modification
  cloud seeding, 127, 311
  evaporation reduction, 310–311
  freeze protection, 310
  irrigation, 308
  mulches, 240–241, 308–309
  shelterbelts, 309, 310
Atmospheric plant association, 201
Autoconvective gradient, 334
Autoconvective lapse rate, 241

Bananas, crop requirements of, 306
Beaufort, Sir James; Beaufort scale, 332
Bell-shaped distributions, 134
  example, 133
  normal curve, 137–138, 142
Bergeron-Findeisen process, 127
Bermuda high, 272
Bilandic, Michael, 317
Binomial distribution, 135–137
  negative, 140
  Poisson distribution, 138–139
  and sign test, 141
  for varying probability, 138
Bioclimatics, 304–305
Biometeorology, human, 317–324
  clothing in, 322
  comfort indices, 320–322
  direct atmospheric effects, 317–318
  food/diet, effects on, 323–324
  health, 293, 296, 322–323
  heat balance, body's, 318–320
Biotemperature, annual mean of ($T_{bio}$), 201
Bivariate linear regression/correlation, 145–146
Blodget, Lorin, 3
Boiling point, decrease in, 323
Bounding conditions for variables, 131
Bowen ratio, 24
Breezes, sea and land, 236–238
Bryson, Reid, 265
Buildings, 324
  climate's effect on, alteration of, 324–327
  effect of, on climate, 245, 246
Buoyancy, 118, 120

Cabbage, crop requirements of, 306
Calder, Nigel, quoted, 259
Carbon dioxide in atmosphere, 264
Cardinal temperatures, 301, 302
Cargoes, ships', atmospheric conditions and, 333
Cattle, climate and, 313
Celsius scale, 29, 30
Cenozoic Era, Tertiary Period of, 254–255
Centers of action, 65, 79
  continental highs, 68–69, 79, 281, 287
  ITCZ, 65–67, 159, 162–163, 283–284, 290
  extreme positions, 207
  monsoon, 69, 218
  subpolar lows, 67–68, 159
  subtropical highs, 67, 68, 79, 159–160, 161, 272
Central limit theorem, 142–143

Central tendency, measures of, 131
  example, 132–133
Chemical industries, climate and, 342
Chi-square test, 142
Chinook winds, 238–239
Circle of illumination, 8–9
Circulation, atmospheric, 26–27, 51–95, 98, 111, 151
  and continentality, 42
  convergence/divergence, 59–60, 120–121, 159, 237–238, 285
  intertropical convergence zone (ITCZ), 65–67, 283–284
  surface-upper air coupling, 92, 93
  forces affecting, 53–59
  general, 84, 93–95
  principal types, North American, 277–278, 282–283
  requirements for, 52–53
  scales of motion, 60–62
  tropical features, 73–74
  seasonal reversals, 163, 165
  surface, 62–84
    for earth of uniform substance, 62–83
    fronts. See Fronts
    land and water, effect of, 63–65
    macroscale (synoptic) phenomena, 60–61
    See also Transient disturbances
    semipermanent features. See Centers of action
  in synoptic climatology, 271–298
  upper-air, 84–95, 285, 287–292, 294–295, 296, 297
  See also Wind
Circulation, ocean, 26, 27, 98–100, 101
Circumpolar trough, 64, 67–68
Citrus fruits, crop requirements of, 306
CLIMAP (Climate: Long-range Investigation Mapping and Prediction) project, 257–258
Climate
  classification. See Climate classification
  definition, 1
    18th century, 2
  elements, 1
  See also Precipitation; Temperature
  energy balance components and, 21–22, 23–24, 25
  and farming. See Agriculture
  forecasting, 269
  land-water proportion, effect of, 95
  marine vs. continental, 37–38, 167, 193–194
  modeling, 267–268
  modification schemes, 269
  ocean circulation, effect of, 100–102
  and precipitation
    intensity, 166
    ratio to rain days, 166, 167
  records, 168, 180–185
  regional. See Regional climates
  scales of atmospheric motion and, 60
  small-scale. See Mesoscale controls; Microscale controls
  of standard continent, 190–192
  and temperature variation, 35–38, 51–52, 152, 157, 177
    continentality, 42–43
    See also Temperature: regional climates
  time variations. See Climatic change
  transient disturbances, effect of, 78–79
  variability. See Variability

350

Climate classification
　and boundary displacement, 180, 181
　factor analysis, by McBoyle, 203
　history, 187, 190
　human comfort and
　　by Maunder, 201–202
　　by Terjung, 202, 203
　by Köppen, 190, 194–197, 198, 199
　　See also Regional climates
　life zone, by Holdridge, 199, 201
　requirements, 193
　by Thornthwaite, 190, 197, 199, 200
Climatic bands, 2
Climatic change
　causes, suspected, 260–267
　　atmospheric composition, 263–264, 267
　　climatic system, linkages in, 260–261
　　solar output, 261–263
　　sun-earth path, 263, 267
　　surface changes, 264–267
　future, prospects for, 269
　geologic time scale
　　Holocene Period, 258–259
　　Mesozoic Era, 254
　　Paleozoic Era, 252–254
　　Precambrian Era, 251, 252
　　table, 253
　　Tertiary Period, 254–255
　history of concern with, 247–249
　last million years, 255–260
　methods for assessing, 249–251, 252, 255–258, 259
　modification, schemes for, 269
Climatic iteration, defined, 176
Climatic optimum, 259
Climatic plant association, 199
Climatological jet stream, 87, 89
　components, 94
Climatology
　definition, 1
　history, 2–3
　　change, concern with, 247–249
　　climate classification, 187, 190
　subdivisions of, 1–2
　synoptic. See Synoptic climatology
Clothing; clo unit, 322
Cloud seeding, 127, 311
Cloudiness
　in general circulation model (GCM), 268
　synoptic weather types, 287
　variations in, 11
Clouds
　constituents, 126–127
　convective, 120, 125, 237
　formation, 116, 126
　mixing and, 123, 125
　and temperature, 32–33
　types, 124, 126
Coalescence of cloud drops, 126–127
Coat sales, weather and, 345
Coefficient of variation, 132
Coffin, James Henry, 190
Coke chemical industry, climate and, 342
Cold (E) climates in Köppen system, 229–230
　frost (EF), 232
　polar marine (EM), 230
　tundra (ET), 230, 232
"Cold" clouds, 127
Collision of cloud drops, 126
Comfort
　cattle, zones for, 313
　human

　climate classifications, 201–203
　indices, 320–322, 327
Commerce, climate/weather and, 342–346
Commodity "futures," weather and, 344–345
Communications, weather and, 346
Compensation point, plants', 301
Condensation, 16, 104, 111, 125–126
　air parcel, 119, 120
　in transient disturbances, 73, 76, 78
Condensation nuclei, 125, 126, 328
Conditional instability, 120
Conduction, 16
　thermal conductivity, 18, 32, 95
Conductive capacity, 32, 41, 95
Consecutive vs. sequential variability, 178
Construction, climate/weather and
　buildings, 245, 246, 324–327
　fences, 245–246
　industry's operations, 341, 343, 344
Continental and maritime air, 69, 83, 84, 95, 174
Continental climates, 37–38, 51–52, 193–194
　Cwa zone, 217–218
　Dfc zone, 224–225
　Dwc zone, 228
　Dwd zone, 229
　precipitation/rain days, 167
　thermoisopleth diagram, 36
Continental highs, 68–69, 79, 281, 287
Continentality, 42–43, 100, 150, 152
Continents
　energy balance, 23–24
　past positions, 264
　　100 MYBP, 254
　　340 MYBP, 249
　seasonal temperature and pressure, 63
　water balance, 109
Continuous vs. discrete variables, 130
Convection, 16, 18, 25, 96–97
　convective activity, 295–296, 297
　eddy diffusivity, 112
　and frictional effects on wind, 57
　geographical variations, 19, 20, 23
　interlatitude imbalances and, 26–27
　thermal, 117–120, 125, 157, 159
　and transient disturbances, 73
Convective activity, 295–296, 297
Convective clouds, 120, 125, 237
Convergence, 59–60, 120–121, 159, 237–238, 285
　and fronts, 70
　intertropical convergence zone (ITCZ), 65–67, 283–284
　latitudinal shifting of zones, 162
　surface-upper air coupling, 92, 93
Cooling degree-days (CDD), 44, 339, 340
Cores, deep-sea, analysis of, 251, 255
Coriolis, G. G., 54
Coriolis force, 54, 55, 59
　friction and, 57
　in tropics, 66, 74
　and water movement, 98
Correlation and regression, 144
　bivariate linear, 145–146
　multivariate linear, 146
　for weather-influenced phenomena, 293
Correlation coefficient, 145–146
Cosmic year, 253
Cotton, crop requirements for, 307
Croll, James, 255
Crop Moisture Index, 116, 117
Crops. See Agriculture

Cryosphere and climatic change, 266–267
Cryptoclimate, 324
　for industries, 341
Cumulus condensation level (CCL); cumulus clouds, 120, 125
Currents, ocean, 26, 27, 98–100, 101
　and standard continent, 191
Curvature effect, 126
Cyclogenesis, 70
　preferred regions, 275–276, 279–281
Cyclone tracks, 276, 281, 283, 284
Cyclones. See Low pressure centers

Dalton, John, 2
Damming of rivers, proposals for, 269
Day length, 7, 8–9, 301
Deforestation, effects of, 265
Degree-days, 302, 304
　freezing/thawing, 334
　heating/cooling, 44, 46, 338, 339, 340
Dendroclimatology, 250, 251
Depression/rise complex, 242–243
Descriptive statistics, 129, 131–135
　graphical descriptors, 132–135
　numerical descriptors, 131–132
Deserts, 165
　in Köppen system, 209, 211–213
　plants, 300
Deviation, standard vs. mean, 132, 137
Dewpoint, 72, 105–106
　extremes, 184
　synoptic weather types, 287
Diet, climate and, 324
Discomfort (Temperature-Humidity) Index, 321–322
　house construction and, 325, 327
Discrete vs. continuous variables, 130
Disease, weather and, 293, 296, 322–323
　of plants, 311, 312
Dispersion of values, measures of, 132, 143
Divergence vs. convergence, 59–60, 120, 159, 237–238
　surface-upper air coupling, 92, 93
Dove, Heinrich Wilhelm, 190
Drizzle, 127
Drought, 304
　and climatic change, idea of, 247
　PDI definition, 116
Drude, Oscar, 190
Drug action, climatic influence on, 323
Dry adiabatic process rate, 118–119, 123
Dry climates
　boundary migration, 180
　and continentality, 42
　extremes, 181
　Köppen system (B climates), 194, 195, 196, 209
　　cold desert (BWk), 212–213
　　cold steppe (BSk), 210–211
　　hot desert (BWh), 211–212
　　hot steppe (BSh), 210
"Dry lines," 174
Duration
　of atmospheric motions, 60–61
　of solar radiation, 7, 8–9, 301
Dust
　in atmosphere, 263
　overgrazing and, 265
Dynamics vs. statistics, Gates's view of, 268

Earth-sun geometry. *See* Orbital characteristics, earth's
Easterly waves, 73, 76
Easterly winds, 87–88
  polar, 68
Eccentricity of earth's orbit, 255, 256, 257
Edaphic plant association, 199
Eddy diffusivity, 112
Effective precipitation, 304
Effective temperature (ET) index, 322
Egg production, climate and, 314
El Niño phenomenon, 341
Electric power, 337, 338
Elevation angle of sun, 33
*Encyclopedia Britannica,* quoted, 2
Energy. *See* Power
Energy balance
  body's heat balance, 318–320
  building, components of, 324
  evapotranspiration calculation, 303–304
  *See also* Energy fluxes
Energy fluxes
  earth-atmosphere system, 24–26, 52
    Hadley cells, 84–85, 94
    interlatitudinal imbalance and, 26–27
    ocean circulation, 26, 27, 98–100, 101
    and transient disturbances, 73
  equations, 17, 111–112
    and temperature, 31
  at interface, 12, 15–17, 25, 33
    conductive capacities and, 32, 41
    geographic and time variations, 19–24
    net radiation, 17, 18–19, 21, 22, 23–24
    oceans, role of, 18, 22, 23, 96–98
    surface variations and, 17–19
Equatorial westerlies, 66–67, 68
Equinoxes, 8, 9, 10
Error curve, 137
Evaporation, 16, 96–97, 100–101, 104, 111–112, 161, 302–304
  body's heat loss, 319–320, 321
  continents' water balance, 109
  and fog, 170, 171
  latitudinal variation, 109
  measurement, 112–113
  reduction, attempted, 310–311
  relative humidity and, 113, 114
  and saturation, 116
  and wet-bulb temperature, 106
  zonally averaged rates, 84, 85
  *See also* Evapotranspiration
Evapotranspiration (ET), 18, 19, 22, 23–24, 112, 302–304
  potential (PET), 113–114, 115, 199, 303, 304
  ratio (P.E.R.), 201
Extraterrestrial solar radiation, 9–10
Extreme value distribution, 140

Factor analysis in climate classification, 203
Fahrenheit scale, 29, 30
Fairbanks, Alaska, Dfc climate of, 224
Fall wind, 239, 243
Farming. *See* Agriculture
Feedback
  atmospheric particles/solar radiation, 263
  negative, situations of, 111, 125, 237, 268
  temperature/ice cover, 266, 268

Fences, climatic effect of, 245–246
Filtering concept, in scales of atmospheric motion, 61
Fires, forest, 314
Fishing industry, weather/climate and, 341–342
Floods, in China, 259
Flux density
  of evaporation, calculating, 111–112
  of radiation
    solar, 5–6
    and temperature, 31
Fluxes, energy. *See* Energy fluxes
Foehn wind, 238–239
Fog, 162, 169–170, 171
  and airport operations, 336
  construction operations affected by, 343
  in Csb zone, 220
  extremes, 185
  ships, problems for, 333
Food, climate and, 323
  industry, 341
Forecasting, climate, 269
Forestry, climate and, 314
Fossil fuels, burning of, and carbon dioxide concentration, 264
Fourier analysis, 147
Freezes
  probability of, 45
    local data, 189
  protection from, 310
Freezing degree-days, 334
Freezing nuclei, 126
Frequency distributions, 133–135, 140
  binomial. *See* Binomial distribution
  and chi-square test, 142
  normal, 137–138, 142
  Poisson, 138–139
Friction layer, 57
  levels above, 54, 56
Friction wind, 57, 62–63
Frictional force/drag, 16, 52–53, 54, 56–57, 121, 237, 240
Frontal fog, 170, 171
Frontal overrunning, 122, 281, 287
Fronts, 70–72, 272
  frequency/zones, 82, 83, 274
    *See also* Polar front zone
  maps, 80–83
    explanation, 79, 82, 83
  and mechanical uplift, 122, 160
  sea-breeze, 237
  and thermal properties of land/water, 84
  weather, effect on, 72–73, 76
  wind speeds above, 86, 91
Frost (EF) climate, 232
Frost hollows, 243
Frost vs. freeze, 310
Fruit, climatic requirements of, 306
Fuel consumption, 337–338, 340
  and carbon dioxide concentration, 264
  heating/cooling degree-days, 44, 46, 338, 339, 340
Fungi and plants, 311
  potato blight, 312

Gamma distribution, 140
Gardening, weather and, 347
Gases, kinetic theory of, 30
Gates, W. Lawrence, quoted, 268–269

Gaussian (normal) distribution, 137–138, 142
General circulation model (GCM), 268
General gas law, 85
Genetic classification of climate, 193, 271
Geostrophic approximation to wind, 55–56, 86
German school of meteoropathology, 293, 296
Glaciation, 255–258
  Little Ice Age, 259
  oxygen-18 and, 251
  Paleozoic, 254
  Precambrian, 252
  requirements for, 266
  Younger Dryas event, 259
Global radiation, 11
Gradient level, 283
Gradient wind approximation, 56
Gradients, 3
Graphical descriptors in statistics, 132–135
Gravity and pressure gradient force, 53
Greenhouse (atmospheric) effect, 25, 264
Grid points in climate modeling, 268
Ground inversions, 42
Growing degree-days, 302
Growing season, 45, 47, 177
Gstettneralm, Austria, frost hollow, 243
Gulf return (weather type), 281, 287
Gyres, 98–99, 100, 101

Hadley cells, 84–85, 94
Hail, 127, 170–173
  extremes, 172, 184
Hale cycle, 262
Halley, Edmund, 2
Hann, Julius, 3
Harbors, 331
Harmonic analysis, 147
Health, weather and, 293, 296, 322–323
Heat balance, body's, 318–320
Heat capacity, 18, 19, 32, 33, 95
Heat hollows, 243
Heat transfer
  water, properties of, 95–96
  *See also* Energy fluxes
Heat units, 302, 304
Heating degree-days (HDD), 44, 46, 338
Hedges, effect of, 326
High latitude (D) climates in Köppen system, 196, 221
  dry summer (Ds), 226
  dry winter
    and cool summer (Dwc), 228–229
    with extreme cold (Dwd), 229
    and hot summer (Dwa), 226, 227
    and warm summer (Dwb), 227–228
  uniform precipitation
    cool summer (Dfc), 224–225
    extremely cold winter (Dfd), 225
    hot summer (Dfa), 221–222
    warm summer (Dfb), 222–224
High pressure centers, 56, 64, 79, 160
  from Canada, 272, 273
  continental, 68–69, 79, 281, 287
  and pollution potential, 293–294, 297
  subtropical, 67, 68, 79, 159–160, 161, 272
  synoptic patterns, 277, 278, 282
  upper levels, 285, 291
Highland (H) climates, 231, 232–233
Hind, J. R., 190

Hipparchus (Greek astronomer), 2, 187, 190
Histogram, 132
Holdridge, L. R., climate classification by, 199, 201
  and forest productivity, 314
Holocene Period, 258–259
Hopkins, A. D., 304–305
Hot, humid (A) climates in Köppen system, 196, 205–206
  short dry season (Am), 207–208
  wet (Af), 206–207
  wet and dry seasons (Aw), 208–209
Houston, Texas, Cfa climate of, 214
Houzeau, Jean Charles, 3
Howard, Luke, 3
Human comfort
  climate classifications based on
    by Maunder, 201–202
    by Terjung, 202, 203
  indices of, 320–322, 327
Humboldt, Alexander von, 190
Humid climates
  boundary migration, 180, 181
  See also Hot, humid (A) climates
Humidity, 100–101, 241
  absolute, 109
    mixing and, 123, 125
  and discomfort, 321
  extremes, 184
  food, effect on, 323
  measures, 104–106
  and precipitation, 120
  relative. See Relative humidity
  vegetation and, 245
    trees, 244
Hurricanes, 76–78
  energy, 104
Hydric plant association, 201
Hydrologic cycle, 96, 104, 106–108
Hydrometeors, 116
Hydrophobic particles, 125
Hydrostatic balance, 53
Hydrostatic equation, 85
Hygroscopic particles, 125
Hypsometric relationship, 85, 86, 92, 285, 291

Ice
  as cryosphere component, 266
  and shipping, 331, 332–333
Ice accretion, 333
Ice ages. See Glaciation
Ice crystals in clouds, 126, 127
Ice pellets, 127
Idrisi (Arab scientist), 190
Inchon, South Korea, Dwa climate of, 226, 227
Inclination of earth, 7, 255, 257
Income, seasonality of, 343–344, 345
Index cycle, 88, 91
Industries, 340–342
  construction, 341, 343, 344
Infiltration/percolation, 107, 302
Infrared flux, 25
Insects
  dispersal patterns, 297–298
  locusts, 298, 312–313
Insolation, 10–12, 13–14, 15, 39–40, 95, 149–150, 241
  on house, 325
  and net radiation, 19
  slopes and, 242

temperature lag, 33, 35, 157
urban climate, 327–328
and vegetation, 243, 245, 299–301
Instability in atmosphere, 118, 119, 174
  of air layers, 122–123
  conditional, 120
Instrument shelters, 31
Insurance, weather and, 345–346
Intensity
  of precipitation, 128, 166
    maximum, 166, 168
  of solar radiation, 7, 8, 9
Interface of atmosphere and earth, 12, 15, 95
  changes in, and climate alteration, 264–267
  See also Energy fluxes: at interface
Intertropical convergence zone (ITCZ), 65–67, 159, 162–163, 283–284, 290
  extreme positions, 207
Inversion, temperature, 41–42, 119
  trade wind, 161–162
Irkutsk, USSR, Dwc climate of, 228
Irrigation, 308
Isobars
  curved, and air motion, 56
  hypsometric relationship and, 85–86
  intertropical convergence zone, 65–66
  maps, 58, 80–81
    explanation, 57, 82
  and pressure gradient force, 53–54, 55
  sea level pressure patterns, 64, 65, 80–82
Isohyets, 279, 284
Isophanes, 304–305
Isopleths, 3
  See also names of individual kinds
Isotachs, 283, 289
Isotherm, Humboldt's introduction of, 190
Isotherm orientations in thermoisopleths, 35–37
Isotopic analysis, 250–251, 255
ITCZ (intertropical convergence zone), 65–67, 159, 162–163, 283–284, 290
  extreme positions, 207
Iterations, defined, 176

Jet-effect wind, 239
Jet stream, 87, 89–91, 92, 93
  STJ/PFJ, 94
Jurin, James, 2

Katabatic (downslope) winds, 239, 243
  foehns, 238–239
Kelvin scale, 29, 30
Kinetic theory of gases, 30
Köppen, Vladimir, climate classification by, 190, 194–197, 198, 199
  See also Regional climates

Lake effects on climate, 239
Land breezes, 236–237
  topography and, 238
Land ice, 266
Land use and climatic change, 265
Landsberg, H. E., 190, 249
Landscaping and air flow, 325, 326
Lapse rates, 40–41, 118, 119, 123, 125, 150, 240
  carbon dioxide concentration and, 264

Large macroscale phenomena, 60, 61
  See also Centers of action
Latent and sensible heat fluxes, 16, 19, 20, 23, 25, 96–97
  interlatitudinal imbalance and, 26–27
  and transient disturbances, 73
Latent heat, kinds of, 104
Latitudinal drift of sites, 249–250
Latitudinal variation
  absolute humidity, 109
  convergence zones, 162
  energy balance, 26
    compensation, 26–27
    components, 21–24, 97–98
  frictional force of wind, 53
  frontal frequency, 83
  hail, 171
  hypsometric relationship and, 85, 86
  precipitation/evaporation, 109
  pressure gradient/wind speed relationship, 59
  snowline, 169
  solar radiation, 8, 9–12, 149–150, 301, 320
  sun's elevation angle, 33
  surface pressure, for uniform earth, 62
  temperature gradients, 39–40
Layer instability, 122–123
Least squares theory, 145
Life zones, 199, 314
Lifting condensation level (LCL), 120, 125
Linear distribution, 134–135
Little climatic optimum, 259
Little Ice Age, 259
Littoral, 100
Livestock, climate and, 313–314
Locusts, 298, 312–313
London, England, Cfb climate of, 215–216
Long (planetary) waves, 88, 91, 288, 289, 292
Low pressure centers, 56, 64, 160
  and convergence, 59, 60, 120–121, 159
  subpolar, 67–68, 159
  synoptic patterns, 274–276, 282
    cyclogenesis, 275–276, 279–281
    cyclone tracks, 276, 281, 283, 284
    height, effect of, 285, 291
  thermal, 159
  and transient disturbances, 70–71
Lysimeters, 303

McBoyle, Geoffrey, 203
Macroscale phenomena, 60–61
  See also Transient disturbances
Mahlmann, Wilhelm, 190
Maps, weather
  cold front, effect of, 76
  explanation, 73
  history of, 2
  insolation, 13–14
    explanation, 11
  surface pressure, 58, 80–81
    explanation, 57, 59, 78–79, 82
  synthesis, synoptic patterns from, 274–284, 290, 294, 296, 297
  synoptic patterns, 57–59, 73, 74–75, 271, 272–274
    extratropical latitudes, 274–287
    frontal frequency, 82, 83, 274
    tropics, 281, 283–285, 289–291
    upper-air, 285, 287–292, 294–295, 296, 297
  upper-level pressure, 89, 90

**353**

explanation, 88, 271
  synthesis, synoptic patterns from, 285, 288–292, 294–295, 296, 297
Marine climate, 37, 193–194
  precipitation/rain days, 167
Maritime and continental air, 69, 83, 84, 95, 174
Maunder, W. J., climate classification by, 201–202
Maunder minimum, 262–263
Maury, Matthew Fontaine, 190
Mean, 131
  and parametric tests, 143–144
  and Poisson distribution, 139
  variability's dependence on, 179
Mean deviation, 132
  binomial distribution, 137
Mean resultant wind, 283, 287, 289
  ITCZ, 66
Mechanical uplift, 122, 160–161
Mechanistic models of climate, 268
Median, 131
Mediterranean climate, 213
  Cs climates, 219–220
Melting, 16
Mesoscale controls of climate, 235–239
  lake effects, 239
  land-sea configuration, 236–238
  topography, 238–239
Mesoscale phenomena, 60, 61–62
Mesozoic Era, 254
Metabolic rate; MET unit, 318, 319
Meteoropathology, 293, 296
Meteorotropic diseases, 323
Microscale controls of climate, 239–246
  bare surface characteristics, 239–241
  constructions, 245–246
  topography, 242–243
  vegetation, 243–245
Microscale phenomena, 60, 61
Mid-latitude climates in Köppen system
  C climates, 196, 213
    dry and hot summer (Csa), 219–220
    dry and warm summer (Csb), 220
    dry winter, hot summer (Cwa), 217–218
    dry winter, warm summer (Cwb), 218–219
    uniform precipitation, cool summer (Cfc), 216–217
    uniform precipitation, hot summer (Cfa), 214–215
    uniform precipitation, warm summer (Cfb), 215–216
  desert (BWk), 212–213
  steppe (BSk), 210–211
Mid-range, 131
Milankovitch, Milutin, 255
Minneapolis-St. Paul, Minn., Dfb climate of, 222–223
Mixing of near-surface air, 41, 123, 125
  and fog, 170
Mixing ratio, 105
Mode, 131
  example, 132–133
MODE (Mid-Ocean Dynamic Experiment), 264
Moist adiabatic process rate, 120, 238–239
Monsoon, 69, 218
Morse, Jedediah, 190
Motion, Newton's first law of, 53
Mountain-valley wind flows, 238

Mountains. See Topography
Mulches, 240–241, 308–309
Multivariate linear regression/correlation, 146

Natural logarithm transformation, 140
Negative binomial distribution, 140
Negative feedback situations, 111, 125, 237, 268
Net radiation, 18–19, 21, 22, 23–24
  equations, 17, 111–112
New Delhi, India, Cwa climate of, 217–218
Newton's first law of motion, 53
Nonparametric tests, 140
  chi-square test, 142
  run test, 141–142
  sign test, 140–141
Normal distribution, 137–138
  and chi-square test, 142
Northers (cold air outbreaks), 192, 215
Numerical descriptors in statistics, 131–132

"Oasis" effects, 20
Obliquity (tilt) of earth's axis, 7, 255, 257
Occlusion, 71
Oceans, 95–102, 150–151
  and breezes, 236–237
  cells, pressure, 67
  and climatic changes, 264
  cores, analysis of, 251, 255
  diurnal precipitation pattern over, 165–166
  and energy fluxes, 18, 23, 96–98
    circulation, 26, 27, 98–100, 101
    table, 22
  and fog formation, 169
  seasonal temperature and pressure, 63–64
  and standard continent, 191
  temperature, 95, 264–265
    surface, 99–100, 101, 150–151, 161, 292
Offshore drilling, weather and, 341
Ogive, 132, 133
Omaha, Nebr., Dfa climate of, 221–222
Orbital characteristics, earth's, 7–8, 255, 256, 257, 263
  and glaciation, 266
  rotation, 7, 53
Orographic effect, 122, 160–161, 168
Orthogonal predictors, 146
Overgrazing, effects of, 265
Overrunning, 122, 281, 287
Oxygen-18 isotopic analysis, 251, 255
Oymyakon, USSR, Dwd climate of, 229

Paleoclimatology. See Climatic change
Paleozoic Era, 252–254
Palmer Drought Index (PDI), 116, 304
Palynology, 250
Pan evaporation, 112, 113, 303
Parallelism of earth's axis, 7–8
Parametric tests, 142–144
Parcels, air, 117–120
Parmenides (Greek philospher), 187
Partial pressure, 104
Particulate matter in atmosphere, 263
  in cities, 328
Path length of solar radiation, 10
Pedestrians, weather and, 335
Penman, H. L., 303

Percolation, infiltration and, 107, 302
Perihelion, 255, 257
Permafrost, 221
PET (potential evapotranspiration), 113–114, 115, 199, 303, 304
  ratio (P. E. R.), 201
Petroleum industry, climate and, 342
Phase changes in water, 16, 96, 104, 105
  See also Evaporation
Phenology, 299, 304–305
Photoperiodism, 299–300, 301
Photosynthesis, 299, 300
Phototropism, 300
Pigs, climate and, 314
Plan Position Indicator (PPI), 295–296
Plane of the ecliptic, 7
Planetary albedos, 15
Planetary scale (large macroscale) phenomena, 60, 61
  See also Centers of action
Planetary temperature, earth's, 6
Planetary (long) waves, 88, 91, 288, 289, 292
Plant associations, 199, 201
Plants
  in climate classification
    by Köppen, 194
    life zone, 199, 201
    by Thornthwaite, 197
  climatic effect
    low-growing, 245
    medium-height, 244–245
    tall, 243–244
  deforestation, effects of, 265
  evaporation from, 112
  hot, humid (A) climates, 207, 208
  pathogens/pests, 311–313
  wind transport, 296–298
  radiation and, 243, 245, 299–301
  temperatures and, 244–245, 301–302
    crop requirements, 306–308
  water and, 302
    irrigation, 308
  See also Agriculture
Poisson distribution, 138–139
Polar easterlies, 68
Polar front jet (PFJ), 94
Polar front zone, 82, 83, 94, 159, 162, 174
  wind speeds above, 86, 91
Polar marine (EM) climate, 230
Pollen studies, 250
Pollution, air, 318, 323, 328
  potential for, 293–294, 297
Populations and parametric tests, 142–144
Positive feedback, 268
Potato blight, 312
Potential evapotranspiration (PET), 113–114, 115, 199, 303, 304
  ratio (P. E. R.), 201
Power
  demand, 337–338, 340
    heating/cooling days, 44, 46, 338, 339, 340
    generation/transmission, 337, 338
Power law, 240
Precambrian Era, 251, 252
Precession of the equinoxes, 255
Precipitable water, 157, 158
Precipitation, 116, 120, 121, 157–173
  characteristics, 127–128
    intensity, 128, 166, 168
    probability, 128, 189

regimes, 162–166
worldwide averages and extremes, 162, 180
continents' water balance, 109
controls
lake effects, 239
ocean surface, 101, 161
of regimes, 162–166, 191–192
temperature, 157
*See also* Vertical motions: mechanisms involved
crop requirements, 306–307
effective vs. actual, 304
enhancement (cloud seeding), 127, 311
formation, processes of, 126–127
forms, 127, 166–173
hydrologic cycle, 96, 107–108
and industries, 342–344
latitudinal variation, 109
in life zone classification, 201
local data, samples of, 188, 189
orographic, 122, 160–161, 168
rain days, 166, 167
records, 168, 181, 183–184
regional climates, in Köppen system, 194, 195, 196, 197
cold (E), 229–230, 232
dry (B), 210–212
high latitude (D), 221–229
highland (H), 231
hot, humid (A), 206, 207, 208–209
mid-latitude (C), 214–220
and road travel, 334–335
and roofs, 325, 327
synoptic patterns, 279–280, 284, 285
upper-air conditions, 288–289, 290, 291, 292
temporal variation
diurnal, 165–166
seasonal, 162–165, 191–192
transient disturbances, 72–73, 78
urban climate, 328
variability, 178–180
as water budget factor, 115
Precipitation effectiveness (PE) index, 197, 199
Pressure, surface. *See* Surface pressure
Pressure, upper-air
features, 56
hypsometric relationship, 85, 86, 92
maps, 89, 90
explanation, 88, 271
synthesis, synoptic patterns from, 285, 288–292, 294–295, 296, 297
Pressure, vapor, 104, 109, 111, 127
Pressure gradient and wind, 57, 59
Pressure gradient force, 53–54, 55, 236
Probability, 130, 135
of freezes, 45, 189
of precipitation, 128, 189
*See also* Frequency distributions; Statistical significance
Process rates, 118–120, 123, 238–239
Psychological responses to weather, 323
Ptolemy (Claudius Ptolemaeus), 190

Quantiles, 132, 133

Radar, PPI scope of, 295–296
Radiation
extremes, 183, 184
solar. *See* Solar radiation
terrestrial, 5, 15–16, 19, 24, 25
and depression/rise complex, 242, 243
Radiation equilibrium temperature, 30
Radiation fog, 169, 170, 171
Radiative exchange by objects, 30
Radiative fluxes, 15–16, 97
infrared, 25
net radiation, 18–19, 21, 22, 23–24
equations, 17, 111–112
Radiosondes, 118
Railroad operations, weather effects on, 335–336
Rain cells, number of, 295–296
Rain days, 166, 167
Rain shadow effect, 122
Raindrops, 126, 127
Rainfall
and construction industry, 343, 344
crop requirements, 306–307
geographical variation, 22
ITCZ and, 162–163
maxima/extremes, 166, 168, 181, 183, 184
and road accidents, 334–335
on roof, 325, 327
synoptic patterns for islands, 279–280, 284, 285
trees and, 244
*See also* Precipitation
Ramage, Colin, 69
Randomness, 136
and chi-square test, 142
Ravenstein, Ernest, 190
Rawinsondes, 118
Réaumur, R. A. F., 304
Recreational activities, weather and, 346, 347
Reflectance, leaf, 300
Regional climates, classification of, 205–233
cold (E), 229–230, 232
dry (B), 194, 195, 196, 209–213
high latitude (D), 196, 221–229
highland (H), 231, 232–233
hot, humid (A), 196, 205–209
mid-latitude (C), 196, 213–220
*See also* Climate classification
Regression and correlation, 144
bivariate linear, 145–146
multivariate linear, 146
for weather-influenced phenomena, 293
Regression coefficients, 145
Relative humidity, 105, 109, 113, 114, 116
and condensation, 125
crop requirements, 306
extremes, 184
industries, optima for, 341
local data, sample of, 188
mixing and, 123
Relative variability, 179
Retail trade, climate/weather and, 342–343, 345
Return period, 139
for precipitation, 128
Revolution of earth, 7
Reykjavik, Iceland, Cfc climate of, 216, 217
Rice, crop requirements of, 307
Ridges, pressure, 56, 283, 288, 290
River travel, 333
Road travel, 333–335

Roofs, precipitation and, 325, 327
Rotation of earth, 7
wind and, 53
Run test, 141–142
Runoff, in continents' water balance, 109

Sales, retail, climate/weather and, 342–343, 345
San Francisco, Calif., Csb climate of, 220
Sand clouds, African, 263
Santa Ana phenomenon, 238, 272, 273
Saturation, attainment of, 116
air parcels, 119–120
ice crystals vs. water, 127
Scalar vs. vector quantities, 53
Scatter diagram, 145
Schwabe cycle, 262
Sea and land breezes, 236–238
Sea ice, 266
Sea level pressure patterns, 64, 65, 80–81, 291
Sea smoke, 170
Sea travel, 331–333
Sensible and latent heat fluxes, 16, 19, 20, 23, 25, 96–97
interlatitudinal imbalance and, 26–27
and transient disturbances, 73
Sequential vs. consecutive variability, 178
"Seven climes" division, 190
Severe weather, defined, 170
Shade species of plants, 301
Shading devices, 246
Shear, James A., 196
Sheep, climate and, 314
Shelterbelts, 309, 310
Ships, climate/weather and, 332–333
Short-wave radiation. *See* Solar radiation
Siberia, USSR, Dwd climate in, 229
Siberian high, 69, 79
Sign test, 140–141
Significance levels, 141
Significance of difference of means, tests for, 143–144
Simulation models of climate, 268
Skewed distribution, 134
binomial, 138
transformation, 140
Slopes, climatic role of, 242
depression/rise complex, 242–243
Small particles in atmosphere, 263
Snow/snowfall, 127, 166, 168–169
and construction industry, 343, 344
as cryosphere component, 266
disruption caused by, 334
high latitude climates, 221, 223
lakes and, 239
and railroad operations, 335
records, 183–184
roof loads, 325
Snowline, 168–169
Soil
and energy fluxes, 18, 19–20
temperature, 240–241
tractionability, 333
water
budgeting, 115–116
infiltration/percolation, 107, 302
Soil moisture, 302
Solar climate classification, 187
Solar declination, 8
Solar energy, 337
Solar parameter *(S)*, earth's, 6, 262
Solar radiation, 5

alterations, 261–262, 263
atmospheric depletion, 10, 11, 24–25, 263–264
in body's heat balance, 318–319, 320
flux density, 5–6
at interface, 15
See also Energy fluxes: at interface; Insolation
long-term variations, calculation of, 255, 256
people, effect on, 317–318
spatial and temporal variations, 6–7
insolation patterns, 10–12, 13–14, 19, 149–150
intensity/duration and, 7, 8–10, 301
load on standing person, 319, 320
orbital characteristics of earth and, 7–8
and temperature of solid, 30
and vegetation, 243, 245, 299–301
wavelength, 5
and plant responses, 299–300
Solstices, 8, 9–10
Solute effect, 126
Sounding of atmosphere, 118
Spectrum analysis, 148
Spörer minimum, 263
Sports, weather and, 347
Spraying of crops, 311
Square root transformation, 140
Stability in atmosphere, 118, 119, 174
conditional instability, 120
Stagnating anticyclones, 294, 297
Standard continent, climate of, 190–192
Standard deviation, 132, 137
and climatic variability, 177, 178, 179
vs. consecutive variability, 178
Standard error (SE), 143, 146
of estimate, 145
Standardization of variables, 137–138
Statistical inference, 129–130
Statistical significance, 129–130
of correlation coefficient, 146
nonparametric tests, 140–142
parametric tests, 142–144
Statistics, 129–148
descriptive, 129
graphical descriptors, 132–135
numerical descriptors, 131–132
dynamics vs., Gates's view of, 268
in forecasting, 269
probability, 130, 135
See also Frequency distributions; Statistical significance
regression/correlation, 144
bivariate linear, 145–146
multivariate linear, 146
for weather-influenced phenomena, 293
terminology, 130–131
tests of statistical significance
nonparametric, 140–142
parametric, 142–144
time series analysis, 147–148
Steam fog, 170, 171
Stefan-Boltzmann equation, 6
Steppe areas, 210–211
Still-air temperature (SAT), 202, 203
Storms
nighttime, lakes and, 239
See also Transient disturbances
Stratiform clouds, 126
Streamlines, 283, 288, 289

Subadiabatic/superadiabatic layers, 119
Subgeostrophic wind, 57
Sublimation, 126
Subpolar lows, 67–68, 159
Subsidence, subtropical highs and, 67
Subtropical highs and, 67, 68, 79, 159–160, 161, 272
Subtropical jet (STJ), 94
Subtropics
boundary variation, 180, 181
evaporation, 23
precipitation/rain days, 167
Sun
elevation angle, 33
radiation. See Solar radiation
Sun species of plants, 301
Sunflecks, 300–301
Sunshine
crop requirements, 307
extremes of, 185
Sunspots, 262–263
Supercooled drops, 126, 127
Surface albedos, 15, 266
Surface pressure
for earth of uniform substance, 62–63
extremes, 183, 185
features. See High pressure centers; Low pressure centers
land and water, effect of, 63–65, 236
maps, 58, 80–81
explanation, 57, 59, 78–79, 82
synthesis, synoptic patterns from, 274–284, 290, 294, 296, 297
and transient disturbances, 70–71, 73, 79
and upper levels, 93
hypsometric relationship, results of, 92, 285, 291
See also Centers of action
Synoptic climatology/meteorology, 271–298
maps. See Maps, weather: synoptic patterns
and weather-influenced phenomena, 293–298
Synoptic scale (macroscale) phenomena, 60–61
See also Transient disturbances

T-test, 143–144
Teleconnections, 292
Temperature, 29–49
adiabatic processes, 117–120
air masses, 174
and air travel, 336, 337
atmospheric composition and, 263, 264
body, 318
buildings' wall materials and, 325, 327
in climate models, 268
in comfort indices, 320–322
continentality, 42–43, 150, 152
depression/rise complex, 242, 243
and electric power utilities, 338
energy balance equations and, 31
food, effect on, 323
human tolerance to, 318
hypsometric relationship, 85, 86
and industries, 341–344
insolation and, 10–11, 12, 14, 39–40, 149–150, 241
time lag, 33, 35, 157
interface substances and, 18–19, 32, 35, 41, 84, 95, 150

inversions, 41–42, 119
trade wind, 161–162
lapse and process rates, 40–41, 118–120, 123, 125, 150, 240
carbon dioxide concentration and, 264
and chinook, 238–239
livestock, thermal-tolerance zones of, 313
mean atmospheric, 39, 40
measurement, height for, 31
oceans, 95, 264–265
surface, 99–100, 101, 150–151, 161, 292
oxygen-18 and, 251
PET estimate based on, 114
physical basis, 30
planetary, earth's, 6
and plants, 244–245, 301–302
crop requirements, 306–308
as precipitation control, 157
and railroad operations, 335–336
record-setting, 181, 182, 212, 232
regional climates, in Köppen system, 194, 195, 196, 197
cold (E), 229–230, 232
dry (B), 210–213
high latitude (D), 221–229
highland (H), 231, 232
hot, humid (A), 206–209
mid-latitude (C), 214–218, 219, 220
and sales of winter coats, 345
scales, 29, 30
and snowfall, 168
soil, 240–241
statistical analysis, examples of, 132–133, 269, 270
still-air (SAT), 202, 203
near surface, 241
synoptic patterns
upper-air conditions, 288, 289–290, 292, 294–295
weather types, 287
thresholds
for excessive heat, 44, 45
for freeze, 45, 47–48, 177, 189
and heating/cooling degree-days, 44, 46, 338, 339, 340
time variations, 7, 32, 33, 34, 82, 86, 150–157, 176
climate and, 35–38, 51–52, 152, 157, 177
lag after insolation, 33, 35, 157
littoral, 100, 152
local data, samples of, 188, 189
long-term, and climatic-change concept, 177, 178, 248, 258, 260
monthly means, 37–38, 39, 150–154, 157, 177, 211, 212, 213
in oceans, 95, 265
and pressure variations, 63–64
standard continent, 191
vegetation and, 244–245, 301–302
weather and, 38
See also regional climates, above
transient disturbances and, 72, 78
variability, 38, 45, 48–49, 176–178
in atmosphere, 39, 40
water vapor and, 32–33, 105–106
Temperature climate, 38
Temperature efficiency (TE) index, 197, 199

356

Temperature-Humidity Index (THI), 321–322
   house construction and, 325, 327
Temperature inversions, 41–42, 119
   trade wind, 161–162
Terjung, W. H., climate classification by, 202, 203
Terrestrial radiation, 5, 15–16, 19, 24, 25
   and depression/rise complex, 242, 243
Tertiary Period, 254–255
Textile industry, climate and, 341
Thawing degree-days, 334
Thermal conductivity, 18, 32, 95
Thermal convection, 117–120, 125, 157, 159
Thermal diffusivity, 240
Thermal lows, 159
Thermal wind, 91
Thermocline, 264–265
Thermodynamic diagrams, 118–119
Thermoisopleth diagrams, 35–37
   for Singapore, 207
Thermometer, working of, 30
Thermoperiodicity, 301–302
Thornthwaite, C. W.
   climate classification, 190, 197,199, 200
   and growth unit concept, 304
   PET measure, 114, 199, 303
Thunderstorms, 172
   hail, 127, 171
Time series analysis, 147–148
Topography
   barriers, 42, 69, 83, 100, 150
      orographic effect, 122, 160–161, 168
      standard continent, 192
   as climate control
      mesoscale, 238–239
      microscale, 242–243
Tornadoes, 215
Tourism, weather and, 346–347
Tower of the Winds, Athens, Greece, 2
Tractionability of soil, 333
Trade wind inversion, 161–162
Trade winds, 63
   convergence zone (ITCZ), 65–67, 284
Trajectory analysis, 297
Transient disturbances, 159, 272
   climate, effect on, 78–79
   and disease, 293, 296
   extratropical, 69–73
   seasonal differences, 79–82
   tropical, 73–74, 76–78
   and upper-level flow, 91, 93
      zone of, 94
   vertical-motion mechanisms and, 121, 122
Transmissivity of leaves, 300
Transpiration, 112, 245
Transportation/travel, climate and
   air, 336–337
   rail, 335–336
   road, 333–335
   water, 331–333
Tree-ring analysis, 250–251
Trees
   climatic effect, 243–244, 326
   as crop, 314
   removal, effects of, 265
Tropical deserts, 211–212
Tropics, 21–22, 35–37, 159
   lapse rate, 40–41
   precipitation/rain days, 167
   vs. subtropics, 23
   synoptic patterns, 281, 283–285, 289–291
   transient disturbances, 73–74, 76–78
   See also Hot, humid (A) climates; Intertropical convergence zone
Tropopause; troposphere, 18, 41, 94
Troughs, pressure, 56, 70, 121, 283, 288, 289–290
   circumpolar, 64, 67–68
Tundra (ET) climate, 230, 232
Two-tailed test, 141

Ultraviolet radiation, effect of, on people, 317, 318
Umbrella sales, weather and, 345
Unbounded variables, 131
Uniform motion, 53
   of wind (geostrophic approximation), 53, 54–56
Uniformitariansim principle, 250
Union of Soviet Socialist Republics, high latitude (D) climates in, 225, 227–228, 229
Upper-air circulation, 84–95, 285, 287–292, 294–295, 296, 297
Upper-air inversions, 41–42
Upslope fog, 170
Upwelling, 99
Urban climates, 327–328

Vacation weather, 346–347
Valleys
   frost hollows, 243
   and wind flows, 238
Vapor gradient, 111, 112
Vapor pressure, 104, 109, 111, 127
Variability, 174, 176
   precipitation, 178–180
   sequential vs. consecutive, 178
   temperature, 38, 45, 48–49, 176–178
   in atmosphere, 39, 40
Varves, 250
Vector vs. scalar quantities, 53
Vegetables, thermal classification of, 308
Vegetation. See Plants
Venitz, Ignatz, 248
Vertical motions
   definition, 116
   mechanisms involved
      convergence, 120–121, 159
      and intensity of precipitation, 166
      layer instability, 122–123
      mechanical uplift, 122, 160–161
      thermal convection, 117–120, 157, 159
Vertical stretching, 122–123
Vertical wind shear, 57, 123
Vladivostok, USSR, Dwb climate of, 227–228
Volcanoes and atmospheric particles, 263

Wall materials, building's, effect of, 325
Water, 103
   balance, continents', 109
   boiling point, decrease in, 323
   budget, 115–116
   and energy fluxes, 18–19, 23, 32
      table of ocean conditions, 22
   fishing industry, temperature effect on, 341–342
   household use, 340
   hydrologic cycle, 96, 106–108
   phase changes, 16, 96, 104, 105
      See also Evaporation
   precipitable, 157, 158
   properties, 95
   roads, effect on, 333–334
   snow equivalent, 168
   supercooled, 126
   surplus/deficit, 108–109, 115–116
   transportation on, 331–333
   and vegetation, 302
      irrigation, 308
Water climate, 115
Water stress, 304
Water vapor, 100–101, 111
   and density of air, 122
   gradient, 111, 112
   hydrologic cycle, 96, 106, 107
   measures of, 104–106
   oxygen-18 and, 251
   pressure, 104, 109, 111, 127
   and relative humidity, 105, 109
   solar radiation, depletion of, 263–264
   and temperature, 32–33, 105–106
      ocean surface, 161
   vertical profile, 241
Wave cyclones. See Transient disturbances
Wavelengths of solar radiation, 5
   and plant responses, 299–300
Waves
   easterly, 73, 76
   long, 88, 91, 288, 289, 292
Weather
   vs. climate, 1
   synoptic climatology and inference of, 272–274
   phenomena influenced by, 293–298
   types, 280–281, 286, 287
   and temperature variability, 38
   transient disturbances and, 72–73, 76
Weather maps. See Maps, weather
Weather routing for ships, 333
Weather stations, 2, 3
   data from, 187, 188–189
   and isobaric analysis, 57
   and scales of motion, 61–62
Weeds, 311
Wegener, Alfred, 264
Westerly winds, 63
   equatorial, 66–67, 68
   upper-level, 86, 87
   waves in, and surface conditions, 93, 292
Wet-bulb temperature, 106
   and THI, 321
Wet months, 165
Wheat, crop requirements of, 307
"White Christmas," probability of, 170
Wien's law, 5
Wind
   barriers to, 122
      shelterbelts, 309
   belts, 63
   construction and, 245–246
   curved flow, 56
   as energy source, 337
   friction wind, 57, 62–63
   frictional force, 52–53, 54, 56–57, 121, 237, 240
   and industries, 342–344
   intertropical convergence zone (ITCZ), 65–67, 284
   katabatic, 239, 243

landscaping, effect of, 325, 326
local data, sample of, 188
mean resultant, 283, 287, 289
   ITCZ, 66
monsoon, criteria for, 69
organisms transported by, 296–298
power law, 240
record speeds, 183, 185
scales of, 332
sea and land breezes, 236–238
and sea travel, 332, 333
stagnation, 294, 297
streamlines, 283, 288, 289

synoptic patterns, 281, 283–284, 287–290
and precipitation, 279, 285
topography and
fall wind, 239
foehn (chinook), 238–239
mountain-valley flows, 238
uniform motion (geostrophic approximation), 53, 54–56
upper-level, 86–88, 89–91, 92, 93, 94
vegetation and, 245
trees, 243–244
weather map, analysis of, 57, 59

Wind shear, 87
  vertical, 57, 123
Windbreaks, 309
Windchill equivalent temperature, 320–321

Younger Dryas event, 259

Zonal climate; zonal soil, 201
Zone of aeration, 302
Zyryanka, USSR, Dfd climate of, 225